Brain Hypoxia and Ischemia

Gabriel G. Haddad • Shan Ping Yu
Editors

Brain Hypoxia and Ischemia

with Special Emphasis on Development

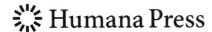

Editors
Gabriel G. Haddad, MD
University of California
San Diego, CA
Rady Children's Hospital
San Diego, CA
ghaddad@ucsd.edu

Shan Ping Yu, MD, PhD
Emory University
Atlanta
GA
yusp@musc.edu

ISBN: 978-1-60327-578-1   e-ISBN: 978-1-60327-579-8
DOI: 10.1007/978-1-60327-579-8

Library of Congress Control Number: 2008939850

© Humana Press 2009, a part of Springer Science+Business Media, LLC
All rights reserved. This work may not be translated or copied in whole or in part without the written permission of the publisher (Humana Press, c/o Springer Science+Business Media, LLC, 233 Spring Street, New York, NY 10013, USA), except for brief excerpts in connection with reviews or scholarly analysis. Use in connection with any form of information storage and retrieval, electronic adaptation, computer software, or by similar or dissimilar methodology now known or hereafter developed is forbidden.
The use in this publication of trade names, trademarks, service marks, and similar terms, even if they are not identified as such, is not to be taken as an expression of opinion as to whether or not they are subject to proprietary rights.
While the advice and information in this book are believed to be true and accurate at the date of going to press, neither the authors nor the editors nor the publisher can accept any legal responsibility for any errors or omissions that may be made. The publisher makes no warranty, express or implied, with respect to the material contained herein.

Printed on acid-free paper

springer.com

Preface

This is a book that has resulted from many discussions and meetings that the editors have had in last few years around mechanisms of cell death and cell survival under stressful conditions in the central nervous system (CNS). Indeed some of the authors of this book participated in a symposium that was organized by the editors at the Society for Neuroscience several years ago. The various aspects of cell death/survival and questions that are addressed here are by and large issues that are crucial for the understanding of the basic mechanisms underlying brain hypoxia and ischemia. The hope is that by understanding these mechanisms we would be better armed to devise much better ways for the diagnosis and treatment of diseases that afflict the brain and potentially other organs when O_2 levels are dysregulated.

Although some overlap between chapters was impossible to prevent, various chapters in this book address specific questions that we believe are critical at this stage of the science. These chapters focus on a panorama of issues, from the role of ion channels/transporters to that of mitochondria and apoptotic mechanisms, to the role of glutamate/NMDA, to the mechanisms in penumbral cells, to the importance of intermittent hypoxia, and certainly to gene regulation under these stressful conditions. While most chapters focus on mammalian models and mechanisms, some chapters center around invertebrate systems since these have proven to be exceedingly important in helping us with mammalian and even human systems and diseases.

Although there is evidence that the newborn or young mammalian CNS is more tolerant than the mature differentiated system, there is also plenty of evidence that the newborn CNS is vulnerable to various stresses including hypoxia and ischemia. Whether the mechanisms are the same or different, whether the vulnerable regions are the same, and how maturation of certain systems can affect survival or injury are all important questions that have major clinical implications at the bedside. This is why this book has a bent toward development.

We believe that this is the first book that has such a panorama of chapters – all in one book – summarizing the state of our knowledge in this area. We are especially excited about the intriguing novel discoveries described in many chapters in this book. The idea is to stimulate and *stir* discussions around these questions by exposing current knowledge and in so doing the state of our ignorance. We believe that neuroscientists, clinicians, and medical/graduate students will find this book useful for both basic research and clinical practice.

San Diego, CA Gabriel G. Haddad
Atlanta, GA Shan Ping Yu

Contents

Preface... v

Contributors.. ix

Part I Ion Channels, Transporters and Excitotoxicity

1 **Regulation of Vulnerability to NMDA Excitotoxicity
    During Postnatal Maturation**................................... 3
    Jeremy D. Marks

2 **Acidosis, Acid-Sensing Ion Channels, and Glutamate
    Receptor-Independent Neuronal Injury**........................ 25
    Zhigang Xiong

3 **Brain Ischemia and Neuronal Excitability**..................... 43
    Ping Deng and Zao C. Xu

4 **Critical Roles of the Na$^+$/K$^+$-ATPase
    in Apoptosis and CNS Diseases**................................. 53
    Adrian Sproul, Xin Zhou, and Shan Ping Yu

5 **Emerging Role of Water Channels in Regulating
    Cellular Volume During Oxygen Deprivation and Cell Death**...... 79
    Thomas James Younts and Francis "Monty" Hughes, Jr.

6 **A Zinc–Potassium Continuum
    in Neuronal Apoptosis**... 97
    Patrick Redman, Megan Knoch, and Elias Aizenman

7 **Mitochondrial Ion Channels in Ischemic Brain**................ 117
    Elizabeth A. Jonas

Part II Reactive Oxygen Species, and Gene Expression to Behavior

8 Perinatal Panencephalopathy in Premature Infants: Is It Due to Hypoxia-Ischemia? 153
Hannah C. Kinney and Joseph J. Volpe

9 Effects of Intermittent Hypoxia on Neurological Function 187
David Gozal

10 Brainstem Sensitivity to Hypoxia and Ischemia 213
Joseph C. LaManna, Paola Pichiule, Kui Xu, and Juan Carlos Chávez

11 Matrix Metalloproteinases in Cerebral Hypoxia-Ischemia 225
Zezong Gu, Jiankun Cui, and Stuart A. Lipton

12 Oxidative Stress in Hypoxic-Ischemic Brain Injury 239
Laura L. Dugan, M. Margarita Behrens, and Sameh S. Ali

13 Postnatal Hypoxia and the Developing Brain: Cellular and Molecular Mechanisms of Injury 255
Robert M. Douglas

14 Hypoxia-Inducible Factor 1 277
Gregg L. Semenza

15 Transcriptional Response to Hypoxia in Developing Brain 289
Dan Zhou

16 Acute Stroke Therapy: Highlighting the Ischemic Penumbra 307
Hang Yao

17 Genes and Survival to Low O_2 Environment: Potential Insights from Drosophila 323
Gabriel G. Haddad

Index 335

Contributors

Elias Aizenman
Department of Neurobiology, University of Pittsburgh School of Medicine, Pittsburgh, PA 15261, USA

Sameh S. Ali
Division of Geriatric Medicine, Department of Medicine, University of California, San Diego, CA 92093, USA

M. Marga Behrens
Division of Geriatric Medicine, Department of Medicine, University of California, San Diego, CA 92093, USA

Juan Carlos Chávez
Case Western Reserve University School of Medicine, Cleveland, OH 44106, USA

Jiankun Cui
Center for Neuroscience, Aging and Stem Cell Research, Burnham Institute for Medical Research, La Jolla, CA 92037, USA

Ping Deng
Department of Anatomy & Cell Biology, Indiana University School of Medicine, Indianapolis, IN 46202, USA

Robert M. Douglas
Department of Pediatrics, University of California, San Diego School of Medicine, La Jolla, CA 92093, USA

Laura L. Dugan
Division of Geriatric Medicine, Department of Medicine, University of California, San Diego, CA 92093, USA
Department of Neurosciences, University of California, San Diego, CA, 92093

David Gozal
Kosair Children's Hospital Research Institute, University of Louisville, Louisville, KY 40202, USA

Zezong Gu
Department of Pathology and Anatomical Sciences, University of Missouri School of Medicine, Columbia, MO 65211, USA
Center for Neuroscience, Aging and Stem Cell Research, Burnham Institute for Medical Research, La Jolla, CA 92037, USA

Gabriel G. Haddad
University of California and Rady Children's Hospital, San Diego, CA

Francis "Monty" Hughes, Jr.
Department of Biology, University of North Carolina at Charlotte, Charlotte, NC 28223, USA

Elizabeth A. Jonas
Department of Internal Medicine (Endocrinology), Yale University School of Medicine, New Haven, CT 06520, USA

Hannah C. Kinney
Department of Pathology, Children's Hospital Boston, Boston, MA 02115, USA

Megan Knoch
Department of Neurobiology, University of Pittsburgh School of Medicine, Pittsburgh, PA 15261

Joseph C. LaManna
Department of Neurology (BRB), Case Western Reserve University School of Medicine, Cleveland, OH 44106, USA

Stuart A. Lipton
Center for Neuroscience, Aging and Stem Cell Research, Burnham Institute for Medical Research, La Jolla, CA 92037, USA
The Salk Institute for Biological Studies, The Scripps Research Institute, and the University of California at San Diego, La Jolla, CA 92037, USA

Jeremy D. Marks
Departments of Pediatrics and Neurology, University of Chicago, Chicago, IL 60637, USA

Paola Pichiule
Case Western Reserve University School of Medicine, Cleveland, OH 44106, USA

Patrick Redman
Department of Neurobiology, University of Pittsburgh School of Medicine, Pittsburgh, PA 15261

Adrian Sproul
Department of Pharmaceutical and Biomedical Sciences, Medical University of South Carolina, Charleston, SC29425

Gregg L. Semenza
The Johns Hopkins University School of Medicine, Baltimore, MD 21205, USA

Contributors

Joseph J. Volpe
Department of Neurology, Children's Hospital Boston, Boston, MA 02115, USA

Zhigang Xiong
Robert S. Dow Neurobiology Laboratories, Legacy Clinical Research Center, Portland, OR 97232, USA

Kui Xu
Case Western Reserve University School of Medicine, Cleveland, OH 44106, USA

Zao C. Xu
Department of Anatomy and Cell Biology, Indiana University School of Medicine, Indianapolis, IN 46202, USA

Hang Yao
Department of Pediatrics, University of California, San Diego, La Jolla, CA 92093, USA

Thomas James Younts
Dominick Purpura Department of Neuroscience, Albert Einstein College of Medicine, Bronx, NY 10461, USA

Shan Ping Yu
Emory University, Atlanta, GA

Dan Zhou
Department of Pediatrics, University of California, San Diego, La Jolla, CA 92093, USA

Xin Zhou
Department of Pharmaceutical and Biomedical Sciences, Medical University of South Carolina, Charleston, SC29425

Part I
Ion Channels, Transporters and Excitotoxicity

Chapter 1
Regulation of Vulnerability to NMDA Excitotoxicity During Postnatal Maturation

Jeremy D. Marks

Abstract Hippocampal and cortical vulnerability to injury following activation of *N*-methyl-D-aspartate (NMDA) receptors increases markedly during development from embryonic life to the adult. The mechanisms underlying this increased vulnerability are multiple, and include developmental regulation of NMDA receptor subunit expression, localization of NMDA to synapses or extrasynaptic locations, and intracellular metabolism of NMDA-induced increases in cytosolic calcium concentrations, particularly by mitochondria. The role of nitric oxide is highlighted, especially with new data demonstrating a role for mitochondrial nitric oxide synthase (NOS) as a primary mediator of the decreased vulnerability of immature neurons to excitotoxicity.

Keywords: Development; NR2A; NR2B; NR1; Nitric oxide; Hypoxia; Ischemia; Hippocampus; Death; Mitochondria; Membrane potential; Calcium; NVP-AAM077; Ifenprodil; NOS; mtNOS

1.1 Postnatal Maturation Alters the Vulnerability of the Brain to Acute Injury

1.1.1 Vulnerability to Hypoxia-Ischemia Changes During Postnatal Development

It has been known since the 1940s that survival of newborn and young animals, as demonstrated by ongoing respiration, is much greater in the face of severe hypoxia (1, 2) and ischemia (3) compared with adults, and that this survival progressively decreases with maturation. Until the 1980s, the primary factors believed to differentiate

---

Jeremy D. Marks
Departments of Pediatrics and Neurology, University of Chicago
5841 S. Maryland Avenue, Chicago, IL 60637, USA
e-mail: jmarks@uchicago.edu

the tolerance of newborn and adult brains for substrate deprivation was the newborn brain's decreased reliance on aerobic metabolism (4).

The discovery of glutamate-induced neurotoxicity (5) and the elaboration of the roles played by N-methyl-D-aspartate (NMDA) and non-NMDA glutamate receptor subtypes in neuronal death (6–8) were followed by the critical observations that excitotoxicity was a primary mediator of hypoxic neuronal death both *in vivo* (9, 10) and *in vitro* (11). Using unilateral carotid ligation and exposure to 6% oxygen as a model of brain hypoxia-ischemia (HI) in rats of increasing maturation, Ikonomidou *et al.* (12) found that the severity and distribution of brain injury 2 h following HI was practically nonexistent in the first 2 days of life, maximal on the sixth postnatal day, and progressively decreased at days 14 and 20. The developmentally regulated regional distribution and magnitude of injury was recapitulated with unilateral NMDA injections into the caudate nucleus when injury was measured 4 h later. Contemporaneously, McDonald *et al.* observed that injections of NMDA (50 mM) into the striatum produced lesions 5 days later that were 21 times larger in 7-day old rats compared with 3-month-old rats (13). It is important to note that this toxicity paradigm resulted in prolonged, generalized tonic-clonic seizures in the 7-day-old rats, compared with ipsilateral turning and rolling behaviors in the adults. Together, these studies gave rise to the idea that (a) the week-old brain (in the rat) is much more vulnerable to HI than the adult; (b) this increased vulnerability is mediated by an increased, developmentally regulated susceptibility to excitotoxicity.

In light of subsequent studies (see below), it is worthwhile to reconcile initial concepts of increasing vulnerability of the brain following HI during postnatal development with the subsequent idea of a window of increased vulnerability at the end of the first week of life. Potential explanations may lie in the timing of observation of neuronal injury following HI. Thus, the brain damage observed 2 h after HI and 4 h after NMDA injection consisted primarily of neuronal edema (12), now known to be due to sodium influx, mediated by both NMDA receptor activation (8) and hypoxia itself. Edema, as such, is not necessarily a harbinger of delayed neuronal death, the hallmark of death from HI or NMDA.

In fact, much more recent studies have corroborated the very early observations of increasing vulnerability to HI during postnatal maturation. Thus, 23-day-old rats require shorter periods of unilateral carotid occlusion and hypoxia compared with 7-day-olds to demonstrate similar cerebral lesions (14). Similarly, in rats between 2- and 30-days old, the incidence and severity of hippocampal lesions seen 72 h following carotid occlusion and hypoxia dramatically increase with postnatal age: Little injury is seen until after 10 days of age; by 20 days of age, the severity of hippocampal lesions surpasses those seen in cortex (15, 16). Furthermore, following HI, the frequency of lesions in cortex, striatum, thalamus, and hippocampus directly correlates with the duration of HI and the postnatal age (15). The preponderance of evidence, therefore, demonstrates that vulnerability of the immature rat brain to standardized periods of hypoxia-ischemia increases during postnatal maturation: at birth, vulnerability is almost negligible; adult levels of vulnerability are reached by about 30 days of age.

1.1.2 Vulnerability to Excitotoxicity During Postnatal Development

Because excitotoxicity has been demonstrated to play a central role in mediating neuronal death following HI, it would not be surprising if the developmental regulation of neuronal vulnerability to excitotoxicity paralleled the development of vulnerability to HI. For the purposes of this review, we shall focus on NMDA toxicity in the forebrain. How postnatal developmental regulates forebrain vulnerability to NMDA has been investigated in whole brain *in vivo*, as well as in brain slices, acutely dissociated neurons, and neurons cultured from animals of different ages.

In the single recent study to assess the development of vulnerability to NMDA *in vivo*, hippocampal injection of a 500 µM solution of NMDA in a 1-µL volume induced minimal lesions in 10-day-old animals when assessed 7 days following injury; injections in progressively more mature animals resulted in progressively larger legions, the largest lesions seen in 30- and 60-day-old animals (17). Importantly, in this model, seizures were not seen. Thus, in the absence of seizures, vulnerability to excitotoxicity does appear to parallel the increased sensitivity to HI, and increases with postnatal age. The differences between this study and the first injection study of McDonald et al. (13) likely stem from a number of factors, including the NMDA concentration [50 mM (13) vs. 500 µM (17)], the presence (13) or absence (17) of seizures following NMDA, and the location of injection [striatum (13) vs. hippocampus (17)].

Assessment of vulnerability to an injected volume of excitotoxin is complicated by developmentally regulated factors that may not be directly related to intrinsic neuronal vulnerability, including diffusion of the volume to neighboring structures (which depends on myelination and abundance of neuropil), activity of neurotransmitter uptake processes by glia, and the possibility of seizures worsening injury. Perfusing brain slices prepared from animals of different ages with known concentrations of excitotoxin has the potential to reveal the nature of developmental regulation of injury without these problems. Using acutely prepared hippocampal slices, Zhou and Baudry have recently compared vulnerability to NMDA in 1–3-week-old rats with that of 2–3-month olds (18). Following 1 h of NMDA (100 µM) perfusion, they observed significant acute increases in LDH release and propidium iodide staining, both markers of acute necrosis, compared with control slices harvested from young rats. However, NMDA did not induce these increases in older rats. It is unclear how prolonged exposure of adult slices to an NMDA concentration sufficiently high to kill hippocampal neurons in other preparations resulted in no neuronal necrosis, especially in light of a wealth of studies identifying the profoundly injurious effects of NMDA receptor activation in adult hippocampus. One possibility may lie in the difficulty of preparing equivalently healthy slices from older animals (19, 20): in this study, LDH release in slices from older animals was elevated at baseline, and Nissl staining following NMDA was decreased.

Vulnerability to excitotoxicity is regulated as a function of maturation even at the level of the single neuron (21–23). In fact, the parallels in vulnerability during maturation between the *in vivo* setting and that seen in several preparations of dissociated neurons strongly suggest that mechanisms of vulnerability intrinsic to the single neuron underlie the developmental pattern *in vivo*. In our laboratory, we have used time-lapse microscopy of hippocampal CA1 neurons acutely dissociated from newborn, 1-week-old and 3-week-old rats to assess morphological and free intracellular calcium ($[Ca^{2+}]_i$) responses to glutamate toxicity (21). In neurons from 3-week-old animals, glutamate induces a persistent $[Ca^{2+}]_i$ increase and necrosis, while in neurons from 1-week-old animals, the $[Ca^{2+}]_i$ increase is less pronounced and not accompanied by morphologic changes leading to death. The absence of any $[Ca^{2+}]_i$ or morphologic response in neurons from 2-day olds likely results from absence of functioning glutamate receptors at that developmental stage.

Acutely dissociated neurons do not lend themselves to the detailed study of mechanisms of cell death since, by nature, they survive *in vitro* for only a few hours and are produced in varying stages of health. To better study cellular mechanisms mediating vulnerability to excitotoxicity, we harvest neurons from postnatal animals of different ages, and maintain them in a monolayer *in vitro*. In this model of postnatal development, hippocampal CA1 neurons are harvested from animals 0–30-days old, and vulnerability is assessed at 4–7 days *in vitro* (22). Since neuronal preparations differ only in the age of the animal from which they are prepared, differences in response can be directly ascribed to maturation. In this model, vulnerability to NMDA (300 µM for 5 min), assessed 48 h later, is minimal in neurons from animals under 7 days old, then progressively rises as the age of the animals from which neurons are harvested increases, up to 18 days old. At this point NMDA kills more than 90% of neurons. NMDA-induced death in neurons from animals up to 30-days old remains at greater than 90%. Most importantly, NMDA-induced increases in free intracellular calcium concentration ($[Ca^{2+}]_i$), reported by high- and low-affinity dyes, are not different in magnitude or kinetics between neurons from older and younger animals, and do not distinguish between neurons that go on to die and those that do not. The conclusion that the magnitude of $[Ca^{2+}]_i$ elevation does not predict the degree of neurodegeneration is supported by multiple observations in which the severity of neuronal death has been dissociated from the calcium loads to which the neurons have been subjected (24–26).

Differential toxicity in the face of similar $[Ca^{2+}]_i$ increases following NMDA receptor activation during development suggests that maturation-regulated, Ca^{2+}-dependent processes and/or intracellular Ca^{2+} metabolism is critical for determining vulnerability to a Ca^{2+} load following NMDA receptor activation. One potential site of developmental regulation is the mitochondrion, a central determinant of both necrotic and apoptotic neuronal death following injury (27–29). Uptake of Ca^{2+} by mitochondria following a $[Ca^{2+}]_i$ elevation is required for subsequent death (30), and the hyperpolarized mitochondrial membrane potential ($\Delta\psi m$) of approximately −220 mV provides the driving force for mitochondrial Ca^{2+} uptake via the mitochondrial Ca^{2+} uniporter (31). Our finding that hippocampal neurons from 19-day-old animals exhibit profound $\Delta\psi m$ dissipation during NMDA-induced

$[Ca^{2+}]_i$ increases while those from 5-day olds exhibit only minimal $\Delta\psi m$ dissipation (22) demonstrate the strong developmental regulation of mitochondrial function.

1.1.3 Increases in Vulnerability of Cultured Neurons with Increasing Time In Vitro

Multiple studies have demonstrated a progressive increase in the vulnerability of neurons cultured from embryonic cortex (7, 32), hippocampus (33, 34), and cerebellum (35, 36) with increasing time *in vitro*. Typically, vulnerability to glutamate is minimal in the (4–7) days after plating (7, 33), appearing at 8–12 days *in vitro* (7, 33), and steadily progressing thereafter. As embryonic neurons remain *in vitro*, they exhibit progressively larger glutamate-induced $[Ca^{2+}]_i$ rises (32), and the magnitude of this $[Ca^{2+}]_i$ rise parallels the rise in vulnerability (37, 38). NMDA receptor expression is similarly low immediately in the first few days *in vitro* (34), but expression of all subunits also increases steadily with time (see later; 35, 38–40). NMDA current amplitude also increases steadily over time (39). Thus, the progressive increase in NMDA receptors leads to greater $[Ca^{2+}]_i$ influx and greater death. This pattern of steadily increasing Ca^{2+} loads following identical excitotoxic stimuli is an important indication that increased vulnerability with maturation *in vitro* is likely not mediated by the same mechanisms as the increased vulnerability seen with maturation *in vivo*. As described earlier, Ca^{2+} loads following excitotoxicity do not increase during *in vivo* maturation (21–23), indicating that mechanisms downstream of the Ca^{2+} elevation that lead to neuronal death are developmentally regulated.

With longer time *in vitro*, embryonic neurons exhibit altered Ca^{2+} homeostasis following excitotoxin-induced $[Ca^{2+}]_i$ elevations. Thus, following a glutamate- or NMDA-induced $[Ca^{2+}]_i$ rise, cortical neurons maintained *in vitro* for several days are able to return $[Ca^{2+}]_i$ homeostatically to baseline levels; in contrast, the same neurons maintained for longer than about 1 week fail to restore $[Ca^{2+}]_i$ to baseline values following glutamate exposure (32). Similarly, the longer embryonic hippocampal or cerebellar neurons are maintained *in vitro*, the more likely a toxic glutamate exposure (100 µM) is to induce a delayed, secondary $[Ca^{2+}]_i$ increase (41, 42), indicative of impending necrosis (43). The decreasing ability of neurons to return elevated $[Ca^{2+}]_i$ levels homeostatically to baseline levels as they age *in vitro* has also been observed following membrane depolarization (36). This observation suggests that failure of Ca^{2+} homeostasis is not specific to Ca^{2+} entry through glutamate receptors, but, instead is a function of the Ca^{2+} load with which the neuron is presented. Failure of Ca^{2+} homeostasis has been shown to be dependent on mitochondrial Ca^{2+} uptake (44), mitochondrial dysfunction (45), increased production of reactive oxygen species (46), and cellular ATP depletion (47). Lack of cellular ATP leads to failure of ATP-dependent Ca-extrusion mechanisms in the plasma membrane (44).

Ongoing excitatory neurotransmitter release following application of NMDA or glutamate also plays an important role in prolonging or potentiating excitotoxic neuronal death (48, 49). With development *in vitro*, KCl-induced glutamate release from cultured embryonic cortical neurons triples between day 10 and day 14 (40). This release is responsible for subsequent delayed neuronal death: death is blocked with prior treatment with tetanus toxin (40) or with subsequent addition of NMDA receptor antagonists (50). This membrane depolarization-induced release steadily increases as neurons are maintained longer *in vitro*. Thus, as cultured neurons remain *in vitro*, excitotoxicity-induced glutamate release participates in a steadily increasing positive feedback loop to increase vulnerability.

1.2 Developmental Regulation of NMDA Receptor Expression

1.2.1 NMDA Receptor Subunits and Functional Properties

With its high permeability to Ca^{2+}, the NMDA subtype of glutamate receptor is a central mediator of excitotoxic death. NMDA receptors are subject to physiologic regulation at multiple levels: by ligands, including Mg^{2+} and glycine (51), Zn^{2+} (52), and polyamines (53), by pH (54) and redox state (55), and by interactions with scaffolding proteins in the postsynaptic density, including PSD-95 (56). The extent of regulation and the functional properties of NMDA receptors are determined by the subunit composition of the receptor [see (57, 58) for recent reviews]. The heteromeric receptor consists of the ubiquitously expressed NR1 subunit and at least one NR2 subunit. Alternatively, the non-NR1 component can be composed of NR2 and NR3 subunits. At least eight functional splice variants of the NR1 subunit have been described, and there is evidence that splice variants that include exon 5 confer susceptibility of the receptor to modulation by pH (58). With the exception of NR2A, each of the NR2 subunits (B–D) and NR3 subunits also exist as splice variants, although the functional significance of these variants remains unclear (57).

Electrophysiologic data indicate that activation of an NMDA receptor requires occupation of two independent glycine binding sites and two independent glutamate binding sites (59, 60). NR1 contains the binding site for glycine, an obligatory coagonist, while NR2 contains the glutamate binding site. Thus, the functional receptor is likely a dimer of two heterodimers of one NR1 subunit and one NR2 subunit (58). Incorporation of an NR3 subunit can replace one NR2 subunit (58), yielding an NR1/NR2/NR3 receptor.

The functional properties of the NMDA receptor are primarily conferred by the NR2 subunit. Functionally, NMDA currents exhibit a slow rise and a very slow decay time (hundreds of ms) (57). The type of NR2 subunit, to which glutamate binds, confers the central determinant of NMDA receptor activity – its inactivation kinetics. The duration of NMDA current following, for example, an identical 1-ms stimulation ranges from 50 ms for NR2A-containing receptors to 1.7 s for NR2D-containing receptors (61). NR2 receptors also differentially confer sensitivity to Mg^{2+} block as well as differences in single channel conductance. Thus, NR2A or NR2B subunits produce higher conductance channels with high sensitivity to Mg^{2+} block; NR2C or NR2D subunits produce lower conductance channels with decreased sensitivity to Mg^{2+} (62). Mg^{2+} sensitivity prevents activation of the NMDA receptor, unless the membrane is depolarized. With membrane depolarization, for example, during AMPA receptor activation, Mg^{2+} block is relieved allowing NMDA receptor activation. Hence, differences in Mg^{2+} sensitivity will strongly affect the timing and magnitude of NMDA receptor activation, converting the NMDA receptor from a coincidence detector to a receptor that functions at more negative membrane potentials.

NR3 receptors do not form functional NMDA receptors when expressed with NR1 alone; alone, they appear to form a function excitatory glycine receptor (63). In NMDA receptor heterodimers, NR3 subunits function as regulatory subunits that decrease NMDAR activity (64–66): when coexpressed with NR1 and NR2, NR3 decreases NMDA current amplitude, single channel conductance, and Ca^{2+} permeability (67). Transgenic mice lacking NR3A subunits exhibit increased NMDA currents and increased spine density (68), providing further evidence that NR3 decreases receptor activity.

NMDA receptor subtypes demonstrate differences in sensitivity to pharmacological antagonists, most effectively with noncompetitive antagonists that act at the NR1 subunit (57). Ifenprodil, for example, which acts by enhancing receptor inhibition by protons (69), inhibits NR1/NR2B receptors with an IC_{50} 200-fold lower than its IC_{50} at NR1/NR2A, NR1/NR2C, or NR1/NR2D (70, 71). Similarly, NVP-AAM077, a phosphonomethyl-quinoxalinedione noncompetitive antagonist, reportedly demonstrates a greater than 100-fold selectivity for NR2A-containing NMDA receptors over receptors containing NR2B (72), although concerns have been raised about this specificity, in rodent neurons, at least (71). Zinc ion demonstrates strong selectivity for blocking NR1/NR2A receptors over other heteromeric receptors at nanomolar concentrations (71, 73). However, its effects on other synaptic mechanisms, as well as its only partial inhibition of NR1/NR2A receptors at its most selective concentrations, make Zn^{2+} problematic as a pharmacological antagonist (71). Finally, triheteromeric receptors, such as NR1/NR2A/NR2B in the cortex, exhibit still different sensitivities to subunit-specific antagonists: in such receptors, the presence of one subtype (NR2B, for example) confers some selectivity for its specific antagonist, but with low efficacy (74).

In addition to the strong developmental regulation of NMDA subunit expression (see later), NR2 subunits have marked differences in regional expression. In the adult rodent brain, NR2A is ubiquitously expressed (75). NR2B is enriched in cerebral cortex and hippocampus, with minimal expression in hindbrain (76). NR2C has highest expression in cerebellum and thalamus, while NR2D is highest in diencephalon and brainstem (77).

1.2.2 Developmental Regulation of NMDA Receptor Subunit Expression

The composition of NMDA receptor subunits is subject to well-described developmental regulation during embryonic and postnatal development. NMDA receptor blockade was first shown to be more effective at blocking visual responses in cortical neurons in kittens than in adult cats (78), suggesting a developmental difference in NMDA receptors. However, Hestrin (79) was the first to demonstrate developmental regulation of NMDA receptor function. He showed that the duration of excitatory postsynaptic current (EPSCs) in response to glutamate using outside-out patches of membrane excised from superior colliculus neurons were several times longer in patches from kittens than from more mature animals (79). In contrast, the amplitude of these currents did not change with development. This initial observation, subsequently borne out by detailed studies of changes in subunit-specific mRNA and protein expression during development, describes the overall picture in the forebrain: during development, changes in subunit expression result in the decay kinetics of NMDA currents becoming progressively shorter.

During postnatal cortical development in the rat, assays of NR1 mRNA transcripts and protein demonstrate that NR1 expression is low at birth, rises to a peak between 2- and 3-weeks postnatal age, and then falls to an intermediate plateau during adulthood (80). Using *in situ* hybridization of NMDA subunit mRNAs, Monyer's now-classic study demonstrated strong NR1 expression at birth, along with strong NR2B expression (81). In contrast, NR2A and NR2C transcripts were undetectable in the embryonic period, appearing around birth (81). NR2D transcripts, mainly in the midbrain, peaked around day 7, decreasing to a plateau of adult levels (81). Subsequent studies employing subunit-specific antibodies have confirmed that NR2A expression in rodent forebrain is low at birth with progressive increases in expression thereafter (82). In contrast, NR2B is expressed at adult levels during forebrain development (82). In human hippocampus, an *in situ* hybridization study of 34 patients at five stages of life demonstrated lower NR1 and higher NR2B transcripts in the neonate compared with older patients, with NR2A transcripts remaining constant, leading to an age-related increase in NR2A/NR2B transcript ratio during development (83). This increase in NR2A-containing receptors in the hippocampus and cortex during early postnatal development has been observed physiologically in heterologous expression systems using mRNA transcripts harvested from animals of different ages (84). In neurons, NMDA receptor-mediated EPSCs have been shown to progressively shorten during postnatal development (85, 86). In addition, ifenprodil block of NMDA receptor currents falls in progressively older animals (85, 87, 88).

1.2.3 Contribution of NR2A and NR2B Subunits to Excitotoxicity

The steady increase in the proportion of NMDA receptors containing NR2A, and the associated progressive shortening of NMDA currents during postnatal development of the forebrain would seem to be unable to directly underlie the progressive vulnerability to excitotoxicity observed over the same time period. Therefore, in assessing the roles played by the developmentally regulated expression of NR2 subunits to the development of vulnerability to excitotoxicity, it is instructive to survey current concepts of the roles played, in the forebrain, by NR2A and NR2B in complex, NMDA-mediated responses. These studies have employed well-characterized antagonists selective for NR2B, ifenprodil and Ro 25–6981, as well as NVP-AAM077, a newer, competitive antagonist reported, at low concentrations (400 nM), to block NR2A-containing receptors selectively (72, 89). Thus, use of ifenprodil or Ro25-6981 to selectively block NR2B-containing receptors can abolish long-term depression (LTD) of hippocampal CA1 field EPSPs, while NVP-AAM077 prevents long-term potentiation (LTP) in the same preparation without effect on LTD (90). Similar observations have been made in perirhinal cortex, focusing on LTP and LTD mediated by extrasynaptic receptors (89). These data suggest that NR2A and NR2B receptors can mediate opposing effects. Deficiencies in LTP in adult mice expressing a C-terminal truncation of NR2A indicate that the differential effects of NR2 subunits on LTP depend, in part, on the integrity of C-terminal-dependent signaling mechanisms (91).

There is now some evidence that NR2B-containing receptors may be the primary mediators of toxicity in several injury models. In contrast, NR2A-containing receptors may provide neuroprotection. For example, mechanical stretch injury of organotypic hippocampal slices, resulting in calpain cleavage and caspase 3 activation in CA3, is prevented by ifenprodil, suggesting a pure NR2B-dependent mechanism (92). Under conditions of extreme stretch, NVP-AAM077 results in increased caspase-3 activation, suggesting that NR2A-containing receptors provide some neuroprotection. Similarly, in cortical neurons cultured from 15-day-old rat embryos, ifenprodil prevented NMDA-induced death at 14 days *in vitro*, while NVP-AAM077 failed to prevent death (93). In marked parallel to the LTP/LTD studies described earlier, at 21 days *in vitro*, NVP-AAM077 exacerbated NMDA-induced death, suggesting NR2A subunit-mediated neuroprotection. Most interestingly, in neurons from mice expressing an NR2A mutant with C-terminal truncation, no evidence of NR2A-mediated neuroprotection was observed (93). Thus, the intracellular tail of the NR2A subunit, a critical determinant of NR2 receptor function (94), appears necessary for any neuroprotective mechanisms mediated by activation of NR2A-containing NMDA receptors (93). Differences in NR2 subunit functions may be mediated through interactions of this cytoplasmic C-terminal tail with different proteins in the receptor-associated protein complex (95). In a similar preparation, in cortical neurons harvested from embryonic day 18 rats and studied at 11–14 days *in vitro*, NMDA-induced death, necrotic or apoptotic, is completely blocked with Ro 25–6981, but not with NVP-AAM077 (96). In this study as well, NVP-AAM077 enhanced NMDA-induced apoptosis, suggesting that activation of NR2A-containing NMDA receptors promotes cell survival in the face of excitotoxicity.

1.2.4 Role of NR2A and NR2B Subunits in Regulating Developmental Regulation of Vulnerability to Excitotoxicity

If NR2B-containing NMDA receptors are primarily responsible for excitotoxicity, how much of the increased vulnerability to NMDA seen during postnatal development can be ascribed to developmental regulation of NR2A and NR2B expression? Using hippocampal slices acutely prepared from 1-week-old and adult rats, Zhou and Baudry (18) employed NR2B-specific antagonists and NVP-AAM077 to dissect NR2A and NR2B contributions to neuronal death following NMDA perfusion (18). Ifenprodil or Ro25-6981 almost completely blocked NMDA toxicity in the immature slices; NVP-AAM077 did not decrease NMDA toxicity. These observations are consistent both with the paucity of NR2A-containing receptors at this age as well as with the studies summarized earlier that suggest that NR2B-containing receptors underlie excitotoxicity *in vitro* (93, 96). This report is the first to link vulnerability of a particular stage of maturation with a particular NMDA receptor subunit. However, in this study, the failure of NMDA to increase hippocampal neuron death in slices from adult rats raises serious concerns regarding the initial health of the adult slices. While the authors interpret this lack of NMDA-induced death in the adult as being mediated by a postnatal decrease in NR2B-containing receptors, these findings fly in the face of a wealth of studies identifying the profoundly injurious effects of NMDA receptor activation in hippocampus.

Studies examining the relative contributions of NR2A- and NR2B-containing receptors to NMDA-induced death in cultured embryonic neurons at different times *in vitro* may shed some light on the role played by these receptor subtypes in mediating developmentally regulated vulnerability *in vivo*, although, for the reasons outlined earlier, mechanisms due to maturation *in vitro* likely bear little relationship to those mechanisms operating *in vivo*. Nonetheless, embryonic day 15 mouse cortical neurons are mostly protected from NMDA toxicity by ifenprodil at 14 days *in vitro*, but not at 21 days *in vitro* (93), suggesting that, at 21 days *in vitro*, a non-NR2B-mediated form of NMDA toxicity supervenes. However, at 21 days *in vitro*, the combination of ifenprodil and NVP-AAM077 (50 nM) does prevent NMDA-induced death; protection is not seen with NVP-AAM077 alone. In cortical cultures derived from embryonic day 18 rats, ifenprodil completely blocks glutamate toxicity at 9 days *in vitro*, but not at 23 days *in vitro* (39). In contrast, increased death following glutamate in embryonic hippocampal neurons with increasing time *in vitro* has also been associated with increased NR2A subunit expression, but not NR2B increases. This pattern has been reported to be associated with a decreased efficacy of ifenprodil to prevent this death (39). To muddy the waters further, at 23 days *in vitro*, neuronal death has been reported to be much more effectively prevented with nimodipine, a blocker of voltage-gated calcium channels. These results do provide some additional support for the importance of NR2B-containing NMDA receptors in mediating excitotoxicity at early stages *in vitro*. However, the role of NR2A-containing receptors cannot be discounted. Finally, the activation of non-NMDA receptors in this paradigm, particularly AMPA receptors, makes comparison of these findings difficult.

1.2.5 *Role of Synaptic and Extrasynaptic NMDA Receptors*

In addition to being localized to synapses, a population of NMDA (and other neurotransmitter) receptors exist extrasynaptically in the neuronal plasma membrane, as demonstrated by lack of colocalization with such presynaptic proteins as synaptophysin. By employing the use-dependent NMDA receptor antagonist, MK-801, during prolonged synaptic activity *in vitro*, NMDA receptors located at synapses can be selectively blocked, allowing molecular dissection of the relative contributions of synaptic and extrasynaptic receptors to NMDA-mediated processes (97, 98). Studies employing this technique have found that these populations differentially modify genome activity through nuclear calcium signaling, resulting in alterations of, for example, cAMP response binding element (CREB)-dependent processes (99). Data now suggest that synaptic and extrasynaptic receptors have distinct roles in neuroprotection and neurodegeneration. Thus, extrasynaptic receptors may mediate the death that follows, for example, oxygen-glucose deprivation, opposing the antiapoptotic effects of synaptic NMDA receptors (98).

In vitro, NMDA receptors have been noted to be expressed first in the absence of presynaptic proteins, before *in vitro* day 7 in embryonic cortical neurons (100), and prior to day 5 in embryonic hippocampal neurons (101). In cortex, NR1/NR2B-containing receptors cluster at extrasynaptic sites on dendrite and somata, while NR2A appearing after day 10 *in vitro* appears to localize only to synapses (100). Furthermore, neurons maintained *in vitro* longer exhibit more synaptic and fewer extrasynaptic receptors.

Electrophysiologically, these extrasynaptic NMDA receptors appear to primarily contain NR2B subunits, since ifenprodil blocks a much larger percentage of whole cell current obtained from CA1 neurons in microisland culture than autaptically driven NMDA currents (97). As these cultures remain *in vitro*, ifenprodil becomes progressively less effective at decreasing synaptic NMDA current, while extrasynaptic currents remain equally sensitive to NR2B selective blockade (97). Thus, extrasynaptic receptors may consist primarily of NR1/NR2B heterodimers. Whether extrasynaptic receptors continue to have this makeup throughout postnatal life is unknown. Interestingly, in hippocampal cultures, Ca^{2+} entry through isolated synaptic NMDA receptors has been reported to cause no toxicity, while entry through extrasynaptic receptors results in mitochondrial dysfunction and neuronal death (98). However, in the identical preparation, other groups have seen no differential contribution to either apoptosis or pro-survival responses between synaptic and extrasynaptic receptors (96). There is currently no information on the relative contributions made by synaptic and extrasynaptic receptors to NMDA-induced neuronal death during postnatal development. However, to the extent that extrasynaptic receptors are activated by spillover of glutamate from pathological glutamate release, such as during hypoxia-ischemia, these differences may prove to underlie some of the differences in vulnerability to excitotoxicity seen during postnatal development.

1.2.6 NMDA Receptor Desensitization During Development

The duration of NMDA currents following identical duration of receptor activation is dependent on NR2 receptor subtype (see earlier sections). In addition, however, there is evidence in cultured neurons that desensitization of the NMDA receptor in response to sustained activation changes with increasing time *in vitro*, perhaps with implications for development *in vivo* (102). Such changes in the magnitude to which NMDA receptor-mediated Ca^{2+} currents decay in the continued presence of NMDA (peak current/steady state current ratio) could alter neuronal predisposition to Ca^{2+} overload. In the only study to address this question, perforated patch recordings were made of NMDA current responses in embryonic day 18 hippocampal neurons, first, in the presence of ifenprodil (to isolate NR2A receptors) and second, in ifenprodil plus MK-801 (to block NR2A-containing receptors) followed by removal of ifenprodil (to recover only NR2B-containing receptors). Under these conditions, NR2B-containing receptors demonstrate greater desensitization than NR2A receptors. Within NR2A- and NR2B-containing receptors, desensitization progressively decreases with greater time *in vitro* (102). Furthermore, extrasynaptic NMDA receptors, isolated using autaptically driven action potentials in the presence of MK-801, exhibit greater desensitization than synaptic NMDA receptors; the decreased desensitization seen with synaptic NMDA receptors appears to depend on association with PSD-95 (102), an integral postsynaptic protein. Activity of extrasynaptic receptors, perhaps associated with excitotoxicity, is more likely to be shut off through desensitization than activity of synaptic receptors.

1.3 Role of Ongoing Synaptic Activity and NMDA Release

In synaptically connected neurons, excitotoxic stimulation can outlast the period during which NMDA receptors are initially activated, whether by agonist application or hypoxia (103); synaptic activity consisting of glutamate release and subsequent Ca^{2+} influx is, therefore, a primary mediator of ongoing neuronal death (102, 104). Consistent with the contributory role of ongoing synaptic activity following an initial excitotoxic stimulus, delayed administration of NMDA receptor antagonists following excitotoxicity can provide near complete neuroprotection (6, 48, 49). Ongoing synaptic activity appears to contribute to the differential vulnerability seen in cultured neurons with increasing time *in vitro*: in contrast to embryonic hippocampal neurons maintained *in vitro* for 1–2 weeks, the persistent $[Ca^{2+}]_i$ elevation seen in neurons maintained *in vitro* for 4–5 weeks is sensitive to NMDA receptor antagonists, indicating that NMDA receptor activation continues after the initial stimulation (50). This persistent elevation in the more mature neurons is accompanied by persistent, NMDA receptor-mediated membrane depolarization, and is not seen in less mature neurons. Use of whole cell voltage clamp, to maintain hyperpolarized membrane potentials, along with glutamate receptor antagonists in these neurons reveals a maturation-dependent increase in non-NMDA receptor EPSCs following glutamate (50). In non-voltage clamped neurons, persistent depolarization coupled with increased non-NMDA receptor activation could lead to persistent NMDA receptor activation and a greater propensity to ongoing excitotoxicity.

Alterations in neurotransmitter release during maturation could contribute also to the increasing vulnerability seen during this period. In embryonic mixed cortical neuron cultures, vulnerability to KCl-induced membrane depolarization increases with increasing time *in vitro* and is sensitive to tetanus toxin (40), an inhibitor of neurotransmitter release (105). Assays of glutamate concentration in the culture medium, performed in the presence of DL-threo-β-benzyloxyaspartate, a nontransportable glutamate transporter inhibitor (106), demonstrate no difference in glutamate release between less vulnerable (10 days *in vitro*) and more vulnerable (14 days *in vitro*) cultures. However, 14-day-old cultures exhibit a threefold increase in glutamate release upon KCl stimulation over 10-day-old cultures (40). These data suggest that, in some settings, transmitter release increases with time *in vitro*, possibly contributing to increasing vulnerability to excitotoxicity. However, astrocytes can also release glutamate (107, 108), and their variable presence in mixed cortical cultures confounds interpretation of these findings.

1.4 Nitric Oxide

1.4.1 Nitric Oxide as a Neurotoxin

Nitric oxide (NO), a gaseous free radical, plays diverse roles in intercellular signaling as a second messenger, in host defense, and in neuronal death and survival (109, 110). In neurons, following NMDA receptor activation, NO production is increased over baseline levels by Ca^{2+}-dependent activation of nitric oxide synthase (NOS) (111), mediated through interactions between the NMDA receptor (25, 26), proteins in the postsynaptic density, including PSD-95 (112), and NOS in the neuronal cytosol. The reduced neuronal death in $NOS^{-/-}$ mice or with neuronal NOS inhibitors following NMDA *in vitro* (113, 114), and following focal cerebral ischemia *in vivo* (115, 116) demonstrates that, under relevant conditions, NMDA-induced NO production is neurotoxic. NO production is linked to neuronal death via several well-described mechanisms: (a) reaction with mitochondrially derived superoxide radical to form peroxynitrite (117), a radical that decomposes to form hydroxyl radical, with subsequent damage to DNA, proteins, and membranes; (b) competition with oxygen at mitochondrial cytochrome *c* oxidase, resulting in inhibition of mitochondrial respiration, dissipation of mitochondrial membrane potential, and energy failure (118–120); (c) S-nitrosylation of such proteins as matrix metalloprotease-9 (121) and glyceraldehyde-3-phosphate dehydrogenase, leading to E3-ubiquitin-ligase-mediated degradation of nuclear proteins and death (109).

1.4.2 Developmental Regulation of NOS Expression and NO Production

NOS expression undergoes strong developmental regulation. In early studies, NOS activity in whole brain was reported to demonstrate a transient peak in the first week of postnatal life in the rat (122). However, no distinction was made between neuronal and nonneuronal NOS. In the guinea pig and rat, studies of the distribution of NOS-positive neurons by immunohistochemistry, NOS expression by western blot of whole brain, and NOS activity all demonstrate that NOS expression is detectable, but is low in hippocampus and cortex just before birth, and then increases to adult levels. In the guinea pig, hippocampal NOS activity approaches adult levels at P13, and is accompanied by increasing expression by Western blot (123). In the rat, NOS-positive neurons are first detected just before birth (124), although RNA transcripts have been reported since embryonic day 10 (125). By immunohistochemistry, the density of NOS-positive neurons progressively increases in the rat to day 14, at which point adult levels are reached (124). In our model of hippocampal neurons cultured from animals of different ages (22), NOS expression by immunohistochemistry changes markedly during development. At 4–7 days *in vitro*, hippocampal CA1 neurons harvested from 5-day-old animals demonstrate essentially no cytosolic NOS staining. In contrast, CA1 neurons harvested from animals older than 18 days exhibit intense neuronal NOS staining (23). In like fashion, microfluorimetry of 3-amino-4-(*N*-methylamino)-2′,7′-difluorofluorescein (DAF-FM) fluorescence to measure NMDA-induced NO production in these neurons over time demonstrates that, while NMDA-induced NO production is present in CA1 neurons harvested from 5-day-old animals, NO production during NMDA receptor activation is significantly larger in neurons from 19 day old animals (23). The importance of this NO increase in mediating NMDA toxicity at this highly vulnerable developmental stage is demonstrated by the profound protection conferred by prior treatment with N_ω-nitro-L-arginine methyl ester (L-NAME), a NOS antagonist (23). Thus, in a variety of experimental preparations, neuronal NOS expression and NO production increases with postnatal age, and this increase contributes to the increased toxicity seen with increased maturation.

1.4.3 Mitochondrial Nitric Oxide Production: A Mediator of Decreased Vulnerability in the Neonatal Period

In recent years, a novel NOS variant localized to mitochondria (mtNOS) has been described in heart, liver, and brain (126–129). Mitochondrial NOS has been localized to mitochondria using biochemical and porphyrinic microsensor techniques (130), as well as by colocalization of NOS and cytochrome oxidase immunofluorescence (23). Mass spectroscopy of proteolytic fragments obtained from affinity purification of liver mitochondrial fractions exhibiting NOS activity demonstrates that mtNOS has 100% homology to neuronal NOS (131). Furthermore, mtNOS is likely the α isoform of neuronal NOS, as demonstrated by RT-PCR of poly(A)$^+$-enriched mRNA of rat liver using primers obtained from spectroscopy analysis (131). Its mitochondrial localization may be due to myristoylation (131). Despite several reports casting doubt on the presence of a NOS isoform within mitochondria (132–134), the preponderance of anatomic, biochemical, and physiologic evidence provides extensive support for the existence and function of mtNOS (127–129, 135–149).

In hippocampal CA1 neurons harvested from 5-day-old rats, we have shown that cytosolic NOS immunostaining is absent (23). Nonetheless, NMDA receptor activation does induce NO production in these neurons. High-resolution confocal microscopy demonstrates exquisite colocalization of NOS immunoreactivity to mitochondria, as imaged with cytochrome oxidase immunoreactivity to colocalize NOS and cytochrome oxidase immunoreactivity (23). Using time-dependent colocalization studies of DAF-FM fluorescence and Mitofluor Red, a potential-independent marker of mitochondria, we have shown physiologically that the source of this NO originates from mitochondria. This NO production depends on uptake of Ca^{2+} into mitochondria, as NO production is abolished with either prior $\Delta\psi m$ dissipation with the protonophore carbonyl cyanide 4-(trifluoromethoxy) phenylhydrazone (FCCP) or pharmacologic blockade of the mitochondrial Ca^{2+} uniporter with Ru-360, a cell-permeable, oxygen-bridged dinuclear ruthenium amine complex (150). Most importantly, mitochondrial NO production in immature neurons plays an important role in mediating the decreased vulnerability of immature neurons to NMDA toxicity: NOS inhibition in 5-day-old neurons increases death following NMDA, from 10 to greater than 60% (23).

How could mitochondrial NO production confer decreased susceptibility to NMDA in immature neurons? NO is well known to regulate mitochondrial respiration, by competing with oxygen for the binuclear Cu_B/cytochrome a_3 site of cytochrome c oxidase (151). By displacing O_2 binding, NO inhibits mitochondrial respiration in the face of ongoing ATP synthesis, thereby dissipating $\Delta\psi m$. Thus, in immature neurons expressing mtNOS, NMDA-induced Ca^{2+} uptake by mitochondria is predicted to dissipate mitochondrial membrane potential; with this dissipation, the driving force for further Ca^{2+} entry into mitochondria is predicted to be reduced. These predictions are upheld by time-lapse measurements of NMDA-induced changes in $\Delta\psi m$, mitochondrial calcium concentrations, and cytosolic calcium concentrations in mtNOS-expressing 5-day-old neurons in the presence and absence of NOS blockade (23). First, NMDA induces a mild $\Delta\psi m$ dissipation, as reported by tetramethylrhodamine methyl ester. This $\Delta\psi m$ dissipation is prevented with NOS blockade. Second, NMDA-induced rises in mitochondrial $[Ca^{2+}]$, as reported by Rhod-2, are larger in the presence of NOS inhibition. Finally, pharmacologic dissipation of mitochondrial depolarization following NMDA, which causes mitochondria to release Ca^{2+} into the cytosol, results in larger $[Ca^{2+}]_i$ increases in the presence of NOS inhibition. Together these data indicate that mild mitochondrial depolarization by NO production in neurons from immature animals decreases mitochondrial Ca^{2+} uptake during NMDA.

The competition of NO and O_2 for binding to cytochrome oxidase also predicts that NO effects on $\Delta\psi m$ and NMDA toxicity will be larger as O_2 tensions decrease. In fact, the peak magnitude of NMDA-induced $\Delta\psi m$ dissipation progressively increases as O_2 tensions fall from 95 to 1% (23). Similarly, NMDA-induced death of 5-day-old neurons is markedly increased in 21% oxygen compared with 3% O_2, an O_2 concentration similar to that measured in rodent hippocampus *in vivo* (152, 153).

The ontogeny of mtNOS in rat brain, as assessed by activity and expression in mitochondrial fractions, parallels these physiologic and immunohistochemical findings. Thus, mtNOS expression, detected as a 144-kDa NOS-immunoreactive band by Western blot in mitochondrial fractions, becomes detectable at embryonic day 15, and increases until postnatal day 1, at which point it persists until about day 8, steadily decreasing thereafter (146). NOS activity in mitochondria exhibits a similarly developmentally regulated time course, with peak activity on postnatal day 0 being sixfold higher than during adulthood. In brains from adult rats, the activity present in the mitochondrial fraction represents about 10% of that of neuronal NOS in the cytosol (146). Thus, mtNOS appears ideally situated within the immature neuron to powerfully modulate mitochondrial function, so that neurons are protected from mitochondrial Ca^{2+} overload leading to death.

References

1. Fazekas JF, Himwich HE. Anaerobic survival of adult animals. Am J Physiol 1943; 139(3): 366–70.
2. Fazekas JF, Alexander FAD, Himwich HE. Tolerance of the newborn to anoxia. Am J Physiol 1941;134(2):281–7.
3. Kabat H. The greater resistance of very young animals to arrest of the brain circulation. Am J Physiol 1940;130(3):588–99.
4. Duffy TE, Kohle SJ, Vannucci RC. Carbohydrate and energy metabolism in perinatal rat brain: Relation to survival in anoxia. J Neurochem 1975;24(2):271–6.
5. Olney JW, Sharpe LG. Brain lesions in an infant rhesus monkey treated with monsodium glutamate. Science 1969;166(903):386–8.
6. Choi DW, Koh J, Peters S. Pharmacology of glutamate neurotoxicity in cortical cell culture: Attenuation by NMDA antagonists. J Neurosci 1988;8:185–96.
7. Choi DW, Maulucci-Gedde M, Kriegstein AR. Glutamate neurotoxicity in cortical cell culture. J Neurosci 1987;7(2):357–68.
8. Choi DW. Ionic dependence of glutamate neurotoxicity. J Neurosci 1987;7(2):369–79.
9. Ikonomidou C, Price MT, Mosinger JL, et al. Hypobaric-ischemic conditions produce glutamate-like cytopathology in infant rat brain. J Neurosci 1989;9(5):1693–700.
10. Olney JW, Ikonomidou C, Mosinger JL, Frierdich G. MK-801 prevents hypobaric-ischemic neuronal degeneration in infant rat brain. J Neurosci 1989;9(5):1701–4.
11. Goldberg MP, Weiss JH, Pham PC, Choi DW. N-methyl-D-aspartate receptors mediate hypoxic neuronal injury in cortical culture. J Pharmacol Exp Ther 1987;243(2):784–91.
12. Ikonomidou C, Mosinger JL, Salles KS, Labruyere J, Olney JW. Sensitivity of the developing rat brain to hypobaric/ischemic damage parallels sensitivity to N-methyl-aspartate neurotoxicity. J Neurosci 1989;9(8):2809–18.
13. McDonald JW, Silverstein FS, Johnston MV. Neurotoxicity of N-methyl-D-aspartate is markedly enhanced in developing rat central nervous system. Brain Res 1988;459(1):200–3.
14. Blumenfeld KS, Welsh FA, Harris VA, Pesenson MA. Regional expression of c-fos and heat shock protein-70 mRNA following hypoxia-ischemia in immature rat brain. J Cereb Blood Flow Metab 1992;12(6):987–95.
15. Towfighi J, Mauger D, Vannucci RC, Vannucci SJ. Influence of age on the cerebral lesions in an immature rat model of cerebral hypoxia-ischemia: A light microscopic study. Dev Brain Res 1997;100(2):149–60.
16. Towfighi J, Mauger D. Temporal evolution of neuronal changes in cerebral hypoxia-ischemia in developing rats: a quantitative light microscopic study. Dev Brain Res 1998;109(2):169–77.
17. Liu Z, Stafstrom CE, Sarkisian M, et al. Age-dependent effects of glutamate toxicity in the hippocampus. Brain Res Dev Brain Res 1996;97(2):178–84.

18. Zhou M, Baudry M. Developmental changes in NMDA neurotoxicity reflect developmental changes in subunit composition of NMDA receptors. J Neurosci 2006;26(11):2956–63.
19. Aitken PG, Breese GR, Dudek FF, et al. Preparative methods for brain slices: A discussion. J Neurosci Methods 1995;59(1):139–49.
20. Lipton P, Aitken PG, Dudek FE, et al. Making the best of brain slices; comparing preparative methods. J Neurosci Methods 1995;59(1):151–6.
21. Marks JD, Friedman JE, Haddad GG. Vulnerability of CA1 neurons to glutamate is developmentally regulated. Brain Res Dev Brain Res 1996;97:194–206.
22. Marks JD, Bindokas VP, Zhang XM. Maturation of vulnerability to excitotoxicity: Intracellular mechanisms in cultured postnatal hippocampal neurons. Brain Res Dev Brain Res 2000;124(1–2):101–16.
23. Marks JD, Boriboun C, Wang J. Mitochondrial nitric oxide mediates decreased vulnerability of hippocampal neurons from immature animals to NMDA. J Neurosci 2005;25(28):6561–75.
24. Dubinsky JM, Rothman SM. Intracellular calcium concentrations during "chemical hypoxia" and excitotoxic neuronal injury. J Neurosci 1991;11(8):2545–51.
25. Tymianski M, Charlton MP, Carlen PL, Tator CH. Source specificity of early calcium neurotoxicity in cultured embryonic spinal neurons. J Neurosci 1993;13(5):2085–104.
26. Sattler R, Charlton MP, Hafner M, Tymianski M. Distinct influx pathways, not calcium load, determine neuronal vulnerability to calcium neurotoxicity. J Neurochem 1998;71(6):2349–64.
27. Ankarcrona M, Dypbukt JM, Bonfoco E, et al. Glutamate-induced neuronal death: A succession of necrosis or apoptosis depending on mitochondrial function. Neuron 1995;15(4):961–73.
28. Budd SL. Mechanisms of neuronal damage in brain hypoxia/ischemia: Focus on the role of mitochondrial calcium accumulation. Pharmacol Ther 1998;80(2):203–29.
29. Nicholls DG, Budd SL. Mitochondria and neuronal glutamate excitotoxicity. Biochim Biophys Acta 1998;1366(1–2):97–112.
30. Stout AK, Raphael HM, Kanterewicz BI, Klann E, Reynolds IJ. Glutamate-induced neuron death requires mitochondrial calcium uptake. Nat Neurosci 1998;1(5):366–73.
31. Nicholls DG, Budd SL. Mitochondria and neuronal survival. Physiol Rev 2000;80(1):316–60.
32. Wahl P, Schousboe A, Honore T, Drejer J. Glutamate-induced increase in intracellular Ca^{2+} in cerebral cortex neurons is transient in immature cells but permanent in mature cells. J Neurochem 1989;53(4):1316–19.
33. Peterson C, Neal JH, Cotman CW. Development of excitotoxicity in cultured hippocampal neurons. Dev Brain Res 1989;48(2):187–95.
34. Mattson MP, Wang H, Michaelis EK. Developmental expression, compartmentalization, and possible role in excitotoxicity of a putative NMDA receptor protein in cultured hippocampal neurons. Brain Res 1991;565(1):94–108.
35. Xia Y, Ragan RE, Seah EE, Michaelis ML, Michaelis EK. Developmental expression of N-methyl-D-aspartate (NMDA)-induced neurotoxicity, NMDA receptor function, and the NMDAR1 and glutamate-binding protein subunits in cerebellar granule cells in primary cultures. Neurochem Res 1995;20(5):617–29.
36. Toescu EC, Verkhratsky A. Neuronal ageing in long-term cultures: Alterations of Ca^{2+} homeostasis. Neuroreport 2000;11(17):3725–9.
37. Eimerl S, Schramm M. The quantity of calcium that appears to induce neuronal death. J Neurochem 1994;62(3):1223–6.
38. Cheng C, Fass DM, Reynolds IJ. Emergence of excitotoxicity in cultured forebrain neurons coincides with larger glutamate-stimulated $[Ca^{2+}]i$ increases and NMDA receptor mRNA levels. Brain Res 1999;849(1–2):97–108.
39. Brewer LD, Thibault O, Staton J, et al. Increased vulnerability of hippocampal neurons with age in culture: Temporal association with increases in NMDA receptor current, NR2A subunit expression and recruitment of L-type calcium channels. Brain Res 2007;1151:20–31.
40. Fogal B, Trettel J, Uliasz TF, Levine ES, Hewett SJ. Changes in secondary glutamate release underlie the developmental regulation of excitotoxic neuronal cell death. Neuroscience 2005;132(4):929–42.

41. Manev H, Favaron M, Guidotti A, Costa E. Delayed increase of Ca^{2+} influx elicited by glutamate: Role in neuronal death. Mol Pharmacol 1989;36(1):106–12.
42. Adamec E, Didier M, Nixon RA. Developmental regulation of the recovery process following glutamate-induced calcium rise in rodent primary neuronal cultures. Dev Brain Res 1998;108(1–2):101–10.
43. Randall RD, Thayer SA. Glutamate-induced calcium transient triggers delayed calcium overload and neurotoxicity in rat hippocampal neurons. J Neurosci 1992;12(5):1882–95.
44. Castilho RF, Hansson O, Ward MW, Budd SL, Nicholls DG. Mitochondrial control of acute glutamate excitotoxicity in cultured cerebellar granule cells. J Neurosci 1998;18(24): 10277–86.
45. Nicholls DG, Vesce S, Kirk L, Chalmers S. Interactions between mitochondrial bioenergetics and cytoplasmic calcium in cultured cerebellar granule cells. Cell Calcium 2003;34(4–5): 407–24.
46. Luetjens CM, Bui NT, Sengpiel B, et al. Delayed mitochondrial dysfunction in excitotoxic neuron death: Cytochrome *c* release and a secondary increase in superoxide production. J Neurosci 2000;20(15):5715–23.
47. Starkov AA, Chinopoulos C, Fiskum G. Mitochondrial calcium and oxidative stress as mediators of ischemic brain injury. Cell Calcium 2004;36(3–4):257–64.
48. Rothman SM, Thurston JH, Hauhart RE. Delayed neurotoxicity of excitatory amino acids in vitro. Neuroscience 1987;22(2):471–80.
49. Hartley DM, Choi DW. Delayed rescue of *N*-methyl-D-aspartate receptor-mediated neuronal injury in cortical culture. J Pharmacol Exp Ther 1989;250(2):752–8.
50. Norris CM, Blalock EM, Thibault O, et al. Electrophysiological mechanisms of delayed excitotoxicity: Positive feedback loop between NMDA receptor current and depolarization-mediated glutamate release. J Neurophysiol 2006;96(5):2488–500.
51. Verdoorn TA, Kleckner NW, Dingledine R. Rat brain *N*-methyl-D-aspartate receptors expressed in Xenopus oocytes. Science 1987;238(4830):1114–16.
52. Rachline J, Perin-Dureau F, Le Goff A, Neyton J, Paoletti P. The micromolar zinc-binding domain on the NMDA receptor subunit NR2B. J Neurosci 2005;25(2):308–17.
53. McGurk JF, Bennett MV, Zukin RS. Polyamines potentiate responses of *N*-methyl-D-aspartate receptors expressed in xenopus oocytes. Proc Natl Acad Sci USA 1990;87(24):9971–4.
54. Traynelis SF, Cull-Candy SG. Pharmacological properties and H^+ sensitivity of excitatory amino acid receptor channels in rat cerebellar granule neurones. J Physiol 1991;433:727–63.
55. Aizenman E, Lipton SA, Loring RH. Selective modulation of NMDA responses by reduction and oxidation. Neuron 1989;2(3):1257–63.
56. Lin Y, Skeberdis VA, Francesconi A, Bennett MVL, Zukin RS. Postsynaptic density protein-95 regulates NMDA channel gating and surface expression. J Neurosci 2004;24(45):10138–48.
57. Cull-Candy S, Brickley S, Farrant M. NMDA receptor subunits: Diversity, development and disease. Curr Opin Neurobiol 2001;11(3):327–35.
58. Kew JN, Kemp JA. Ionotropic and metabotropic glutamate receptor structure and pharmacology. Psychopharmacology 2005;179(1):4–29.
59. Benveniste M, Mayer ML. Kinetic analysis of antagonist action at *N*-methyl-D-aspartic acid receptors. Two binding sites each for glutamate and glycine. Biophys J 1991;59(3): 560–73.
60. Clements JD, Westbrook GL. Activation kinetics reveal the number of glutamate and glycine binding sites on the *N*-methyl-D-aspartate receptor. Neuron 1991;7(4):605–13.
61. Vicini S, Wang JF, Li JH, et al. Functional and pharmacological differences between recombinant *N*-methyl-D-aspartate receptors. J Neurophysiol 1998;79(2):555–66.
62. Behe P, Colquhoun D, Wyllie DJA. Activation of single AMPA- and NMDA-type glutamate receptor channels. In: Jonas P, Monyer H, eds. Ionotropic glutamate receptors in the CNS. Berlin: Springer; 1999:175–218.
63. Chatterton JE, Awobuluyi M, Premkumar LS, et al. Excitatory glycine receptors containing the NR3 family of NMDA receptor subunits. Nature 2002;415(6873):793–8.

64. Sucher NJ, Akbarian S, Chi CL, et al. Developmental and regional expression pattern of a novel NMDA receptor-like subunit (NMDAR-L) in the rodent brain. J Neurosci 1995;15(10):6509–20.
65. Ciabarra AM, Sullivan JM, Gahn LG, Pecht G, Heinemann S, Sevarino KA. Cloning and characterization of chi-1: A developmentally regulated member of a novel class of the ionotropic glutamate receptor family. J Neurosci 1995;15(10):6498–508.
66. Nishi M, Hinds H, Lu H-P, Kawata M, Hayashi Y. Motoneuron-specific expression of NR3B, a novel NMDA-type glutamate receptor subunit that works in a dominant-negative manner. J Neurosci 2001;21(23):185RC.
67. Matsuda K, Fletcher M, Kamiya Y, Yuzaki M. Specific assembly with the NMDA receptor 3B subunit controls surface expression and calcium permeability of NMDA receptors. J Neurosci 2003;23(31):10064–73.
68. Das S, Sasaki YF, Rothe T, et al. Increased NMDA current and spine density in mice lacking the NMDA receptor subunit NR3A. Nature 1998;393(6683):377–81.
69. Mott DD, Doherty JJ, Zhang S, et al. Phenylethanolamines inhibit NMDA receptors by enhancing proton inhibition. Nat Neurosci 1998;1(8):659–67.
70. Williams K. Ifenprodil discriminates subtypes of the N-methyl-D-aspartate receptor: Selectivity and mechanisms at recombinant heteromeric receptors. Mol Pharmacol 1993;44(4):851–9.
71. Neyton J, Paoletti P. Relating NMDA receptor function to receptor subunit composition: Limitations of the pharmacological approach. J Neurosci 2006;26(5):1331–3.
72. Auberson YP, Allgeier H, Bischoff S, Lingenhoehl K, Moretti R, Schmutz M. 5-Phosphonomethylquinoxalinediones as competitive NMDA receptor antagonists with a preference for the human 1A/2A, rather than 1A/2B receptor composition. Bioorg Med Chem Lett 2002; 12(7):1099–102.
73. Chen N, Moshaver A, Raymond LA. Differential sensitivity of recombinant N-methyl-D-aspartate receptor subtypes to zinc inhibition. Mol Pharmacol 1997;51(6):1015–23.
74. Hatton CJ, Paoletti P. Modulation of triheteromeric NMDA receptors by N-terminal domain ligands. Neuron 2005;46(2):261–74.
75. Köhr G. NMDA receptor function: Subunit composition versus spatial distribution. Cell Tissue Res 2006;326(2):439–46.
76. Wang YH, Bosy TZ, Yasuda RP, et al. Characterization of NMDA receptor subunit-specific antibodies: Distribution of NR2A and NR2B receptor subunits in rat brain and ontogenic profile in the cerebellum. J Neurochem 1995;65(1):176–83.
77. Wenzel A, Scheurer L, Kunzi R, Fritschy JM, Mohler H, Benke D. Distribution of NMDA receptor subunit proteins NR2A, 2B, 2C and 2D in rat brain. Neuroreport 1995;7(1):45–8.
78. Tsumoto T, Hagihara K, Sato H, Hata Y. NMDA receptors in the visual cortex of young kittens are more effective than those of adult cats. Nature 1987;327(6122):513–14.
79. Hestrin S. Developmental regulation of NMDA receptor-mediated synaptic currents at a central synapse. Nature 1992;357(6380):686–9.
80. Sheng M, Cummings J, Roldan LA, Jan YN, Jan LY. Changing subunit composition of heteromeric NMDA receptors during development of rat cortex. Nature 1994;368(6467):144–7.
81. Monyer H, Burnashev N, Laurie DJ, Sakmann B, Seeburg PH. Developmental and regional expression in the rat brain and functional properties of four NMDA receptors. Neuron 1994; 12(3):529–40.
82. Portera-Cailliau C, Price DL, Martin LJ. N-methyl-D-aspartate receptor proteins NR2A and NR2B are differentially distributed in the developing rat central nervous system as revealed by subunit-specific antibodies. J Neurochem 1996;66(2):692–700.
83. Law AJ, Weickert CS, Webster MJ, Herman MM, Kleinman JE, Harrison PJ. Expression of NMDA receptor NR1, NR2A and NR2B subunit mRNAs during development of the human hippocampal formation. Eur J Neurosci 2003;18(5):1197–205.
84. Kleckner NW, Dingledine R. Regulation of hippocampal NMDA receptors by magnesium and glycine during development. Mol Brain Res 1991;11(2):151–9.
85. Barth AL, Malenka RC. NMDAR EPSC kinetics do not regulate the critical period for LTP at thalamocortical synapses. Nat Neurosci 2001;4(3):235–6.

86. Lu H-C, Gonzalez E, Crair MC. Barrel cortex critical period plasticity is independent of changes in NMDA receptor subunit composition. Neuron 2001;32(4):619–34.
87. Townsend M, Yoshii A, Mishina M, Constantine-Paton M. Developmental loss of miniature N-methyl-D-aspartate receptor currents in NR2A knockout mice. Proc Natl Acad Sci USA 2003;100(3):1340–5.
88. Williams K, Russell SL, Shen YM, Molinoff PB. Developmental switch in the expression of NMDA receptors occurs in vivo and in vitro. Neuron 1993;10(2):267–78.
89. Massey PV, Johnson BE, Moult PR, et al. Differential roles of NR2A and NR2B-containing NMDA receptors in cortical long-term potentiation and long-term depression. J Neurosci 2004;24(36):7821–8.
90. Liu L, Wong TP, Pozza MF, et al. Role of NMDA receptor subtypes in governing the direction of hippocampal synaptic plasticity. Science 2004;304(5673):1021–4.
91. Kohr G, Jensen V, Koester HJ, et al. Intracellular domains of NMDA receptor subtypes are determinants for long-term potentiation induction. J Neurosci 2003;23(34):10791–9.
92. DeRidder MN, Simon MJ, Siman R, Auberson YP, Raghupathi R, Meaney DF. Traumatic mechanical injury to the hippocampus in vitro causes regional caspase-3 and calpain activation that is influenced by NMDA receptor subunit composition. Neurobiol Dis 2006;22(1):165–76.
93. von Engelhardt J, Coserea I, Pawlak V, et al. Excitotoxicity in vitro by NR2A- and NR2B-containing NMDA receptors. Neuropharmacology 2007;53(1):10–17.
94. Sprengel R, Suchanek B, Amico C, et al. Importance of the intracellular domain of NR2 subunits for NMDA receptor function in vivo. Cell 1998;92(2):279–89.
95. Bliss T, Schoepfer R. Neuroscience. Controlling the ups and downs of synaptic strength. Science 2004;304(5673):973–4.
96. Liu Y, Wong TP, Aarts M, et al. NMDA receptor subunits have differential roles in mediating excitotoxic neuronal death both in vitro and in vivo. J Neurosci 2007;27(11):2846–57.
97. Tovar KR, Westbrook GL. The incorporation of NMDA receptors with a distinct subunit composition at nascent hippocampal synapses in vitro. J Neurosci 1999;19(10):4180–8.
98. Hardingham GE, Fukunaga Y, Bading H. Extrasynaptic NMDARs oppose synaptic NMDARs by triggering CREB shut-off and cell death pathways. Nat Neurosci 2002;5(5):405–14.
99. Zhang SJ, Steijaert MN, Lau D, et al. Decoding NMDA receptor signaling: Identification of genomic programs specifying neuronal survival and death. Neuron 2007;53(4):549–62.
100. Li JH, Wang YH, Wolfe BB, et al. Developmental changes in localization of NMDA receptor subunits in primary cultures of cortical neurons. Eur J Neurosci 1998;10(5):1704–15.
101. Rao A, Kim E, Sheng M, Craig AM. Heterogeneity in the molecular composition of excitatory postsynaptic sites during development of hippocampal neurons in culture. J Neurosci 1998;18(4):1217–29.
102. Abele AE, Scholz KP, Scholz WK, Miller RJ. Excitotoxicity induced by enhanced excitatory neurotransmission in cultured hippocampal pyramidal neurons. Neuron 1990;4(3):413–19.
103. Meldrum B. Protection against ischaemic neuronal damage by drugs acting on excitatory neurotransmission. Cerebrovasc Brain Metab Rev 1990;2(1):27–57.
104. Limbrick DD, Jr., Sombati S, DeLorenzo RJ. Calcium influx constitutes the ionic basis for the maintenance of glutamate-induced extended neuronal depolarization associated with hippocampal neuronal death. Cell Calcium 2003;33(2):69–81.
105. Humeau Y, Doussau F, Grant NJ, Poulain B. How botulinum and tetanus neurotoxins block neurotransmitter release. Biochimie 2000;82(5):427–46.
106. Shimamoto K, Lebrun B, Yasuda-Kamatani Y, et al. DL-Threo-beta-benzyloxyaspartate, a potent blocker of excitatory amino acid transporters. Mol Pharmacol 1998;53(2):195–201.
107. Montana V, Ni Y, Sunjara V, Hua X, Parpura V. Vesicular glutamate transporter-dependent glutamate release from astrocytes. J Neurosci 2004;24(11):2633–42.
108. Bezzi P, Gundersen V, Galbete JL, et al. Astrocytes contain a vesicular compartment that is competent for regulated exocytosis of glutamate. Nat Neurosci 2004;7(6):613–20.
109. Hara MR, Snyder SH. Cell signaling and neuronal death. Annu Rev Pharmacol Toxicol 2007;47(1):117–41.

110. Dalkara T, Yoshida T, Irikura K, Moskowitz MA. Dual role of nitric oxide in focal cerebral ischemia. Neuropharmacology 1994;33(11):1447–52.
111. Mayer B, Klatt P, Bohme E, Schmidt K. Regulation of neuronal nitric oxide and cyclic GMP formation by Ca^{2+}. J Neurochem 1992;59(6):2024–9.
112. Cui H, Hayashi A, Sun H-S, et al. PDZ protein interactions underlying NMDA receptor-mediated excitotoxicity and neuroprotection by PSD-95 inhibitors. J Neurosci 2007;27(37):9901–15.
113. Dawson VL, Dawson TM, London ED, Bredt DS, Snyder SH. Nitric oxide mediates glutamate neurotoxicity in primary cortical cultures. Proc Natl Acad Sci USA 1991;88(14):6368–71.
114. Dawson V, Kizushi V, Huang P, Snyder S, Dawson T. Resistance to neurotoxicity in cortical cultures from neuronal nitric oxide synthase-deficient mice. J Neurosci 1996;16(8):2479–87.
115. Yoshida T, Limmroth V, Irikura K, Moskowitz MA. The NOS inhibitor, 7-nitroindazole, decreases focal infarct volume but not the response to topical acetylcholine in pial vessels. J Cereb Blood Flow Metab 1994;14(6):924–9.
116. Huang Z, Huang PL, Panahian N, Dalkara T, Fishman MC, Moskowitz MA. Effects of cerebral ischemia in mice deficient in neuronal nitric oxide synthase. Science 1994;265(5180):1883–5.
117. Huie RE, Padmaja S. The reaction of NO with superoxide. Free Radic Res Commun 1993;18(4):195–9.
118. Brorson JR, Schumacker PT, Zhang H. Nitric oxide acutely inhibits neuronal energy production. J Neurosci 1999;19(1):147–58.
119. Ushmorov A, Ratter F, Lehmann V, Droge W, Schirrmacher V, Umansky V. Nitric-oxide-induced apoptosis in human leukemic lines requires mitochondrial lipid degradation and cytochrome *c* release. Blood 1999;93(7):2342–52.
120. Koivisto A, Matthias A, Bronnikov G, Nedergaard J. Kinetics of the inhibition of mitochondrial respiration by NO. FEBS Lett 1997;417(1):75–80.
121. Gu Z, Kaul M, Yan B, et al. S-nitrosylation of matrix metalloproteinases: Signaling pathway to neuronal cell death. Science 2002;297(5584):1186–90.
122. Matsumoto T, Pollock JS, Nakane M, Forstermann U. Developmental changes of cytosolic and particulate nitric oxide synthase in rat brain. Brain Res Dev Brain Res 1993;73(2):199–203.
123. Kimura KA, Reynolds JN, Brien JF. Ontogeny of nitric oxide synthase I and III protein expression and enzymatic activity in the guinea pig hippocampus. Dev Brain Res 1999;116(2):211–16.
124. Terada H, Nagai T, Okada S, Kimura H, Kitahama K. Ontogenesis of neurons immunoreactive for nitric oxide synthase in rat forebrain and midbrain. Dev Brain Res 2001;128(2):121–37.
125. Keilhoff G, Seidel B, Noack H, Tischmeyer W, Stanek D, Wolf G. Patterns of nitric oxide synthase at the messenger RNA and protein levels during early rat brain development. Neuroscience 1996;75(4):1193–201.
126. Ghafourifar P, Richter C. Nitric oxide synthase activity in mitochondria. FEBS Lett 1997;418(3):291–6.
127. Lacza Z, Puskar M, Figueroa JP, Zhang J, Rajapakse N, Busija DW. Mitochondrial nitric oxide synthase is constitutively active and is functionally upregulated in hypoxia. Free Radic Biol Med 2001;31(12):1609–15.
128. Bringold U, Ghafourifar P, Richter C. Peroxynitrite formed by mitochondrial NO synthase promotes mitochondrial Ca^{2+} release. Free Radic Biol Med 2000;29(3–4):343–8.
129. Giulivi C. Characterization and function of mitochondrial nitric-oxide synthase. Free Radic Biol Med 2003;34(4):397–408.
130. Kanai AJ, Pearce LL, Clemens PR, et al. Identification of a neuronal nitric oxide synthase in isolated cardiac mitochondria using electrochemical detection. Proc Natl Acad Sci 2001;98(24):14126–31.
131. Elfering SL, Sarkela TM, Giulivi C. Biochemistry of mitochondrial nitric-oxide synthase. J Biol Chem 2002;277(41):38079–86.
132. Lacza Z, Snipes JA, Zhang J, et al. Mitochondrial nitric oxide synthase is not eNOS, nNOS or iNOS. Free Radic Biol Med 2003;35(10):1217–28.

133. Lacza Z, Horn TF, Snipes JA, et al. Lack of mitochondrial nitric oxide production in the mouse brain. J Neurochem 2004;90(4):942–51.
134. Gao S, Chen J, Brodsky SV, et al. Docking of eNOS to the mitochondrial outer membrane: A pentabasic amino acid sequence in the autoinhibitory domain of eNOS targets a proteinase K-cleavable peptide on the cytoplasmic face of mitochondria. J Biol Chem 2004;279: 15968–74.
135. Leavesley HB, Li L, Prabhakaran K, Borowitz JL, Isom GE. Interaction of cyanide and nitric oxide with cytochrome *c* oxidase: Implications for acute cyanide toxicity. Toxicol Sci 2008;101(1):101–11.
136. Giusti S, Converso DP, Poderoso JJ, Fiszer de Plazas S. Hypoxia induces complex I inhibition and ultrastructural damage by increasing mitochondrial nitric oxide in developing CNS. Eur J Neurosci 2008;27(1):123–31.
137. Giulivi C, Kato K, Cooper CE. Nitric oxide regulation of mitochondrial oxygen consumption I: Cellular physiology. Am J Physiol Cell Physiol 2006;291(6):C1225–C1231.
138. Persichini T, Mazzone V, Polticelli F, et al. Mitochondrial type I nitric oxide synthase physically interacts with cytochrome *c* oxidase. Neurosci Lett 2005;384:254–9.
139. Lores-Arnaiz S, Perazzo JC, Prestifilippo JP, et al. Hippocampal mitochondrial dysfunction with decreased mtNOS activity in prehepatic portal hypertensive rats. Neurochem Int 2005;47(5):362–8.
140. Zanella B, Giordano E, Muscari C, Zini M, Guarnieri C. Nitric oxide synthase activity in rat cardiac mitochondria. Basic Res Cardiol 2004;99(3):159–64.
141. Traaseth N, Elfering S, Solien J, Haynes V, Giulivi C. Role of calcium signaling in the activation of mitochondrial nitric oxide synthase and citric acid cycle. Biochim Biophys Acta (BBA) Bioenergetics 2004;1658(1–2):64–71.
142. Lores-Arnaiz S, D'Amico G, Czerniczyniec A, Bustamante J, Boveris A. Brain mitochondrial nitric oxide synthase: In vitro and in vivo inhibition by chlorpromazine. Arch Biochem Biophys 2004;430(2):170–7.
143. Dedkova EN, Ji X, Lipsius SL, Blatter LA. Mitochondrial calcium uptake stimulates nitric oxide production in mitochondria of bovine vascular endothelial cells. Am J Physiol Cell Physiol 2003;286(2):C406–C415.
144. Boveris A, Valdez LB, Alvarez S, Zaobornyj T, Boveris AD, Navarro A. Kidney mitochondrial nitric oxide synthase. Antioxid Redox Signal 2003;5(3):265–71.
145. Alvarez S, Valdez LB, Zaobornyj T, Boveris A. Oxygen dependence of mitochondrial nitric oxide synthase activity. Biochem Biophys Res Commun 2003;305(3):771–5.
146. Riobo NA, Melani M, Sanjuan N, et al. The modulation of mitochondrial nitric-oxide synthase activity in rat brain development. J Biol Chem 2002;277(45):42447–55.
147. Sarkela TM, Berthiaume J, Elfering S, Gybina AA, Giulivi C. The modulation of oxygen radical production by nitric oxide in mitochondria. J Biol Chem 2001;276(10):6945–9.
148. Batista CMC, de Paula KC, Cavalcante LA, Mendez-Otero R. Subcellular localization of neuronal nitric oxide synthase in the superficial gray layer of the rat superior colliculus. Neurosci Res 2001;41(1):67–70.
149. Lopez-Figueroa MO, Caamano C, Morano MI, Ronn LC, Akil H, Watson SJ. Direct evidence of nitric oxide presence within mitochondria. Biochem Biophys Res Commun 2000;272(1):129–33.
150. Matlib MA, Zhou Z, Knight S, et al. Oxygen-bridged dinuclear ruthenium amine complex specifically inhibits Ca^{2+} uptake into mitochondria in vitro and in situ in single cardiac myocytes. J Biol Chem 1998;273(17):10223–31.
151. Brown GC. Regulation of mitochondrial respiration by nitric oxide inhibition of cytochrome c oxidase. Biochim Biophys Acta 2001;1504(1):46–57.
152. Feng ZC, Roberts EL, Jr., Sick TJ, Rosenthal M. Depth profile of local oxygen tension and blood flow in rat cerebral cortex, white matter and hippocampus. Brain Res 1988;445(2):280–8.
153. Buerk DG, Nair P. $PtiO_2$ and $CMRO_2$ changes in cortex and hippocampus of aging gerbil brain. J Appl Physiol 1993;74(4):1723–8.

Chapter 2
Acidosis, Acid-Sensing Ion Channels, and Glutamate Receptor-Independent Neuronal Injury

Zhigang Xiong

Abstract Stroke/brain ischemia, caused predominantly by an occlusion of cerebral blood flow to the brain tissue, is a leading cause of death and long-term disabilities in the developed countries. Unfortunately, there is no effective treatment for stroke patients other than the use of thrombolitic agents, which have limited therapeutic time window of less than 3 h and a potential side effect of intracranial hemorrhage. Therefore, searching for new cell injury mechanism(s) and effective therapeutic strategies has been a major challenge. It was recognized several decades ago that excessive intracellular Ca^{2+} accumulation and subsequent Ca^{2+} toxicity is critical for neuronal injury associated with brain ischemia. However, the exact pathway(s) underlying toxic Ca^{2+} loading responsible for ischemic brain injury remained elusive. For many years, accumulation of glutamate in the extracellular space and overactivation of postsynaptic glutamate receptors have been the central focus for Ca^{2+} toxicity in brain ischemia. However, the recent large-scale multicenter clinical trials using the antagonists of glutamate receptors as the neuroprotective agents have failed to demonstrate a satisfactory effect. Although multiple factors may have contributed to the failure of the trials, studies in the last 5 years have suggested that Ca^{2+} loading through several glutamate receptor-independent pathways, e.g., Ca^{2+}-permeable acid-sensing ion channels and TRPM7 channels, may contribute equally to the pathological loading of Ca^{2+} in ischemic brain. This review focuses on the role of acid-sensing ion channels in glutamate receptor-independent neuronal injury.

Keywords: Brain; Ischemia; Acidosis; Acid-sensing ion channel; Calcium; Neuron; Injury

Z. Xiong
Robert S. DowNeurobiology Laboratories, Legacy Clinical Research Center,
1225 NE 2nd Ave, Portland, OR 97232, USA
e-mail: zxiong@downeurobiology.org

2.1 Glutamate Excitotoxicity

Glutamate is the principal excitatory neurotransmitter in the central nervous system (CNS) (1–3). Activation of the postsynaptic glutamate receptors is critical for normal neurological functions such as learning and memory, movement and sensation (1, 4). One subtype of the glutamate receptors, the NMDA receptor, is a ligand-gated ion channel highly permeable to Ca^{2+} in addition to Na^+ and K^+ ions. The resultant increase in the concentration of intracellular Ca^{2+} is essential for normal physiological function of the receptor, and for Ca^{2+} toxicity in ischemic conditions (5–9).

Following the onset of ischemia, extracellular glutamate is rapidly elevated. The mechanisms of the elevation include enhanced efflux of glutamate and a reduction of glutamate uptake. Deprivation of oxygen/glucose and the shortage of energy supply results in depolarization of neurons and glial cells. Depolarization of these cells leads to synaptic release of glutamate from neurons and reversed transport of glutamate from astrocytes (10–12). Prolonged accumulation of glutamate in the extracellular space causes overactivation of the postsynaptic NMDA receptors, resulting in uncontrolled Ca^{2+} influx and overload of neurons with toxic amount of Ca^{2+} (13, 14). Excessive intracellular Ca^{2+} overload leads to inappropriate activation of several enzyme systems including nitric oxide synthase (NOS), proteases, phospholipase A2 (PLA2), and endonucleases (15–17), which cause final destruction of cells. Despite the factor that a number of studies, both in vitro and in vivo, demonstrated clear neuroprotection by the antagonists of glutamate receptors (18–20), recent clinical trials have failed to show a satisfactory protection by these agents (15, 21–23). Although the intolerance of glutamate antagonists and the inability to initiate treatment at an early stage of ischemia may have contributed to the failure of the trials, studies in the last 5 years have suggested that Ca^{2+} loading through several glutamate receptor-independent pathways, e.g., Ca^{2+}-permeable acid-sensing ion channels (24), TRPM7 channels (25), gap junction channels (26), and newly expressed SUR1-regulated NC_{Ca-ATP} channels (27) may also contribute to the pathological loading of Ca^{2+} in ischemic brain. This review focuses on the role of acid-sensing ion channels in glutamate receptor-independent ischemic neuronal injury.

2.2 Brain Acidosis Activates Acid-Sensing Ion Channels

2.2.1 Brain Acidosis

Since the activities of a large number of enzymes and membrane ion channels are highly pH-sensitive, a stable pH is vital for normal cellular function of all tissues (28). In the brain, electrical activity can elicit rapid pH change in both intracellular and extracellular compartments; mechanisms that govern H^+ homeostasis are therefore especially important. In physiological conditions, extracellular pH (pH_o) and intracellular pH (pH_i) in the brain are well maintained at ~7.3 and ~7.0 through a number of H$^+$-transporting mechanisms including Na$^+$/H$^+$ exchange, Na$^+$-driven Cl$^-$/HCO$_3^-$ exchange, Na$^+$–HCO$_3^-$ cotransport, and passive Cl$^-$/HCO$_3^-$ exchange (28). In the pathological conditions such as ischemic stroke, neurotrauma, and epileptic seizure, however, dramatic decrease of tissue pH, a condition termed acidosis, may take place due to various factors. In ischemia, for example, shortage of oxygen supply enhances anaerobic glucose metabolism resulting in accumulation of lactic acid (29, 30). Energy shortage and increased ATP hydrolysis also induces release of H$^+$. In addition, increased acid secretion from depolarized glial cells likely contributes to the tissue acidosis (28). In ischemic brain, extracellular pH typically falls to ~6.5, and during severe ischemia or under hyperglycemic conditions, it can fall below 6.0 (31, 32).

2.2.2 Acidosis Induces Neuronal Injury

Brain acidosis has long been known to cause neuronal injury (30, 33, 34). However, the detailed cellular and molecular mechanisms underlying acid-induced injury remained uncertain, multifactorial, and vague. Low tissue pH has been suggested to cause a nonselective denaturation of proteins and nucleic acids (35). It may also stimulate the Na$^+$/H$^+$ and Cl$^-$/HCO$_3^-$ exchangers, which leads to cellular edema and osmolysis (36). Low pH may hinder postischemic metabolic recovery by inhibiting mitochondrial energy metabolism and by impairing postischemic blood flow via vascular edema (29). In addition, the stimulation of pathologic free radical formation by acidosis has been shown in some systems (37). At the neurotransmitter level, profound acidosis inhibits astrocytic glutamate uptake, which may contribute to excitatory neuronal injury (38). Marked acidosis, with tissue pH < 5.5, may influence neuronal vulnerability indirectly by damaging glial cells (39–41). It has also been suggested that acidosis may potentiate AMPA/kinate receptor-mediated neuronal injury (42).

In contrast to these harmful effects, some in vitro studies have indicated that mild acidosis may in fact be beneficial by protecting neurons from excitotoxic injury (43–45). A good explanation is that decreasing pH inhibits NMDA receptor function (46, 47), thus attenuating excitatory neuronal injury. This finding may suggest that during ischemia where acidosis takes place, Ca^{2+} entry through NMDA channels might not play an essential role in neuronal injury since the activity of NMDA channels is largely inhibited by the acidosis. In addition to inhibiting the activity of NMDA channels, acidosis also inhibits several other voltage-gated and ligand-gated ion channels (48–56).

2.2.3 Acid-Sensing Ion Channels

In contrast to its modulating effect on other ion channels, which is predominantly inhibitory, it has been demonstrated by early studies that H$^+$ can induce a direct excitation of neurons (57–59). Further studies suggest that this acid-induced neuronal excitation was mediated by proton activation of a Na⁺-selective cation channel (60–62), which was late cloned and named as the proton-gated or acid-sensing ion channel – ASIC (63).

Up to date, four genes (*ASIC1–ASIC4*) encoding six different ASIC subunits have been identified. Each ASIC subunit contains about 500 amino acids. All ASICs belong to amiloride-sensitive epithelial Na⁺-channel/degenerin (ENaC/Deg) superfamily (64–67). Based on the biochemical analysis of ENaC (63, 68) and glycosylation studies of ASIC2a subunits (69), it has been proposed that each ASIC subunit consists of two transmembrane domains (TM I and TM II), linked by a large extracellular cystein-rich loop, and intracellular N and C termini.

The total number of subunits that are required to form a functional ASIC has been rather controversial. It was orginally believed that functional ASICs are tetrameric assemblies of homomeric or heteromeric subunits (70), which was based on some studies of ENaC (71) and supported by fluorescence resonance energy transfer analysis (72). Other studies of ENaC have suggested that this superfamily of ion channels may be assembled with as many as nine subunits (73–76). Very recent crystal structure analysis, however, demonstrated that ASICs might be, in fact, trimeric assemblies (77).

2.2.4 Tissue Distribution and Electrophysiological Properties of ASICs

Distribution of ASICs has been extensively studied using in situ hybridization, immunohistochemistry, and confirmed by electrophysiological recordings, whereas the properties of individual ASICs have been analyzed largely in heterologous expression systems with transient expression of individual ASIC subunit, and in native neurons by gene knockout approaches. ASIC1a is the most common subunit expressed in the neurons of peripheral sensory and CNS (63, 78, 79). Homomeric ASIC1a channels respond to extracellular H⁺ by mediating a fast and transient inward current with a threshold pH of ~7.0 and a $pH_{0.5}$ (the pH for half maximal activation of the channel) of ~6.2 (63). In addition to conducting Na⁺ ions, homomeric ASIC1a channels are permeable to Ca^{2+} (24, 63, 80). ASIC1b (ASIC1β) is a splice variant of ASIC1a with restricted expression in sensory neurons (81). It differs from the ASIC1a subunit in the first 175 amino acids including the first transmembrane region. Homomeric ASIC1b channels are activated by extracellular H⁺ with a similar transient current and a $pH_{0.5}$ of ~5.9 (81, 82). Unlike ASIC1a, however, homomeric ASIC1b channels do not show detectable Ca^{2+} permeability. ASIC2a is expressed broadly in peripheral sensory and CNS neurons. It shares 67% sequence identity with ASIC1a (83). Homomeric ASIC2a channels have a low sensitivity to H⁺ with a threshold pH for activation at ~5.5 and a $pH_{0.5}$ of ~4.4 (83, 84). ASIC2b is a splice variant of ASIC2a. Although widely expressed in peripheral sensory and central neurons, these subunits do not seem to form functional homomeric channels. However, they have been shown to be associated with other ASIC subunits (e.g., ASIC2a or ASIC3) to form heteromeric channels with distinct kinetics, pH dependences, and ion selectivities (83). ASIC3 is predominantly expressed in dorsal root ganglia (85). Homomeric ASIC3 channels are activated by H⁺ with a biphasic response – a fast desensitizing current followed by a sustained component (85, 86). These channels have a high sensitivity to H⁺ with a $pH_{0.5}$ of 6.7 (86). ASIC4 subunits show high level of expression in pituitary gland. Similar to ASIC2b, they do not seem to form functional homomeric channels (87, 88).

The relative contribution by ASIC1a, ASIC2a, or ASIC3 subunits to acid-evoked currents in native sensory and CNS neurons has been examined (24, 78, 89, 90). In medium-sized DRG neurons, for example, acid-activated currents match those recorded from heterologous cells expressing a mix of ASIC1, ASIC2, and ASIC3 subunits (89). Deletion of any one subunit did not abolish acid-activated currents, but altered currents in a manner consistent with heteromultimerization of the two remaining subunits. In rodent cortical and hippocampal neurons, however, knockout of *ASIC1* gene alone almost completely eliminated the acid-activated current (24, 78, 90), suggesting that ASIC1a is a predominant functional ASIC subunit in CNS neurons. Further studies suggested that the acid-activated currents in CNS neurons are mediated largely by a combination of ASIC1a homomeric channels and ASIC1a/ASIC2a heteromeric channels (90, 91).

2.3 Pharmacology of ASICs

2.3.1 Amiloride

The potassium-sparing natriuretic agent amiloride is a commonly used nonspecific blocker for ASICs. It inhibits the ASIC current and acid-induced increase of $[Ca^{2+}]_i$ with an IC_{50} of 10–50 µM (24, 63, 80, 81, 85, 92–94). Unlike the currents mediated by other ASICs, however, the sustained current mediated by ASIC3 channels is insensitive to amiloride (85, 95).

Based on the studies of ENaC, it is believed that amiloride inhibits ASICs by a direct blockade of the channel (96). The pre-TM II region of the channel is critical for the effect of amiloride. Mutation of Gly-430 in this region, for example, dramatically changed the sensitivity of ASIC2a current to amiloride (97). Consistent with its inhibition on the ASIC current, amiloride has been shown to suppress acid-induced pain in peripheral sensory system (98–101), and acidosis-mediated injury of CNS neurons (24, 80). However, because of its nonspecificity for other ion channels (e.g., ENaC and T-type Ca^{2+} channels) and ion exchange systems (e.g., Na^+/H^+ and Na^+/Ca^{2+} exchanger), it is less likely that amiloride will be used as a future neuroprotective agent in human subjects. It is worth mentioning that the normal activity of Na^+/Ca^{2+} exchanger, for example, is critical for maintaining the cellular Ca^{2+} homeostasis and the survival of neurons against delayed calcium deregulation caused by glutamate receptor activation (102). Inhibition of Na^+/Ca^{2+} exchange by amiloride may therefore compromise normal neuronal Ca^{2+} handling, thus potentiating the glutamate toxicity (102).

2.3.2 A-317567

A-317567, a small molecule structurally unrelated to amiloride, is another nonselective ASIC blocker (101). It inhibits the ASIC1a, ASIC2a, and ASIC3-like currents with an IC_{50} of 2–30 µM. Unlike amiloride, which has no effect on the slow component of the ASIC3 current, A-317567 blocks both the fast and the sustained ASIC3 currents. Also different from amiloride, A-317567 does not show diuresis or natriuresis activity (101), suggesting that it is more specific for ASICs than amiloride. Its inhibition of sustained ASIC3 current suggests that it might be potent in reducing acidosis-mediated chronic pain. Indeed, A-317567 has been shown to be effective in suppressing the pain in a rat model of thermal hyperalgesia at a dose tenfold lower than amiloride (101).

2.3.3 Psalmotoxin 1 (PcTX1)

Being a peptide toxin isolated from venom of the South American tarantula *Psalmopoeus cambridgei*, PcTX1 is a potent and specific inhibitor for homomeric ASIC1a channels (103). This toxin contains 40 amino acids cross-linked by three disulfide bridges. In heterologous expression systems, PcTX1 specifically inhibits the acid-activated current mediated by homomeric ASIC1a subunits with an IC_{50} of ~1 nM (103). At concentrations that effectively inhibit the ASIC1a current, it has no effect on the currents mediated by other configurations of ASICs (103), or known voltage-gated Na^+, K^+, Ca^{2+} channels as well as several ligand-gated ion channels (24). Unlike amiloride, which directly blocks the ASICs, PcTX1 acts as a gating modifier. It shifts the channel from its resting state toward the inactivated state by increasing its apparent affinity for protons (104).

2.3.4 APETx2

Being a peptide toxin isolated from sea anemone *Anthopleura elegantissima*, APETx2 is a potent and selective inhibitor for homomeric ASIC3 and ASIC3 containing channels (105). The toxin contains 42 amino acids, also cross-linked by three disulfide bridges. It reduces transient peak acid-evoked currents mediated by homomeric ASIC3 channels (105). In contrast to the peak ASIC3 current, the sustained component of the ASIC3 current is insensitive to APETx2. In addition to homomeric ASIC3 channels (IC_{50} = 63 nM for rat and 175 nM for human), APETx2 inhibits heteromeric ASIC3/1a (IC_{50} = 2 µM), ASIC3/1b (IC_{50} = 900 nM), and ASIC3/2b (IC_{50} = 117 nM). Homomeric ASIC1a, ASIC1b, ASIC2a, and heteromeric ASIC3/2a channels, on the other hand, are not sensitive to APETx2.

2.3.5 Nonsteroid Anti-inflammatory Drugs (NSAIDs)

NSAIDs are the most commonly used anti-inflammatory and analgesic agents. They inhibit the synthesis of prostaglandins (PGs), a main tissue inflammatory substance. A recent study demonstrated that NSAIDs also inhibit the activity of ASICs at their therapeutic doses for analgesic effects (106). Ibuprofen and flurbiprofen, for example, inhibit ASIC1a containing channels with an IC_{50} of 350 µM. Aspirin and salicylate inhibit ASIC3 containing channels with an IC_{50} of 260 µM, whereas diclofenac inhibits the same channels with an IC_{50} of 92 µM. In addition to a direct inhibition of the ASIC activity, NSAIDs also prevent inflammation-induced increase of ASIC expression in sensory neurons (106).

2.4 Modulation of ASIC Activity by Ischemia-Related Signals

In addition to a decreased pH, which activates the ASICs, various other biochemical changes take place in neurological diseases including brain ischemia. The majority of these changes have been shown to modulate the activities of several voltage-gated and ligand-gated channels. Similarly, the activities of ASICs are modulated by the ischemia-related signaling molecules (107).

2.4.1 Proteases

Brain ischemia is accompanied by increased protease activity (108). Following ischemia, blood-derived proteases such as thrombin, tissue plasminogen activator, and plasmin can have access to the interstitial space from compromised blood–brain barrier (108, 109). Previous studies have demonstrated that proteases modulate the activities of some ion channels, including ENaC (110–112). Recent studies by Poirot et al. demonstrated an ASIC1a-specifc modulation of the ASIC activity by serine proteases (113). Exposure of cells to trypsin, proteinase K, or chymotrypsin leads to a decreased ASIC1a current when the channel is activated by a pH drop from 7.4. Interestingly, if the channel is activated from a basal pH of 7, protease exposure increases, rather than decreases, the ASIC1a activity (113). The effects of proteases involve proteolytic activity on the channel protein, as the inhibition of trypsin by soybean trypsin inhibitor or modification of its catalytic site by TLCK prevents trypsin modulation of the ASIC1a current. Further studies demonstrate that trypsin modulates the ASIC1a function by cleaving ASIC1a subunits at Arg-145, which is in the N-terminal part of the extracellular loop (114).

It is likely that the bidirectional regulation of ASIC1a function in neurons may allow ASIC1a to adapt its gating to situations of changed extracellular pH, e.g., in ischemia. Since activation of ASIC1a is involved in acidosis-mediated ischemic brain injury (24, 80), modulation of ASIC1a by proteases may play some role in the pathology of brain ischemia.

2.4.2 Arachidonic Acid

Arachidonic acid is a major metabolite of membrane phospholipids. It has be shown to be involved in the physiological processes (115, 116) and pathology of various neurological disorders (116–118). During brain ischemia, the rise of $[Ca^{2+}]_i$ leads to the activation of PLA2, which results in an increased production of arachidonic acid (116, 117, 119). Earlier studies have shown that arachidonic acid modulates the activity of a variety of voltage-gated and ligand-gated ion channels (120–126). For example, it potentiates the opening of NMDA-gated channels (120, 126, 127). Similarly, a recent study demonstrated that arachidonic acid also enhances the ASIC current recorded in rat cerebellar Purkinje neurons (128). The potentiation of the ASIC currents appears to be produced directly by arachidonic acid and not by its derivatives, since including an agent known to block the breakdown of arachidonic acid does not affect the potentiation (128). The mechanism for arachidonic acid potentiation of ASICs is not fully understood. One explanation is that insertion of arachidonic acid into the membrane induces membrane stretch and that the ASICs are stretch-sensitive, as has been proposed for NMDA channels (127). In support of this argument, perfusion of neurons with hypotonic saline, which causes cell swelling and membrane stretch, mimics the potentiation of the ASIC current by arachidonic acid (128).

2.4.3 Lactate

The switch from aerobic to anaerobic metabolism of glucose during ischemia leads to increased production of lactic acid, which is further converted to lactate and H^+. In brain ischemia, concentrations of lactate between 12 and 20 mM have been reported in the extracellular space, which is dramatically higher than the ~1 mM lactate in the normal conditions (129, 130). In sensory neurons that innervate the heart, Immke and McCleskey demonstrated that addition of 15 mM lactate dramatically increased the amplitude of the ASIC current (131). Application of the same concentration of lactate at pH values that do not activate the ASICs (8.0 or 7.4) has no response. Thus, lactate acts by potentiating but not activating the ASICs. In COS-7 cells transfected with different subunit of ASICs, both ASIC1a and ASIC3 currents are potentiated by lactate (131). The effect of lactate does not require second messenger or signaling cascade because the potentiation persists in excised membrane patches. Since lactate has the ability to chelate the divalent cations including Ca^{2+} and Mg^{2+}, and the concentration of extracellular divalent cations, particularly Ca^{2+}, has a modulatory role on various membrane receptors and ion channels (132–134), the authors hypothesized that potentiation of the ASICs may be due to the chelation of divalent cations. Indeed, adjusting the concentrations of Ca^{2+} and Mg^{2+} eliminates the effect of lactate, whereas reducing the divalent concentrations mimics the effect of lactate (131). Similar to the cardiac sensory neurons, potentiation of the ASIC current by lactate has been reported in cerebellar Purkinje neurons (128).

2.4.4 Glucose

Depletion of glucose, along with oxygen, is a major neurochemical change associated with ischemic brain injury. However, it was unclear whether a change in glucose concentration affects the activity of ASICs. A recent report suggests that glucose depletion/reduction potentiates the ASIC current and acid-induced increase of $[Ca^{2+}]_i$ (135). The potentiation of ASIC currents is associated with a significant leftward shift in pH dose-response curve, or an increase in the affinity of ASICs to H^+. The potentiation of ASIC current by glucose depletion/reduction is independent of the level of extracellular Ca^{2+}, Zn^{2+}, and intracellular ATP. With ASICs expressed in Chinese hamster ovary cells, glucose depletion potentiates the currents mediated by homomeric ASIC1a channels, without affecting the currents mediated by homomeric ASICb, ASIC2a, and ASIC3 channels. Current-clamp recording and calcium-imaging experiments demonstrate that glucose depletion increases both acid-induced membrane depolarization and increase of $[Ca^{2+}]_i$. These findings suggest that the activities of ASICs and ASIC-mediated injury of CNS neurons are tightly modulated by the concentrations of glucose. The exact mechanism underlying glucose modulation of ASICs is unclear (135).

2.5 Activation of ASICs Induces Neuronal Excitation and Increased Intracellular Ca^{2+}

ASICs are Na^+-selective channels, which have a reversal potential near Na^+ equilibrium potential (~+60 mV). Activations of ASICs at normal resting potentials produce large inward currents, which result in membrane depolarization and the excitation of neurons (91, 136). This ASIC-mediated membrane depolarization may, in turn, facilitate the activation of voltage-gated Ca^{2+} channels and NMDA receptor-gated channels (78), further promoting neuronal excitation and $[Ca^{2+}]_i$ accumulation. For homomeric ASIC1a channels, acid activation also induces Ca^{2+} influx directly through these channels (24, 63, 80).

The Ca^{2+} permeability of ASICs in CNS neurons has been characterized using fluorescent Ca^{2+} imaging and ion-substitution protocols (24, 80). In mouse cortical and hippocampal neurons, activation of ASICs by decreasing pH_o induces elevated $[Ca^{2+}]_i$. This acid-induced increase of $[Ca^{2+}]_i$ could be recorded in the presence of blockers of common voltage-gated and ligand-gated Ca²⁺ channels (24), indicating Ca²⁺ entry directly through ASICs. This argument is further supported by the finding that acid-induced increase of [Ca²⁺]$_i$ is eliminated by specific and nonspecific ASIC1a blockade, or by *ASIC1* gene knockout (24, 80). Consistent with the finding of fluorescent imaging studies, patch-clamp recording shows acid-activated inward current when extracellular solution contains Ca²⁺ as the only conducting cation (24). Thus, homomeric ASIC1a channels constitute an important Ca²⁺ entry pathway for CNS neurons.

2.6 ASIC1a Activation Plays an Important Role in Acidosis-Induced Neuronal Injury

The widespread expression of ASIC1a in the brain, its activation by pH drop to the level commonly seen in ischemic brain, and its demonstrated permeability to Ca²⁺ suggest a role for ASICs in the pathology of brain ischemia. Indeed, a series of recent studies have demonstrated a clear link between ASIC1a activation, [Ca²⁺]$_i$ accumulation, and acidosis-mediated brain injury (24, 80, 137, 138) (also see Fig. 2.1). In cultured mouse cortical neurons, for example, activation of ASICs by brief acid incubation induces glutamate receptor-independent Ca²⁺-dependent neuronal injury inhibited by specific and nonspecific ASIC1a blockade, and by *ASIC1* gene knockout (24). Consistent with ASIC1a activation, reducing [Ca²⁺]$_o$, which lowers the driving force for Ca²⁺ entry through ASICs, also decreases the acid-induced neuronal injury. In an in vivo rodent model of focal ischemia, intracerebroventricular injection of the ASIC1a blocker, or *ASIC1* gene knockout, reduces the infarct volume induced by transient or permanent focal ischemia by up to 60% (24, 138). Attenuating brain acidosis by intracerebroventricular administration of NaHCO₃ is

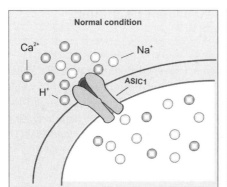
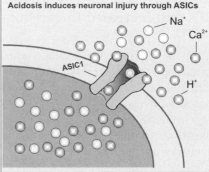

Fig. 2.1 Role of ASIC activation in acidosis-induced neuronal injury. Left panel represents neurons in nonischemic conditions where the concentration of extracelullar proton is low and the ASIC channels remain closed. Right panel represents neurons in ischemic conditions where the concentration of extracellular proton rises, resulting in activation of ASIC channels and the flux of Na⁺ and Ca²⁺ ions into neurons, which leads to neuronal injury (*See Color Plates*)

also protective. Thus, Ca^{2+}-permeable ASIC1a represents a novel pharmacological target for ischemic brain injury.

2.7 Evidence of a Developmental Change of ASICs

Subunit expression and the density of various ion channels may undergo developmental changes. For example, different isoforms of GABA receptors are expressed in neonate and adult animal, and the properties of the GABAergic inhibitory postsynaptic currents change with development (139, 140). The expression of NR2A and NR2B subunits of NMDA receptors also changes dramatically with development (141).

Since individual ASIC subunits have dramatic difference in their pH sensitivity, Ca^{2+} permeability, and pharmacological properties, changing the level of individual ASIC subunit or relative ratio of different subunits may have significant influence on the electrophysiological/pharmacological properties of these channels in native neurons and their role in both physiological and pathological conditions. Similar to other ion channels, the expression of ASIC1a, ASIC2a, and the electrophysiological/pharmacological properties of ASICs in CNS neurons appear to change at different developmental stages (142). In cultured and acutely dissociated mouse cortical neurons, for example, the amplitude and the density of ASIC current increase with neuronal maturation. A shift in H^+ dose-response relationship toward more acidic pH and a decrease in the value of $pH_{0.5}$ are observed in mature neurons. In addition, the desensitization of ASIC currents becomes slower whereas the rate of recovery from the current desensitization becomes faster in mature neurons. These findings suggest an increase in ASIC2a subunit expression with neuronal maturation. Consistent with this notion, an increase in the sensitivity of ASIC currents to Zn^{2+} potentiation, a reduction in the currents to PcTX1 blockade, and a decrease in acid-evoked increase of intracellular $[Ca^{2+}]_i$ are observed in mature neurons. Indeed, RT-PCR reveals an increase in the relative ratio of ASIC2a/ASIC1a mRNA in mature neurons (142). These results suggest that relative contributions of ASIC1a and ASIC2a subunits to overall ASIC-mediated responses in CNS neurons may undergo significant developmental changes. Future studies will determine how this developmental change in ASIC subunit expression affects the function of these channels in physiological and pathological conditions.

References

1. Nakanishi S. Molecular diversity of glutamate receptors and implications for brain function. Science 1992; 258:597–603.
2. Curtis DR, Watkins JC. Acidic amino acids with strong excitatory actions on mammalian neurones. J Physiol 1960; 166:1–14.
3. Krnjevic K. Glutamate and gamma-aminobutyric acid in brain. Nature 1970; 228(267):119–124.
4. Gasic GP, Hollmann M. Molecular neurobiology of glutamate receptors. Annu Rev Physiol 1992; 54:507–536.

5. Mori H, Mishina M. Structure and function of the NMDA receptor channel. Neuropharmacology 1995; 34(10):1219–1237.
6. Sucher NJ, Awobuluyi M, Choi YB, Lipton SA. NMDA receptors: from genes to channels. Trends Pharmacol Sci 1996; 17(10):348–355.
7. Tymianski M. Cytosolic calcium concentrations and cell death in vitro. Adv Neurol 1996; 71:85–105.
8. Rothman SM, Olney JW. Excitotoxicity and the NMDA receptor –still lethal after eight years. Trends Neurosci 1995; 18:57–58.
9. Choi DW. Calcium-mediated neurotoxicity: relationship to specific channel types and role in ischemic damage. Trends Neurosci 1988; 11(10):465–469.
10. Nicholls D, Attwell D. The release and uptake of excitatory amino acids. Trends Pharmacol Sci 1990; 11:462–467.
11. Siesjo BK. Pathophysiology and treatment of focal cerebral ischemia. II. Mechanisms of damage and treatment. J Neurosurg 1992; 77(3):337–354.
12. Benveniste H, Drejer J, Schousboe A, Diemer NH. Elevation of the extracellular concentrations of glutamate and aspartate in rat hippocampus during transient cerebral ischemia monitored by intracerebral microdialysis. J Neurochem 1984; 43(5):1369–1374.
13. Choi DW. Glutamate neurotoxicity and diseases of the nervous system. Neuron 1988; 1(8):623–634.
14. Choi DW. Excitotoxic cell death. J Neurobiol 1992; 23(9):1261–1276.
15. Lee JM, Zipfel GJ, Choi DW. The changing landscape of ischaemic brain injury mechanisms. Nature 1999; 399(6738 Suppl):A7–A14.
16. Simonian NA, Coyle JT. Oxidative stress in neurodegenerative diseases. Annu Rev Pharmacol Toxicol 1996; 36:83–106.
17. Coyle JT, Puttfarcken P. Oxidative stress, glutamate, and neurodegenerative disorders. Science 1993; 262(5134):689–695.
18. Simon RP, Swan JH, Griffiths T, Meldrum BS. Blockade of *N*-methyl-D-aspartate receptors may protect against ischemic damage in the brain. Science 1984; 226(4676):850–852.
19. Wieloch T. Hypoglycemia-induced neuronal damage prevented by an *N*-methyl-D-aspartate antagonist. Science 1985; 230(4726):681–683.
20. Tymianski M, Charlton MP, Carlen PL, Tator CH. Source specificity of early calcium neurotoxicity in cultured embryonic spinal neurons. J Neurosci 1993; 13(5):2085–2104.
21. Muir KW. Glutamate-based therapeutic approaches: clinical trials with NMDA antagonists. Curr Opin Pharmacol 2006; 6(1):53–60.
22. Hoyte L, Barber PA, Buchan AM, Hill MD. The rise and fall of NMDA antagonists for ischemic stroke. Curr Mol Med 2004; 4(2):131–136.
23. Ikonomidou C, Turski L. Why did NMDA receptor antagonists fail clinical trials for stroke and traumatic brain injury? Lancet Neurol 2002; 1(6):383–386.
24. Xiong ZG, Zhu XM, Chu XP, Minami M, Hey J, Wei WL et al. Neuroprotection in ischemia: blocking calcium-permeable acid-sensing ion channels. Cell 2004; 118(6):687–698.
25. Aarts M, Iihara K, Wei WL, Xiong ZG, Arundine M, Cerwinski W et al. A key role for TRPM7 channels in anoxic neuronal death. Cell 2003; 115(7):863–877.
26. Thompson RJ, Zhou N, MacVicar BA. Ischemia opens neuronal gap junction hemichannels. Science 2006; 312(5775):924–927.
27. Simard JM, Chen M, Tarasov KV, Bhatta S, Ivanova S, Melnitchenko L et al. Newly expressed SUR1-regulated NC(Ca-ATP) channel mediates cerebral edema after ischemic stroke. Nat Med 2006; 12(4):433–440.
28. Chesler M. Regulation and modulation of pH in the brain. Physiol Rev 2003; 83(4): 1183–1221.
29. Hillered L, Smith ML, Siesjo BK. Lactic acidosis and recovery of mitochondrial function following forebrain ischemia in the rat. J Cereb Blood Flow Metab 1985; 5(2):259–266.
30. Siesjo BK. Acidosis and ischemic brain damage. Neurochem Pathol 1988; 9:31–88.
31. Nedergaard M, Kraig RP, Tanabe J, Pulsinelli WA. Dynamics of interstitial and intracellular pH in evolving brain infarct. Am J Physiol 1991; 260(3 Part 2):R581–R588.

32. Rehncrona S. Brain acidosis. Ann Emerg Med 1985; 14(8):770–776.
33. Siesjo BK, Katsura KI, Kristian T, Li PA, Siesjo P. Molecular mechanisms of acidosis-mediated damage. Acta Neurochir Suppl 1996; 66:8–14.
34. Kristian T, Katsura K, Gido G, Siesjo BK. The influence of pH on cellular calcium influx during ischemia. Brain Res 1994; 641(2):295–302.
35. Kalimo H, Rehncrona S, Soderfeldt B, Olsson Y, Siesjo BK. Brain lactic acidosis and ischemic cell damage. II. Histopathology. J Cereb Blood Flow Metab 1981; 1(3): 313–327.
36. Kimelberg HK, Barron KD, Bourke RS, Nelson LR, Cragoe EJ. Brain anti-cytoxic edema agents. Prog Clin Biol Res 1990; 361:363–385.
37. Rehncrona S, Hauge HN, Siesjo BK. Enhancement of iron-catalyzed free radical formation by acidosis in brain homogenates: differences in effect by lactic acid and CO_2. J Cereb Blood Flow Metab 1989; 9(1):65–70.
38. Swanson RA, Farrell K, Simon RP. Acidosis causes failure of astrocyte glutamate uptake during hypoxia. J Cereb Blood Flow Metab 1995; 15(3):417–424.
39. Tombaugh GC, Sapolsky RM. Evolving concepts about the role of acidosis in ischemic neuropathology. J Neurochem 1993; 61(3):793–803.
40. Goldman SA, Pulsinelli WA, Clarke WY, Kraig RP, Plum F. The effects of extracellular acidosis on neurons and glia in vitro. J Cereb Blood Flow Metab 1989; 9(4):471–477.
41. Giffard RG, Monyer H, Choi DW. Selective vulnerability of cultured cortical glia to injury by extracellular acidosis. Brain Res 1990; 530(1):138–141.
42. McDonald JW, Bhattacharyya T, Sensi SL, Lobner D, Ying HS, Canzoniero LM et al. Extracellular acidity potentiates AMPA receptor-mediated cortical neuronal death. J Neurosci 1998; 18(16):6290–6299.
43. Giffard RG, Monyer H, Christine CW, Choi DW. Acidosis reduces NMDA receptor activation, glutamate neurotoxicity, and oxygen-glucose deprivation neuronal injury in cortical cultures. Brain Res 1990; 506(2):339–342.
44. Kaku DA, Giffard RG, Choi DW. Neuroprotective effects of glutamate antagonists and extracellular acidity. Science 1993; 260:1516–1518.
45. Sapolsky RM, Trafton J, Tombaugh GC. Excitotoxic neuron death, acidotic endangerment, and the paradox of acidotic protection. Adv Neurol 1996; 71:237–244.
46. Tang CM, Dichter M, Morad M. Modulation of the *N*-methyl-D-aspartate channel by extracellular H$^+$. Proc Natl Acad Sci USA 1990; 87(16):6445–6449.
47. Traynelis SF, Cull-Candy SG. Proton inhibition of *N*-methyl-D-aspartate receptors in cerebellar neurons. Nature 1990; 345(6273):347–350.
48. Chu XP, Zhu XM, Wei WL, Li GH, Simon RP, MacDonald JF et al. Acidosis decreases low Ca(2$^+$)-induced neuronal excitation by inhibiting the activity of calcium-sensing cation channels in cultured mouse hippocampal neurons. J Physiol 2003; 550 (Part 2):385–399.
49. Chen XH, Bezprozvanny I, Tsien RW. Molecular basis of proton block of L-type Ca^{2+} channels. J Gen Physiol 1996; 108(5):363–374.
50. Lopes CM, Gallagher PG, Buck ME, Butler MH, Goldstein SA. Proton block and voltage gating are potassium-dependent in the cardiac leak channel Kcnk3. J Biol Chem 2000; 275(22):16969–16978.
51. Tombaugh GC, Somjen GG. Effects of extracellular pH on voltage-gated Na$^+$ K$^+$ and Ca^{2+} currents in isolated rat CA1 neurons. J Physiol 1996; 493 (Part 3):719–732.
52. Kaibara M, Kameyama M. Inhibition of the calcium channel by intracellular protons in single ventricular myocytes of the guinea-pig. J Physiol 1988; 403:621–640.
53. Khoo C, Helm J, Choi HB, Kim SU, McLarnon JG. Inhibition of store-operated Ca(2$^+$) influx by acidic extracellular pH in cultured human microglia. Glia 2001; 36(1):22–30.
54. Claydon TW, Boyett MR, Sivaprasadarao A, Ishii K, Owen JM, O'Beirne HA et al. Inhibition of the K$^+$ channel kv1.4 by acidosis: protonation of an extracellular histidine slows the recovery from N-type inactivation. J Physiol 2000; 526(2):253–264.
55. Zhu G, Chanchevalap S, Cui N, Jiang C. Effects of intra- and extracellular acidifications on single channel Kir2.3 currents. J Physiol 1999; 516 (Part 3):699–710.

56. Zhai J, Peoples RW, Li C. Proton inhibition of GABA-activated current in rat primary sensory neurons. Pflugers Arch 1998; 435(4):539–545.
57. Steen KH, Reeh PW, Anton F, Handwerker HO. Protons selectively induce lasting excitation and sensitization to mechanical stimulation of nociceptors in rat skin, in vitro. J Neurosci 1992; 12(1):86–95.
58. Gruol DL, Barker JL, Huang LY, MacDonald JF, Smith TG, Jr. Hydrogen ions have multiple effects on the excitability of cultured mammalian neurons. Brain Res 1980; 183(1): 247–252.
59. Frederickson RC, Jordan LM, Phillis JW. The action of noradrenaline on cortical neurons: effects of pH. Brain Res 1971; 35(2):556–560.
60. Krishtal OA, Pidoplichko VI. A receptor for protons in the nerve cell membrane. Neuroscience 1980; 5(12):2325–2327.
61. Ueno S, Nakaye T, Akaike N. Proton-induced sodium current in freshly dissociated hypothalamic neurones of the rat. J Physiol 1992; 447:309–327.
62. Kovalchuk Y, Krishtal OA, Nowycky MC. The proton-activated inward current of rat sensory neurons includes a calcium component. Neurosci Lett 1990; 115(2–3):237–242.
63. Waldmann R, Champigny G, Bassilana F, Heurteaux C, Lazdunski M. A proton-gated cation channel involved in acid-sensing. Nature 1997; 386(6621):173–177.
64. Alvarez dLR, Canessa CM, Fyfe GK, Zhang P. Structure and regulation of amiloride-sensitive sodium channels. Annu Rev Physiol 2000; 62:573–594.
65. Corey DP, Garcia-Anoveros J. Mechanosensation and the DEG/ENaC ion channels. Science 1996; 273(5273):323–324.
66. Waldmann R, Lazdunski M. H(+)-gated cation channels: neuronal acid sensors in the ENaC/DEG family of ion channels. Curr Opin Neurobiol 1998; 8(3):418–424.
67. Waldmann R, Champigny G, Lingueglia E, De Weille J, Heurteaux C, Lazdunski M. H(+)-gated cation channels. Ann N Y Acad Sci 1999; 868:67–76.
68. Renard S, Lingueglia E, Voilley N, Lazdunski M, Barbry P. Biochemical analysis of the membrane topology of the amiloride-sensitive Na$^+$ channel. J Biol Chem 1994; 269(17): 12981–12986.
69. Saugstad JA, Roberts JA, Dong J, Zeitouni S, Evans RJ. Analysis of the membrane topology of the acid-sensing ion channel 2a. J Biol Chem 2004; 279(53):55514–55519.
70. Krishtal O. The ASICs: signaling molecules? Modulators? Trends Neurosci 2003; 26(9): 477–483.
71. Dijkink L, Hartog A, van Os CH, Bindels RJ. The epithelial sodium channel (ENaC) is intracellularly located as a tetramer. Pflugers Arch 2002; 444(4):549–555.
72. Gao Y, Liu SS, Qiu S, Cheng W, Zheng J, Luo JH. Fluorescence resonance energy transfer analysis of subunit assembly of the ASIC channel. Biochem Biophys Res Commun 2007; 359(1):143–150.
73. Eskandari S, Snyder PM, Kreman M, Zampighi GA, Welsh MJ, Wright EM. Number of subunits comprising the epithelial sodium channel. J Biol Chem 1999; 274(38):27281–27286.
74. Firsov D, Gautschi I, Merillat AM, Rossier BC, Schild L. The heterotetrameric architecture of the epithelial sodium channel (ENaC). EMBO J 1998; 17(2):344–352.
75. Staruschenko A, Medina JL, Patel P, Shapiro MS, Booth RE, Stockand JD. Fluorescence resonance energy transfer analysis of subunit stoichiometry of the epithelial Na$^+$ channel. J Biol Chem 2004; 279(26):27729–27734.
76. Snyder PM, Cheng C, Prince LS, Rogers JC, Welsh MJ. Electrophysiological and biochemical evidence that DEG/ENaC cation channels are composed of nine subunits. J Biol Chem 1998; 273(2):681–684.
77. Jasti J, Furukawa H, Gonzales EB, Gouaux E. Structure of acid-sensing ion channel 1 at 1.9 A resolution and low pH. Nature 2007; 449(7160):316–323.
78. Wemmie JA, Chen J, Askwith CC, Hruska-Hageman AM, Price MP, Nolan BC et al. The acid-activated ion channel ASIC contributes to synaptic plasticity, learning, and memory. Neuron 2002; 34(3):463–477.

79. Alvarez dLR, Krueger SR, Kolar A, Shao D, Fitzsimonds RM, Canessa CM. Distribution, subcellular localization and ontogeny of ASIC1 in the mammalian central nervous system. J Physiol 2003; 546(Part 1):77–87.
80. Yermolaieva O, Leonard AS, Schnizler MK, Abboud FM, Welsh MJ. Extracellular acidosis increases neuronal cell calcium by activating acid-sensing ion channel 1a. Proc Natl Acad Sci USA 2004; 101(17):6752–6757.
81. Chen CC, England S, Akopian AN, Wood JN. A sensory neuron-specific, proton-gated ion channel. Proc Natl Acad Sci USA 1998; 95(17):10240–10245.
82. Bassler EL, Ngo-Anh TJ, Geisler HS, Ruppersberg JP, Grunder S. Molecular and functional characterization of acid-sensing ion channel (ASIC) 1b. J Biol Chem 2001; 276(36):33782–33787.
83. Lingueglia E, De Weille JR, Bassilana F, Heurteaux C, Sakai H, Waldmann R et al. A modulatory subunit of acid sensing ion channels in brain and dorsal root ganglion cells. J Biol Chem 1997; 272(47):29778–29783.
84. Waldmann R, Champigny G, Voilley N, Lauritzen I, Lazdunski M. The mammalian degenerin MDEG, an amiloride-sensitive cation channel activated by mutations causing neurodegeneration in *Caenorhabditis elegans*. J Biol Chem 1996; 271(18): 10433–10436.
85. Waldmann R, Bassilana F, de Weille J, Champigny G, Heurteaux C, Lazdunski M. Molecular cloning of a non-inactivating proton-gated Na$^+$ channel specific for sensory neurons. J Biol Chem 1997; 272(34):20975–20978.
86. Sutherland SP, Benson CJ, Adelman JP, McCleskey EW. Acid-sensing ion channel 3 matches the acid-gated current in cardiac ischemia-sensing neurons. Proc Natl Acad Sci USA 2001; 98(2):711–716.
87. Akopian AN, Chen CC, Ding Y, Cesare P, Wood JN. A new member of the acid-sensing ion channel family. Neuroreport 2000; 11(10):2217–2222.
88. Grunder S, Geissler HS, Bassler EL, Ruppersberg JP. A new member of acid-sensing ion channels from pituitary gland. Neuroreport 2000; 11(8):1607–1611.
89. Benson CJ, Xie J, Wemmie JA, Price MP, Henss JM, Welsh MJ et al. Heteromultimers of DEG/ENaC subunits form H$^+$-gated channels in mouse sensory neurons. Proc Natl Acad Sci USA 2002; 99(4):2338–2343.
90. Askwith CC, Wemmie JA, Price MP, Rokhlina T, Welsh MJ. ASIC2 modulates ASIC1 H$^+$-activated currents in hippocampal neurons. J Biol Chem 2004; 279(18):18296–18305.
91. Baron A, Waldmann R, Lazdunski M. ASIC-like, proton-activated currents in rat hippocampal neurons. J Physiol 2002; 539 (Part 2):485–494.
92. Bassilana F, Champigny G, Waldmann R, De Weille JR, Heurteaux C, Lazdunski M. The acid-sensitive ionic channel subunit ASIC and the mammalian degenerin MDEG form a heteromultimeric H$^+$-gated Na$^+$ channel with novel properties. J Biol Chem 1997; 272(46):28819–28822.
93. Vukicevic M, Kellenberger S. Modulatory effects of acid-sensing ion channels on action potential generation in hippocampal neurons. Am J Physiol Cell Physiol 2004; 287(3):C682–C690.
94. Wu LJ, Duan B, Mei YD, Gao J, Chen JG, Zhuo M et al. Characterization of acid-sensing ion channels in dorsal horn neurons of rat spinal cord. J Biol Chem 2004; 279(42): 43716–43724.
95. Benson CJ, Eckert SP, McCleskey EW. Acid-evoked currents in cardiac sensory neurons: a possible mediator of myocardial ischemic sensation. Circ Res 1999; 84(8):921–928.
96. Schild L, Schneeberger E, Gautschi I, Firsov D. Identification of amino acid residues in the alpha, beta, and gamma subunits of the epithelial sodium channel (ENaC) involved in amiloride block and ion permeation. J Gen Physiol 1997; 109(1):15–26.
97. Champigny G, Voilley N, Waldmann R, Lazdunski M. Mutations causing neurodegeneration in *Caenorhabditis elegans* drastically alter the pH sensitivity and inactivation of the mammalian H$^+$-gated Na$^+$ channel MDEG1. J Biol Chem 1998; 273(25):15418–15422.

98. Ugawa S, Ueda T, Ishida Y, Nishigaki M, Shibata Y, Shimada S. Amiloride-blockable acid-sensing ion channels are leading acid sensors expressed in human nociceptors. J Clin Invest 2002; 110(8):1185–1190.
99. Jones NG, Slater R, Cadiou H, McNaughton P, McMahon SB. Acid-induced pain and its modulation in humans. J Neurosci 2004; 24(48):10974–10979.
100. Sluka KA, Price MP, Breese NM, Stucky CL, Wemmie JA, Welsh MJ. Chronic hyperalgesia induced by repeated acid injections in muscle is abolished by the loss of ASIC3, but not ASIC1. Pain 2003; 106(3):229–239.
101. Dube GR, Lehto SG, Breese NM, Baker SJ, Wang X, Matulenko MA et al. Electrophysiological and in vivo characterization of A-317567, a novel blocker of acid sensing ion channels. Pain 2005; 117(1–2):88–96.
102. Bano D, Young KW, Guerin CJ, Lefeuvre R, Rothwell NJ, Naldini L et al. Cleavage of the plasma membrane Na^+/Ca^{2+} exchanger in excitotoxicity. Cell 2005; 120(2):275–285.
103. Escoubas P, De Weille JR, Lecoq A, Diochot S, Waldmann R, Champigny G et al. Isolation of a tarantula toxin specific for a class of proton-gated Na^+ channels. J Biol Chem 2000; 275(33):25116–25121.
104. Chen X, Kalbacher H, Grunder S. The tarantula toxin psalmotoxin 1 inhibits acid-sensing ion channel (ASIC) 1a by increasing its apparent H^+ affinity. J Gen Physiol 2005; 126(1): 71–79.
105. Diochot S, Baron A, Rash LD, Deval E, Escoubas P, Scarzello S et al. A new sea anemone peptide, APETx2, inhibits ASIC3, a major acid-sensitive channel in sensory neurons. EMBO J 2004; 23(7):1516–1525.
106. Voilley N, de Weille J, Mamet J, Lazdunski M. Nonsteroid anti-inflammatory drugs inhibit both the activity and the inflammation-induced expression of acid-sensing ion channels in nociceptors. J Neurosci 2001; 21(20):8026–8033.
107. Xu TL, Xiong ZG. Dynamic regulation of acid-sensing ion channels by extracellular and intracellular modulators. Curr Med Chem 2007; 14(16):1753–1763.
108. Gingrich MB, Traynelis SF. Serine proteases and brain damage – is there a link?. Trends Neurosci 2000; 23(9):399–407.
109. Vivien D, Buisson A. Serine protease inhibitors: novel therapeutic targets for stroke?. J Cereb Blood Flow Metab 2000; 20(5):755–764.
110. Chraibi A, Vallet V, Firsov D, Hess SK, Horisberger JD. Protease modulation of the activity of the epithelial sodium channel expressed in Xenopus oocytes. J Gen Physiol 1998; 111(1):127–138.
111. Holt JC, Lioudyno M, Athas G, Garcia MM, Perin P, Guth PS. The effect of proteolytic enzymes on the alpha9-nicotinic receptor-mediated response in isolated frog vestibular hair cells. Hear Res 2001; 152(1–2):25–42.
112. Nicole O, Docagne F, Ali C, Margaill I, Carmeliet P, MacKenzie ET et al. The proteolytic activity of tissue-plasminogen activator enhances NMDA receptor-mediated signaling. Nat Med 2001; 7(1):59–64.
113. Poirot O, Vukicevic M, Boesch A, Kellenberger S. Selective regulation of acid-sensing ion channel 1 by serine proteases. J Biol Chem 2004; 279(37):38448–38457.
114. Vukicevic M, Weder G, Boillat A, Boesch A, Kellenberger S. Trypsin cleaves acid-sensing ion channel 1a in a domain that is critical for channel gating. J Biol Chem 2006; 281(2):714–722.
115. Sang N, Chen C. Lipid signaling and synaptic plasticity. Neuroscientist 2006; 12(5): 425–434.
116. Farooqui AA, Horrocks LA. Phospholipase A2-generated lipid mediators in the brain: the good, the bad, and the ugly. Neuroscientist 2006; 12(3):245–260.
117. Muralikrishna AR, Hatcher JF. Phospholipase A2, reactive oxygen species, and lipid peroxidation in cerebral ischemia. Free Radic Biol Med 2006; 40(3):376–387.
118. Farooqui AA, Ong WY, Horrocks LA. Inhibitors of brain phospholipase A2 activity: their neuropharmacological effects and therapeutic importance for the treatment of neurologic disorders. Pharmacol Rev 2006; 58(3):591–620.
119. Rehncrona S, Westerberg E, Akesson B, Siesjo BK. Brain cortical fatty acids and phospholipids during and following complete and severe incomplete ischemia. J Neurochem 1982; 38(1):84–93.

120. Miller B, Sarantis M, Traynelis SF, Attwell D. Potentiation of NMDA receptor currents by arachidonic acid. Nature 1992; 355(6362):722–725.
121. Nagano N, Imaizumi Y, Watanabe M. Modulation of calcium channel currents by arachidonic acid in single smooth muscle cells from vas deferens of the guinea-pig. Br J Pharmacol 1995; 116(2):1887–1893.
122. Mignen O, Thompson JL, Shuttleworth TJ. Arachidonate-regulated Ca^{2+}-selective (ARC) channel activity is modulated by phosphorylation and involves an A-kinase anchoring protein. J Physiol 2005; 567 (Part 3):787–798.
123. Angelova P, Muller W. Oxidative modulation of the transient potassium current IA by intracellular arachidonic acid in rat CA1 pyramidal neurons. Eur J Neurosci 2006; 23(9):2375–2384.
124. Hu HZ, Xiao R, Wang C, Gao N, Colton CK, Wood JD et al. Potentiation of TRPV3 channel function by unsaturated fatty acids. J Cell Physiol 2006; 208(1):201–212.
125. Keros S, McBain CJ. Arachidonic acid inhibits transient potassium currents and broadens action potentials during electrographic seizures in hippocampal pyramidal and inhibitory interneurons. J Neurosci 1997; 17(10):3476–3487.
126. Paoletti P, Ascher P. Mechanosensitivity of NMDA receptors in cultured mouse central neurons. Neuron 1994; 13:645–655.
127. Casado M, Ascher P. Opposite modulation of NMDA receptors by lysophospholipids and arachidonic acid: common features with mechanosensitivity. J Physiol 1998; 513 (Part 2): 317–330.
128. Allen NJ, Attwell D. Modulation of ASIC channels in rat cerebellar Purkinje neurons by ischemia-related signals. J Physiol 2002; 543(2):521–529.
129. Schurr A, Rigor BM. Brain anaerobic lactate production: a suicide note or a survival kit? Dev Neurosci 1998; 20(4–5):348–357.
130. Schurr A. Lactate, glucose and energy metabolism in the ischemic brain (Review). Int J Mol Med 2002; 10(2):131–136.
131. Immke DC, McCleskey EW. Lactate enhances the acid-sensing Na^+ channel on ischemia-sensing neurons. Nat Neurosci 2001; 4(9):869–870.
132. Xiong ZG, MacDonald JF. Sensing of extracellular calcium by neurones. Can J Physiol Pharmacol 1999; 77(9):715–721.
133. Hess P, Lansman JB, Tsien RW. Calcium channel selectivity for divalent and monovalent cations. Voltage and concentration dependence of single channel current in ventricular heart cells. J Gen Physiol 1986; 88(3):293–319.
134. Zhou W, Jones SW. Surface charge and calcium channel saturation in bullfrog sympathetic neurons. J Gen Physiol 1995; 105(4):441–462.
135. Chu XP, Simon RP, Xiong ZG. Modulation of acid-sensing ion channels by glucose. Soc Neurosci Abstr 2006; 83:3.
136. Lilley S, LeTissier P, Robbins J. The discovery and characterization of a proton-gated sodium current in rat retinal ganglion cells. J Neurosci 2004; 24(5):1013–1022.
137. Gao J, Duan B, Wang DG, Deng XH, Zhang GY, Xu L et al. Coupling between NMDA receptor and acid-sensing ion channel contributes to ischemic neuronal death. Neuron 2005; 48(4):635–646.
138. Pignataro G, Simon RP, Xiong ZG. Prolonged activation of ASIC1a and the time window for neuroprotection in cerebral ischaemia. Brain 2007; 130 (Part 1):151–158.
139. Dunning DD, Hoover CL, Soltesz I, Smith MA, O'Dowd DK. GABA(A) receptor-mediated miniature postsynaptic currents and alpha-subunit expression in developing cortical neurons. J Neurophysiol 1999; 82(6):3286–3297.
140. Taketo M, Yoshioka T. Developmental change of GABA(A) receptor-mediated current in rat hippocampus. Neuroscience 2000; 96(3):507–514.
141. de Lopez A, Sah P. Development and subunit composition of synaptic NMDA receptors in the amygdala: NR2B synapses in the adult central amygdala. J Neurosci 2003; 23(17):6876–6883.
142. Li M, Xiong ZG. Developmental changes of acid-sensing ion channels in cultured mouse cortical neurons. Soc Neurosci Abstr 2005; 957:9.

Chapter 3
Brain Ischemia and Neuronal Excitability

Ping Deng and Zao C. Xu

Abstract Selective neuronal death in certain brain regions has long been recognized as a consequence of transient cerebral ischemia; however, its mechanisms remain unclear. Growing evidence indicates that an increase in neuronal excitability may contribute to this process. Both excitatory synaptic inputs and voltage-dependent potassium currents are important for regulating neuronal excitability. Recent studies demonstrate that the activities of excitatory synaptic inputs and potassium channels are differentially altered in ischemia-sensitive and ischemia-resistant neurons after ischemic insults. It is suggested that a suppression of excitatory neurotransmission or an enhancement of voltage-dependent potassium currents may protect neurons against cerebral ischemia.

Keywords: Cerebral ischemia; Potassium channels; Hippocampal CA3 pyramidal neurons; Striatal cholinergic neurons; Excitability; Synaptic transmission

Transient cerebral ischemia results in neuronal death in certain brain regions. For instance, CA1 pyramidal cells in the hippocampus and medium-sized spiny neurons in the neostriatum are highly sensitive to ischemic insults (1). In contrast, other neurons in the same regions, such as hippocampal CA3 pyramidal neurons and striatal cholinergic interneurons, are relatively resistant to transient ischemia (1, 2). Although the mechanisms underlying this selective neuronal damage following ischemia are not fully understood, accumulating evidence indicates that differential alterations of neuronal excitability may be critical to the ischemia-induced neuronal injury in particular type of neurons. Increase in neuronal excitability may result in excessive Ca^{2+} influx and subsequent intracellular Ca^{2+} overload, which triggers the cascades of cell death. Both intrinsic membrane properties (which are dynamically regulated by the activity of voltage-dependent ion channels) and synaptic inputs

---

P.Deng and Z. C. Xu (✉)
Department of Anatomy and Cell Biology, Indiana University School of Medicine, 635 Barnhill Drive, MS 507 Indianapolis, IN 46202, USA
e-mail: zxu@anatomy.iupui.edu

(especially excitatory neurotransmission) contribute to the neuronal excitability, and thus influence the neuronal responses to ischemic insults. It has been demonstrated that the neuronal excitability of ischemia-sensitive neurons is increased, whereas that of ischemia-resistant neurons is decreased. Here, we briefly summarize recent advances in understanding the postischemic changes in neuronal excitability and their possible mechanisms, with emphases on the voltage-dependent potassium currents and the excitatory synaptic transmission.

3.1 Excitatory Neurotransmission After Ischemia

Excitotoxicity has long been recognized as a major cause of ischemia-induced neuronal damage. According to this hypothesis, over stimulation of glutamate receptors by increased concentration of extracellular glutamate causes extensive Ca^{2+} influx, which triggers a wide variety of cytoplasmic and nuclear processes that promote neuronal death (3, 4). The extracellular levels of glutamate are dramatically increased during ischemia and rapidly return to control levels following reperfusion (5, 6). However, both in vivo and in vitro studies demonstrate that, during ischemia, excitatory neurotransmission in the hippocampus and the neostriatum is suppressed (7–9). Thus, it is possible that postischemic changes in excitatory synaptic transmission may be more important for the occurrence of selective neuronal damage.

The changes of excitatory synaptic transmission after ischemic insults have been extensively investigated in brain regions that are sensitive to ischemia. It is consistently revealed that these changes are opposite between ischemia-sensitive and ischemia-resistant neurons. In the hippocampus, evoked excitatory postsynaptic potentials (EPSPs) in CA1 pyramidal neurons are potentiated after ischemia/hypoxia (10, 11). In contrast, the slope and amplitude of EPSPs in CA3 neurons and dentate granule cells are reduced 12–36h after transient forebrain ischemia (12). In the neostriatum, transient cerebral ischemia in vivo induces a depression of the intrastriatally evoked fast excitatory postsynaptic currents (EPSCs) in ischemia-resistant cholinergic interneurons (13), and an enhancement of the EPSCs in ischemia-sensitive medium-sized spiny neurons (14). Studies using intracellular recording in vivo technique find that the incidence of cortically evoked polysynaptic EPSPs is also significantly increased after ischemia (15). Furthermore, deprivation of oxygen and glucose in brain slices is capable of inducing long-term potentiation in striatal spiny neurons but not in cholinergic interneurons (16). Therefore, the enhancement of excitatory neurotransmission might be associated with ischemic cell death, whereas the depression of synaptic excitation might be involved in neuroprotection after ischemia. This possibility is further supported by the findings that the synaptic transmission in CA1 pyramidal cells and striatal spiny neurons is depressed after mild cerebral ischemia (with shorter duration of ischemia) that does not produce neuronal death (17, 18).

Both pre- and postsynaptic mechanisms influence the strength of synaptic transmission, which has been shown to be involved in the postischemic changes of excitatory neurotransmission. In the hippocampus, it is believed that the potentiation of excitatory neurotransmission after ischemia results from the enhanced postsynaptic responsiveness. The responses of both NMDA and non-NMDA receptors to exogenous application of receptor agonist are enhanced after ischemia, causing larger currents and higher intracellular Ca^{2+} concentrations than those of control ones (19). The enhanced responses of glutamate receptors might be associated with the modulation of receptor activity by protein kinase phosphorylation. A variety of protein kinases, including PKA, PKC, protein tyrosine kinases, and calcium/calmodulin-dependent kinase II, profoundly modulate glutamate receptors (NMDA receptors and AMPA receptors). It has been shown that transient forebrain ischemia induces phosphorylation of NMDA receptor 2A (NR2A) subunit in CA1 pyramidal cells, which is catalyzed by cyclin-dependent kinase 5 (Cdk5). Prevention of NMDA receptor phosphorylation by inhibiting Cdk5 activity protects CA1 pyramidal cells against ischemia (20). In addition, transient global ischemia induces an increase in the tyrosine phosphorylation of NR2A and NMDA receptor 2B (NR2B) subunits, and enhances the interaction of NR2B subunits with phosphatidylinositol 3-kinase (21). On the other hand, presynaptic mechanisms may also be invovled in the potentiation of excitatory synaptic transmission in CA1 pyramidal neurons after transient cerebral ischemia. For example, intracellular Ca^{2+} buffering is capable of protecting hippocampal neurons against anoxia/aglycemia, but has no protective effect on NMDA-induced neuronal damage, suggesting an involvement of presynaptic mechanisms in the neuroprotection (22).

In the neostriatum, most of the studies support the conclusion that the postischemic changes in excitatory neurotransmission are due to the presynaptic mechanisms (13, 14, 23). In cholinergic interneurons, the amplitude of evoked EPSCs (mainly mediated by AMPA receptors) is significantly decreased within 24 h after transient ischemia, whereas the responses to exogenous application of glutamate remain the same as that before ischemia. During high-frequency stimulation, the fifth/first EPSC ratio in cholinergic interneurons early (4–6 h) after ischemia is significantly higher than those of controls and at later time points (24 h) after reperfusion, indicating that the release probability is transiently decreased shortly after ischemia. More detailed studies suggest that the suppression of EPSCs shortly after ischemia might be attributable to the tonic inhibitory function of adenosine A1 receptors on synaptic release of glutamate, and the depression at late time points might result from the degeneration of presynaptic terminals (13). Adenosine is a potential modulator that has broad actions on neuronal activities, with inhibitory effects on the release of neurotransmitter, especially in glutamatergic systems (24). The extracellular levels of adenosine remarkably increase after ischemia/hypoxia. Activation of A1 receptors has also been demonstrated to play an important role in the postischemic depression of excitatory synaptic transmission in the hippocampus, which might be one of the mechanisms underlying neuroprotection against ischemic insults (9, 25). In contrast to that of cholinergic interneurons, the excitatory synaptic transmission in striatal spiny neurons is potentiated after ischemia. This ischemia-induced potentiation in synaptic efficacy is caused by an enhancement of presynaptic release, because the frequency of miniature EPSCs is increased and the paired-pulse ratio is decreased. Furthermore, exogenous glutamate-evoked inward currents exhibit no significant changes in ischemic neurons, suggesting that postsynaptic mechanisms may not be involved. The ischemia-induced increase in the frequency of miniature EPSCs is not affected by blockade of voltage-gated calcium channels, but it is eliminated in the absence of extracellular calcium. Blockade of ionotropic ATP receptors (P2X) almost completely reverses the increase of mEPSC frequency in ischemic neurons, but had no effect in the control neurons. These findings indicate that the tonic enhancement of P2X receptor function after ischemia, which directly mediates Ca^{2+} influx, is responsible for the increased glutamate release from presynaptic terminals (14). This conclusion is also supported by other experimental observations that extracellular concentration of ATP is increased in pathophysiological conditions including ischemia (26), and that the increased expression of P2X receptors after ischemia is mainly located on the presynaptic components of synapses (27). Based on these studies, the postischemic alterations of excitatory neurotransmission in striatal neurons (inhibition in ischemia-resistant neurons, and enhancement in ischemia-sensitive neurons) are mainly mediated by presynaptic modulators that differentially regulate the release probability of neurotransmitters. However, postsynaptic mechanisms could not be completely excluded.

In addition to regulating neuronal excitability, the Ca^{2+} permeability of glutamate receptors is also important for determining ischemia-induced cell death. Although NMDA receptors are highly Ca^{2+}-permeable, recent studies indicate that Ca^{2+} permeable AMPA receptors may play important roles in neuronal damage after ischemia (28, 29). The AMPA receptors are composed of subunits (GluR1-4) as homo- or heterotetramers, and the functional properties of AMPA receptors are largely dictated by their subunit composition. AMPA receptors lacking GluR2 subunits are Ca^{2+}-permeable. After transient cerebral ischemia, GluR2 mRNA and GluR2 subunit expression are significantly downregulated in CA1 pyramidal neurons, but remain unchanged in CA3 neurons and dentate granule cells (29). These findings suggest that changes in AMPA receptor composition, with increased Ca^{2+} permeability, may be related to selective neuronal death in the hippocampus. Indeed, reducing Ca^{2+} permeability of AMPA receptors, either through viral transfection of functional GluR2 subunits or by treatment with selective blockers of GluR2-lacking AMPA receptors, protects CA1 pyramidal neurons from damage induced by forebrain ischemia (30, 31). On the contrary, overexpression of Ca^{2+}-permeable AMPA receptors promotes cell death of CA3 neurons and dentate granule cells that are normally resistant to transient ischemia (30). The postischemic downregulation of GluR2 subunits may result from a suppression of GluR2 promoter activity or a reduction in the expression of the nuclear enzyme that is responsible for GluR2 pre-mRNA editing (31, 32). Additionally, a variety of intracellular proteins, such as *N*-ethylmaleimide-sensitive factor, protein interacting with C-kinase 1, and glutamate receptor-interacting protein/AMPA receptor binding protein, are involved in the trafficking and anchoring of Ca^{2+}-permeable AMPA receptors (28). In cultured hippocampal neurons, ischemic insults induce targeting of GluR2-lacking AMPA receptors to synapses, by promoting clathrin-dependent endocytosis of

GluR2-containing AMPA receptors and facilitating *N*-ethylmaleimide-sensitive factor attachment protein receptor-dependent exocytosis of GluR2-lacking AMPA receptors. The postischemic alteration in subunit composition of AMPA receptors is also dependent on a dissociation of GluR2 subunits from AMPA receptor binding protein, and an association with protein interacting with C-kinase 1 (33). Therefore, modulation of intracellular proteins that interact with AMPA receptors may influence the neuronal sensitivity to ischemic insults.

The subunit composition of NMDA receptors may also be important for neuronal damage after ischemia. The NMDA receptor subunits (NR1, NR2A-D, NR3A) are capable of forming homo- or heteromeric channels. Among these subunits, NR1 subunit is fundamental for function of NMDA receptors. The electrophysiological and pharmacological properties of NMDA receptors largely depend on the composition of NR2 subunits. In mature hippocampal and cortical neurons, NR2A-containing and NR2B-containing NMDA receptors are preferentially expressed at synaptic and extrasynaptic sites, respectively (34). It has been shown that the location of NMDA receptors may be associated with the glutamate-induced neuronal damage. Selective stimulation of synaptic NMDA receptors has neuroprotective effects, whereas activation of extrasynaptic NMDA receptors mediates neuronal death (35). However, it is also possible that the opposite effects of synaptic and extrasynaptic NMDA receptors may be related to the different subunit composition of NMDA receptors (i.e., NR2A-containing vs. NR2B-containing). Indeed, more recent experimental evidence demonstrates that selective stimulation of either synaptic or extrasynaptic NR2B-containing NMDA receptors leads to an increase in neuronal apoptosis of cortical neurons, whereas activation of either synaptic or extrasynaptic NR2A-containing NMDA receptors significantly promotes neuronal survival and protects neurons against ischemic cell death (36, 37). Thus, activation of NMDA receptors may have opposite effects on ischemia-induced neuronal damage, which is, most likely, dependent on the subunit composition of receptors. Most importantly, these experiments suggest that specific blockade of NR2B-containing NMDA receptors and selective activation of NR2A-containing NMDA receptors may provide a novel strategy for the treatment of stroke (36). However, it remains unclear whether this mechanism is involved in selective neuronal damage after transient cerebral ischemia.

3.2 Voltage-Dependent Potassium Currents After Ischemia

Voltage-dependent potassium (Kv) channels are widely expressed in CNS. The activity of these channels is important for regulating neuronal excitability, including the resting membrane potential, the membrane input resistance, the shape of action potential, and the integration of synaptic inputs (38). Based on the biophysical characteristics, Kv channels mediate different types of potassium currents, including transient A-type potassium currents (I_A) and delayed rectifier potassium currents (I_K). In most neurons, Kv4.2 subunits are the major component of channels conducting somatodendritic I_A. Somatic I_K channels are mainly composed of Kv2.1 subunits.

It is clear that potassium currents in neurons are profoundly altered in response to ischemia/hypoxia. Furthermore, it has been shown that alterations of I_A and I_K in ischemia-sensitive neurons may differ from those in ischemia-resistant neurons. For example, in hippocampal CA1 pyramidal neurons, I_A amplitude is increased 12–14 h after transient cerebral ischemia, accompanied by a hyperpolarizing shift of the steady-state activation curve, and an increase in the time constant of the recovery from inactivation. At 36–38 h after ischemia, the amplitude of I_A returns to control levels, and the activation curve shifts to the depolarizing direction. In addition, the rising slope and decay time constant of I_A are increased accordingly after ischemia (39, 40). In dentate granule cells, which are relatively resistant to ischemia, the amplitude and rising slope of I_A are significantly increased 38 h after ischemia, and the inactivation curve shifts toward the depolarizing direction. In CA3 pyramidal neurons, the I_A remains unaltered after ischemia. The I_K in hippocampal CA1 pyramidal neurons is also changed after ischemia, with a progressive increase in the I_K amplitude. However, the I_K shows no significant change in either CA3 neurons or granule cells at any time points after reperfusion (40). These studies demonstrate that the voltage-dependent potassium currents (both I_A and I_K) in hippocampal neurons are differentially altered after transient cerebral ischemia.

An increase in the activity of Kv channels may lead to a hyperpolarization of membrane potential, and thus a depression of neuronal excitability. In hippocampal pyramidal cells, activation of I_A channels on distal dendrites also attenuates the excitatory synaptic inputs. Thus, the postischemic enhancement of I_A might have neuroprotective effects. However, excessive K^+ efflux and intracellular K^+ depletion mediated by I_K have been implicated in triggering neuronal apoptosis (41, 42). Several studies suggest that the activation of Kv2.1 channels is necessary for neuronal apoptosis and is sufficient to increase sensitivity to a sublethal dose of apoptotic stimulus (43, 44). Therefore, a sustained increase of I_K in CA1 neurons might be associated with the neuronal damage after ischemia. This hypothesis is supported by the findings that application of tetraethylammonium, a potential blocker of I_K channels, attenuates hypoxia- and ischemia-induced damage in ischemia-vulnerable neurons, such as hippocampal CA1 neurons and neocortical neurons (45). Interestingly, a transient increase in I_K is detected in ischemia-resistant neurons following ischemia. The amplitude of I_K in striatal cholinergic interneurons is significantly increased 24 h, and returns to control levels within 72 h after cerebral ischemia. The increase in I_K dramatically shortens the spike duration. Because Ca^{2+} entry is primarily triggered by spikes in these neurons, the shortened spike duration will certainly reduce Ca^{2+} influx, and hence prevent the activation of cell death pathways that is due to the intracellular Ca^{2+} overload (46). More recent studies suggest that transient upregulation of Kv2.1 channel-mediated I_K can protect hippocampal neurons against ischemic insults, by reducing neuronal excitability (47). Therefore, the functional significance of the postischemic alterations in Kv channel activity may depend on the nature of potassium currents (I_A vs. I_K) and the temporal profile of the changes (transient vs. sustained).

It should be mentioned that, in addition to Kv channels, other types of potassium channels, such as Ca^{2+}-activated potassium channels, ATP-sensitive potassium channels, and two-pore domain potassium channels, are also altered during and after ischemic insults. Studies both in vivo and in vitro support the conclusion that upregulation of these currents depresses neuronal excitability, and confers neuroprotective effects (48–50).

The mechanisms of the postischemic alterations of Kv channels are still unclear. Changes in the functional channel number might be one of the reasons. However, it is frequently observed that, in addition to the amplitude, the voltage dependence and the gating kinetics of Kv currents are altered in response to ischemic insults, suggesting that other mechanisms must be involved in this process. Several lines of evidence have shown that Kv channels are a target for protein kinase phosphorylation, which profoundly modulate the voltage dependence and gating kinetics (51). For example, in CA1 pyramidal neurons, activation of protein kinase A (PKA) and PKC results in a depolarizing shift of the activation curve and a reduction in the amplitude of the Kv4.2 subunits mediated dendritic I_A, through a common downstream pathway involving extracellular signal-regulated kinases (52). In striatal cholinergic interneurons, the somatic I_K is also suppressed by activation of PKA and protein tyrosine kinases (46). Moreover, the activity and the distribution of PKA and PKC in neurons have been shown to be altered following ischemia (53, 54). In fact, the recovery of increased I_K in cholinergic interneurons 72 h after ischemia is due to the tonic activation of PKA (46). It is interesting that, in hippocampal neurons, the activity and the location of Kv4.2 and Kv2.1 channels are dependent on the neuronal activity (55, 56). A brief treatment with glutamate rapidly causes dephopholarylation and translocation of Kv2.1 channels from clusters to a more uniform location. At the same time, the activation curve of I_K shifts toward a hyperpolarization direction. Upon glutamate receptor stimulation, internalization of Kv4.2 channels is also rapidly induced in hippocampal spines and dendrites. These observations provide another possibility that postischemic modulation of Kv channels might be dependent on the neuronal activity. Supporting evidence comes from the findings that, under ischemic conditions, the enhancement of I_K and the hyperpolarizing shift in the voltage-dependent activation are caused by the dephosphorylation and the translocation of surface Kv2.1 channels from clusters to a uniform localization, which is dependent on intracellular Ca^{2+}- and metabolic state (47). Taken together, multiple mechanisms might contribute to the postischemic alterations of Kv channel activity in both ischemia-sensitive and ischemia-resistant neurons.

3.3 Conclusion

Selective neuronal death in certain brain regions after transient cerebral ischemia is mediated by multiple factors. Differential alterations of neuronal excitability have been implicated to be an important mechanism underlying these processes.

The activity of Kv channels and excitatory synaptic inputs, which dynamically regulates neuronal excitability, is differentially altered in ischemia-sensitive and ischemia-resistant neurons after ischemic insults. A reduction of neuronal excitability might be neuroprotective against cerebral ischemia.

References

1. Pulsinelli WA, Brierley JB. A new model of bilateral hemispheric ischemia in the unanesthetized rat. Stroke 1979;10:267–72.
2. Francis A, Pulsinelli W. The response of GABAergic and cholinergic neurons to transient cerebral ischemia. Brain Res 1982;243:271–8.
3. Choi DW, Rothman SM. The role of glutamate neurotoxicity in hypoxic-ischemic neuronal death. Annu Rev Neurosci 1990;13:171–82.
4. Rothman SM, Olney JW. Glutamate and the pathophysiology of hypoxic-ischemic brain damage. Ann Neurol 1986;19:105–11.
5. Benveniste H, Drejer J, Schousboe A, Diemer NH. Elevation of the extracellular concentrations of glutamate and aspartate in rat hippocampus during transient cerebral ischemia monitored by intracerebral microdialysis. J Neurochem 1984;43:1369–74.
6. Globus MY, Busto R, Dietrich WD, Martinez E, Valdes I, Ginsberg MD. Effect of ischemia on the in vivo release of striatal dopamine, glutamate, and gamma-aminobutyric acid studied by intracerebral microdialysis. J Neurochem 1988;51:1455–64.
7. Gervitz LM, Lutherer LO, Davies DG, Pirch JH, Fowler JC. Adenosine induces initial hypoxic-ischemic depression of synaptic transmission in the rat hippocampus in vivo. Am J Physiol Regul Integr Comp Physiol 2001;280:R639–R645.
8. Sugahara M, Asai S, Zhao H, et al. Extracellular glutamate changes in rat striatum during ischemia determined by a novel dialysis electrode and conventional microdialysis. Neurochem Int 2001;39:65–73.
9. Tanaka E, Yasumoto S, Hattori G, Niiyama S, Matsuyama S, Higashi H. Mechanisms underlying the depression of evoked fast EPSCs following in vitro ischemia in rat hippocampal CA1 neurons. J Neurophysiol 2001;86:1095–103.
10. Gao TM, Pulsinelli WA, Xu ZC. Prolonged enhancement and depression of synaptic transmission in CA1 pyramidal neurons induced by transient forebrain ischemia in vivo. Neuroscience 1998;87:371–83.
11. Urban L, Neill KH, Crain BJ, Nadler JV, Somjen GG. Postischemic synaptic physiology in area CA1 of the gerbil hippocampus studied in vitro. J Neurosci 1989;9:3966–75.
12. Gao TM, Howard EM, Xu ZC. Transient neurophysiological changes in CA3 neurons and dentate granule cells after severe forebrain ischemia in vivo. J Neurophysiol 1998;80:2860–9.
13. Pang ZP, Deng P, Ruan YW, Xu ZC. Depression of fast excitatory synaptic transmission in large aspiny neurons of the neostriatum after transient forebrain ischemia. J Neurosci 2002;22:10948–57.
14. Zhang Y, Deng P, Li Y, Xu ZC. Enhancement of excitatory synaptic transmission in spiny neurons after transient forebrain ischemia. J Neurophysiol 2006;95:1537–44.
15. Gajendiran M, Ling GY, Pang Z, Xu ZC. Differential changes of synaptic transmission in spiny neurons of rat neostriatum following transient forebrain ischemia. Neuroscience 2001;105:139–52.
16. Calabresi P, Saulle E, Centonze D, Pisani A, Marfia GA, Bernardi G. Post-ischaemic long-term synaptic potentiation in the striatum: a putative mechanism for cell type-specific vulnerability. Brain 2002;125:844–60.

17. Xu ZC. Neurophysiological changes of spiny neurons in rat neostriatum after transient forebrain ischemia: an in vivo intracellular recording and staining study. Neuroscience 1995;67: 823–36.
18. Xu ZC, Pulsinelli WA. Electrophysiological changes of CA1 pyramidal neurons following transient forebrain ischemia: an in vivo intracellular recording and staining study. J Neurophysiol 1996;76:1689–97.
19. Mitani A, Namba S, Ikemune K, Yanase H, Arai T, Kataoka K. Postischemic enhancements of N-methyl-D-aspartic acid (NMDA) and non-NMDA receptor-mediated responses in hippocampal CA1 pyramidal neurons. J Cereb Blood Flow Metab 1998;18:1088–98.
20. Wang J, Liu S, Fu Y, Wang JH, Lu Y. Cdk5 activation induces hippocampal CA1 cell death by directly phosphorylating NMDA receptors. Nat Neurosci 2003;6:1039–47.
21. Takagi N, Sasakawa K, Besshoh S, Miyake-Takagi K, Takeo S. Transient ischemia enhances tyrosine phosphorylation and binding of the NMDA receptor to the Src homology 2 domain of phosphatidylinositol 3-kinase in the rat hippocampus. J Neurochem 2003;84:67–76.
22. Abdel-Hamid KM, Tymianski M. Mechanisms and effects of intracellular calcium buffering on neuronal survival in organotypic hippocampal cultures exposed to anoxia/aglycemia or to excitotoxins. J Neurosci 1997;17:3538–53.
23. Calabresi P, Centonze D, Pisani A, Bernardi G. Endogenous adenosine mediates the presynaptic inhibition induced by aglycemia at corticostriatal synapses. J Neurosci 1997;17:4509–16.
24. Dunwiddie TV, Masino SA. The role and regulation of adenosine in the central nervous system. Annu Rev Neurosci 2001;24:31–55.
25. Hsu SS, Newell DW, Tucker A, Malouf AT, Winn HR. Adenosinergic modulation of CA1 neuronal tolerance to glucose deprivation in organotypic hippocampal cultures. Neurosci Lett 1994;178:189–92.
26. Volonte C, Amadio S, Cavaliere F, D'Ambrosi N, Vacca F, Bernardi G. Extracellular ATP and neurodegeneration. Curr Drug Targets CNS Neurol Disord 2003;2:403–12.
27. Franke H, Gunther A, Grosche J, et al. P2X7 receptor expression after ischemia in the cerebral cortex of rats. J Neuropathol Exp Neurol 2004;63:686–99.
28. Cull-Candy S, Kelly L, Farrant M. Regulation of Ca^{2+}-permeable AMPA receptors: synaptic plasticity and beyond. Curr Opin Neurobiol 2006;16:288–97.
29. Tanaka H, Grooms SY, Bennett MV, Zukin RS. The AMPAR subunit GluR2: still front and center-stage. Brain Res 2000;886:190–207.
30. Liu S, Lau L, Wei J, et al. Expression of Ca(2+)-permeable AMPA receptor channels primes cell death in transient forebrain ischemia. Neuron 2004;43:43–55.
31. Noh KM, Yokota H, Mashiko T, Castillo PE, Zukin RS, Bennett MV. Blockade of calcium-permeable AMPA receptors protects hippocampal neurons against global ischemia-induced death. Proc Natl Acad Sci USA 2005;102:12230–5.
32. Peng PL, Zhong X, Tu W, et al. ADAR2-dependent RNA editing of AMPA receptor subunit GluR2 determines vulnerability of neurons in forebrain ischemia. Neuron 2006;49:719–33.
33. Liu B, Liao M, Mielke JG, et al. Ischemic insults direct glutamate receptor subunit 2-lacking AMPA receptors to synaptic sites. J Neurosci 2006;26:5309–19.
34. Tovar KR, Westbrook GL. The incorporation of NMDA receptors with a distinct subunit composition at nascent hippocampal synapses in vitro. J Neurosci 1999;19:4180–8.
35. Hardingham GE, Fukunaga Y, Bading H. Extrasynaptic NMDARs oppose synaptic NMDARs by triggering CREB shut-off and cell death pathways. Nat Neurosci 2002;5: 405–14.
36. Liu Y, Wong TP, Aarts M, et al. NMDA receptor subunits have differential roles in mediating excitotoxic neuronal death both in vitro and in vivo. J Neurosci 2007;27:2846–57.
37. von Engelhardt J, Coserea I, Pawlak V, et al. Excitotoxicity in vitro by NR2A- and NR2B-containing NMDA receptors. Neuropharmacology 2007;53:10–17.
38. Rudy B. Diversity and ubiquity of K channels. Neuroscience 1988;25:729–49.
39. Chi XX, Xu ZC. Differential changes of potassium currents in CA1 pyramidal neurons after transient forebrain ischemia. J Neurophysiol 2000;84:2834–43.

40. Zou B, Li Y, Deng P, Xu ZC. Alterations of potassium currents in ischemia-vulnerable and ischemia-resistant neurons in the hippocampus after ischemia. Brain Res 2005;1033: 78–89.
41. Yu SP, Yeh C, Strasser U, Tian M, Choi DW. NMDA receptor-mediated K+ efflux and neuronal apoptosis. Science 1999;284:336–9.
42. Yu SP, Yeh CH, Sensi SL, et al. Mediation of neuronal apoptosis by enhancement of outward potassium current. Science 1997;278:114–17.
43. Pal S, Hartnett KA, Nerbonne JM, Levitan ES, Aizenman E. Mediation of neuronal apoptosis by Kv2.1-encoded potassium channels. J Neurosci 2003;23:4798–802.
44. Redman PT, He K, Hartnett KA, et al. Apoptotic surge of potassium currents is mediated by p38 phosphorylation of Kv2.1. Proc Natl Acad Sci USA 2007;104:3568–73.
45. Wei L, Yu SP, Gottron F, Snider BJ, Zipfel GJ, Choi DW. Potassium channel blockers attenuate hypoxia- and ischemia-induced neuronal death in vitro and in vivo. Stroke 2003;34:1281–6.
46. Deng P, Pang ZP, Zhang Y, Xu ZC. Increase of delayed rectifier potassium currents in large aspiny neurons in the neostriatum following transient forebrain ischemia. Neuroscience 2005;131:135–46.
47. Misonou H, Mohapatra DP, Menegola M, Trimmer JS. Calcium- and metabolic state-dependent modulation of the voltage-dependent Kv2.1 channel regulates neuronal excitability in response to ischemia. J Neurosci 2005;25:11184–93.
48. Gribkoff VK, Starrett JE, Jr., Dworetzky SI, et al. Targeting acute ischemic stroke with a calcium-sensitive opener of maxi-K potassium channels. Nat Med 2001;7:471–7.
49. Heron-Milhavet L, Xue-Jun Y, Vannucci SJ, et al. Protection against hypoxic-ischemic injury in transgenic mice overexpressing Kir6.2 channel pore in forebrain. Mol Cell Neurosci 2004;25:585–93.
50. Heurteaux C, Guy N, Laigle C, et al. TREK-1, a K+ channel involved in neuroprotection and general anesthesia. EMBO J 2004;23:2684–95.
51. Jonas EA, Kaczmarek LK. Regulation of potassium channels by protein kinases. Curr Opin Neurobiol 1996;6:318–23.
52. Yuan LL, Adams JP, Swank M, Sweatt JD, Johnston D. Protein kinase modulation of dendritic K+ channels in hippocampus involves a mitogen-activated protein kinase pathway. J Neurosci 2002;22:4860–8.
53. Bright R, Mochly-Rosen D. The role of protein kinase C in cerebral ischemic and reperfusion injury. Stroke 2005;36:2781–90.
54. Tanaka K. Alteration of second messengers during acute cerebral ischemia – adenylate cyclase, cyclic AMP-dependent protein kinase, and cyclic AMP response element binding protein. Prog Neurobiol 2001;65:173–207.
55. Kim J, Jung SC, Clemens AM, Petralia RS, Hoffman DA. Regulation of dendritic excitability by activity-dependent trafficking of the A-type K+ channel subunit Kv4.2 in hippocampal neurons. Neuron 2007;54:933–47.
56. Misonou H, Mohapatra DP, Park EW, et al. Regulation of ion channel localization and phosphorylation by neuronal activity. Nat Neurosci 2004;7:711–18.

Chapter 4
Critical Roles of the Na⁺/K⁺-ATPase in Apoptosis and CNS Diseases

Adrian Sproul, Xin Zhou, and Shan Ping Yu

Abstract The Na⁺/K⁺-ATPase is a transmembrane protein that serves as the primary electrogenic ion transporter in the plasma membrane of all mammalian cells. In addition to this critical ubiquitous function, the Na⁺/K⁺-ATPase is also essential for several important tissue-specific functions, which include Na⁺ resorption in the kidney, muscular contraction in cardiac and skeletal myocytes, and membrane repolarization. Although the Na⁺/K⁺-ATPase was first discovered in neuronal tissue, its function has been much more extensively characterized in cardiac and renal tissues, and the roles of the Na⁺/K⁺-ATPase as a novel drug target in these systems have been recently reviewed (1, 2). The goal of this chapter is to review the function and dysfunction of the plasma membrane Na⁺/K⁺-ATPase in the context of neuronal diseases. More specifically, this chapter will outline the structure, function, and regulation of the Na⁺/K⁺-ATPase under normal as well as hypoxic/ischemic conditions. We will review recent evidence on the relationship between the Na⁺/K⁺-ATPase and K⁺ homeostasis in neuronal apoptosis, outline the emerging role of ouabain and other Na⁺/K⁺-ATPase inhibitors as endogenous hormones, which activate specific signal transduction cascades by binding to the Na⁺/K⁺-ATPase, and evaluate the impact of these signal transduction functions on the role of Na⁺/K⁺-ATPase in CNS diseases.

Keywords: Na⁺/K⁺-ATPase; Apoptosis; Hybrid death; Phosphorylation; Src kinases; Hypoxia; Neurodegenerative diseases

4.1 Introduction

The Na⁺/K⁺-ATPase, or Na⁺/K⁺ pump, belongs to the P-type class of ATPases, which hydrolyzes ATP to power ion transport across the plasma membrane. The Na⁺/K⁺-ATPase exports Na⁺ ions out of the cell and imports K⁺ ions into the

---

A. Sproul, X. Zhou, and S.P. Yu(✉)
Emory University, Atlanta, GA
e-mail: yusp@musc.edu

cell, thus allowing for the generation and maintenance of high extracellular Na$^+$ concentration and high intracellular K$^+$ concentration. These ionic gradients in turn serve several essential cellular functions including the maintenance of the resting membrane potential and cell excitability. The Na$^+$ gradient created by the Na$^+$/K$^+$-ATPase is used to power multiple secondary transport systems; these include the uptake of glucose and amino acids, and the export of Ca^{2+} and H$^+$ via the Na$^+$/Ca^{2+} and Na$^+$/H$^+$ exchanger (3–5). Na$^+$/K$^+$-ATPase activity is subjected to physiological regulation by endogenous modulators and by alterations in the phosphorylation states of the protein. Impairment of the Na$^+$/K$^+$-ATPase can also result from a diverse set of stimuli, including energy deficiency, increased reactive oxygen species (ROS), and different stress signals.

4.1.1 Molecular Structure of the Na$^+$/K$^+$-ATPase

The Na$^+$/K$^+$-ATPase is a heterodimeric protein containing two subunits designated as α and β subunits (Fig. 4.1). There are four different α and three β subunit isoforms selectively expressed in diverse tissues.

4.1.1.1 Structure and Function of α Subunit

The α subunit serves the transport function of the enzyme and carries the binding sites for ATP and the specific inhibitor ouabain. It is composed of approximately

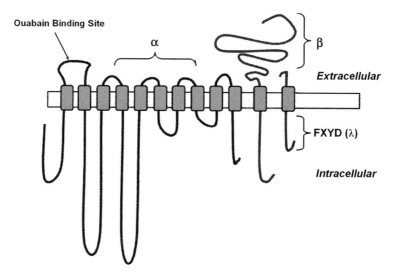

Fig. 4.1 Molecular structure of the Na$^+$/K$^+$-ATPase. The diagraph illustrates the α, β, and λ subunits in the Na$^+$/K$^+$-ATPase protein (*See Color Plates*)

1,000 amino acids and includes 10 transmembrane domains (Fig. 4.1). There are four distinct isoforms of the catalytic α subunit (α1–α4) which share 90% or greater sequence homology that is conserved across species. The major conserved sequences of the α subunit are located at the hydrophobic transmembrane domains, cytoplasmic phosphorylation sites, and the C-terminus. Variant sequences within the α subunit occur most frequently at the N-terminus (6–8).

The α1 isoform is expressed ubiquitously, whereas the α2 isoform is expressed in the brain, heart, muscle, eye, and some other tissues. The α3 isoform appears to be the most abundant form in the nervous system, and it is also present in the heart. The α4 subunit is exclusively expressed in the testis and regulates sperm motility (9). In the CNS, the α2 isoform is usually expressed in glial cells, while the α3 isoform is only located in neurons (10).

4.1.1.2 Structure and Function of β Subunit

The β subunit is a small polypeptide of 300 residues with only a single transmembrane segment that serves to regulate the conformational stability and activity of the α subunit (Fig. 4.1). Among the group of ATPase ion transporters, the presence of a β subunit appears to be required only by those transporters that export K^+ from the cell (11). There are three different isoforms (β1, β2, and β3), and similar to the α subunit, these are also highly conserved across species. β1 and β2 share 94–96% similarity in amino acid content while the sequence homology of β3 is approximately 75% (8). The most conserved regions of the β subunit occur in the transmembrane segments and in the six cysteines of the extracellular region that form intramolecular disulfide bonds (6–8). While the β1 subunit is expressed ubiquitously, the β2 subunit is localized primarily in the nervous system and skeletal muscle. The β3 subunit is found in the liver, lungs, and testis (12). The β subunits are necessary for proper trafficking of the Na^+/K^+-ATPase from the ER to the plasma membrane, and are further required for proper orientation of the Na^+/K^+-ATPase within the plasma membrane. They additionally modulate the K^+ and Na^+ affinity of the enzyme (12). For a more extensive review of β subunit functions, see the recent review by Geering (11).

4.1.1.3 Structure and Function of γ Subunit (FXYD Proteins)

A third small polypeptide, initially termed the γ subunit, was first identified by its association with the Na^+/K^+-ATPase in renal tissue (13), and was subsequently characterized in Xenopus oocytes as a modulator of Na^+/K^+-ATPase currents (14). It is now known that the γ subunit belongs to a class of proteins containing FXYD amino acid motifs in their extracellular N-terminus, and the γ subunit is thus known as FXYD2 (15). The seven FXYD family members identified to date exhibit tissue-specific expression patterns. Five have been found to modulate the function of the Na^+/K^+-ATPase. While the γ subunit/FXYD2 is not required for

Na⁺/K⁺-ATPase function, it appears to increase the ATP-binding affinity of the enzyme and to reduce the enzyme's affinity for both Na⁺ and K⁺ (13, 16, 17).

Although most tissues do not express FXYD2 under physiological conditions, it has been hypothesized that upregulation of FXYD2 may function across cell types to preserve Na⁺/K⁺-ATPase function under conditions of reduced ATP (15). For example, increased osmolarity or temperature may trigger expression of the renal-specific FXYD2 in a variety of cell types including hippocampal neurons (18, 19).

Of the known FXYD proteins, FXYD7 is the only family member whose expression is restricted to neuronal tissue (14). In contrast to other FXYD proteins, FXYD7 is extensively O-glycosylated, expressed to a much greater extent in neurons than in glia, and associates specifically with the α1β1 isoform of Na⁺/K⁺-ATPase (14). When expressed in Xenopus, FXYD7 enhances the affinity of the Na⁺/K⁺-ATPase for K⁺ (14), and it has thus been hypothesized that FXYD7 may function to ensure efficient repolarization following the neuronal action potential by facilitating K⁺ import (15). FXYD6, which is also expressed in many nonneuronal tissues, has recently been identified as playing a CNS-specific role in regulating the homeostasis of endolymph fluid in the inner ear (20, 21).

4.1.2 Physiological Function of the Na⁺/K⁺-ATPase

The primary role of the Na⁺/K⁺-ATPase is to maintain the homeostasis of Na⁺ and K⁺ ions across the plasma membrane. The pump exerts its function by extruding three Na⁺ ions from the cell and moving two K⁺ ions into the cell using one ATP according to the following equation (12):

$$ATP + 3Na^+_{intracellular} + 2K^+_{extracellular} \rightarrow ADP + Pi + 3Na^+_{extracellular} + 2K^+_{intracellular}$$

The Na⁺/K⁺-ATPase, like all P-type-ATPases, has two conformations designated-E1 and E2. In the E1 conformational state, the Na⁺/K⁺-ATPase binds Na⁺ and ATP and is phosphorylated by the γ phosphate group of the ATP molecule. This phosphorylation induces the conformational change to the E2 state, which allows for the release of three Na⁺ ions to the extracellular space. The E2 conformational state exhibits high-affinity binding for K⁺, and the binding of K⁺ to the enzyme is followed by dephosphorylation, reversion to the E1 state, and the release of K⁺ ions to the intracellular space (22).

Variations in levels of intracellular Na⁺ can affect other cellular functions, such as regulation of intracellular free Ca²⁺ ($[Ca^{2+}]_i$). Slight increases in cytoplasmic Na⁺ favor import of extracellular Ca²⁺ via the Na⁺/Ca²⁺ exchanger, which can in turn impact multiple downstream Ca²⁺-dependent signal transduction events. Such an effect is of particular relevance in tissues such as cardiac myocytes and neurons, which couple their essential functions to variations in $[Ca^{2+}]_i$. Because cell volume is largely controlled by ionic gradients and water movements, Na⁺/K⁺-ATPase-mediated

K^+ and Na^+ homeostasis is an important determinant of the osmotic regulation of cell volume (23, 24).

Under normal conditions, the Na^+/K^+-ATPase operates at about one-third of its maximal capacity (8). This allows room for significant increases in pump function under conditions of cellular stress. Thus, Na^+/K^+-ATPase activity can be upregulated even when its expression is decreased, providing an important means of maintaining normal ion transport during physiological and pathological conditions.

4.1.3 Physiologic Regulation of the Na^+/K^+-ATPase

4.1.3.1 Dual Effects of Ouabain and Cardiac Glycosides

The best characterized modulators of the Na^+/K^+-ATPase are the cardiac glycosides (CGs), which are a family of cholesterol-derived steroids either isolated from plant and seed extracts, or from the skin and saliva glands of amphibians (25). The specific downstream effects of the various CGs are distinct, tissue specific, and far from fully characterized. For a current review of the differences between known species of CGs, please see the review by Dvela et al. (25). The specific inhibitory effect of CGs on the Na^+/K^+-ATPase has been the predominant rationale for both their clinical efficacy and for their use as experimental tools for the study of Na^+/K^+-ATPase. More recent evidence indicates that ouabain, digitalis, and several other CGs exist in vivo (1, 26, 27). The exact identities, synthetic pathways, and specific effects for these hormones have not been fully characterized and so they have been referred to collectively as endogenous ouabain-like substances (EOLS). The synthesis of EOLS shows a partial overlap with the mineralocorticoid synthesis pathway (28). Following synthesis, EOLS are released from the hypothalamus and adrenal glands in to the blood stream. The specific stimuli that trigger release of EOLS are diverse, including stress, physical exercise, and hypoxia (27, 29, 30). The nature of these stimuli and consequent cellular events suggest that EOLS may function as stress hormones that act in conjunction with hypothalamic-pituitary axis.

4.1.3.2 Ouabain-Mediated Signal Transduction

Xie et al. suggested that CGs act not only as inhibitors of Na^+/K^+-ATPase activity, but also exhibit Src-dependent activation of PLC/IP3/PKC and RAS/RAF/ERK, with the latter occurring via transactivation of the EGF receptor (31–34). Because the pump-current inhibition and signal transduction functions of CGs both serve to increase $[Ca^{2+}]_i$, it is difficult to distinguish their specific resultant effects. There is no reason, however, to expect that these dual mechanisms need act exclusively of each other. At the subnanomolar levels of circulating CGs in conjunction with their known receptor potencies, it is suggested that the predominant normal physiological effect of EOLS is that of signal transduction, since much higher doses would be necessary to significantly block Na/K$^+$ ATPase currents (2, 35, 36). EOLS signal transduction may even, in some instances, occur through nonconducting variants of the Na$^+$/K$^+$-ATPase, a possibility that is supported by the presence of non-conducting α-1 containing Na$^+$ K$^+$-ATPases at the cell surface in renal proximal tubular cell lines (37).

4.1.3.3 Protein Kinase Regulation of the Na$^+$/K$^+$-ATPase

The effects of PKA and PKG on the Na$^+$/K$^+$-ATPase vary depending upon which α subunit isoform is present, with the neuronally expressed α3 subunit displaying a uniquely upregulated profile. On the other hand, PKA or PKG phosphorylation of α1- or α2-containing Na$^+$/K$^+$-ATPases tends to decrease the pump activity (12). PKC, in contrast, phosphorylates Ser11 and Ser18 induces either endocytosis or exocytosis of the Na$^+$/K$^+$-ATPase, depending on whether only Ser18 or both Ser11 and Ser18 are phosphorylated (12, 38). Since PKC has been characterized as a downstream effector of ouabain (32, 34), one can thus envision a homeostatic mechanism in which circulating EOLS induces either increased or decreased surface expression of the Na$^+$/K$^+$-ATPase, depending upon the degree of ouabain-dependent PKC activation.

4.1.3.4 Src Family Kinases as Regulators of the Na$^+$, K$^+$-ATPase

Several Src family members are expressed predominantly in highly differentiated tissues. In neurons, the Src family kinases Lck, Lyn, Yes, and Src are localized to the postsynaptic density (39). Our laboratory has reported a novel regulatory role of the Src family kinase member Lyn on the Na$^+$/K$^+$-ATPase in cortical neurons (40). The electrogenic activity of the Na$^+$/K$^+$ pump was measured using whole-cell voltage clamp. The pump activity-associated inward and outward currents were attenuated by the tyrosine kinase inhibitor genistein, herbimycin A, or lavendustin A, while blocking tyrosine phosphatases increased the pump currents. Downregulation of the pump current was also seen with the Src inhibitor PP1 and with intracellularly applied anti-lyn antibody. Consistently, intracellular application of lyn kinase upregulated the pump current. Immunoprecipitation and Western blotting showed tyrosine phosphorylation and a direct interaction between lyn and the α3 subunit of the Na$^+$/K$^+$-ATPase. The tyrosine phosphorylation of the α3 subunit was reduced by the apoptotic insult of serum deprivation. This data suggests that Na$^+$, K$^+$-ATPase activity in central neurons is regulated by specific Src tyrosine kinases via a protein–protein mechanism and may play a role in apoptotic cell death (40).

The observed regulation of α3 by Src family kinases is consistent with the known unique upregulation of α3-containing Na$^+$/K$^+$-ATPases by PKA, which could conceivably constitute a downstream signaling cascade leading to the lyn-dependent increase of α3 containing Na$^+$/K$^+$-ATPase currents. The neuron-specific expression of the α3 subunit, coupled with the observation that α3 appears to be uniquely upregulated as compared with other α subunits, suggests that α3 may be particularly well suited to facilitate a more rapid repolarization following the neuronal action potential. From a cell death perspective, the presence of a rapidly upregulatable Na$^+$/K$^+$-ATPase in neurons would also be consistent with the need to favor preservation of ionic homeostasis over progression to apoptosis (5, 23, 24).

4.1.3.5 AMP-Kinase Regulation of Na$^+$/K$^+$-ATPase

AMP-kinase (AMPK) is a highly conserved heterotrimeric serine/threonine kinase, whose downstream effects can generally be summarized as inhibiting anabolism and stimulating catabolism. As intracellular stores of ATP are depleted, the AMP/ATP ratio increases, allowing for the binding to and activation of AMPK by AMP. Several epithelial channels (e.g., CFTR, MDR, ENaC, and Na$^+$/H$^+$ exchanger) decrease conductance during conditions of cellular energy depletion (41, 42), and AMPK has recently emerged as a regulator of some of these ion channels, with specific AMPK effects demonstrated on chloride channels (CFTR), epithelial Na$^+$ channels, and voltage-gated Na$^+$ channels (42). Recent evidence suggests that AMPK may also function to inhibit the Na$^+$/K$^+$-ATPase and inhibit passive K$^+$ efflux through other channels. In H441 lung cells, AMPK activation with phenformin or 5-aminoimidazole-4-carboxamide ribonucleoside (AICAR) inhibits Na$^+$ export via the Na$^+$/K$^+$-ATPase (43, 44).

4.2 The Na$^+$/K$^+$-ATPase and Cell Death

4.2.1 Regulation of the Na$^+$/K$^+$-ATPase Under Pathological Conditions

4.2.1.1 Hypoxic and Ischemic Regulation of the Na$^+$/K$^+$-ATPase

Given sufficient duration and severity, decreases in blood supply and available O$_2$ will eventually lead to decreases in available ATP. Thus, it is not surprising that cells might evolve mechanisms that preserve ATP during conditions of hypoxia or ischemia. As the primary consumer of cellular ATP, the Na$^+$/K$^+$-ATPase would constitute an ideal target for such energy conservation. In primary hepatocyte cultures exposed to hypoxia, Na$^+$/K$^+$-ATPase currents in a dose-dependent fashion as O$_2$ levels decreased (45). In isolated perfused kidneys, ouabain reduces the extent of hypoxic damage as measured by accumulation of pimonidazole adducts (46).

In addition to reducing Na$^+$/K$^+$-ATPase activity, hypoxic and ischemic insults in neuronal and non-neuronal cells also generally lead to decreases in the expression level of Na$^+$/K$^+$-ATPase (47–50). The decrease of the Na$^+$/K$^+$-ATPase in the plasma membrane is most likely due to degradation and endocytosis of the protein. For example, hypoxia appears to trigger endocytosis of the α1 subunit of the Na+/K+-ATPase in alveolar epithelial cells (51–53). The Na+/K+-ATPase endocytosis and ubiquitination were prevented when the Ser-18 in the PKC phosphorylation motif of the α1 subunit was mutated to an alanine, suggesting that phosphorylation at Ser-18 is required for ubiquitination. Mutation of the four lysines surrounding Ser-18 to arginine prevented Na+/K+-ATPase ubiquitination and endocytosis during hypoxia (53). Phosphorylation of the adaptor protein-2 (AP-2) mu2 subunit may also mediate Na+/K+-ATPase endocytosis in response to a variety of signals, such as dopamine and ROS (52). In human alveolar epithelial cells, ROS-mediated RhoA/ROCK activation is necessary for internalization of surface Na+/K+-ATPase (54).

Von Hippel Lindau protein (pVHL) is a key mediator in cellular adaptation to hypoxia; in normoxia, pVHL mediates degradation of hypoxia inducible factor (HIF). A recent report by Zhou et al. suggests that pVHL also plays a role in degrading plasma membrane Na+/K+-ATPases, which is independent of its role in HIF degradation (50). In pVHL-deficient renal carcinoma cells, hypoxia did not affect Na+/K+-ATPase activity, and the degradation of plasma membrane Na+/K+-ATPase was prevented (50). ROS generation leads to pVHL-mediated Na+/K+-ATPase degradation during hypoxia. Overexpression of ROS scavenger proteins such as superoxide dismutase (SOD) prevented the hypoxia-mediated degradation of Na+/K+-ATPase in different cells (55, 56). Transport and hydrolytic activity of the Na+/K+-ATPase in cerebellar granule cells was equally suppressed under hypoxic and hyperoxic conditions. Interestingly, experiments with the NO synthesis inhibitor 1-NAME suggested that NO and its derivatives were involved in maintenance of high Na+/K+-ATPase activity under physiological conditions (47).

Consistent with a regulatory mechanism operating across tissue types, hypoxic regulation examined in cerebellar granule cells in vitro appears to be specific to Na+/K+-ATPase isoforms containing the ubiquitously expressed α subunit, while the tissue-specific isoforms α2 and α3 may not exhibit hypoxic regulation of transport activities (47). Na+/K+-ATPase activity in vivo also appears to be subjected to regulation of EOLS that are upregulated upon hypoxia and ischemia (30, 57). In animal studies, Na+/K+-ATPase activity is often reduced in the hypoxic/ischemic brain and heart (58–61). While the exact mechanisms and tissue specificity of hypoxic regulation of Na+/K+-ATPase function are far from being fully characterized, experimental evidence to date suggests that multiple mechanisms are involved. As stated, one regulatory mechanism of the Na+/K+-ATPase during hypoxia appears to be phosphorylation-dependent endocytosis and ubiquitination of the ATPase (53). In renal epithelia alterations in endogenous lipid metabolism during hypoxia produce acyl carnitines and phosphocholine species, which are capable of inhibiting approximately two-third of the Na+/K+-ATPase current (62).

Astrocytes, which are increasingly being recognized as playing an important role in postischemic neuronal pathology, appear to have a profile of posthypoxic Na+/K+-ATPase activation, which is distinct from that of neurons. Astrocyte recordings in acute hippocampal slices exposed to combined oxygen and glucose deprivation (OGD) show depolarization primarily due to increased extracellular K⁺, which is followed by an increase in ouabain-sensitive hyperpolarization with cessation of OGD (63). This suggests that the Na⁺/K⁺-ATPase in astrocytes may show enhanced activity after OGD, which serves as a compensatory mechanism for maintaining ionic homeostasis in the microenvironment of the CNS.

4.2.2 K⁺ Homeostasis and Apoptosis

Programmed cell death or apoptosis plays an important role in development and many pathological conditions. Apoptosis is defined by specific morphological features that include cell shrinkage, nuclear condensation, caspase activation, cytochrome *c* release, DNA fragmentation, and formation of apoptotic bodies. In addition to regulation by multiple genes, apoptosis is also under regulation by a K⁺ homeostasis-based ionic mechanism (64–68). Thus, activity of the Na⁺/K⁺-ATPase appears to have direct and significant impacts on apoptotic cell death.

As contrasted with other major ion species, K⁺ is predominant inside of the cell (~140 mM), while the intracellular concentrations of Na⁺, Ca⁺, and Cl⁻ are around 12 mM, 100 nM, and 4–60 mM, respectively (Fig. 4.2). As the most abundant intracellular cation, K⁺ acts as a dominant regulator of cytosolic osmolarity. Moreover, the K⁺ electrochemical gradient is created and maintained by the Na⁺/K⁺-ATPase (69). Dysfunction of the Na⁺/K⁺-ATPase in combination with passive K⁺ efflux via K⁺-permeable channels will lead to cellular K⁺ depletion (24). In apoptotic cells, intracellular K⁺ concentrations may decrease from well over 100 to 30–50 mM (70, 71). The K⁺ efflux ultimately results in water loss; the process is believed to be a major determinant of cell volume decrease in physiological and pathological conditions (12). Simultaneous efflux of Cl⁻ prevents excessive hyperpolarization from K⁺ efflux and may also contribute to cell volume regulation (67). Compelling evidence agrees that the large K⁺ reduction is likely a prerequisite signal for executing the apoptotic cascade, including cytochrome *c* release from mitochondria (24), apoptosome formation, and activation of apoptotic enzymes such as caspase-3 (24, 72) and endonucleases (73).

Early studies, using toxins and selective K⁺ ionophores, linked K⁺ efflux to cell shrinkage and some other apoptotic events (74). Findings from our lab and other works demonstrates that K⁺ ionophores, such as valinomycin, or overexpression of certain K⁺ channels can cause K⁺ depletion and lead to apoptosis in primary cortical neurons and peripheral cell types, whereas the addition of extracellular solutions with high K⁺ concentrations (e.g., 25–35 mM) attenuates K⁺ efflux and cell death (64, 75, 76). A major focus of our group has been on the discriminating roles of K⁺ channels and the Na⁺/K⁺-ATPase in neuronal apoptosis. Apoptosis in cortical neurons is associated with an early increase (during 3–12 h exposure) in outward I_K currents, followed in the next few hours by impairment of Na⁺/K⁺-ATPase and depletion of intracellular K⁺(23). More significantly, preventing K⁺ depletion, either by increasing extracellular K⁺ or by application of K⁺ channel blockers such as tetraethylammonium (TEA) attenuates apoptosis (64, 65, 75). K+ loss occurs before production of ROS, apoptosome formation, activation of caspase-3, and nuclear condensation, and it also associates with changes in mitochondrial membrane potential (73). Proapoptotic K+ efflux can be mediated by delayed rectifier K+ channels including Kv channels, A-type I_A channels, Ca^{2+}-activated K+ channels, the ATP-sensitive K+ (K_{ATP}) channel, and also by the NMDA or AMPA subtypes of ionotropic glutamate receptors (64, 66, 75, 77–82). Although routes of K+ efflux may vary depending on cell types, a dysfunction of the Na+/K+-ATPase likely develops in apoptotic cells as a universal result from ATP reduction, ROS production, and altered phosphorylation states of the pump protein (40, 83, 84).

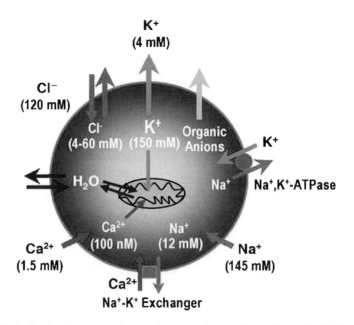

Fig. 4.2 Ionic distributions across the membranes and regulation pathways for K+ homeostasis. Driven by their concentration gradients as well as electrical forces, K+, Cl−, and organic anions flow from the cytoplasm to extracellular space via ion channels and exchangers. Active ionic movements can be achieved by corresponding transporters/pumps. K+ may also move into the mitochondria upon activation of mitochondrial K+ channels such as the mK_{ATP} channel. Water moves following the osmolarity changes in the cytoplasm and mitochondria. The K+, Cl−, and water efflux are most likely responsible for apoptotic cell volume decrease. Because of excessive K+ and Cl− efflux over the water movement, there are significant decreases in intracellular concentrations of K+ and Cl− in apoptotic cells. Moving of K+ and water into the mitochondria may cause mitochondrial swelling, loss of mitochondrial potential, disruption of the outer membrane, and probably the release of some apoptotic factors such as cytochrome *c* into the cytoplasm (see text for detail). Activation of Ca^{2+} and Na+ channels in the plasma membrane results in influx of Ca^{2+} and Na+, which is believed to be a trigger for necrosis. The homeostasis of K+, Ca^{2+}, and Na+ is principally maintained by normal function of the Na+, K+-ATPase. The cytoskeleton and other membrane channels/transporters may contribute to the cell volume control and various apoptotic events; which are not the topics of this review and are not illustrated in the figure [adopted from (23) with permission] (*See Color Plates*)

4.2.3 Na+/K+-ATPase, Apoptosis, and Hybrid Cell Death

With respect to the relationship between cellular K+ loss and apoptosis, it is reasonable to argue that the K+ efflux mediated by K+-permeable channels/receptors can be balanced by sufficient K+ uptake carried by an increased Na+/K+-ATPase activity. Inhibition of the Na+/K+-ATPase must thus be an essential event in proapoptotic K+ depletion and apoptosis (23, 83) (Fig. 4.3). Consistent with this prediction, 50–80% K+ loss has been observed in ouabain-treated T-lymphocyte cell lines and in primary cortical neuronal cultures (5, 24, 72, 85). In Jurkat cells, Western blot analysis of the α subunit of the Na+/K+-ATPase showed the loss of its expression in apoptotic cells (86). We reported that in neuronal cells undergoing early apoptotic process, there was a progressive suppression of Na+/K+-ATPase currents. Up to 80–90% of the pump currents were suppressed after 8–10h of an apoptotic insult, preceded by increased K+ efflux via overactivated outward delayed rectifier K+ channels (23, 64, 83).

The potential interplay between Na+/K+-ATPase and apoptosis is further demonstrated by B-cell lymphoma cell lines overexpressing the antiapoptotic protein Bcl-2. Following Bcl-2 overexpression, cells exhibit a hyperpolarized membrane potential (−100 mV) and increased Na+/K+-ATPase currents (87, 88). While the mechanism for such enhanced Na+/K+-ATPase activity is obscure, one plausible explanation is that a tonic ROS-induced inhibition of the Na+/K+-ATPase is prevented with Bcl-2 overexpression. Such an ROS-based regulatory mechanism would be consistent with the previously mentioned ROS-dependence of pVHL-mediated degradation of

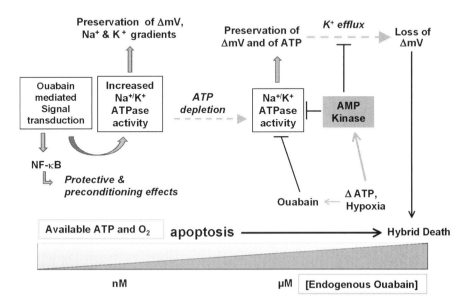

Fig. 4.3 Strategies for modulation of Na+/K+-ATPase under pathological conditions (*See Color Plates*)

the Na⁺/K⁺-ATPase, and suggests that alterations in ROS may act as a fundamental regulator of the Na⁺/K⁺-ATPase in multiple settings.

More robust evidence linking the Na⁺/K⁺-ATPase to apoptotic cell death comes from studies showing that apoptosis can be induced by administration of CGs. A demonstration of this effect came from a study using bufadieneolides isolated as the principal active components from traditional Chinese analgesic medicines, which were shown to be capable of inducing apoptosis in human leukemia cell lines (89). Ouabain-induced apoptosis was subsequently shown to occur in a variety of cell lines, including lymphocytes, human androgen-independent prostate cancer cells, human prostatic smooth muscle cells, human monocytic leukemia, colon adenocarcinoma cells, and neurons (12). Interestingly, some classical K⁺ channel blockers such as 4-aminopyridine and clofilium have been identified as potent inhibitors of the Na⁺/K⁺-ATPase, which may account for their long-recognized neurotoxicity (84, 90).

4.2.3.1 Synergistic Effects of Low Concentrations of Ouabain and Sublethal Apoptotic Insults

In addition to induction of apoptosis, ouabain can also sensitize cells to other apoptotic insults, including TNF-induced cell death in rodent tumor cell lines, Fas ligand-induced shrinkage and apoptosis in Jurkat T lymphocytes, and radiation-induced cell death in several human cell lines (5). We reported that application of ouabain at sublethal low µM concentrations significantly increased the vulnerability of cortical neurons to other apoptotic insults. For example, ouabain at 0.1 µM alone had a minimal effect on Na⁺/K⁺ pump activity, and did not change $[K^+]_i$ cell viability. Likewise, applications of C_2-ceramide at the low concentration of 5 µM or amyloid beta (Aβ) peptides 1–42 at 5 µM caused little cell death. Coapplication of ouabain and C_2-ceramide or Aβ 1–42 at above concentrations, however, led to approximately 50% depletion of $[K^+]_i$, and substantial apoptotic death (72). We proposed that even slight impairment of the pump activity might impair the ability of the pump to efficiently recruit its reserve capacity and thus prevent it from maintaining K⁺ homeostasis in the face of apoptotic stresses. The sensitizing effect of low-concentration ouabain may also involve the ouabain-induced intracellular signaling pathways, although it remains to be elucidated. Whatever the mechanisms involved, it appears clinically relevant that even slight dysfunction of the Na⁺/K⁺-ATPase may markedly increase the susceptibility of neurons to an apoptotic stress.

4.2.3.2 Blocking Na⁺/K⁺-ATPase Induces Hybrid Cell Death with Both Apoptotic and Necrotic Features

While slight and moderate inhibition of the Na⁺/K⁺-ATPase allows the time-dependent development of cellular K⁺ depletion and activation of apoptotic cascade, application of high concentrations of ouabain (80–100 µM) typically elicits a mixed form of cell death (Fig. 4.3). Cell volume is initially increased during the first 2–3 h of exposure to ouabain, but is then followed by gradual cell shrinkage in the next 5–10 h (24). To identify the specific type of cell death involved, we examined morphological and biochemical events in these cells and confirmed that concurrent necrosis and apoptotic processes were evolving within the same cells. Examination of neuronal ultrastructural features with electron microscopy further supported a mixed form of cell death in which both necrotic and apoptotic alterations occur simultaneously from early stages of the neuronal injury. In addition to several necrotic features such as cytoplasmic vacuolization, degradation of subcellular organelles, and deterioration of membranes, these cells showed features of cytochrome *c* release, caspase activation, DNA fragmentation, and chromatin condensation. This mixed form of concurrent necrosis and apoptosis is associated with cellular Ca^{2+} increase and K^+ depletion and has been named *hybrid death* or *necrapoptosis* (24, 91). Supporting the involvement of both $[Ca^{2+}]_i$ and $[K^+]_i$ changes in hybrid cell death, ouabain-induced cell death was only partially blocked by the K^+ inhibitor TEA, the Ca^{2+} channel blocker nifedipine, or by the caspase inhibitor Z-VAD, but was virtually prevented by coapplication of Z-VAD and nifedipine (24). In contrast, no protective effect of nifedipine was observed when typical apoptosis was induced in neurons with serum deprivation, staurosporine, ceramide, or the K^+ ionophore valinomycin (64, 65, 75, 92).

4.3 Neuronal Function of the Na^+/K^+-ATPase and Roles in CNS Diseases

4.3.1 Emerging Neuronal Functions of the Na^+/K^+-ATPase

While the Na^+/K^+-ATPase is a major player for ionic homeostasis and resting membrane potential it is also emerging as an active participant in signal transduction, a necessary component for proper neuronal development, and a specific regulator of synaptic activity.

4.3.1.1 The Na^+/K^+-ATPase as a Receptor for Agrin

Recently, the neuron-specific α3 subunit of the Na^+/K^+-ATPase has been identified as a specific receptor for agrin (93). Agrin is a large proteoglycan secreted into the extracellular matrix at peripheral nerve terminals, which functions to cluster postsynaptic acetylcholine receptors at the neuromuscular junction (94, 95). Agrin also appears to play an analogous role in synapse formation in the CNS (96). While agrin knockout mice do not survive beyond birth, CNS-specific deletion of agrin results in a viable phenotype (97), which allows for a closer examination of agrin's role in the CNS. Consistent with agrin's specific localization at excitatory synapses, agrin-CNS-deficient mice exhibit decreased frequency of spontaneous miniature postsynaptic currents (sEPSC) at excitatory but not inhibitory synapses (97). These mice additionally exhibit a significant reduction in total synapse number (97). Agrin decreases Na$^+$/K$^+$-ATPase activity, subsequently leading to neuronal depolarization and increased firing frequency (93). Such an effect suggests that Na$^+$/K$^+$-ATPase may participate in selectively strengthening active synapses required by Hebbian plasticity (98). Membrane currents generated by α3 containing Na$^+$/K$^+$-ATPase have also been found to mediate presynaptic hyperpolarization following high-frequency stimulation (99). Thus the agrin-mediated regulation of Na$^+$/K$^+$-ATPase may serve as a selection mechanism to increase the probability of depolarization.

4.3.1.2 Na$^+$/K$^+$-ATPase α Subunit Knockout Phenotypes

Distinct neuronal pathologies have been identified in mice deficient in either the α2 or α3 subunits, and these phenotypes may help to explain the human diseases that have specific mutations in human α2 and α3 subunits. Deletion or mutation of Na$^+$/K$^+$-ATPase α subunits in drosophila results in a phenotype of seizure-like hyperexcitability or age-dependent neurodegeneration (100). Deletion of the α2 subunit results in a nonviable phenotype with extensive apoptotic neuronal loss in the amygdala, or in the case of single allele deletion, a viable phenotype, which exhibits excessive fear responses and impaired learning (10, 101). Mice deficient in the α3 subunit also exhibit impaired learning and memory, and have significant reductions in hippocampal NMDA receptor expression (10).

4.3.2 The Na$^+$/K$^+$-ATPase and CNS Diseases

The following section discusses the roles of the Na$^+$/K$^+$-ATPase in several CNS disorders.

4.3.2.1 Stroke

A role for Na$^+$/K$^+$-ATPase in conditions involving acute neuronal insults such as stroke is fairly well supported by consistent evidence for hypoxia/ischemia-induced degradation of the Na$^+$/K$^+$-ATPase and the impairment of Na$^+$/K$^+$-ATPase in neuronal cell death (5, 47–50). Consistent with a critical role of the Na$^+$/K$^+$-ATPase in ischemic injury, its activation induces significant positive inotropic effect in animal models of cardiac ischemia (102). Beneficial effects of Na$^+$/K$^+$-ATPase activation have also been reported in the ischemic brain, with several studies showing a neuroprotective effect of Na$^+$/K$^+$-ATPase activation (103–105). While direct genetic linkages to stroke susceptibility have not been established for the Na$^+$/K$^+$-ATPase, a strong link has been established between hypertension and the Na$^+$/K$^+$-ATPase (106, 107), with hypertension representing the greatest single primary risk factor for stroke. Thus, alterations in renal Na$^+$/K$^+$-ATPase function that impair Na$^+$ resorption create an increased risk of stroke.

Multiple lines of evidence point to the involvement of the Na$^+$/K$^+$-ATPase as a critical participant in the process of neuronal cell death. More recent studies have also begun to highlight the validity of the Na$^+$/K$^+$-ATPase as a novel pharmacological target for preventing neuronal damage following stroke. As mentioned earlier, dysfunction of the Na$^+$/K$^+$ ATPase directly results in intracellular K$^+$ depletion and Ca^{2+}/Na$^+$ accumulation, which are likely the ionic basis for hybrid cell death after ischemic stroke (24, 108, 109).

While inhibition of the Na$^+$/K$^+$ ATPase is generally associated with decreased cellular viability, pretreatment with low doses of ouabain and other CGs may constitute a viable protective strategy. Brief exposure of primary cortical neurons to ouabain mimics the protective effects of hypoxic preconditioning to subsequent oxygen-glucose deprivation (OGD) in vitro (110). In vivo cardiac preconditioning with low-dose ouabain is also protective and acts via Src and PLC-dependent signal transduction (111). Thus, it is possible that CGs may be explored as a preconditioning strategy for preventing ischemic stroke.

4.3.2.2 Alzheimer Disease/Alzheimer's Disease?

Plaques consisting of aggregates of Aβ peptides are a cardinal feature of Alzheimer's disease (AD) and have been proposed to be central to the pathogenesis of AD. Aβ fragments have been shown to decrease Na$^+$/K$^+$-ATPase currents in Xenopus oocytes (112). Application of Aβ to primary cultures of hippocampal neurons similarly leads to an impaired Na$^+$/K$^+$-ATPase activity, which is followed by elevation of [Ca^{2+}]$_i$ and cell death (113). Consistent with these in vitro findings, mouse models of AD that express hippocampal amyloid plaques exhibit deficits in Na$^+$/K$^+$-ATPase activity, which are restricted to amyloid-expressing brain areas (114). Thus, while Na$^+$/K$^+$-ATPase dysfunction in AD does not appear to represent a primary pathological mechanism, the correlation between impaired Na$^+$/K$^+$-ATPase and brain regions exhibiting AD pathology does raise the possibility that agents aimed at improving Na$^+$/K$^+$-ATPase function might provide clinical benefit in the treatment of AD.

4.3.2.3 Parkinson's Disease

Potential involvement of the Na$^+$/K$^+$-ATPase in the pathogenesis of Parkinson's disease (PD) has recently emerged. It was noticed that the Na$^+$/K$^+$-ATPase activity was low in patients with PD and some other neurodegenerative diseases (115). Dopamine (50–200 µM) caused dose-dependent inhibition of Na$^+$/K$^+$-ATPase activity of rat brain crude synaptosomal-mitochondrial fraction during in vitro incubation. The inactivation of Na$^+$/K$^+$-ATPase in this system might involve both H$_2$O$_2$ and metal ions. The authors suggested that the inactivation of neuronal Na$^+$/K$^+$-ATPases by dopamine might give rise to various toxic sequelae with potential implications for dopaminergic cell death in Parkinson's disease (116). More compelling evidence comes from molecular biological studies showing that mutations of the *DJ-1* gene cause early-onset, familial Parkinson's disease (117). *DJ-1* gene has been shown to have a potential neuroprotective role in oxidative stress, protein folding, and degradation pathways. In DJ-1$^{-/-}$ mice, OGD-induced membrane hyperpolarization was significantly enhanced. Moreover, rotenone caused a shorter hyperpolarization followed by an irreversible depolarization. Compared with WT mice, in dopaminergic neurons from DJ-1$^{-/-}$ mice, ouabain induced rapid and irreversible membrane potential changes. It was suggested that loss of function of DJ-1 enhances vulnerability to energy metabolism alterations, and that nigral neurons are particularly sensitive to Na$^+$/K$^+$-ATPase impairment (118).

4.3.2.4 Rapid Onset Dystonia-Parkinsonism

The hypothesis that Na$^+$/K$^+$-ATPase dysfunction underlies Parkinsonian neuronal degeneration is significantly strengthened by the finding that patients with rapid onset dystonia-parkinsonism (RDP) have been found to have several missense mutations in the α3 subunit of the Na$^+$/K$^+$-ATPase (119, 120, 121). RPD is inherited in an autosomal dominant fashion, with disease onset occurring in the teens or twenties, frequently in association with an external physiological stressor. In addition to typical Parkinsonian symptoms (shuffling gate, hand tremor, mask-like facies), RPD is typified by involuntary muscle contractions and twisting and writhing movements (dystonia) (119, 120). Although a specific pathological basis for RPD is far from established, one possibility is that impaired signaling between agrin and the α3 subunit of Na$^+$/K$^+$-ATPase (93) is impaired in RPD, leading to impaired Hebbian maintenance of critical circuitry during the period of cortical pruning, which takes place during adolescence. Alternatively, impaired postfiring hyperpolarization (99) could lead to unregulated and excessive neuronal activation and excitotoxic cell death of dopaminergic neurons.

4.3.2.5 Bipolar Disorder

Bipolar disorder (BD) is a mood disorder in which patients alternate between severe depressive and manic episodes. The manic phase of bipolar disease can be conceived of as a prolonged period of excessive cortical excitability, and the involvement of Na$^+$/K$^+$-ATPase activity in bipolar disease has thus been suggested because of its role in regulating neuronal repolarization (122, 123). Patients with BD show decreased Na$^+$/K$^+$-ATPase activity during the manic phase of BD and increased activity during the depressive phase (124). Consistent with Na$^+$/K$^+$-ATPase dysfunction as a relevant aspect of BD pathology, the mainstay of treatment for BD (lithium) has been shown to reverse stress-induced neuronal deficits of Na$^+$/K$^+$-ATPase activity in vivo (125). Such differences could, however, be secondary to system-wide variation in metabolic rate occurring in BD, and there does not appear to be any significant genetic linkage between mutations in α1 or β3 subunit isoforms and bipolar disease (126). However, linkage analysis of the potentially more relevant α2 and α3 subunits in susceptible populations has not been reported. The emerging role of the Na+/K+-ATPase as a specific regulator of neuronal excitability and hybrid cell injury suggests that further investigation of the potential role of the Na+/K+-ATPase dysfunction in BD is warranted.

4.3.2.6 Familial Hemiplegic Migraine Type II

Mutations in the α2 subunit of the Na+/K+-ATPase have been associated with infantile seizures (127) and common familial migraine (128), while the strongest linkage evidence is for familial hemiplegic migraine type II (FHMII) (119, 129). FHMII is an autosomal dominant inherited propensity for migraine headaches in which weakness occurs on one side of the body during the aura phase of the migraine. Genetic studies have identified multiple mutations within the α2 subunit of the Na+/K+-ATPase in FHMII patients (129). Mutations at L764P and W887R exhibit a dominant negative phenotype (130). Other identified mutations encode a functional Na+/K+-ATPase with altered ion affinity or kinetics. Mutations at sites R689Q and M731T exhibit altered affinity for ATP, which consequently decreases the rate of ATP hydrolysis by the pump (131). Mutation at site T345A appears to decrease the affinity of the Na+/K+-ATPase for extracellular K+ and consequently decreases capacity for K+ import (132). In most instances the final effect of identified mutations appears to be to decrease the activity of the Na+/K+-ATPase.

The specific linkage of the Na+/K+-ATPase with FHMII raises the possibility that Na+/K+-ATPase dysfunction may represent an underlying pathological mechanism in nonfamilial migraine as well, and a role for Na+/K+-ATPase in migraine has been hypothesized prior to the discovery of the FHMII linkage. Recent evidence suggests that the primary pathological phenomenon thought to cause aura in migraine (cortical spreading depression) causes a localized neuronal hypoxia in vivo (133). Such an unexpected common mechanism may also represent a clinically relevant link between migraine and stroke, and may help to explain why young women presenting with migraine with aura exhibit a 1.5-fold increased risk of subsequent stroke (134, 135).

4.4 Na+/K+-ATPases as Drug Targets

4.4.1 Naturally Occurring Na+/K+-ATPase Inhibitors

The validity of targeting the Na+/K+-ATPase is highlighted by its wide distribution in different cells with cell-specific subunits and its vital roles in cellular activities. More interestingly, the existence of EOLS strongly suggests that the hormonal regulation of the Na+/K+-ATPase activity may represent an endogenous defense mechanism capable of significantly affecting physiological and pathophysiological responses. In addition to the classical plant-derived cardenolides (digitalis, ouabain) and the amphibian-derived bufadienolides (bufalin, marinobufagenin), the bloodroot-derived inhibitor sanguinarine and the highly toxic coral-derived channel-opener palytoxin also show inhibitory effects on the Na$^+$/K$^+$-ATPase. Several human cancer cell lines are sensitive to cytotoxic effects of the cardenolides isolated from milkweed and novel cardenolide species isolated from plants growing in endangered habitats such as the alpine regions of China and the rainforests of Madagascar (63, 103, 136).

4.4.2 Na$^+$/K$^+$-ATPase as Therapeutics for Cancer

While targeting cancerous cells with Na$^+$/K$^+$-ATPase inhibitors raises the concern of insulting nonmalignant tissues, the activity of the Na$^+$/K$^+$-ATPase differs between normal and malignant cells, as does their sensitivity to Na$^+$/K$^+$-ATPase inhibitors. This may be due to the numbers of enzymes on the plasma membrane and different isoforms expressed in the cells (8). It has been reported that the β1 subunit is downregulated in epithelial cancer cells, whereas the α subunit is upregulated in some malignant cells (8). Many studies report increases in Na$^+$/K$^+$-ATPase activity during the course of malignant cell transformation (137), which is consistent with our notion that cells with lower Na$^+$/K$^+$-ATPase activity are more vulnerable to apoptosis (72). The Na$^+$/K$^+$-ATPase is also involved in cell adhesion and migration, processes which are both essential to the growth and metastases of malignant cells (22, 49). A number of Na$^+$/K$^+$-ATPase inhibitors have accordingly been used against cancer as proapoptotic drugs, and can also restore sensitivity to chemotherapeutic drugs and decrease rates of metastasis (8), which agrees with the notion that sublethal ouabain sensitizes neuronal cells to apoptotic insults (72).

4.4.3 Targeting the Na$^+$/K$^+$-ATPase for Cytoprotection

In addition to the potential proapoptotic applications of Na$^+$/K$^+$-ATPase inhibition in cancer therapy, antiapoptotic effects might also be achieved by stimulating signal transduction through the Na$^+$/K$^+$-ATPase (138). At concentrations well below the threshold for significant inhibition of the Na$^+$/K$^+$-ATPase activity, ouabain can activate the PLC/IP3 component of Na$^+$/K$^+$-ATPase signal transduction, triggering increases in [Ca^{2+}]$_i$ and leading to the activation of AKT with downstream activation of the antiapoptotic transcription factor NF-κB (35, 139, 140). The activation of NF-κB may explain the antiapoptotic effect of ouabain in some cell types (35, 141, 142). Ouabain at very low doses has also been shown to enhance cell growth without affecting the Na$^+$/K$^+$-ATPase activity (142). Additional cytoprotective strategies might include enhancing Na$^+$ pump activity with antioxidants or with substrates of ATP synthesis (87, 88), as has been demonstrated with zinc-desferrioxamine in bovine lens cultures (143) and with pyruvate/succinate in cortical neurons, respectively (83). As previously mentioned, low and moderate concentrations of Na$^+$/K$^+$-ATPase inhibitors may be used as preconditioning agents to increase the tolerance of cells to consequent ischemic insults.

4.5 Conclusion

Despite being initially discovered and characterized in neuronal tissue, the Na$^+$/K$^+$-ATPase has received substantially less scrutiny in the CNS than in cardiac and renal systems. Hopefully this chapter has served to highlight the role of this ancient molecule as an intriguing and multifaceted modulator of specific neuronal functions. As the Na$^+$/K$^+$ pump is enjoying something of a renaissance as a drug target in peripheral systems, it should also be thoroughly investigated as a relevant target for the treatment of CNS diseases. The Na$^+$/K$^+$-ATPase possesses the distinct advantage of a class of specific inhibitors that have seen extensive prior clinical and experimental uses. As the CNS diseases discussed here generally tend to exhibit decreased pump activity, the development of agents that preserve or stimulate Na$^+$/K$^+$-ATPase activity may represent an effective therapeutic strategy. Hopefully, both inhibitors and stimulators will be further explored and as the multiple functions and points of regulation of the Na$^+$/K$^+$-ATPase are better understood, more targeted and effective pharmacological modulation of this ancient molecule will also result.

References

1. Schoner, W. and G. Scheiner-Bobis, *Endogenous and exogenous cardiac glycosides: their roles in hypertension, salt metabolism, and cell growth.* Am J Physiol Cell Physiol, 293: C509–536, 2007.
2. Xie, Z. and J. Xie, *The Na/K-ATPase-mediated signal transduction as a target for new drug development.* Front Biosci, 10:3100–3109, 2005.
3. Geering, K., *Na,K-ATPase.* Curr Opin Nephrol Hypertens, 6:434–439, 1997.
4. Pavlov, K.V. and V.S. Sokolov, *Electrogenic ion transport by Na$^+$,K$^+$-ATPase.* Membr Cell Biol, 13:745–788, 2000.
5. Yu, S.P., *Na(+), K(+)-ATPase: the new face of an old player in pathogenesis and apoptotic/hybrid cell death.* Biochem Pharmacol, 66:1601–1609, 2003.
6. Blanco, G., *Na,K-ATPase subunit heterogeneity as a mechanism for tissue-specific ion regulation.* Semin Nephrol, 25:292–303, 2005.
7. Blanco, G., *The NA/K-ATPase and its isozymes: what we have learned using the baculovirus expression system.* Front Biosci, 10:2397–2411, 2005.
8. Mijatovic, T., E. Van Quaquebeke, B. Delest, O. Debeir, F. Darro, and R. Kiss, *Cardiotonic steroids on the road to anti-cancer therapy.* Biochim Biophys Acta, 1776:32–57, 2007.
9. Dostanic-Larson, I., J.N. Lorenz, J.W. Van Huysse, J.C. Neumann, A.E. Moseley, and J.B. Lingrel, *Physiological role of the alpha1- and alpha2-isoforms of the Na$^+$-K$^+$-ATPase and biological*

significance of their cardiac glycoside binding site. Am J Physiol Regul Integr Comp Physiol, 290:R524–R528, 2006.
10. Moseley, A.E., M.T. Williams, T.L. Schaefer, C.S. Bohanan, J.C. Neumann, M.M. Behbehani, C.V. Vorhees, and J.B. Lingrel, *Deficiency in Na,K-ATPase alpha isoform genes alters spatial learning, motor activity, and anxiety in mice.* J Neurosci, 27:616–626, 2007.
11. Geering, K., *The functional role of beta subunits in oligomeric P-type ATPases.* J Bioenerg Biomembr, 33:425–438, 2001.
12. Panayiotidis, M.I., C.D. Bortner, and J.A. Cidlowski, *On the mechanism of ionic regulation of apoptosis: would the Na+/K+-ATPase please stand up?* Acta Physiol, 187:205–215, 2006.
13. Therien, A.G., R. Goldshleger, S.J. Karlish, and R. Blostein, *Tissue-specific distribution and modulatory role of the gamma subunit of the Na,K-ATPase.* J Biol Chem, 272:32628–32634, 1997.
14. Beguin, P., G. Crambert, F. Monnet-Tschudi, M. Uldry, J.D. Horisberger, H. Garty, and K. Geering, *FXYD7 is a brain-specific regulator of Na,K-ATPase alpha 1-beta isozymes.* EMBO J, 21:3264–3273, 2002.
15. Geering, K., *FXYD proteins: new regulators of Na-K-ATPase.* Am J Physiol Renal Physiol, 290:F241–F250, 2006.
16. Therien, A.G., S.J. Karlish, and R. Blostein, *Expression and functional role of the gamma subunit of the Na, K-ATPase in mammalian cells.* J Biol Chem, 274:12252–12256, 1999.
17. Arystarkhova, E., R.K. Wetzel, N.K. Asinovski, and K.J. Sweadner, *The gamma subunit modulates Na(+) and K(+) affinity of the renal Na,K-ATPase.* J Biol Chem, 274:33183–33185, 1999.
18. Wetzel, R.K., J.L. Pascoa, and E. Arystarkhova, *Stress-induced expression of the gamma subunit (FXYD2) modulates Na,K-ATPase activity and cell growth.* J Biol Chem, 279:41750–41757, 2004.
19. Kassed, C.A., T.L. Butler, G.W. Patton, D.D. Demesquita, M.T. Navidomskis, S. Memet, A. Israel, and K.R. Pennypacker, *Injury-induced NF-kappaB activation in the hippocampus: implications for neuronal survival.* FASEB J, 18:723–724, 2004.
20. Delprat, B., D. Schaer, S. Roy, J. Wang, J.L. Puel, and K. Geering, *FXYD6 is a novel regulator of Na,K-ATPase expressed in the inner ear.* J Biol Chem, 282:7450–7456, 2007.
21. Delprat, B., J.L. Puel, and K. Geering, *Dynamic expression of FXYD6 in the inner ear suggests a role of the protein in endolymph homeostasis and neuronal activity.* Dev Dyn, 236:2534–2540, 2007.
22. Geibel, S., J.H. Kaplan, E. Bamberg, and T. Friedrich, *Conformational dynamics of the Na+/K+-ATPase probed by voltage clamp fluorometry.* Proc Natl Acad Sci USA, 100:964–969, 2003.
23. Yu, S.P., *Regulation and critical role of potassium homeostasis in apoptosis.* Prog Neurobiol, 70:363–386, 2003.
24. Xiao, A.Y., L. Wei, S. Xia, S. Rothman, and S.P. Yu, *Ionic mechanism of ouabain-induced concurrent apoptosis and necrosis in individual cultured cortical neurons.* J Neurosci, 22:1350–1362, 2002.
25. Dvela, M., H. Rosen, T. Feldmann, M. Nesher, and D. Lichtstein, *Diverse biological responses to different cardiotonic steroids.* Pathophysiology, 14:159–166, 2007.
26. Mathews, W.R., D.W. DuCharme, J.M. Hamlyn, D.W. Harris, F. Mandel, M.A. Clark, and J.H. Ludens, *Mass spectral characterization of an endogenous digitalis-like factor from human plasma.* Hypertension, 17:930–935, 1991.
27. Schoner, W., N. Bauer, J. Muller-Ehmsen, U. Kramer, N. Hambarchian, R. Schwinger, H. Moeller, H. Kost, C. Weitkamp, T. Schweitzer, U. Kirch, H. Neu, and E.G. Grunbaum, *Ouabain as a mammalian hormone.* Ann N Y Acad Sci, 986:678–684, 2003.
28. Hamlyn, J.M., J. Laredo, J.R. Shah, Z.R. Lu, and B.P. Hamilton, *11-hydroxylation in the biosynthesis of endogenous ouabain: multiple implications.* Ann N Y Acad Sci, 986:685–693, 2003.
29. Goto, A., K. Yamada, H. Nagoshi, Y. Terano, and M. Omata, *Stress-induced elevation of ouabainlike compound in rat plasma and adrenal.* Hypertension, 26:1173–1176, 1995.
30. De Angelis, C. and G.T. Haupert, Jr., *Hypoxia triggers release of an endogenous inhibitor of Na(+)-K(+)-ATPase from midbrain and adrenal.* Am J Physiol, 274:F182–F188, 1998.

31. Tian, J., T. Cai, Z. Yuan, H. Wang, L. Liu, M. Haas, E. Maksimova, X.Y. Huang, and Z.J. Xie, *Binding of Src to Na+/K+-ATPase forms a functional signaling complex.* Mol Biol Cell, 17: 317–326, 2006.
32. Mohammadi, K., P. Kometiani, Z. Xie, and A. Askari, *Role of protein kinase C in the signal pathways that link Na+/K+-ATPase to ERK1/2.* J Biol Chem, 276:42050–42056, 2001.
33. Yuan, Z., T. Cai, J. Tian, A.V. Ivanov, D.R. Giovannucci, and Z. Xie, *Na/K-ATPase tethers phospholipase C and IP3 receptor into a calcium-regulatory complex.* Mol Biol Cell, 16: 4034–4045, 2005.
34. Xie, Z., *Molecular mechanisms of Na/K-ATPase-mediated signal transduction.* Ann N Y Acad Sci, 986:497–503, 2003.
35. Aperia, A., *New roles for an old enzyme: Na,K-ATPase emerges as an interesting drug target.* J Intern Med, 261:44–52, 2007.
36. Ihenetu, K., H.M. Qazzaz, F. Crespo, R. Fernandez-Botran, and R. Valdes, Jr., *Digoxin-like immunoreactive factors induce apoptosis in human acute T-cell lymphoblastic leukemia.* Clin Chem, 53:1315–1322, 2007.
37. Liang, M., J. Tian, L. Liu, S. Pierre, J. Liu, J. Shapiro, and Z.J. Xie, *Identification of a pool of non-pumping Na/K-ATPase.* J Biol Chem, 282:10585–10593, 2007.
38. Lopina, O.D., *Interaction of Na,K-ATPase catalytic subunit with cellular proteins and other endogenous regulators.* Biochemistry, 66:1122–1131, 2001.
39. Kalia, L.V. and M.W. Salter, *Interactions between Src family protein tyrosine kinases and PSD-95.* Neuropharmacology, 45:720–728, 2003.
40. Wang, X.Q. and S.P. Yu, *Novel regulation of Na, K-ATPase by Src tyrosine kinases in cortical neurons.* J Neurochem, 93:1515–1523, 2005.
41. Rajasekaran, S.A., L.G. Palmer, K. Quan, J.F. Harper, W.J. Ball, Jr., N.H. Bander, A. Peralta Soler, and A.K. Rajasekaran, *Na,K-ATPase beta-subunit is required for epithelial polarization, suppression of invasion, and cell motility.* Mol Biol Cell, 12:279–295, 2001.
42. Hallows, K.R., *Emerging role of AMP-activated protein kinase in coupling membrane transport to cellular metabolism.* Curr Opin Nephrol Hypertens, 14:464–471, 2005.
43. Woollhead, A.M., J.W. Scott, D.G. Hardie, and D.L. Baines, *Phenformin and 5-aminoimidazole-4-carboxamide-1-beta-D-ribofuranoside (AICAR) activation of AMP-activated protein kinase inhibits transepithelial Na+ transport across H441 lung cells.* J Physiol, 566:781–792, 2005.
44. Woollhead, A.M., J. Sivagnanasundaram, K.K. Kalsi, V. Pucovsky, L.J. Pellatt, J.W. Scott, K.J. Mustard, D.G. Hardie, and D.L. Baines, *Pharmacological activators of AMP-activated protein kinase have different effects on Na+ transport processes across human lung epithelial cells.* Br J Pharmacol, 151:1204–1215, 2007.
45. Bogdanova, A., B. Grenacher, M. Nikinmaa, and M. Gassmann, *Hypoxic responses of Na+/K+ ATPase in trout hepatocytes.* J Exp Biol, 208:1793–1801, 2005.
46. Rosenberger, C., S. Rosen, A. Shina, W. Bernhardt, M.S. Wiesener, U. Frei, K.U. Eckardt, and S.N. Heyman, *Hypoxia-inducible factors and tubular cell survival in isolated perfused kidneys.* Kidney Int, 70:60–70, 2006.
47. Petrushanko, I.Y., N.B. Bogdanov, N. Lapina, A.A. Boldyrev, M. Gassmann, and A.Y. Bogdanova, *Oxygen-induced Regulation of Na/K ATPase in cerebellar granule cells.* J Gen Physiol, 130: 389–398, 2007.
48. Jung, Y.W., I.J. Choi, and T.H. Kwon, *Altered expression of sodium transporters in ischemic penumbra after focal cerebral ischemia in rats.* Neurosci Res, 59:152–159, 2007.
49. Chen, C.M., S.H. Liu, and S.Y. Lin-Shiau, *Honokiol, a neuroprotectant against mouse cerebral ischaemia, mediated by preserving Na+, K+-ATPase activity and mitochondrial functions.* Basic Clin Pharmacol Toxicol, 101:108–116, 2007.
50. Zhou, G., L.A. Dada, N.S. Chandel, K. Iwai, E. Lecuona, A. Ciechanover, and J.I. Sznajder, *Hypoxia-mediated Na-K-ATPase degradation requires von Hippel Lindau protein.* FASEB J, 2007.
51. Dada, L.A., N.S. Chandel, K.M. Ridge, C. Pedemonte, A.M. Bertorello, and J.I. Sznajder, *Hypoxia-induced endocytosis of Na,K-ATPase in alveolar epithelial cells is mediated by mitochondrial reactive oxygen species and PKC-zeta.* J Clin Invest, 111:1057–1064, 2003.

52. Chen, Z., R.T. Krmar, L. Dada, R. Efendiev, I.B. Leibiger, C.H. Pedemonte, A.I. Katz, J.I. Sznajder, and A.M. Bertorello, *Phosphorylation of adaptor protein-2 mu2 is essential for Na+,K+-ATPase endocytosis in response to either G protein-coupled receptor or reactive oxygen species*. Am J Respir Cell Mol Biol, 35:127–132, 2006.
53. Dada, L.A., L.C. Welch, G. Zhou, R. Ben-Saadon, A. Ciechanover, and J.I. Sznajder, *Phosphorylation and ubiquitination are necessary for Na,K-ATPase endocytosis during hypoxia*. Cell Signal, 19:1893–1898, 2007.
54. Dada, L.A., E. Novoa, E. Lecuona, H. Sun, and J.I. Sznajder, *Role of the small GTPase RhoA in the hypoxia-induced decrease of plasma membrane Na,K-ATPase in A549 cells*. J Cell Sci, 120:2214–2222, 2007.
55. Litvan, J., A. Briva, M.S. Wilson, G.R. Budinger, J.I. Sznajder, and K.M. Ridge, *Beta-adrenergic receptor stimulation and adenoviral overexpression of superoxide dismutase prevent the hypoxia-mediated decrease in Na,K-ATPase and alveolar fluid reabsorption*. J Biol Chem, 281:19892–19898, 2006.
56. Comellas, A.P., L.A. Dada, E. Lecuona, L.M. Pesce, N.S. Chandel, N. Quesada, G.R. Budinger, G.J. Strous, A. Ciechanover, and J.I. Sznajder, *Hypoxia-mediated degradation of Na,K-ATPase via mitochondrial reactive oxygen species and the ubiquitin-conjugating system*. Circ Res, 98:1314–1322, 2006.
57. Schoner, W., *Endogenous cardiac glycosides, a new class of steroid hormones*. Eur J Biochem, 269:2440–2448, 2002.
58. Wyse, A.T., E.L. Streck, S.V. Barros, A.M. Brusque, A.I. Zugno, and M. Wajner, *Methylmalonate administration decreases Na+,K+-ATPase activity in cerebral cortex of rats*. Neuroreport, 11:2331–2334, 2000.
59. Qiao, M., K.L. Malisza, M.R. Del Bigio, and U.I. Tuor, *Transient hypoxia-ischemia in rats: changes in diffusion-sensitive MR imaging findings, extracellular space, and Na+-K+-adenosine triphosphatase and cytochrome oxidase activity*. Radiology, 223:65–75, 2002.
60. Nagafuji, T., T. Koide, and M. Takato, *Neurochemical correlates of selective neuronal loss following cerebral ischemia: role of decreased Na+,K(+)-ATPase activity*. Brain Res, 571:265–271, 1992.
61. Dzurba, A., A. Ziegelhoffer, L. Okruhlicova, N. Vrbjar, and J. Styk, *Salutary effect of tedisamil on post-ischemic recovery rat heart: involvement of sarcolemmal (Na,K)-ATPase*. Mol Cell Biochem, 215:129–133, 2000.
62. Schonefeld, M., S. Noble, A.M. Bertorello, L.J. Mandel, M.H. Creer, and D. Portilla, *Hypoxia-induced amphiphiles inhibit renal Na+, K(+)-ATPase*. Kidney Int, 49:1289–1296, 1996.
63. Xie, M., W. Wang, H.K. Kimelberg, and M. Zhou, *Oxygen and glucose deprivation-induced changes in astrocyte membrane potential and their underlying mechanisms in acute rat hippocampal slices*. J Cereb Blood Flow Metab, 28:456–467, 2008.
64. Yu, S.P., C.H. Yeh, S.L. Sensi, B.J. Gwag, L.M. Canzoniero, Z.S. Farhangrazi, H.S. Ying, M. Tian, L.L. Dugan, and D.W. Choi, *Mediation of neuronal apoptosis by enhancement of outward potassium current*. Science, 278:114–117, 1997.
65. Yu, S.P., Z.S. Farhangrazi, H.S. Ying, C.H. Yeh, and D.W. Choi, *Enhancement of outward potassium current may participate in beta-amyloid peptide-induced cortical neuronal death*. Neurobiol Dis, 5:81–88, 1998.
66. Yu, S.P., C. Yeh, U. Strasser, M. Tian, and D.W. Choi, *NMDA receptor-mediated K^+ efflux and neuronal apoptosis*. Science, 284:336–339, 1999.
67. Yu, S.P. and D.W. Choi, *Ions, cell volume, and apoptosis*. Proc Natl Acad Sci USA, 97:9360–9362, 2000.
68. Yu, S.P., L.M. Canzoniero, and D.W. Choi, *Ion homeostasis and apoptosis*. Curr Opin Cell Biol, 13:405–411, 2001.
69. Robinson, J.D. and M.S. Flashner, *The (Na+ + K+)-activated ATPase. Enzymatic and transport properties*. Biochim Biophys Acta, 549:145–176, 1979.
70. Barbiero, G., F. Duranti, G. Bonelli, J.S. Amenta, and F.M. Baccino, *Intracellular ionic variations in the apoptotic death of L cells by inhibitors of cell cycle progression*. Exp Cell Res, 217:410–418, 1995.

71. Hughes, F.M. Jr., and J.A. Cidlowski, *Potassium is a critical regulator of apoptotic enzymes in vitro and in vivo*. Adv Enzyme Regul, 39:157–171, 1999.
72. Xiao, A.Y., X.Q. Wang, A. Yang, and S.P. Yu, *Slight impairment of Na+,K+-ATPase synergistically aggravates ceramide- and beta-amyloid-induced apoptosis in cortical neurons*. Brain Res, 955:253–259, 2002.
73. Dallaporta, B., T. Hirsch, S.A. Susin, N. Zamzami, N. Larochette, C. Brenner, I. Marzo, and G. Kroemer, *Potassium leakage during the apoptotic degradation phase*. J Immunol, 160: 5605–5615, 1998.
74. Jonas, D., I. Walev, T. Berger, M. Liebetrau, M. Palmer, and S. Bhakdi, *Novel path to apoptosis: small transmembrane pores created by staphylococcal alpha-toxin in T lymphocytes evoke internucleosomal DNA degradation*. Infect Immun, 62:1304–1312, 1994.
75. Yu, S.P., C.H. Yeh, F. Gottron, X. Wang, M.C. Grabb, and D.W. Choi, *Role of the outward delayed rectifier K+ current in ceramide-induced caspase activation and apoptosis in cultured cortical neurons*. J Neurochem, 73:933–941, 1999.
76. Lang, F., S.M. Huber, I. Szabo, and E. Gulbins, *Plasma membrane ion channels in suicidal cell death*. Arch Biochem Biophys, 462:189–194, 2007.
77. Xiao, A.Y., M. Homma, X.Q. Wang, X. Wang, and S.P. Yu, *Role of K(+) efflux in apoptosis induced by AMPA and kainate in mouse cortical neurons*. Neuroscience, 108:61–67, 2001.
78. Ekhterae, D., O. Platoshyn, S. Krick, Y. Yu, S.S. McDaniel, and J.X. Yuan, *Bcl-2 decreases voltage-gated K+ channel activity and enhances survival in vascular smooth muscle cells*. Am J Physiol Cell Physiol, 281:C157–C165, 2001.
79. Norman, D.J., L. Feng, S.S. Cheng, J. Gubbay, E. Chan, and N. Heintz, *The lurcher gene induces apoptotic death in cerebellar Purkinje cells*. Development, 121:1183–1193, 1995.
80. Krick, S., O. Platoshyn, M. Sweeney, H. Kim, and J.X. Yuan, *Activation of K+ channels induces apoptosis in vascular smooth muscle cells*. Am J Physiol Cell Physiol, 280:C970–C979, 2001.
81. Fan, Z. and R.A. Neff, *Susceptibility of ATP-sensitive K+ channels to cell stress through mediation of phosphoinositides as examined by photoirradiation*. J Physiol, 529 (Part 3):707–721, 2000.
82. Pal, S., K. He, and E. Aizenman, *Nitrosative stress and potassium channel-mediated neuronal apoptosis: is zinc the link?* Pflugers Arch, 448:296–303, 2004.
83. Wang, X.Q., A.Y. Xiao, C. Sheline, K. Hyrc, A. Yang, M.P. Goldberg, D.W. Choi, and S.P. Yu, *Apoptotic insults impair Na+, K+-ATPase activity as a mechanism of neuronal death mediated by concurrent ATP deficiency and oxidant stress*. J Cell Sci, 116:2099–2110, 2003.
84. Wang, X.Q., A.Y. Xiao, A. Yang, L. LaRose, L. Wei, and S.P. Yu, *Block of Na+,K+-ATPase and induction of hybrid death by 4-aminopyridine in cultured cortical neurons*. J Pharmacol Exp Ther, 305:502–506, 2003.
85. Nobel, C.S., J.K. Aronson, D.J. van den Dobbelsteen, and A.F. Slater, *Inhibition of Na+/K(+)-ATPase may be one mechanism contributing to potassium efflux and cell shrinkage in CD95-induced apoptosis*. Apoptosis, 5:153–163, 2000.
86. Bortner, C.D., M. Gomez-Angelats, and J.A. Cidlowski, *Plasma membrane depolarization without repolarization is an early molecular event in anti-Fas-induced apoptosis*. J Biol Chem, 276:4304–4314, 2001.
87. Gilbert, M.S., A.H. Saad, B.A. Rupnow, and S.J. Knox, *Association of BCL-2 with membrane hyperpolarization and radioresistance*. J Cell Physiol, 168:114–122, 1996.
88. Gilbert, M. and S. Knox, *Influence of Bcl-2 overexpression on Na+/K(+)-ATPase pump activity: correlation with radiation-induced programmed cell death*. J Cell Physiol, 171:299–304, 1997.
89. Numazawa, S., M.A. Shinoki, H. Ito, T. Yoshida, and Y. Kuroiwa, *Involvement of Na+,K(+)-ATPase inhibition in K562 cell differentiation induced by bufalin*. J Cell Physiol, 160: 113–120, 1994.
90. Yang, A., X.Q. Wang, C.S. Sun, L. Wei, and S.P. Yu, *Inhibitory effects of clofilium on membrane currents associated with Ca channels, NMDA receptor channels and Na+, K+-ATPase in cortical neurons*. Pharmacology, 73:162–168, 2005.
91. Jaeschke, H. and J.J. Lemasters, *Apoptosis versus oncotic necrosis in hepatic ischemia/reperfusion injury*. Gastroenterology, 125:1246–1257, 2003.

92. Abdalah, R., L. Wei, K. Francis, and S.P. Yu, *Valinomycin-induced apoptosis in Chinese hamster ovary cells.* Neurosci Lett, 405:68–73, 2006.
93. Hilgenberg, L.G., H. Su, H. Gu, D.K. O'Dowd, and M.A. Smith, *Alpha3Na+/K+-ATPase is a neuronal receptor for agrin.* Cell, 125:359–369, 2006.
94. Gautam, M., P.G. Noakes, L. Moscoso, F. Rupp, R.H. Scheller, J.P. Merlie, and J.R. Sanes, *Defective neuromuscular synaptogenesis in agrin-deficient mutant mice.* Cell, 85:525–535, 1996.
95. McMahan, U.J., *The agrin hypothesis.* Cold Spring Harb Symp Quant Biol, 55:407–418, 1990.
96. Bowe, M.A. and J.R. Fallon, *The role of agrin in synapse formation.* Annu Rev Neurosci, 18:443–462, 1995.
97. Ksiazek, I., C. Burkhardt, S. Lin, R. Seddik, M. Maj, G. Bezakova, M. Jucker, S. Arber, P. Caroni, J.R. Sanes, B. Bettler, and M.A. Ruegg, *Synapse loss in cortex of agrin-deficient mice after genetic rescue of perinatal death.* J Neurosci, 27:7183–7195, 2007.
98. Hebb, D.O., *The Organization of Behavior: A Neuropsychological Theory.* Wiley: New York 437, 1949.
99. Kim, J.H., I. Sizov, M. Dobretsov, and H. von Gersdorff, *Presynaptic Ca^{2+} buffers control the strength of a fast post-tetanic hyperpolarization mediated by the alpha3 Na(+)/K(+)-ATPase.* Nat Neurosci, 10:196–205, 2007.
100. Palladino, M.J., J.E. Bower, R. Kreber, and B. Ganetzky, *Neural dysfunction and neurodegeneration in Drosophila Na+/K+ ATPase alpha subunit mutants.* J Neurosci, 23:1276–1286, 2003.
101. Ikeda, K., T. Onaka, M. Yamakado, J. Nakai, T.O. Ishikawa, M.M. Taketo, and K. Kawakami, *Degeneration of the amygdala/piriform cortex and enhanced fear/anxiety behaviors in sodium pump alpha2 subunit (Atp1a2)-deficient mice.* J Neurosci, 23:4667–4676, 2003.
102. Xu, K.Y., E. Takimoto, and N.S. Fedarko, *Activation of (Na+ + K+)-ATPase induces positive inotropy in intact mouse heart in vivo.* Biochem Biophys Res Commun, 349:582–587, 2006.
103. Cao, D., B. Yang, L. Hou, J. Xu, R. Xue, L. Sun, C. Zhou, and Z. Liu, *Chronic daily administration of ethyl docosahexaenoate protects against gerbil brain ischemic damage through reduction of arachidonic acid liberation and accumulation.* J Nutr Biochem, 18:297–304, 2007.
104. Zhan, C. and J. Yang, *Protective effects of isoliquiritigenin in transient middle cerebral artery occlusion-induced focal cerebral ischemia in rats.* Pharmacol Res, 53:303–309, 2006.
105. Imaizumi, S., K. Kurosawa, H. Kinouchi, and T. Yoshimoto, *Effect of phenytoin on cortical Na(+)-K(+)-ATPase activity in global ischemic rat brain.* J Neurotrauma, 12:231–234, 1995.
106. Hamlyn, J.M., R. Ringel, J. Schaeffer, P.D. Levinson, B.P. Hamilton, A.A. Kowarski, and M.P. Blaustein, *A circulating inhibitor of (Na+ + K+)ATPase associated with essential hypertension.* Nature, 300:650–652, 1982.
107. Blaustein, M.P., J.M. Hamlyn, and T.L. Pallone, *Sodium pumps: ouabain, ion transport, and signaling in hypertension.* Am J Physiol Renal Physiol, 293:F438; author reply F439, 2007.
108. Wei, L., B.H. Han, Y. Li, C.L. Keogh, D.M. Holtzman, and S.P. Yu, *Cell death mechanism and protective effect of erythropoietin after focal ischemia in the whisker-barrel cortex of neonatal rats.* J Pharmacol Exp Ther, 317:109–116, 2006.
109. Wei, L., D.J. Ying, L. Cui, J. Langsdorf, and S.P. Yu, *Necrosis, apoptosis and hybrid death in the cortex and thalamus after barrel cortex ischemia in rats.* Brain Res, 1022:54–61, 2004.
110. Bruer, U., M.K. Weih, N.K. Isaev, A. Meisel, K. Ruscher, A. Bergk, G. Trendelenburg, F. Wiegand, I.V. Victorov, and U. Dirnagl, *Induction of tolerance in rat cortical neurons: hypoxic preconditioning.* FEBS Lett, 414:117–121, 1997.
111. Pierre, S.V., C. Yang, Z. Yuan, J. Seminerio, C. Mouas, K.D. Garlid, P. Dos-Santos, and Z. Xie, *Ouabain triggers preconditioning through activation of the Na+,K+-ATPase signaling cascade in rat hearts.* Cardiovasc Res, 73:488–496, 2007.
112. Gu, Q.B., J.X. Zhao, J. Fei, and W. Schwarz, *Modulation of Na(+),K(+) pumping and neurotransmitter uptake by beta-amyloid.* Neuroscience, 126:61–67, 2004.
113. Mark, R.J., K. Hensley, D.A. Butterfield, and M.P. Mattson, *Amyloid beta-peptide impairs ion-motive ATPase activities: evidence for a role in loss of neuronal Ca^{2+} homeostasis and cell death.* J Neurosci, 15:6239–6249, 1995.

114. Dickey, C.A., M.N. Gordon, D.M. Wilcock, D.L. Herber, M.J. Freeman, and D. Morgan, *Dysregulation of Na+/K+ ATPase by amyloid in APP+ PS1 transgenic mice.* BMC Neurosci, 6:7, 2005.
115. Kurup, R.K. and P.A. Kurup, *Hypothalamic digoxin-mediated model for Parkinson's disease.* Int J Neurosci, 113:515–536, 2003.
116. Khan, F.H., T. Sen, and S. Chakrabarti, *Dopamine oxidation products inhibit Na+, K+-ATPase activity in crude synaptosomal-mitochondrial fraction from rat brain.* Free Radic Res, 37:597–601, 2003.
117. Bonifati, V., P. Rizzu, M.J. van Baren, O. Schaap, G.J. Breedveld, E. Krieger, M.C. Dekker, F. Squitieri, P. Ibanez, M. Joosse, J.W. van Dongen, N. Vanacore, J.C. van Swieten, A. Brice, G. Meco, C.M. van Duijn, B.A. Oostra, and P. Heutink, *Mutations in the DJ-1 gene associated with autosomal recessive early-onset parkinsonism.* Science, 299:256–259, 2003.
118. Pisani, A., G. Martella, A. Tscherter, C. Costa, N.B. Mercuri, G. Bernardi, J. Shen, and P. Calabresi, *Enhanced sensitivity of DJ-1-deficient dopaminergic neurons to energy metabolism impairment: role of Na+/K+ ATPase.* Neurobiol Dis, 23:54–60, 2006.
119. de Carvalho Aguiar, P., K.J. Sweadner, J.T. Penniston, J. Zaremba, L. Liu, M. Caton, G. Linazasoro, M. Borg, M.A. Tijssen, S.B. Bressman, W.B. Dobyns, A. Brashear, and L.J. Ozelius, *Mutations in the Na+/K+-ATPase alpha3 gene ATP1A3 are associated with rapid-onset dystonia parkinsonism.* Neuron, 43:169–175, 2004.
120. Brashear, A., W.B. Dobyns, P. de Carvalho Aguiar, M. Borg, C.J. Frijns, S. Gollamudi, A. Green, J. Guimaraes, B.C. Haake, C. Klein, G. Linazasoro, A. Munchau, D. Raymond, D. Riley, R. Saunders-Pullman, M.A. Tijssen, D. Webb, J. Zaremba, S.B. Bressman, and L.J. Ozelius, *The phenotypic spectrum of rapid-onset dystonia-parkinsonism (RDP) and mutations in the ATP1A3 gene.* Brain, 130:828–835, 2007.
121. Kamphuis, D.J., H. Koelman, A.J. Lees, and M.A. Tijssen, *Sporadic rapid-onset dystonia-parkinsonism presenting as Parkinson's disease.* Mov Disord, 21:118–119, 2006.
122. el-Mallakh, R.S. and R.J. Wyatt, *The Na,K-ATPase hypothesis for bipolar illness.* Biol Psychiatry, 37:235–244, 1995.
123. Goldstein, I., T. Levy, D. Galili, H. Ovadia, R. Yirmiya, H. Rosen, and D. Lichtstein, *Involvement of Na(+), K(+)-ATPase and endogenous digitalis-like compounds in depressive disorders.* Biol Psychiatry, 60:491–499, 2006.
124. Kurup, A.R. and P.A. Kurup, *Membrane Na(+)-K+ ATPase mediated cascade in bipolar mood disorder, major depressive disorder, and schizophrenia – relationship to hemispheric dominance.* Int J Neurosci, 112:965–982, 2002.
125. de Vasconcellos, A.P., A.I. Zugno, A.H. Dos Santos, F.B. Nietto, L.M. Crema, M. Goncalves, R. Franzon, A.T. de Souza Wyse, E.R. da Rocha, and C. Dalmaz, *Na+,K(+)-ATPase activity is reduced in hippocampus of rats submitted to an experimental model of depression: effect of chronic lithium treatment and possible involvement in learning deficits.* Neurobiol Learn Mem, 84:102–110, 2005.
126. Philibert, R.A., D. Cheung, N. Welsh, P. Damschroder-Williams, B. Thiel, E.I. Ginns, and H.K. Gershenfeld, *Absence of a significant linkage between Na(+),K(+)-ATPase subunit (ATP1A3 and ATP1B3) genotypes and bipolar affective disorder in the old-order Amish.* Am J Med Genet, 105:291–294, 2001.
127. Vanmolkot, K.R., E.E. Kors, J.J. Hottenga, G.M. Terwindt, J. Haan, W.A. Hoefnagels, D.F. Black, L.A. Sandkuijl, R.R. Frants, M.D. Ferrari, and A.M. van den Maagdenberg, *Novel mutations in the Na+, K+-ATPase pump gene ATP1A2 associated with familial hemiplegic migraine and benign familial infantile convulsions.* Ann Neurol, 54:360–366, 2003.
128. Todt, U., M. Dichgans, K. Jurkat-Rott, A. Heinze, G. Zifarelli, J.B. Koenderink, I. Goebel, V. Zumbroich, A. Stiller, A. Ramirez, T. Friedrich, H. Gobel, and C. Kubisch, *Rare missense variants in ATP1A2 in families with clustering of common forms of migraine.* Hum Mutat, 26:315–321, 2005.
129. De Fusco, M., R. Marconi, L. Silvestri, L. Atorino, L. Rampoldi, L. Morgante, A. Ballabio, P. Aridon, and G. Casari, *Haploinsufficiency of ATP1A2 encoding the Na+/K+ pump alpha2 subunit associated with familial hemiplegic migraine type 2.* Nat Genet, 33:192–196, 2003.

130. Koenderink, J.B., G. Zifarelli, L.Y. Qiu, W. Schwarz, J.J. De Pont, E. Bamberg, and T. Friedrich, *Na,K-ATPase mutations in familial hemiplegic migraine lead to functional inactivation*. Biochim Biophys Acta, 1669:61–68, 2005.
131. Segall, L., A. Mezzetti, R. Scanzano, J.J. Gargus, E. Purisima, and R. Blostein, *Alterations in the alpha2 isoform of Na,K-ATPase associated with familial hemiplegic migraine type 2*. Proc Natl Acad Sci USA, 102:11106–11111, 2005.
132. Segall, L., R. Scanzano, M.A. Kaunisto, M. Wessman, A. Palotie, J.J. Gargus, and R. Blostein, *Kinetic alterations due to a missense mutation in the Na,K-ATPase alpha2 subunit cause familial hemiplegic migraine type 2*. J Biol Chem, 279:43692–43696, 2004.
133. Takano, T., G.F. Tian, W. Peng, N. Lou, D. Lovatt, A.J. Hansen, K.A. Kasischke, and M. Nedergaard, *Cortical spreading depression causes and coincides with tissue hypoxia*. Nat Neurosci, 10:754–762, 2007.
134. Etminan, M., B. Takkouche, F.C. Isorna, and A. Samii, *Risk of ischaemic stroke in people with migraine: systematic review and meta-analysis of observational studies*. BMJ, 330:63, 2005.
135. MacClellan, L.R., W. Giles, J. Cole, M. Wozniak, B. Stern, B.D. Mitchell, and S.J. Kittner, *Probable migraine with visual aura and risk of ischemic stroke: the stroke prevention in young women study*. Stroke, 38:2438–2445, 2007.
136. Roy, M.C., F.R. Chang, H.C. Huang, M.Y. Chiang, and Y.C. Wu, *Cytotoxic principles from the formosan milkweed, Asclepias curassavica*. J Nat Prod, 68:1494–1499, 2005.
137. Weidemann, H., *Na/K-ATPase, endogenous digitalis like compounds and cancer development – a hypothesis*. Front Biosci, 10:2165–2176, 2005.
138. Scheiner-Bobis, G. and W. Schoner, *A fresh facet for ouabain action*. Nat Med, 7:1288–1289, 2001.
139. Aizman, O., P. Uhlen, M. Lal, H. Brismar, and A. Aperia, *Ouabain, a steroid hormone that signals with slow calcium oscillations*. Proc Natl Acad Sci USA, 98:13420–13424, 2001.
140. Miyakawa-Naito, A., P. Uhlen, M. Lal, O. Aizman, K. Mikoshiba, H. Brismar, S. Zelenin, and A. Aperia, *Cell signaling microdomain with Na,K-ATPase and inositol 1,4,5-trisphosphate receptor generates calcium oscillations*. J Biol Chem, 278:50355–50361, 2003.
141. Li, J., S. Zelenin, A. Aperia, and O. Aizman, *Low doses of ouabain protect from serum deprivation-triggered apoptosis and stimulate kidney cell proliferation via activation of NF-kappaB*. J Am Soc Nephrol, 17:1848–1857, 2006.
142. Aydemir-Koksoy, A., J. Abramowitz, and J.C. Allen, *Ouabain-induced signaling and vascular smooth muscle cell proliferation*. J Biol Chem, 276:46605–46611, 2001.
143. Schaal, S., I. Beiran, E. Bormusov, M. Chevion, and A. Dovrat, *Zinc-desferrioxamine reduces damage to lenses exposed to hyperbaric oxygen and has an ameliorative effect on catalase and Na, K-ATPase activities*. Exp Eye Res, 84:455–463, 2007.

Chapter 5
Emerging Role of Water Channels in Regulating Cellular Volume During Oxygen Deprivation and Cell Death

Thomas James Younts and Francis "Monty" Hughes, Jr.

Abstract In this review, we will discuss how proteinaceous water channels, termed aquaporins (AQPs), regulate water fluxes across plasma membranes within various physiological and pathophysiological contexts. Particular emphasis has been assigned to changes in aquaporin expression in brain matter in response to conditions where oxygen deprivation, such as hypoxia or ischemia, has been experimentally induced. We also discuss the specific role AQPs play in apoptosis, also known as programmed cell death, with special interest paid to AQPs in conjunction with potassium channels, and their seemingly interdependent function in regulating downstream apoptotic cascades. Finally, we argue for the possibility of connected aquaporin and potassium channel translocation during apoptotic insults.

Keywords: Apoptosis; Aquaporin; Brain; Expression; Hypoxia; Ischemia; Potassium channel; Translocation; Volume regulation; Water channel

5.1 Volume Regulatory Mechanisms

Under physiological conditions, extracellular fluids are subject to fluctuations in ionic content. Such fluctuations bring about osmotic gradients that can induce severe damage to cells, possibly even leading to cell death (1). Clearly, preventing or minimizing excessive cellular shrinkage and expansion is critical if cells are to successfully adapt to their microenvironment. To counteract such stresses, cells possess two primary mechanisms to regulate physiological volume. Both mechanisms function by altering ionic content within the cell, which ultimately drives water to its respective compartment.

To restore cellular volume when presented with hypertonic challenges, cells undergo the regulatory volume increase (RVI). Normal cells accomplish the RVI by

---

T.J. Younts and F. "Monty" Hughes, Jr. (✉)
Department of Biology, University of North Carolina at Charlotte, 9201 University City Blvd, Charlotte, NC 28223, USA
e-mail: mhughes@uncc.edu

G.G. Haddad, S.P. Yu (eds.), *Brain Hypoxia and Ischemia*,
© 2009 Humana Press, a Part of Springer Science + Business Media, LLC

several plasma membrane ion channels acting alone or in concert. While the channels may vary, the result is essentially to recruit Na⁺ and Cl⁻ into the cell, creating an osmotic gradient responsible for drawing water out of the extracellular fluid and restoring cellular volume (2, 3). Conversely, the regulatory volume decrease (RVD) reestablishes cellular volume when cells are exposed to hypotonic conditions by mechanisms designed to expel K⁺ and Cl⁻ (4, 5). The RVI and RVD have primarily been described in vitro where solute changes can be made to occur rapidly, and pharmacological inhibitors that are specific to various volume regulatory ion channels have identified their roles in individual cells (6–9). However, in vivo, cells more or less isovolumetrically utilize the RVI and RVD, and hence their apparent volume does not appear to change as dramatically as what can be measured in vitro.

One in vivo physiological condition in which the cell's size does change dramatically is during the process of apoptosis. In virtually every model in which it has been examined, with one nonphysiological exception (1), cells undergoing apoptosis lose water and shrink in a process that has become known as the apoptotic volume decrease (AVD). Similar to the RVI and RVD, the AVD also makes use of ionic fluxes across the plasma membrane to create the osmotic gradients necessary to bring about volume change.

Much work has been done on the various ionic fluxes constituting these volume regulatory mechanisms, but until recently, little attention was paid to the pathway(s) by which water traversed the plasma membrane. We now know that protein water channels termed aquaporins (AQPs) are critical in mediating this water movement in a large number of physiological systems, and defects in AQP function have been implicated in numerous diseases and pathological conditions such as stroke as a result of hypoxia/ischemia (10), nephrogenic diabetes insipidus (11, 12), congestive heart failure (13), obesity (14), loss of vision and cataracts (13, 15, 16), HIV (17), malaria (18), and pulmonary obstruction (19). In this chapter, we will discuss the expression and importance of these water channels in neuronal tissues focusing on the role of AQPs in the onset of brain edema in response to hypoxia and ischemia. Furthermore, we will present evidence for the emerging role and importance of AQPs in apoptosis.

5.2 Properties of AQP and Their Neuronal Expression

AQPs are a subset of the Major Intrinsic Protein (MIP) family of proteins, of which over 84 have been identified in many different organisms ranging from bacteria to humans (20). AQPs were discovered in 1992 [for a history of AQP discovery, see (21)] by Peter Agre and colleagues as the long-sought water channels in red blood cell membranes (21, 22), and it was given the name CHIP28 (channel-forming integral membrane protein of 28 kDa). Since that time, 13 AQPs (designated AQP0-12) have been identified and cloned in mammals. Development of AQP knockout mice revealed many unexpected cellular roles ranging from reduced tumor angiogenesis and cellular migration to providing energy substrates such as glycerol to epidermis and fat tissues for skin hydration and fat metabolism, respectively (23). Taken together, the functional roles of AQPs are vast.

5 Emerging Role of Water Channels in Regulating Cellular Volume

Monomeric AQP polypeptides exhibit six membrane-spanning α-helical domains with intracellular carboxyl and amino termini (24) (Fig. 5.1). While embedded in the plasma membrane, AQPs are known to form homotetramers; however, each monomer appears capable of mediating water movement independent of other subunits (26). Originally, all AQPs were thought to exclusively mediate water movement, but careful study has shown that, in fact, two families of AQPs exist: one specific to water and a second class more promiscuous in their transport. The latter class, consisting of AQP3, 7, 9, and 10, has been termed aquaglyceroporins (27) because the major solute transported is glycerol. However, these aquaglyceroporins are also capable of transporting other small uncharged molecules such as urea (28).

In general, AQPs are expressed in fluid-transporting tissues such as kidney, liver, and epithelia; however, nonfluid-transporting tissues such as adipose tissue, neurons, and astroglia also express AQPs, suggesting that these channels play fundamental roles in cellular physiology. It is important to point out that water movement across the plasma membrane is a passive event. Although some water can be moved concomitantly with ions by various other transporters (29–31), this

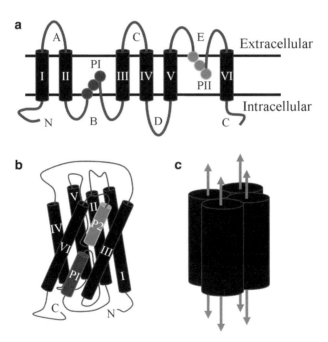

Fig. 5.1 Structural model of AQP water channel. (**a**) Generalized topological map illustrating one subunit of an AQP and indicating the six transmembrane α-helical spanning regions (labeled I–VI), five loop regions (labeled A–E), and intracellular amino (N) and carboxyl (C) termini. Two putative phosphorylation loops or P-loops (PI, PII) are shown, each of which contain a specific NPA (asparagine, proline, alanine) motif responsible for water selectivity. (**b**) Representation of one folded AQP subunit from the generalized topological structure as described in (25). (**c**) Four individual subunits homotetramerize giving the complete AQP structure. Each AQP subunit facilitates bidirectional water passage as indicated by double-headed arrows (*See Color Plates*)

is not thought to contribute to the basal plasma membrane water permeability of most cells. Permeability is generally accepted to be a result of the combined effect of simple diffusion of water molecules through the lipid bilayer and movement through AQPs. Contribution of each type of water transport to a cell's permeability will understandably vary according to the amounts and ratios of various lipids present as well as the number, type, and location of water channels. Nevertheless, in most cell types it appears that water transport through AQPs constitutes the major permeability pathway and that the presence and/or availability of AQPs is the rate-limiting step in this process (32–34).

5.2.1 AQP in the Brain

Various AQP subtypes are expressed in brain tissues, consistent with specific functional roles of AQPs. For example, aquaporin-1, a pure water channel, is restricted to the choroid plexus of the lateral ventricles where it appears to be responsible for providing the water component of the cerebral spinal fluid (35, 36). This water channel is also expressed in primary sensory neurons presumably responsible for nociception. Interestingly, no phenotypic differences in nociception were detected in a AQP1–/– mouse model as compared with wild-type mouse model (37), suggesting that redundant mechanisms may be present. At early stages of Alzheimer's disease, increased expression of AQP1 has been observed (38). AQP1 has recently been localized to astrocytes (39), and mutations in AQP1 have been linked to migrating malignant astrocytes (40); a correlation between AQP1 mutations and the neoplastic state exists. Likewise, we have published data suggesting that changes in AQP expression may contribute to the cancerous state by imparting resistance to apoptosis (41).

Aquaporin-4 is by far the predominant water channel in the brain and can be found in ependyma and pial surfaces in contact with cerebrospinal fluid. There is also distinct expression in periventricular areas, namely, astrocytic endfeet adjacent to blood vessels (42–45). Thus, AQP4 can facilitate water exchange between blood and brain, a crucial interface for homeostatic volume regulation. The roles of AQP4 in brain pathologies are explored in greater detail in a following section. Although some evidence exists for aquaporin-5 expression in brain (46, 47), the information is limited and this AQP will not be discussed in this chapter.

Aquaporin-9 protein is strongly expressed in brain as well as liver, testes, epididymis, and spleen (48–51). This AQP channel is considered promiscuous in that it is permeable to water and several solutes including carbamides, polyols, purines, and pyrimidines (52). AQP9 expression in the CNS is primarily restricted to catecholaminergic neurons (53), which are not thought to be directly involved in osmoregulation. In this context, it is believed that brain AQP9 may function as a metabolic transporter of glycerol and lactate. Indeed, it has been suggested that AQP9 might participate in the *lactate shuttle model* (54). Under pathological conditions such as hypoxia/ischemia, AQP9 may participate in the clearance of excess lactate from the extracellular environment.

5.3 Changes in Aquaporin Expression During Hypoxia and Ischemia

Before discussing how AQP expression is affected in response to lack of oxygenated blood, it is important to understand how hypoxia and ischemia often lead to apoptosis, necrosis, and subsequent brain edema (55). Although apoptosis is typified by cellular shrinkage and necrosis by cellular swelling, differences between apoptosis and necrosis are not always clear (56), and the methodology for analyzing hypoxia/ischemia must be carefully considered (57). Importantly though, both apoptosis and necrosis often result from hypoxia/ischemia, and edema is a classical postindicator of this type of injury within the brain. Two types of edema have been identified, although the lines are often blurred. Vasogenic edema is defined by deterioration of the blood brain barrier (BBB) leading to expanded extracellular space whereas cytotoxic edema is classified by disruptions in cellular metabolism, without BBB breakdown, leading to expanded intracellular space (58).

During experimental cerebral focal ischemia, which resembles ischemic stroke, blood flow control is disrupted creating an infarct zone consisting of necrotic tissue at the core surrounded by a penumbra (59–61). Necrotic tissue exhibits membrane failure and subsequent ion gradient disruption whereas cells of the penumbra exhibit acidosis and edema (62). After hypoxia or ischemia, decreased extracellular space has been reported (63), corresponding to an increase in astrocytic and neuronal swelling (64). Decreased extracellular space leads to concentration of harmful substances (65) and altered ionic gradients (65–67). Astrocytes lacking energy in pathophysiological circumstances are unable to maintain Na^+ and K^+ gradients because the Na^+/K^+ ATP-fueled pump fails, resulting in cellular swelling (68). Arachidonic acid, when released into the extracellular environment during ischemia, has several consequences including inhibition of the RVD (69), further leading to cytotoxic edema. Neuronal overactivity, common when excitatory amino acids are released during hypoxia/ischemia (70, 71), is also known to increase local concentrations of K^+, leading to blood vessel dilation (72, 73). All of these failures drive edema formation.

5.3.1 Nonapoptotic Roles for AQP During Hypoxia/Ischemia

Globally, it is well established that AQP4 plays a critical role in brain edema. However, since alterations in AQP4 expression during hypoxia/ischemia are time and edema-type dependent, seemingly disparate roles for AQP4 in vasogenic and cytotoxic brain edema have emerged. Regardless, astrocyte-specific adaptation (74) appears to be an important mechanism in edema formation and/or resolution given the unique localization of AQP4 protein to astrocytic endfeet apposed to perivascular regions of the BBB (42, 43). Using AQP4 −/− mice several roles for this AQP have been found, including their involvement in cytotoxic and vasogenic brain edema, neural signal transduction, and cellular migration (75). In several models of cytotoxic edema, AQP4 −/− mice showed reduced brain swelling and improved neurological outcome (76). On the other hand, brain swelling and neurological outcome were worse in models of vasogenic edema. This dichotomy is being explored for therapeutic possibilities.

Downregulation of AQP4 can decrease water permeability of the BBB, thereby preventing cytotoxic edema, consistent with the AQP4 −/− data (76). Several groups have shown marked decreases in astrocytic AQP4 mRNA and protein expression in response to hypoxic/ischemic insults (47, 77–79). However, other researchers have demonstrated increases in astrocytic AQP4 expression under oxygen-deprived conditions (46, 80, 81). In other cases still, overall AQP4 protein expression did not appreciably change, although more intense immunoreactivity was detected at perivascular regions (82, 83). It should not go without mention that each of these studies used various hypoxia/ischemia-inducing methodologies to stimulate edema (cytotoxic vs. vasogenic) in various regions of the brain (focal vs. global, for example), which are variable in their hypoxia/ischemia vulnerabilities (57). Taken together, such differences may account for the apparent variability of results.

Findings from mice lacking AQP4 are consistent with protection against cytotoxic brain edema; however, these studies revealed an unanticipated role for AQP4 in vasogenic edema. Previous models of vasogenic edema suggested extracellular fluid moved out of the brain through a bulk flow mechanism that does not involve cellular water channels (23). One possible explanation for the unanticipated role of AQP4 in vasogenic edema is borne out in older experiments (84) where the effect of neuronal protection was studied in the cerebellum of water-loaded rats, a condition mimicking vasogenic edema. In those studies it was found that taurine-containing neurons redistributed the location of taurine to nearby nontaurine-containing astrocytes. This produces an osmotic gradient that draws water into the astrocytes and away from the extracellular space, thereby sparing the nearby neuronal cells. Although the following has not been experimentally verified, these astrocytes could discharge their water into the vasculature through AQP4. Thus, a *bidirectional role* for AQP4 has been established (Fig. 5.2), which is consistent with downregulation of AQP4 during hypoxia/ischemia as a protective factor against water influx into astrocytes early during edema whereas upregulation of AQP4 in later phases may assist in water clearance from astrocytes (85). Experiments are needed to clarify the role of AQP4 in cytotoxic and vasogenic edema in response to hypoxia/ischemia.

Data regarding changes in AQP9 expression in response to oxygen deprivation have varied less than that of AQP4 with AQP9 tending toward increased expression in specific neuronal cells, namely, catecholaminergic neurons (86–88). Recall that catecholaminergic neurons are not thought to be osmosensors; rather, it is likely that they act as energy balance sensors (89). During pathological conditions such hypoxia/ischemia, lactate is released into the extracellular regions of the infarct zone. Given AQP9's permeability to monocarboxylates such as lactate, AQP9 likely has a function in lactate clearance.

5 Emerging Role of Water Channels in Regulating Cellular Volume   85

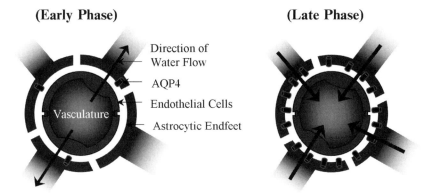

Fig. 5.2 Proposed model of AQP4-mediated resolution of brain edema. Early phase: Early during edema formation in response to a hypoxic/ischemic insult, AQP4 expression may decrease at, or be redistributed away from, astrocytic endfeet surrounding perivascular regions. Water flow would be reduced in astrocytes and the neighboring syncytium, staving-off cytotoxic edema. However, reduced water permeability results in increased blood pressure resulting in either, or a combination of, the following: leakage from the vasculature into the extracellular space due to hydrostatic pressure, or breakdown of the BBB leading to massive perturbations of osmotic homeostasis. Water clearance may be accomplished by neuronal release of taurine. Osmotically obliged water follows taurine into astrocytes and away from neurons and extracellular regions. Late phase: With upregulated expression or relocalization of AQP4 to astrocytic endfeet surrounding perivascular regions, water can be cleared from the system. Thus, a bidirectional, time-dependent function for AQP4 may be important for resolving vasogenic and cytotoxic brain edema. Emboldened arrows reflect the direction of water flow through AQP4 during early- and late-phase edema (*See Color Plates*)

In addition to the nonapoptotic roles played by AQPs, hypoxia/ischemia also leads to apoptosis and subsequent changes in cellular volume. Ionic mechanisms of neuronal volume changes in response to hypoxia/ischemia have been well characterized (55, 90–93). However, the role of AQPs in hypoxia/ischemia-induced apoptotic cellular shrinkage has not been investigated.

5.4 AVD: Role and Significance of AQP

The AVD is facilitated by a well-documented decrease in intracellular K^+ and Na^+, which is critical for creating an environment conducive to the activation of caspases and nucleases (8, 9, 94–96). Coupled with this ion movement is an inhibition of both the Na^+/K^+ ATPase and the Na^+/H^+ exchanger (97–99), activation of Cl^- channels (100, 101), and release of taurine and other molecules. While no study has defined the absolute contribution of each of these events to the AVD, what is clear is that release of these agents creates an osmotic gradient where isotonicity had previously existed. This gradient then serves to draw water out of the cell to create the characteristic cell shrinkage, or the AVD. Because efflux of water rapidly follows the loss of osmolytes, the process is often called normotonic cell shrinkage (6).

AQPs are thought to mediate the vast majority of the plasma membrane's water permeability. Since the AVD involves loss of a considerable amount of water, one would expect the bulk of water leaving a cell during this process to exit via water channels. Indeed, we have recently shown that AQPs are responsible for the majority of this flow (102). The absolute conserved nature of the AVD also led us to consider that the expression and/or availability of AQPs may be a regulated step in apoptosis. We demonstrated that overexpression of AQP1 in CHO cells increased their rate of apoptosis (102), suggesting that loss of water may be a rate-limiting step in the AVD. When we performed the opposite experiment and blocked water flow through AQPs, we found this treatment blocked apoptotic cell shrinkage (102).

5.5 Regulation of Aquaporin Expression and Function After (and Before) the AVD

Apoptotic loss of K^+ is known to have at least two very important effects. The first is to drive the loss of water that brings about the characteristic shrinkage associated with the AVD (96, 103); the other is to create an environment conducive to activation of apoptotic cascades (94, 95). We previously demonstrated that resting intracellular concentrations of K^+ (in the range of 140mM) completely block DNA degradation and caspase activation (96) and that activation of these enzymes required K^+ concentrations below 50mM (95). Measurements of shrunken cell size and intracellular K^+ content confirmed that following the AVD, K^+ levels decreased to approximately 35mM. Logically, if water loss was equivalent to K^+ loss, an environment would be created with normal levels of intracellular K^+, a situation not allowing apoptotic enzyme activity. Thus, to create the lowered K^+ concentration observed in the post-AVD cell, water loss, at some point, must be curtailed. Indeed, we have shown AQPs are inactivated following the AVD (102). Although mechanisms underlying AQP inhibition are still not entirely clear, we have presented evidence that association with caveolins may mediate this event (104). Other possibilities of AQP inactivation include *pinching* and *capping* gating mechanisms via phosphorylation (105).

While mechanisms underlying AQP inactivation during the AVD remain a mystery, one might predict that if AQPs were blocked prior to the initiation of apoptosis, the cell would actually die faster because the concentration of intracellular K^+ would begin to fall immediately without the additional step of cell shrinkage. However, much to our surprise, blocking AQPs not only blocked the AVD but also downstream events such as caspase activation and DNA degradation (102). Thus, in addition to decreased intracellular K^+ concentration, water loss through AQPs is essential for the cell death process to proceed, and AQP activity could control the rate by which a cell dies, or perhaps if it dies at all.

Recently, we have presented data in support of AQP-mediated resistance to apoptosis. When H4IIE liver cancer cells are growing in vivo or in culture immediately

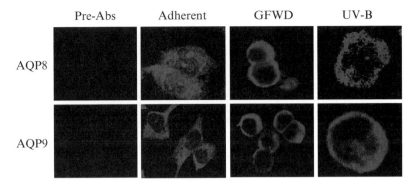

Fig. 5.3 Aquaporin 8 and 9 translocation in RUCA-1 cells (106) during growth factor withdrawal (GFWD) and ultraviolet-B (UV-B)-induced apoptosis. Cells were plated overnight in 24-well plates and apoptosis was then stimulated by GFWD (for 48 h) or exposure to UV-B (25 mJ cm^{-2} followed by 24-h incubation). The localization of AQP8 and 9 was examined in the apoptotic (nonadherent) population by standard immunocytochemical methods. Controls include staining for nonapoptotic (adherent) cells and separately, use of a preabsorbed (Pre-Abs) primary antibody in nonadherent populations (*See Color Plates*)

after being removed from rats, they express little or no AQP8 or 9 and are highly resistant to apoptosis (41). However, when grown in culture for as little as 24 h, expression of these two AQP homologs significantly increases and these cells become highly sensitive to apoptotic insults. These results demonstrate a strong relationship between AQP expression and functionality (sensitivity to apoptotic stimuli), suggesting the intriguing possibility that decreased expression of AQPs may contribute to a cancerous cell's resistance to apoptosis and thus identify a novel therapeutic target for cancer. Studies are underway to explore this exciting possibility.

A reasonable extension of the idea that the number of AQPs present in the plasma membrane may control the rate of cell death states that if a cell expresses AQPs, but they are not functionally available (for example, not present on the plasma membrane), that cell will not be able to die as rapidly as possible. In the previous studies with H4IIE cells, we noticed that cells in culture expressed AQP8 and 9 predominately in the cytoplasm in a punctate manner, a distribution consistent with their being localized to intracellular vesicles (41). Moreover, a similar pattern has been observed in a rat uterine cancer adenocarcinoma cell line (RUCA-1) (Fig. 5.3) and primary rat ovarian granulosa cells (Fig. 5.4). Since inefficient death can potentially lead to secondary necrosis, inflammation, and perhaps autoimmune diseases, AQPs located intracellularly might be expected to translocate to the plasma membrane to facilitate the death process. As hypothesized, induction of apoptosis in these cells stimulated AQP translocation to the plasma membrane where they could mediate loss of water during the AVD (Fig. 5.3 and 5.4). Signals and mechanisms of this translocation process are currently under investigation.

Aquaporin translocation from subcellular compartments is not a new concept; aquaporin-2 translocation in the kidney collecting duct has been studied extensively

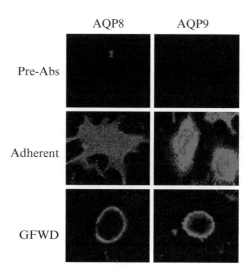

Fig. 5.4 Aquaporin translocation during apoptosis in ovarian granulosa cells. Granulosa cells were harvested from ovaries of immature rats; apoptosis was later induced using GFWD in cells previously plated for 24 h (102). Localization of AQP8 and 9 was examined in the apoptotic (nonadherent) population by standard immunocytochemical methods. Controls include staining of nonapoptotic (adherent) cells and separately, use of a Pre-Abs primary antibody in nonadherent populations (*See Color Plates*)

(107–110). For example, it is known that vasopressin (antidiuretic hormone, ADH) regulates AQP expression and translocation in kidney, facilitating water resorption into the body. A similar role for ADH in brain was confirmed (111), suggesting a possible role of ADH-induced AQP translocation in the brain.

Other AQP homologs such as AQP6, 11, and 12 are expressed subcellularly, where they are restricted to intracellular vesicles (112, 113); AQP11 and 12 may comprise a novel class of AQPs since they failed to transport water, glycerol, urea, or ions. Although appearing to lack function, AQP11 knockout mice were lethal (114). AQP11 and 12 likely comprise a new AQP subfamily whose possible role in apoptosis should be investigated.

5.6 Colocalization of AQP and Potassium Channels

Efficient water flow into and out of cellular compartments should be greatest when AQPs are tightly coupled to ion channels. Arrays of particles (115) or *assemblies* (116), today known as orthogonal arrays of particles (OAPs), are positioned to mediate substance flow given their unique distribution. Interestingly, AQP4 developmental expression increases parallel to the formation of OAPs as recognized by freeze-fracture techniques (117), and solid evidence that AQP4 is the OAP was revealed when OAPs were not detected in mice with genetic deletion of AQP4 (118, 119). Pathophysiologically, disordering of OAPs on astrocytic processes under hypoxic/ischemic conditions may account for edema formation (120). Moreover, if AQP4 were spatially coupled to ion channels, one might imagine a more efficient system for maintaining osmotic perturbations (121, 122). Investigating spatial relationships between AQP4 and the glial inwardly rectifying potassium channel $K_{ir}4.1$ revealed enrichment of both channels in plasma membrane domains near the vitreous body and blood vessels of Müller cells (123). Observations (after transient ischemia) that glial cells exhibited less K$^+$ conductance due to decreased expression of $K_{ir}4.1$ at blood vessels suggested that downregulation of K$^+$ channels caused an increased concentration of ions in the glial cells, which osmotically drove water away from the blood, inducing cytotoxic edema (124).

Interestingly, transgenic mice lacking dystrophin, a protein normally responsible for securing the extracellular matrix and membrane proteins to the cytoskeleton, demonstrated unaltered total AQP4 protein; however, a large reduction of AQP4 localization to astrocytic endfeet indicated a disruption in AQP4 trafficking to these regions, suggesting that dystrophin acts as one trafficking mediator of AQP4 protein (125). Coimmunoprecipitations and cross-linking experiments revealed association of AQP4 with other components of the dystrophin complex including α-syntrophin. Similarly, total AQP4 expression was unaltered in brains of α-Syn −/− mice, but normal subcellular localization to astrocytic endfeet surrounding blood vessels was disrupted, and reversed polarization of AQP4 to neuropil was observed (126). Although $K_{ir}4.1$ expression and localization was largely unchanged between wild type and α-Syn −/− mice (127), $K_{ir}4.1$ is known to bind directly to α-syntrophin (128). Further testing of whether AQP4 and K$^+$ channels cooperate with one another to restore osmotic imbalances after neural activation showed a twofold prolonged K$^+$ clearance in α-Syn −/− mice (127). As may be expected, prolonged K$^+$ clearance in nearly half of the α-Syn −/− mice leads to increased epileptic seizures. Furthermore, AQP4 −/− mice had higher seizure thresholds (129).

Delocalized AQP4 from astroglial endfeet implicates dystrophin and its architectural components in the pathophysiology of brain edema in response to hypoxia and ischemia (130). Given alterations in AQP expression and distribution during hypoxia/ischemia, it is possible that disruption of dystrophin complexes occurs through cytoskeletal and/or extracellular matrix rearrangements. Aquaporin-4, $K_{ir}4.1$, and the dystrophin complex are logical targets for therapeutics designed to alleviate symptoms of hypoxia/ischemia and the classical postindicator, brain edema.

We have been vague in the identification of specific K$^+$ channels involved in the AVD because it appears that different cells utilize different K$^+$ channels, and perhaps individual cells utilize different channels in response to different stimuli. Since virtually all examples of the AVD require large increases in K$^+$ efflux in response to the apoptotic insult, specific experiments are needed to determine which K$^+$ channels are involved in a particular model. For example, one model implements the Kv2.1 channel in the death of cortical neurons (131). Intriguingly, these experiments revealed increased K$^+$ efflux due to the translocation of vesicles containing Kv2.1 to the plasma membrane (131). Given the colocalization of potassium and water channels in certain tissues and the necessary role of each channel in facilitating the AVD, we postulate the existence of yet-to-be-defined vesicles containing both K$^+$ and AQP channels that play critical roles in apoptosis. These vesicles would translocate to the plasma membrane in response to apoptotic stimuli, and therefore, we suggest that they be designated *translocating apoptotic vesicles* (TAVs).

5.7 Concluding Remarks

In researching the role of AQPs during hypoxia/ischemia and apoptosis, complex functional roles for water channels have emerged. In the physiological setting, AQPs passively mediate water flux in response to changing ionic conditions, facilitating changes in cellular volumes necessary for maintaining volume homeostasis. In pathophysiological settings, AQP expression levels and localization are subject to change. During hypoxia/ischemia, massive changes in the extracellular environment are observed; it is not surprising that certain AQPs like AQP4 work to reestablish the local environment. Further evidence for the role of water channels during apoptosis was discussed along with the existence of a novel mechanism requiring translocating apoptotic vesicles thought to contain K^+ and AQP channels. Given the nuances of AQPs to the field of edema formation as a result of hypoxia/ischemia and separately, their possible role in apoptosis, future work designed to probe details of this channel's absolute contribution to each is necessary. And, given the high relevance of apoptosis and edema to the clinic, therapeutic targeting of water channels will likely lead to enhanced survivability.

References

1. Bortner CD, Cidlowski JA. Absence of volume regulatory mechanisms contributes to the rapid activation of apoptosis in thymocytes. Am J Physiol 1996;271(3, Part 1):C950–C961.
2. Wehner F, Olsen H, Tinel H, Kinne-Saffran E, Kinne RK. Cell volume regulation: osmolytes, osmolyte transport, and signal transduction. Rev Physiol Biochem Pharmacol 2003;148: 1–80.
3. Okada Y. Ion channels and transporters involved in cell volume regulation and sensor mechanisms. Cell Biochem Biophys 2004;41(2):233–258.
4. Lippmann BJ, Yang R, Barnett DW, Misler S. Pharmacology of volume regulation following hypotonicity-induced cell swelling in clonal N1E115 neuroblastoma cells. Brain Res 1995; 686(1):29–36.
5. Knoblauch C, Montrose MH, Murer H. Regulatory volume decrease by cultured renal cells. Am J Physiol 1989;256(2, Part 1):C252–C259.
6. Maeno E, Ishizaki Y, Kanaseki T, Hazama A, Okada Y. Normotonic cell shrinkage because of disordered volume regulation is an early prerequisite to apoptosis. Proc Natl Acad Sci USA 2000;97(17):9487–9492.
7. O'Reilly N, Xia Z, Fiander H, Tauskela J, Small DL. Disparity between ionic mediators of volume regulation and apoptosis in N1E 115 mouse neuroblastoma cells. Brain Res 2002; 943(2):245–256.
8. Yu SP, Yeh CH, Sensi SL, Gwag BJ, Canzoniero LM, Farhangrazi ZS, Ying HS, Tian M, Dugan LL, Choi DW. Mediation of neuronal apoptosis by enhancement of outward potassium current. Science 1997;278(5335):114–117.

9. Yu SP, Yeh C, Strasser U, Tian M, Choi DW. NMDA receptor-mediated K⁺ efflux and neuronal apoptosis. Science 1999;284(5412):336–339.
10. King LS, Kozono D, Agre P. From structure to disease: the evolving tale of aquaporin biology. Nat Rev Mol Cell Biol 2004;5(9):687–698.
11. Iolascon A, Aglio V, Tamma G, D'Apolito M, Addabbo F, Procino G, Simonetti MC, Montini G, Gesualdo L, Debler EW, Svelto M, Valenti G. Characterization of two novel missense mutations in the *AQP2* gene causing nephrogenic diabetes insipidus. Nephron Physiol 2007;105(3):33–41.
12. Sohara E, Rai T, Yang SS, Uchida K, Nitta K, Horita S, Ohno M, Harada A, Sasaki S, Uchida S. Pathogenesis and treatment of autosomal-dominant nephrogenic diabetes insipidus caused by an aquaporin 2 mutation. Proc Natl Acad Sci USA 2006;103(38):14217–14222.
13. Bolignano D, Coppolino G, Criseo M, Campo S, Romeo A, Buemi M. Aquaretic agents: what's beyond the treatment of hyponatremia? Curr Pharm Des 2007;13(8):865–871.
14. Wintour EM, Henry BA. Glycerol transport: an additional target for obesity therapy? Trends Endocrinol Metab 2006;17(3):77–78.
15. Shiels A, Bassnett S. Mutations in the founder of the *MIP* gene family underlie cataract development in the mouse. Nat Genet 1996;12(2):212–215.
16. Agre P, Kozono D. Aquaporin water channels: molecular mechanisms for human diseases. FEBS Lett 2003;555(1):72–78.
17. St Hillaire C, Vargas D, Pardo CA, Gincel D, Mann J, Rothstein JD, McArthur JC, Conant K. Aquaporin 4 is increased in association with human immunodeficiency virus dementia: implications for disease pathogenesis. J Neurovirol 2005;11(6):535–543.
18. Liu Y, Promeneur D, Rojek A, Kumar N, Frokiaer J, Nielsen S, King LS, Agre P, Carbrey JM. Aquaporin 9 is the major pathway for glycerol uptake by mouse erythrocytes, with implications for malarial virulence. Proc Natl Acad Sci USA 2007;104(30):12560–12564.
19. Wang K, Feng YL, Wen FQ, Chen XR, Ou XM, Xu D, Yang J, Deng ZP. Decreased expression of human aquaporin-5 correlated with mucus overproduction in airways of chronic obstructive pulmonary disease. Acta Pharmacol Sin 2007;28(8):1166–1174.
20. Kruse E, Uehlein N, Kaldenhoff R. The aquaporins. Genome Biol 2006;7(2):206.
21. Benga G. Water channel proteins: from their discovery in 1985 in Cluj-Napoca, Romania, to the 2003 Nobel Prize in Chemistry. Cell Mol Biol 2006;52(7):10–19.
22. Preston GM, Carroll TP, Guggino WB, Agre P. Appearance of water channels in Xenopus oocytes expressing red cell CHIP28 protein. Science 1992;256(5055):385–387.
23. Verkman AS. More than just water channels: unexpected cellular roles of aquaporins. J Cell Sci 2005;118 (Part 15):3225–3232.
24. Gonen T, Walz T. The structure of aquaporins. Q Rev Biophys 2006;39(4):361–396.
25. Hiroaki Y, Tani K, Kamegawa A, Gyobu N, Nishikawa K, Suzuki H, Walz T, Sasaki S, Mitsuoka K, Kimura K, Mizoguchi A, Fujiyoshi Y. Implications of the aquaporin-4 structure on array formation and cell adhesion. J Mol Biol 2006;355(4):628–639.
26. Verkman AS, Mitra AK. Structure and function of aquaporin water channels. Am J Physiol Renal Physiol 2000;278(1):F13–F28.
27. Hara-Chikuma M, Verkman AS. Physiological roles of glycerol-transporting aquaporins: the aquaglyceroporins. Cell Mol Life Sci 2006;63(12):1386–1392.
28. Ishibashi K, Kuwahara M, Gu Y, Kageyama Y, Tohsaka A, Suzuki F, Marumo F, Sasaki S. Cloning and functional expression of a new water channel abundantly expressed in the testis permeable to water, glycerol, and urea. J Biol Chem 1997;272(33):20782–20786.
29. MacAulay N, Hamann S, Zeuthen T. Water transport in the brain: role of cotransporters. Neuroscience 2004;129(4):1031–1044.
30. Gwan JF, Baumgaertner A. Cooperative transport in a potassium ion channel. J Chem Phys 2007;127(4):045103.
31. Nilius B. Is the volume-regulated anion channel VRAC a "water-permeable" channel? Neurochem Res 2004;29(1):3–8.

32. de Groot BL, Grubmuller H. Water permeation across biological membranes: mechanism and dynamics of aquaporin-1 and GlpF. Science 2001;294(5550):2353–2357.
33. Solenov EI, Vetrivel L, Oshio K, Manley GT, Verkman AS. Optical measurement of swelling and water transport in spinal cord slices from aquaporin null mice. J Neurosci Methods 2002;113(1):85–90.
34. Papahadjopoulos D, Kimelberg HK. Phospholipid vesicles (liposomes) as models for biological membranes: their properties and interactions with cholesterol and proteins: progress in surface scenarios. Oxford:Pergamon. 141–232; 1974.
35. Nielsen S, Smith BL, Christensen EI, Agre P. Distribution of the aquaporin CHIP in secretory and resorptive epithelia and capillary endothelia. Proc Natl Acad Sci USA 1993;90(15): 7275–7279.
36. Boassa D, Stamer WD, Yool AJ. Ion channel function of aquaporin-1 natively expressed in choroid plexus. J Neurosci 2006;26(30):7811–7819.
37. Shields SD, Mazario J, Skinner K, Basbaum AI. Anatomical and functional analysis of aquaporin 1, a water channel in primary afferent neurons. Pain 2007;131(1–2):8–20.
38. Perez E, Barrachina M, Rodriguez A, Torrejon-Escribano B, Boada M, Hernandez I, Sanchez M, Ferrer I. Aquaporin expression in the cerebral cortex is increased at early stages of Alzheimer disease. Brain Res 2007;1128(1):164–174.
39. Satoh J, Tabunoki H, Yamamura T, Arima K, Konno H. Human astrocytes express aquaporin-1 and aquaporin-4 in vitro and in vivo. Neuropathology 2007;27(3):245–256.
40. McCoy E, Sontheimer H. Expression and function of water channels (aquaporins) in migrating malignant astrocytes. Glia 2007;55(10):1034–1043.
41. Jablonski EM, Mattocks MA, Sokolov E, Koniaris LG, Hughes FM, Jr., Fausto N, Pierce RH, McKillop IH. Decreased aquaporin expression leads to increased resistance to apoptosis in hepatocellular carcinoma. Cancer Lett 2007;250(1):36–46.
42. Nagelhus EA, Veruki ML, Torp R, Haug FM, Laake JH, Nielsen S, Agre P, Ottersen OP. Aquaporin-4 water channel protein in the rat retina and optic nerve: polarized expression in Muller cells and fibrous astrocytes. J Neurosci 1998;18(7):2506–2519.
43. Nielsen S, Nagelhus EA, Amiry-Moghaddam M, Bourque C, Agre P, Ottersen OP. Specialized membrane domains for water transport in glial cells: high-resolution immunogold cytochemistry of aquaporin-4 in rat brain. J Neurosci 1997;17(1):171–180.
44. Jung JS, Bhat RV, Preston GM, Guggino WB, Baraban JM, Agre P. Molecular characterization of an aquaporin cDNA from brain: candidate osmoreceptor and regulator of water balance. Proc Natl Acad Sci USA 1994;91(26):13052–13056.
45. Venero JL, Vizuete ML, Machado A, Cano J. Aquaporins in the central nervous system. Prog Neurobiol 2001;63(3):321–336.
46. Yamamoto N, Yoneda K, Asai K, Sobue K, Tada T, Fujita Y, Katsuya H, Fujita M, Aihara N, Mase M, Yamada K, Miura Y, Kato T. Alterations in the expression of the AQP family in cultured rat astrocytes during hypoxia and reoxygenation. Brain Res Mol Brain Res 2001; 90(1):26–38.
47. Fujita Y, Yamamoto N, Sobue K, Inagaki M, Ito H, Arima H, Morishima T, Takeuchi A, Tsuda T, Katsuya H, Asai K. Effect of mild hypothermia on the expression of aquaporin family in cultured rat astrocytes under hypoxic condition. Neurosci Res 2003;47(4):437–444.
48. Elkjaer M, Vajda Z, Nejsum LN, Kwon T, Jensen UB, Amiry-Moghaddam M, Frokiaer J, Nielsen S. Immunolocalization of AQP9 in liver, epididymis, testis, spleen, and brain. Biochem Biophys Res Commun 2000;276(3):1118–1128.
49. Nicchia GP, Frigeri A, Nico B, Ribatti D, Svelto M. Tissue distribution and membrane localization of aquaporin-9 water channel: evidence for sex-linked differences in liver. J Histochem Cytochem 2001;49(12):1547–1556.
50. Loitto VM, Huang C, Sigal YJ, Jacobson K. Filopodia are induced by aquaporin-9 expression. Exp Cell Res 2007;313(7):1295–1306.

51. Loitto VM, Forslund T, Sundqvist T, Magnusson KE, Gustafsson M. Neutrophil leukocyte motility requires directed water influx. J Leukoc Biol 2002;71(2):212–222.
52. Tsukaguchi H, Shayakul C, Berger UV, Mackenzie B, Devidas S, Guggino WB, van Hoek AN, Hediger MA. Molecular characterization of a broad selectivity neutral solute channel. J Biol Chem 1998;273(38):24737–24743.
53. Badaut J, Petit JM, Brunet JF, Magistretti PJ, Charriaut-Marlangue C, Regli L. Distribution of Aquaporin 9 in the adult rat brain: preferential expression in catecholaminergic neurons and in glial cells. Neuroscience 2004;128(1):27–38.
54. Magistretti PJ, Pellerin L, Rothman DL, Shulman RG. Energy on demand. Science 1999;283(5401):496–497.
55. Pasantes-Morales H, Tuz K. Volume changes in neurons: hyperexcitability and neuronal death. Contrib Nephrol 2006;152:221–240.
56. Nakajima W, Ishida A, Lange MS, Gabrielson KL, Wilson MA, Martin LJ, Blue ME, Johnston MV. Apoptosis has a prolonged role in the neurodegeneration after hypoxic ischemia in the newborn rat. J Neurosci 2000;20(21):7994–8004.
57. Hossmann KA. Experimental models for the investigation of brain ischemia. Cardiovasc Res 1998;39(1):106–120.
58. Klatzo I. Presidental address. Neuropathological aspects of brain edema. J Neuropathol Exp Neurol 1967;26(1):1–14.
59. Lee TY, Murphy BD, Aviv RI, Fox AJ, Black SE, Sahlas DJ, Symons S, Lee DH, Pelz D, Gulka IB, Chan R, Beletsky V, Hachinski V, Hogan MJ, Goyal M, Demchuk AM, Coutts SB. Cerebral blood flow threshold of ischemic penumbra and infarct core in acute ischemic stroke: a systematic review. Stroke 2006;37(9):2201; author reply 2203.
60. Astrup J, Siesjo BK, Symon L. Thresholds in cerebral ischemia – the ischemic penumbra. Stroke 1981;12(6):723–725.
61. Symon L, Lassen NA, Astrup J, Branston NM. Thresholds of ischaemia in brain cortex. Adv Exp Med Biol 1977;94:775–782.
62. Weinstein PR, Hong S, Sharp FR. Molecular identification of the ischemic penumbra. Stroke 2004;35(11, Suppl 1):2666–2670.
63. van Harreveld A, Collewijn H, Malhotra SK. Water, electrolytes, and extracellular space in hydrated and dehydrated brains. Am J Physiol 1966;210(2):251–256.
64. Garcia JH, Kalimo H, Kamijyo Y, Trump BF. Cellular events during partial cerebral ischemia. I. Electron microscopy of feline cerebral cortex after middle-cerebral-artery occlusion. Virchows Arch B Cell Pathol 1977;25(3):191–206.
65. Kimelberg HK. Astrocytic swelling in cerebral ischemia as a possible cause of injury and target for therapy. Glia 2005;50(4):389–397.
66. Walz W, Wuttke WA. Independent mechanisms of potassium clearance by astrocytes in gliotic tissue. J Neurosci Res 1999;56(6):595–603.
67. Kempski O, Staub F, Jansen M, Baethmann A. Molecular mechanisms of glial cell swelling in acidosis. Adv Neurol 1990;52:39–45.
68. Macknight AD, Leaf A. Regulation of cellular volume. Physiol Rev 1977;57(3):510–573.
69. Lambert IH. Effect of arachidonic acid on conductive Na, K and anion transport in Ehrlich ascites tumor cells under isotonic and hypotonic conditions. Cell Physiol Biochem 1991;1: 177–194.
70. Eddleston M, Mucke L. Molecular profile of reactive astrocytes – implications for their role in neurologic disease. Neuroscience 1993;54(1):15–36.
71. Davalos A, Shuaib A, Wahlgren NG. Neurotransmitters and pathophysiology of stroke: evidence for the release of glutamate and other transmitters/mediators in animals and humans. J Stroke Cerebrovasc Dis 2000;9(6, Part 2):2–8.
72. Kuschinsky W, Wahl M, Bosse O, Thurau K. Perivascular potassium and pH as determinants of local pial arterial diameter in cats. A microapplication study. Circ Res 1972;31(2):240–247.

73. Filosa JA, Bonev AD, Straub SV, Meredith AL, Wilkerson MK, Aldrich RW, Nelson MT. Local potassium signaling couples neuronal activity to vasodilation in the brain. Nat Neurosci 2006;9(11):1397–1403.
74. Vizuete ML, Venero JL, Vargas C, Ilundain AA, Echevarria M, Machado A, Cano J. Differential upregulation of aquaporin-4 mRNA expression in reactive astrocytes after brain injury: potential role in brain edema. Neurobiol Dis 1999;6(4):245–258.
75. Verkman AS, Binder DK, Bloch O, Auguste K, Papadopoulos MC. Three distinct roles of aquaporin-4 in brain function revealed by knockout mice. Biochim Biophys Acta 2006; 1758(8):1085–1093.
76. Manley GT, Fujimura M, Ma T, Noshita N, Filiz F, Bollen AW, Chan P, Verkman AS. Aquaporin-4 deletion in mice reduces brain edema after acute water intoxication and ischemic stroke. Nat Med 2000;6(2):159–163.
77. Ke C, Poon WS, Ng HK, Pang JC, Chan Y. Heterogeneous responses of aquaporin-4 in oedema formation in a replicated severe traumatic brain injury model in rats. Neurosci Lett 2001;301(1):21–24.
78. Kiening KL, van Landeghem FK, Schreiber S, Thomale UW, von Deimling A, Unterberg AW, Stover JF. Decreased hemispheric Aquaporin-4 is linked to evolving brain edema following controlled cortical impact injury in rats. Neurosci Lett 2002;324(2):105–108.
79. Meng S, Qiao M, Lin L, Del Bigio MR, Tomanek B, Tuor UI. Correspondence of AQP4 expression and hypoxic-ischaemic brain oedema monitored by magnetic resonance imaging in the immature and juvenile rat. Eur J Neurosci 2004;19(8):2261–2269.
80. Taniguchi M, Yamashita T, Kumura E, Tamatani M, Kobayashi A, Yokawa T, Maruno M, Kato A, Ohnishi T, Kohmura E, Tohyama M, Yoshimine T. Induction of aquaporin-4 water channel mRNA after focal cerebral ischemia in rat. Brain Res Mol Brain Res 2000;78(1–2): 131–137.
81. Mao X, Enno TL, Del Bigio MR. Aquaporin 4 changes in rat brain with severe hydrocephalus. Eur J Neurosci 2006;23(11):2929–2936.
82. Aoki K, Uchihara T, Tsuchiya K, Nakamura A, Ikeda K, Wakayama Y. Enhanced expression of aquaporin 4 in human brain with infarction. Acta Neuropathol 2003;106(2):121–124.
83. Nesic O, Lee J, Ye Z, Unabia GC, Rafati D, Hulsebosch CE, Perez-Polo JR. Acute and chronic changes in aquaporin 4 expression after spinal cord injury. Neuroscience 2006;143(3): 779–792.
84. Nagelhus EA, Lehmann A, Ottersen OP. Neuronal-glial exchange of taurine during hypoosmotic stress: a combined immunocytochemical and biochemical analysis in rat cerebellar cortex. Neuroscience 1993;54(3):615–631.
85. Fu X, Li Q, Feng Z, Mu D. The roles of aquaporin-4 in brain edema following neonatal hypoxia ischemia and reoxygenation in a cultured rat astrocyte model. Glia 2007;55(9): 935–941.
86. Hwang IK, Yoo KY, Li H, Lee BH, Suh HW, Kwon YG, Won MH. Aquaporin 9 changes in pyramidal cells before and is expressed in astrocytes after delayed neuronal death in the ischemic hippocampal CA1 region of the gerbil. J Neurosci Res 2007;85(11): 2470–2479.
87. Dibas A, Yang MH, Bobich J, Yorio T. Stress-induced changes in neuronal aquaporin-9 (AQP9) in a retinal ganglion cell-line. Pharmacol Res 2007;55(5):378–384.
88. Badaut J, Hirt L, Granziera C, Bogousslavsky J, Magistretti PJ, Regli L. Astrocyte-specific expression of aquaporin-9 in mouse brain is increased after transient focal cerebral ischemia. J Cereb Blood Flow Metab 2001;21(5):477–482.
89. Yang XJ, Kow LM, Funabashi T, Mobbs CV. Hypothalamic glucose sensor: similarities to and differences from pancreatic beta-cell mechanisms. Diabetes 1999;48(9):1763–1772.
90. Lang F, Shumilina E, Ritter M, Gulbins E, Vereninov A, Huber SM. Ion channels and cell volume in regulation of cell proliferation and apoptotic cell death. Contrib Nephrol 2006;152:142–160.
91. Yu SP, Canzoniero LM, Choi DW. Ion homeostasis and apoptosis. Curr Opin Cell Biol 2001;13(4):405–411.

92. Wei L, Han BH, Li Y, Keogh CL, Holtzman DM, Yu SP. Cell death mechanism and protective effect of erythropoietin after focal ischemia in the whisker-barrel cortex of neonatal rats. J Pharmacol Exp Ther 2006;317(1):109–116.
93. Wei L, Yu SP, Gottron F, Snider BJ, Zipfel GJ, Choi DW. Potassium channel blockers attenuate hypoxia- and ischemia-induced neuronal death in vitro and in vivo. Stroke 2003;34(5):1281–1286.
94. Hughes FM, Jr., Bortner CD, Purdy GD, Cidlowski JA. Intracellular K$^+$ suppresses the activation of apoptosis in lymphocytes. J Biol Chem 1997;272(48):30567–30576.
95. Hughes FM, Jr., Cidlowski JA. Potassium is a critical regulator of apoptotic enzymes in vitro and in vivo. Adv Enzyme Regul 1999;39:157–171.
96. Bortner CD, Hughes FM, Jr., Cidlowski JA. A primary role for K$^+$ and Na$^+$ efflux in the activation of apoptosis. J Biol Chem 1997;272(51):32436–32442.
97. Orlov SN, Thorin-Trescases N, Pchejetski D, Taurin S, Farhat N, Tremblay J, Thorin E, Hamet P. Na$^+$/K$^+$ pump and endothelial cell survival: [Na$^+$]i/[K$^+$]i-independent necrosis triggered by ouabain, and protection against apoptosis mediated by elevation of [Na$^+$]i. Pflugers Arch 2004;448(3):335–345.
98. Scholz W, Albus U. Na$^+$/H$^+$ exchange and its inhibition in cardiac ischemia and reperfusion. Basic Res Cardiol 1993;88(5):443–455.
99. Yu SP. Na(+), K(+)-ATPase: the new face of an old player in pathogenesis and apoptotic/hybrid cell death. Biochem Pharmacol 2003;66(8):1601–1609.
100. Okada Y, Shimizu T, Maeno E, Tanabe S, Wang X, Takahashi N. Volume-sensitive chloride channels involved in apoptotic volume decrease and cell death. J Membr Biol 2006;209(1): 21–29.
101. Okada Y, Maeno E. Apoptosis, cell volume regulation and volume-regulatory chloride channels. Comp Biochem Physiol A Mol Integr Physiol 2001;130(3):377–383.
102. Jablonski EM, Webb AN, McConnell NA, Riley MC, Hughes FM, Jr. Plasma membrane aquaporin activity can affect the rate of apoptosis but is inhibited after apoptotic volume decrease. Am J Physiol Cell Physiol 2004;286(4):C975–C985.
103. Bortner CD, Cidlowski JA. Cell shrinkage and monovalent cation fluxes: role in apoptosis. Arch Biochem Biophys 2007;462(2):176–188.
104. Jablonski EM, Hughes FM, Jr. The potential role of caveolin-1 in inhibition of aquaporins during the AVD. Biol Cell 2006;98(1):33–42.
105. Hedfalk K, Tornroth-Horsefield S, Nyblom M, Johanson U, Kjellbom P, Neutze R. Aquaporin gating. Curr Opin Struct Biol 2006;16(4):447–456.
106. Vollmer G. Endometrial cancer: experimental models useful for studies on molecular aspects of endometrial cancer and carcinogenesis. Endocr Relat Cancer 2003;10(1):23–42.
107. Petrovic MM, Vales K, Stojan G, Basta-Jovanovic G, Mitrovic DM. Regulation of selectivity and translocation of aquaporins: an update. Folia Biol 2006;52(5):173–180.
108. van Balkom BW, Graat MP, van Raak M, Hofman E, van der Sluijs P, Deen PM. Role of cytoplasmic termini in sorting and shuttling of the aquaporin-2 water channel. Am J Physiol Cell Physiol 2004;286(2):C372–C379.
109. Nielsen S, Marples D, Birn H, Mohtashami M, Dalby NO, Trimble M, Knepper M. Expression of VAMP-2-like protein in kidney collecting duct intracellular vesicles. Colocalization with aquaporin-2 water channels. J Clin Invest 1995;96(4):1834–1844.
110. Noda Y, Sasaki S. Trafficking mechanism of water channel aquaporin-2. Biol Cell 2005;97(12):885–892.
111. Niermann H, Amiry-Moghaddam M, Holthoff K, Witte OW, Ottersen OP. A novel role of vasopressin in the brain: modulation of activity-dependent water flux in the neocortex. J Neurosci 2001;21(9):3045–3051.
112. Beitz E, Liu K, Ikeda M, Guggino WB, Agre P, Yasui M. Determinants of AQP6 trafficking to intracellular sites versus the plasma membrane in transfected mammalian cells. Biol Cell 2006;98(2):101–109.
113. Ishibashi K. Aquaporin subfamily with unusual NPA boxes. Biochim Biophys Acta 2006;1758(8):989–993.

114. Morishita Y, Matsuzaki T, Hara-chikuma M, Andoo A, Shimono M, Matsuki A, Kobayashi K, Ikeda M, Yamamoto T, Verkman A, Kusano E, Ookawara S, Takata K, Sasaki S, Ishibashi K. Disruption of aquaporin-11 produces polycystic kidneys following vacuolization of the proximal tubule. Mol Cell Biol 2005;25(17):7770–7779.
115. Kreutziger GO. Freeze-etching of intercellular junction of mouse liver. In: Proceedings of the 26th Meeting of the Electron Microscope Society of America; 1968. Claitor's Publishing Division, Baton Rouge, LA. 234; 1968.
116. Landis DM, Reese TS. Arrays of particles in freeze-fractured astrocytic membranes. J Cell Biol 1974;60(1):316–320.
117. Nico B, Frigeri A, Nicchia GP, Quondamatteo F, Herken R, Errede M, Ribatti D, Svelto M, Roncali L. Role of aquaporin-4 water channel in the development and integrity of the blood–brain barrier. J Cell Sci 2001;114 (Part 7):1297–1307.
118. Rash JE, Yasumura T, Hudson CS, Agre P, Nielsen S. Direct immunogold labeling of aquaporin-4 in square arrays of astrocyte and ependymocyte plasma membranes in rat brain and spinal cord. Proc Natl Acad Sci USA 1998;95(20):11981–11986.
119. Verbavatz JM, Ma T, Gobin R, Verkman AS. Absence of orthogonal arrays in kidney, brain and muscle from transgenic knockout mice lacking water channel aquaporin-4. J Cell Sci 1997;110 (Part 22):2855–2860.
120. Landis DM, Reese TS. Astrocyte membrane structure: changes after circulatory arrest. J Cell Biol 1981;88(3):660–663.
121. Amedee T, Robert A, Coles JA. Potassium homeostasis and glial energy metabolism. Glia 1997;21(1):46–55.
122. Kofuji P, Connors NC. Molecular substrates of potassium spatial buffering in glial cells. Mol Neurobiol 2003;28(2):195–208.
123. Nagelhus EA, Horio Y, Inanobe A, Fujita A, Haug FM, Nielsen S, Kurachi Y, Ottersen OP. Immunogold evidence suggests that coupling of K^+ siphoning and water transport in rat retinal Muller cells is mediated by a coenrichment of Kir4.1 and AQP4 in specific membrane domains. Glia 1999;26(1):47–54.
124. Pannicke T, Iandiev I, Uckermann O, Biedermann B, Kutzera F, Wiedemann P, Wolburg H, Reichenbach A, Bringmann A. A potassium channel-linked mechanism of glial cell swelling in the postischemic retina. Mol Cell Neurosci 2004;26(4):493–502.
125. Vajda Z, Pedersen M, Fuchtbauer EM, Wertz K, Stodkilde-Jorgensen H, Sulyok E, Doczi T, Neely JD, Agre P, Frokiaer J, Nielsen S. Delayed onset of brain edema and mislocalization of aquaporin-4 in dystrophin-null transgenic mice. Proc Natl Acad Sci USA 2002; 99(20):13131–13136.
126. Neely JD, Amiry-Moghaddam M, Ottersen OP, Froehner SC, Agre P, Adams ME. Syntrophin-dependent expression and localization of aquaporin-4 water channel protein. Proc Natl Acad Sci USA 2001;98(24):14108–14113.
127. Amiry-Moghaddam M, Williamson A, Palomba M, Eid T, de Lanerolle NC, Nagelhus EA, Adams ME, Froehner SC, Agre P, Ottersen OP. Delayed K^+ clearance associated with aquaporin-4 mislocalization: phenotypic defects in brains of alpha-syntrophin-null mice. Proc Natl Acad Sci USA 2003;100(23):13615–13620.
128. Connors NC, Adams ME, Froehner SC, Kofuji P. The potassium channel Kir4.1 associates with the dystrophin–glycoprotein complex via alpha-syntrophin in glia. J Biol Chem 2004; 279(27):28387–28392.
129. Binder DK, Yao X, Zador Z, Sick TJ, Verkman AS, Manley GT. Increased seizure duration and slowed potassium kinetics in mice lacking aquaporin-4 water channels. Glia 2006; 53(6):631–636.
130. Frigeri A, Nicchia GP, Nico B, Quondamatteo F, Herken R, Roncali L, Svelto M. Aquaporin-4 deficiency in skeletal muscle and brain of dystrophic mdx mice. FASEB J 2001;15(1):90–98.
131. Pal SK, Takimoto K, Aizenman E, Levitan ES. Apoptotic surface delivery of K^+ channels. Cell Death Differ 2006;13(4):661–667.

Chapter 6
A Zinc–Potassium Continuum in Neuronal Apoptosis

Patrick Redman, Megan Knoch, and Elias Aizenman

Abstract Intraneuronal zinc homeostasis is tightly regulated by a variety of specialized ionic transport mechanisms, metal binding proteins, and organelle compartmentalization. Dysregulation of these processes by pathophysiological conditions, especially those in which reactive oxygen and nitrogen species (ROS, RNS) have been implicated, is central to the progression of neuronal injury. In addition to disruption of zinc homeostatic mechanisms, a reduction of intracellular ionic strength via enhanced potassium efflux is a critical and perhaps requisite mediator of apoptotic cell death induced by lethal insults to the brain. In this chapter, we describe the manner by which the cellular dysregulation of these two essential ions, zinc and potassium, lies along a signaling continuum leading to the demise of central neurons. By reviewing the current status of the field and presenting new information, we detail the molecular mechanisms linking intracellular zinc liberation to cellular potassium efflux in acute neuronal injury, such as ischemia. The complete characterization of this potentially ubiquitous cell death signaling cascade will likely provide novel therapeutic targets for cerebral ischemia and, possibly, other forms of neurodegeneration.

Keywords: Zinc; Oxidation; Kv2.1; Potassium channel; p38; Apoptosis

6.1 Introduction

Zinc is a ubiquitous, essential trace element required for many biological processes, from regulating protein structure and function (1), to serving as an ionic second messenger (2, 3), to modulating neurotransmission (4). Although Zn^{2+} is present in all eukaryotic cell types, the concentration of Zn^{2+} in the mammalian brain is particularly high (~150 µM) (5). However, intracellular free, or *chelatable* Zn^{2+} concentrations are in picomolar to low nanomolar levels under normal conditions,

---

P. Redman, M. Knoch, and E. Aizenman (✉)
Department of Neurobiology, University of Pittsburgh School of Medicine E1456 BST
3500 Terrace Street, Pittsburgh, PA 15261, USA
e-mail: redox@pitt.edu

due to the precise regulation of this metal by ionic transport mechanisms, compartmentalization in organelles, and the presence of a large number of metal binding proteins (6, 7). Nonetheless, pathological conditions, such as ischemic generation of reactive oxygen and nitrogen species (ROS, RNS), can disrupt intracellular Zn^{2+} homeostasis, generating excess free intracellular Zn^{2+} and triggering the activation of a myriad of cell death cascades in neurons (8–10).

Another essential ion under tight cellular regulation is potassium. K^+ is the single most abundant free intracellular cation (~140 mM), while its extracellular concentration is relatively low (~4 mM). This concentration gradient is maintained primarily by the Na^+/K^+-ATPase pump, in addition to various other ionic transport mechanisms (11). This concentration gradient drives the efflux of intracellular potassium when permeable routes for ionic movement, such as voltage-gated K^+ channels, are made available. Increasing evidence supports the notion that these gateways for K^+ efflux are indispensable mediators of the execution phase of neuronal apoptosis (12). As with Zn^{2+} dysregulation, when the intracellular K^+ concentration is dramatically reduced during pathophysiological conditions, cell death ensues (13). Until recently, dysregulation of either Zn^{2+} or K^+ was often considered in isolation within the context of neuronal injury. However, it is now known that the sequential disruption in the homeostatic regulation of both these ions is part of a continuum along a common apoptotic signaling pathway (10). Here, by reviewing previous findings, in addition to presenting more recent results, we illustrate the relationship between zinc and potassium dysregulation that governs neuronal cell death programs in which oxidative and nitrative mechanisms have been implicated.

6.2 Role of Zn^{2+} in Neuronal Injury

Zn^{2+} has been referred to as the *calcium of the twenty-first century* (1), and indeed, the many functions of Zn^{2+} in the central nervous system have just begun to be fully appreciated, including its newly recognized role as a second messenger (2, 3). However, the association of Zn^{2+} with neurodegeneration now spans nearly two decades since this metal was first shown to be directly toxic to neurons (14, 15). In nearly all of the initial studies in this field, exogenous Zn^{2+}-mediated neuronal cell death was proposed to be dependent upon the influx of the metal into cells following its release from presynaptic terminals (16). Indeed, Zn^{2+}-induced injury was shown to be blocked by antagonists of potential entry sites for this metal, such as Ca^{2+}-permeable voltage- and glutamate-gated ion channels (7, 16–20), as well as by cell-impermeant metal chelators (21).

Cellular Zn^{2+} influx has been associated with neuronal damage resulting from ischemia, trauma, and epilepsy, among other disorders. Free (chelatable) Zn^{2+} is present in synaptic vesicles (6, 22), and is generally believed to be synaptically released [(23–30), although see (31) and (32)] and enter postsynaptic neurons (18, 21, 26, 33–35). Indeed, the significance of zinc-mediated neuronal injury is best exemplified by the observation that zinc chelation is neuroprotective in in vivo models of ischemia (21, 36, 37) and head trauma (38).

6.3 Mechanism of Zn^{2+} Neurotoxicity

The mechanisms by which an increase in intracellular free Zn^{2+} induce neuronal cell death in neurodegenerative disorders and aging are areas of current, intense study (1, 9, 39, 40). Zn^{2+} is redox inactive (41, 42) and, as such, is likely to be relatively nontoxic by itself (43). Several mechanisms have been suggested to account for the toxic actions of this metal. Manev et al. (44) and Sensi et al. (45, 46) first proposed that Zn^{2+} can enter mitochondria and induce the release of oxygen-derived free radicals. Zn^{2+} also affects several key components of mitochondrial function (47–49), in addition to inducing permeability transition (50, 51), inhibiting mitochondrial movement (52), and activating mitochondrial multiconductance channels associated with cell death signaling molecules (53). Furthermore, Zn^{2+} can inhibit glycolysis and deplete NAD^+ (54). Although each of these studies has contributed significantly to our understanding of Zn^{2+} toxicity, the specific signaling pathways that contribute to Zn^{2+}-induced injuries are incomplete, and remain important and critical areas of study.

Zn^{2+}-associated cellular injury has been implicated in both caspase-dependent and -independent cell death pathways (55–61). The activation of these divergent pathways has linked to the intensity of Zn^{2+} exposure (44, 55, 56). Other factors, such as regulation of Zn^{2+}-dependent transcription factors (62, 63), activation of kinases (57–59, 61, 64, 65), and generation of ROS (56, 65–67), as well as the release of apoptosis-inducing molecules (51, 68–71) and activation of death induction proteins (53, 72, 73), likely determine the cellular processes by which neurons die following Zn^{2+} exposure. Importantly, Zn^{2+} can also *prevent* apoptosis under certain circumstances (74–80), and the depletion of this metal can be deleterious to cells (81–86). Therefore, the balance between the physiological and pathophysiological actions of Zn^{2+} closely resembles those which have been described for calcium (15, 87–89). Zn^{2+}, like Ca^{2+}, is a selective ligand for a metabotropic zinc-sensing receptor (ZnR; 90, 91), can be sequestered by intracellular organelles (44–46, 53, 92), is bound to metal chaperone proteins such as metallothionein (MT) (42), and is regulated by plasma membrane transporters (89, 93–95), all of which work in unison to maintain intracellular Zn^{2+} homeostasis, similar to Ca^{2+} regulation (42, 89). Additionally, like Ca^{2+} (96), Zn^{2+} is an important component of a variety of molecular signaling pathways (1, 5). Interestingly, a recent study has gone as far as to suggest that some of the previously proposed neurotoxic consequences of intracellular Ca^{2+} dysregulation may be actually due to Zn^{2+} (97).

6.4 Intracellular Release of Zn^{2+}

Of the total Zn^{2+} content in the brain, only 10% is free and restricted to a population of glutamate-containing synaptic vesicles (6). The majority of the cellular Zn^{2+}, however, is complexed to proteins such as metallothionein (MT), Zn^{2+} finger-containing transcription factors (1, 6, 42, 98, 99), and other functional proteins, such as protein kinases and phosphatases (100, 101). Nonetheless, Zn^{2+} can be liberated from these stores by oxidative stimuli and NO-related species under oxidative conditions (2, 10, 59, 60, 102–110). Moreover, aldehyde products of lipid peroxidation have also been recently shown to trigger intracellular Zn^{2+} release, thereby linking a common oxidative stress product directly to Zn^{2+}-mediated signaling (111). Maret and Valle (102) have pointed out that MT-bound Zn^{2+} is readily released by thiol oxidants (42, 112, 113), because of MT's relatively low redox potential (−366 mV) (102). Hence, even though this metalloprotein has a relatively high affinity for Zn^{2+} ($K_d = 1.4 \times 10^{-13}$ M), oxidants acting as cellular signals can rapidly and effectively induce the transfer of Zn^{2+} from MT to other proteins with lower affinities for this metal (42, 103). Under injurious oxidant conditions, however, the liberated Zn^{2+} is a highly significant source of the metal for triggering neuronal injury (10, 35, 104, 114, 115).

6.5 Cell Death Signaling Events Following Liberation of Intracellular Zn^{2+}

Incubation of MT with the cell-permeant thiol oxidant 2,2′-dithiodipyridine (DTDP) results in the oxidation of all 20 cysteines contained in the protein and release of all bound Zn^{2+} atoms (102). Our laboratory utilized DTDP to induce the release of Zn^{2+} from intracellular metalloproteins in cortical neurons in vitro (116). We observed that DTDP-liberated Zn^{2+} was an upstream mediator of neuronal apoptosis via a p38 kinase-dependent process (57, 117, 118). Our studies have described in detail a cell death pathway (Fig. 6.1) that is activated by the liberated metal in neurons: the intracellular release of Zn^{2+} is followed by generation of ROS from mitochondria (119) and 12-lipoxygenase (12-LOX) (59) leading to activation of p38 via the upstream kinase ASK-1 (57, 117). Importantly, the generation of ROS likely provides a feedback-like stimulus for additional Zn^{2+} release from metallothionein or related proteins. The activation of p38 is followed by an enhancement of delayed rectifying K^+ currents (Fig. 6.2) (121), and finally caspase activation (57, 59, 117, 120). We have demonstrated that the observed enhancement of K^+ currents is mediated by the N-ethylmaleimide-sensitive factor attachment protein receptor (SNARE)-dependent membrane insertion of Kv2.1-encoded channels following their direct phosphorylation by p38 (120, 122, 123).

To reveal the direct link between p38 activation and Kv2.1 channel insertion, we made use of various biochemical and electrophysiological experimental approaches following identification of the amino acid residue serine 800 (S800) in Kv2.1 as a putative p38 phosphorylation site (124). Substitution of S800 with a nonphosphorylatable alanine residue (Kv2.1S800A) completely blocked the apoptotic K^+ current enhancement, while substitution of S800 with phospho-mimicking glutamate (Kv2.1S800E) or aspartate (Kv2.1S800D) residues resulted in significant K^+ current increases analogous to apoptotic cells expressing wild-type Kv2.1. Also, a thiol-reactive functional inhibitor ([2-trimethylammoniumethyl] methanethiosulfate or MTSET) (124) was used in conjunction with cysteine-containing mutant

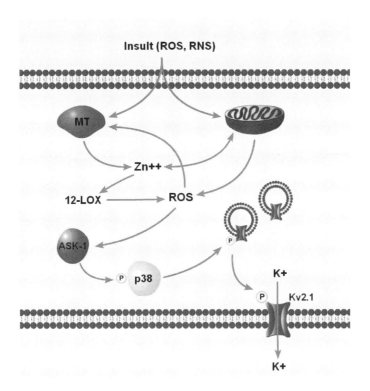

Fig. 6.1 Schematic of proposed pathway linking ROS/RNS-mediated liberation of intracellular Zn^{2+} and a subsequent enhancement of Kv2.1-mediated K^+ currents. Increases in intracellular Zn^{2+} lead to activation of p38 via a ROS-dependent activation of ASK-1. Kv2.1 channels are inserted in the cell membrane following their direct phosphorylation by p38. Channel membrane insertion is a SNARE-dependent process (*See Color Plates*)

Kv2.1 channels (Kv2.1I379C) to demonstrate that S800 is essential for apoptotic membrane insertion. As a 10-min treatment with MTSET functionally silenced all membrane-resident channels, plasma membrane insertion during apoptosis could be easily measured by the appearance of new voltage-gated currents. These new currents never appeared when S800 was mutated to alanine in Kv2.1I379C. Furthermore, a new phospho-S800-specific antibody revealed a p38-dependent increase in Kv2.1 phosphorylation in apoptotic neurons. This antibody also detected cell-free phosphorylation of immunopurified Kv2.1, but not Kv2.1(S800A), by active p38. Most importantly, channels lacking a phosphorylatable S800 residue could not sustain apoptosis mediated by wild-type Kv2.1 in a recombinant expression system.

Until these recent findings, the well-described phosphorylation and dephosphorylation events reported for Kv2.1 had been shown to strongly influence channel gating and neuronal excitability (125, 126), but not apoptosis. Indeed, none of the previously established functional effects of phosphorylation of Kv2.1

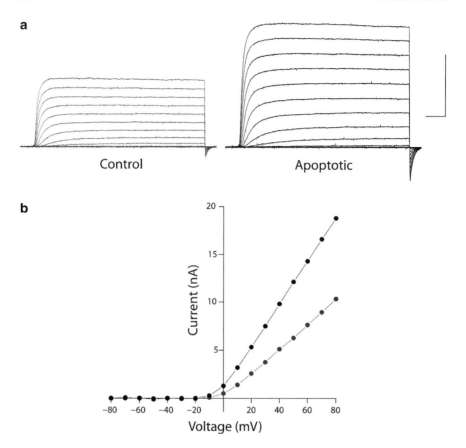

Fig. 6.2 DTDP induces characteristic apoptotic K⁺ current surge. (**a**) Representative whole-cell K⁺ currents from Kv2.1-expressing Chinese hamster ovary (CHO) cells recorded under control (*red*) and DTDP treatment (*black*) conditions (25 μM for 5 min). Cells were cotreated with butoxy-carbonyl-aspartate-fluoromethyl ketone (BAF, 10 μM), a broad-spectrum cysteine protease inhibitor in both vehicle and DTDP treatment conditions. Cells were also subsequently maintained in BAF- (10 μM) containing posttreatment medium until recordings were performed. In our experience, caspase inhibition via BAF is required to maintain cell viability for electrophysiological recordings, as Kv2.1-expressing CHO cells readily undergo apoptosis when challenged with DTDP (120). Currents were obtained 24-h posttransfection, 3-h postoxidative injury, and evoked by sequential 10-mV voltage steps to +80 mV from a holding potential of −80 mV. Calibration: 10 nA, 20 ms. (**b**) Steady state current/voltage relationship of the vehicle-treated, Kv2.1-expressing CHO cell (*red*) and Kv2.1-expressing CHO cell exposed to 25-μM DTDP (*black*) shown earlier (*See Color Plates*)

had involved S800. Importantly, in addition to identifying Kv2.1 as a novel substrate for direct phosphorylation by the MAPK p38, our studies exposed a previously unknown mechanism of Kv2.1 channel modulation essential not for channel gating and neuronal excitability, but for completion of an apoptotic program. Following an apoptotic insult, we do not observe alterations in the voltage dependence of membrane-resident channels (Fig. 6.2b), a phenomenon that has been reported by others to accompany nonlethal neuronal injury and other stimuli (127, 128). In those cases, calcineurin-dependent dephosphorylation of multiple intracellular sites on Kv2.1 participates in the graded regulation of channel activation. Kv2.1 does not generally participate in neuronal repolarization during an action potential, due to an elevated voltage-dependent activation threshold and slow activation kinetics (126, 129). However, the dynamic regulation of calcineurin-sensitive sites on Kv2.1 serves to suppress neuronal firing during periods of hyperexcitability (130), as dephosphorylation of these sites both lowers the voltage threshold for channel activation threshold and hastens channel activation kinetics (131). S800, on the other hand, is calcineurin resistant (125) and therefore does not modulate Kv2.1 gating during sublethal stimuli, but rather is uniquely subjected to phosphorylation by p38 following a more injurious stimulus, leading to channel trafficking and completion of the neuronal cell death program (123). Studies are ongoing to determine the molecular mechanism linking p38 phosphorylation of Kv2.1 to apoptotic plasma membrane insertion. As mentioned earlier, channel membrane insertion during apoptosis is a SNARE-dependent process (122). Interestingly, both SNAP-25 and syntaxin have been shown to directly associate with Kv2.1 (132). Therefore, channel phosphorylation by p38 may facilitate membrane insertion by changing the physical association with SNARE proteins with the channel, but this has yet to be experimentally demonstrated.

6.6 Potassium Efflux and Apoptosis

The enhancement of voltage-gated K$^+$ channel activity, resulting in K$^+$ efflux, has been shown to be a crucial element in various apoptotic paradigms (12). As such, several models of apoptosis have demonstrated K$^+$ efflux as a required event in the cell death, albeit through gateways other than Kv2.1 (12, 13). In most instances, the molecular mechanism by which the K$^+$ current enhancement and/or K$^+$ efflux is achieved in these models has yet to be elucidated. Regardless, the increased K$^+$ efflux seems to be the driving factor for cell death in pathological circumstances. Interestingly, blocking K$^+$ efflux via channel *inhibition* can also lead to apoptosis under certain circumstances (133, 134). Drosophila *Ether-a-go-go* (EAG) K$^+$ channels have been shown to *inhibit* apoptosis and increase proliferation (135). However, the antiapoptotic actions in this instance were attributed to intracellular signaling cascades governed by channel conformation, as nonconducting channel mutants produced similar results.

It has been proposed that the enhancement of potassium currents during apoptosis leads to a decrease in the concentration of this cation in the cytoplasm (136). It is this decrease of intracellular potassium that acts as a permissive apoptotic signal (13, 137), as caspases and apoptotic nucleases are activated most efficiently at significantly reduced potassium concentrations (138). Furthermore, normal intracellular potassium levels hinder oligomerization of Apaf-1 and assembly of the apoptosome (139). It must be noted that inhibition of both Na^+/K^+-ATPase and potassium leak channels has also been shown to contribute to reduced intracellular K^+ concentrations during apoptosis (11, 12, 140). As such, maintaining physiological intracellular potassium levels by blocking K^+ channels or increasing the extracellular K^+ concentration effectively attenuates cell death in many models of neuronal apoptosis (57, 60, 116, 121, 136, 141–145).

6.7 The Zinc–Potassium Continuum in Ischemia

Cell death pathways linking intracellular Zn^{2+} liberation and enhanced K^+ currents may be widespread (10). In our laboratory, we have provided a direct link between these two phenomena following the release of Zn^{2+} by thiol oxidants (57) and peroxynitrite (10, 59). More recently we have also reported that neurons exposed to microglia-derived ROS and RNS have observable Zn^{2+} release and a very pronounced K^+ current surge preceding cell death (146). Intraneuronal Zn^{2+} liberation has been associated with several acute neurodegenerative conditions such as epileptic seizures (32, 35), cerebral ischemia (36), and target deprivation (114). Relevant to our work (146), microglial activation is a likely contributor to the neuropathalogical changes accompanying these conditions (147–152). Microglia are also activated during normal brain development (153), where they play an important role in programmed neural cell death (154). Not surprisingly, both intracellular Zn^{2+} release and increased K^+ currents have been suggested to be indelible markers of developmental neuronal cell death (115, 155).

As mentioned earlier, alterations in potassium channel activity have also been seen in transient cerebral ischemia models, which are known to be similarly associated with ROS production (156, 157). Along these lines, we recently detected the liberation of intracellular Zn^{2+} in cortical neurons exposed to 30 min of oxygen-glucose deprivation (OGD) (Fig. 6.3), which was followed by an apoptotic surge in K^+ currents (Fig. 6.4). To verify that the increase in intracellular Zn^{2+} was essential in the K^+ current enhancement following OGD, neurons were cotreated with the membrane-permeant Zn^{2+} chelating agent tetrakis-(2-pyridylmethyl)ethylenediamine (TPEN; 20 µM) during OGD. The coexposure to TPEN effectively abrogated OGD-induced K^+ current enhancement (Fig. 6.4). Our models documenting intracellular release of Zn^{2+} have reported a subsequent ROS-dependent activation of p38 MAPK prior to K^+ efflux (57, 117). Therefore, we also coexposed OGD-treated cortical neurons with the highly selective p38 MAPK inhibitor SB293063 (SB; 20 µM) (57, 121, 123, 158, 159). Similar to TPEN coexposure, SB293063 cotreatment abolished the characteristic K^+ current enhancement (Fig. 6.4). Thus, it seems that neuronal injury provoked by various injurious stimuli, including transient ischemia in vitro, is associated with amplified potassium currents following intracellular Zn^{2+} release.

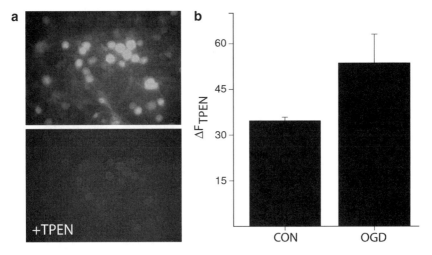

Fig. 6.3 (**a**) Oxidative injury induces intracellular zinc release. Cortical neurons were loaded with the Zn^{2+}-sensitive fluorescent reporter Fluozin-3 AM (5 μM for 30 min) and then perfused with DTDP (100 μM for 5 min). Images were acquired with a 485-nm excitation light *(top)*. Following DTDP exposure, cells were treated with 20-μM TPEN to chelate-free intracellular Zn^{2+} and images were again acquired at 485 nm *(bottom)*. (**b**) Zn^{2+} release following an ischemic insult. Data represents TPEN-sensitive intraneuronal Zn^{2+} 4 h after 30 min of oxygen-glucose deprivation (OGD; $n = 3$) or 30 min non-OGD control exposure (glucose-containing salt solution; $n = 3$). Following OGD, neurons were loaded with the Zn^{2+}-sensitive fluorescent reporter Fluozin-3 AM (5 μM for 30 min). Images were then acquired with 485-nm excitation light. Baseline images were acquired and then cells were treated with 20-μM TPEN to chelate free intracellular Zn^{2+}. Relative Zn^{2+} fluorescence for all neurons was determined by subtracting the Fluozin-3 AM signal after TPEN application from the baseline Fluozin-3 AM signal (ΔF_{TPEN}). Data represent pooled (Mean ± SEM) TPEN-sensitive Zn^{2+} fluorescence measurements from three coverslips, each containing 5–15 cortical neurons

6.8 Alternative Zn^{2+} Signaling Pathways

Zn^{2+}, like Ca^{2+}, is an intracellular second messenger (3, 160). As such, it is not altogether surprising that intracellular Zn^{2+} release is associated with divergent cell death signaling cascades. For example, we have described a noncaspase, non-K^+ current-dependent ERK-mediated cell death pathway triggered in neurons after the liberation of Zn^{2+} by isothiazole-derived biocides (58, 118). In addition, although the exogenous addition of toxic concentrations of $ONOO^-$ can recruit a Zn^{2+} and p38-dependent cell death pathway in neurons (10, 60), $ONOO^-$ can also induce ERK-dependent cell death in oligodendrocytes by stimulating Zn^{2+} release (118) through a pathway that is nearly identical to that of isothiazole toxicity in neurons (58). This ERK-dependent cell death cascade is also very reminiscent to that triggered during oxidative glutamate toxicity in oligodendrocytes and immature neurons and neuronal cell lines (161, 162). This ERK-dependent pathway relies on

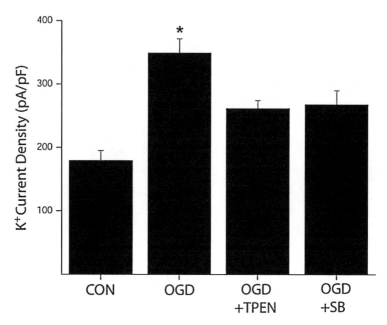

Fig. 6.4 OGD induces an apoptosis-associated K⁺ current enhancement. Whole cell potassium currents were obtained from non-OGD exposed control cortical neurons ($n = 10$), cortical neurons following 105 min of OGD ($n = 22$), cortical neurons following 105 min of OGD in 25-μM TPEN ($n = 12$), and cortical neurons following 105 min of OGD in 20-μM SB293063 ($n = 15$). Recordings were performed 3 h after each treatment condition, as at this time point the apoptotic K⁺ current enhancement is reliably detected. Potassium currents were evoked by a single voltage step to +10 mV from a holding potential of −80 mV. Current amplitudes were normalized to cell capacitance. Mean ± SEM; $^*P < 0.001$ (ANOVA followed by Tukey–Kramer multiple comparisons test)

the activation of NADPH oxidase, leading to ROS-mediated DNA damage and overactivation of the DNA repair enzyme poly(ADP-ribose) polymerase, resulting in a depletion of reducing equivalents and noncaspase-dependent neuronal cell death (58). How Zn^{2+} signals can initiate divergent cell death signaling pathways is an area of current work in our laboratory.

6.9 Intracellular Zn^{2+} Release in Chronic Models of Neurodegeneration

Evidence indicates that Zn^{2+} may play a crucial role in various chronic neurodegenerative conditions. *Alzheimer's disease (AD)* – Zn^{2+} colocalizes with beta amyloid (Aβ) plaques in transgenic mouse models of AD (163) and in human diseased brain tissue (38, 164), while chelation of Zn^{2+} facilitates plaque dissolution (165–167) and solubilizes Aβ from deposits in postmortem tissue from Alzheimer's disease patients (168). In addition, Zn^{2+} also induces aggregation of synthetic Aβ in vitro (169) and enables Aβ-mediated H_2O_2 formation (170). Interestingly, potassium channel dysfunction has also been implicated in AD. In rat cortical neurons exposed to Aβ(1–40), a selective enhancement in transient voltage-gated potassium current is detected prior to completion of the apoptotic program (142, 171). Very recently, another group provided evidence implicating a nuclear factor κB-dependent transcriptional coupregulation of Kv3.4 and accessory subunit MiRP2 as a probable mechanism responsible for the increased K^+ current observed prior to caspase activation and cell death (172). Thus, increased K^+ channel activity may also be responsible for facilitating the apoptotic cascade downstream of Aβ exposure in cortical neurons. *Parkinson's disease* – ATP-dependent potassium (K-ATP) channels mediate selective degeneration of dopaminergic neurons in the substantia nigra of mice following treatment with 1-methyl-4-phenyl-1,2,3,6-tetrahydropyridine (MPTP) (173). K-ATP channels are activated in response to the convergence of cellular energy depletion and increased ROS generation induced by mitochondrial complex I inhibition. In addition, metallothionein levels are significantly reduced in the substantia nigra of mice following a single neurotoxic injection of MPTP (174). Thus, a potential rise in intracellular Zn^{2+} resulting from removal of this metal-buffering protein would in all probability contribute to subsequent ROS generation and K-ATP channel activation. It has yet to be established, however, which downstream events connect K-ATP channel activation to dopaminergic cell death. It is quite possible that K-ATP channels are facilitating an apoptotic K^+ efflux similar to the requisite event observed in dopaminergic neurons exposed to the oxidative neurotoxin 6-hydroxydopamine in vitro (121).

6.10 Concluding Remarks

We propose that increases in intraneuronal Zn^{2+} represent a ubiquitous signal for neuronal injury in neurodegenerative disorders where the generation of ROS and RNS has been implicated. In addition, intracellular dysregulation of a second ionic species, K^+, seems to be an equally critical cell death catalyst downstream of the initial oxidant-induced Zn^{2+} release. Here, we have summarized the key molecular components that link the Zn^{2+} liberation to the loss of cytoplasmic K^+ in injured neurons destined to die. By characterizing the signaling pathway linking these two events, we therefore provide novel therapeutic targets for preventing neuronal cell death in ischemia and other neurodegenerative disorders.

Acknowledgements This work was supported by an American Heart Association Predoctoral Fellowship to Patrick Redman, an American Heart Association Postdoctoral Fellowship to Megan Knoch, and by NIH grant NS043227 to Elias Aizenman. The technical expertise of Karen Hartnett is gratefully acknowledged.

References

1. Frederickson CJ, Koh JY, Bush AI. The neurobiology of zinc in health and disease. Nat Rev Neurosci 2005;6:449–62.
2. Pearce LL, Wasserloos K, St Croix CM, Gandley R, Levitan ES, Pitt BR. Metallothionein, nitric oxide and zinc homeostasis in vascular endothelial cells. J Nutr 2000;130:1467S–70S.
3. Yamasaki S, Sakata-Sogawa K, Hasegawa A, et al. Zinc is a novel intracellular second messenger. J Cell Biol 2007;177:637–45.
4. Frederickson CJ, Bush AI. Synaptically released zinc: physiological functions and pathological effects. Biometals 2001;14:353–66.
5. Weiss JH, Sensi SL, Koh JY. Zn(2+): a novel ionic mediator of neural injury in brain disease. Trends Pharmacol Sci 2000;21:395–401.
6. Frederickson CJ, Hernandez MD, McGinty JF. Translocation of zinc may contribute to seizure-induced death of neurons. Brain Res 1989;480:317–21.
7. Sensi SL, Canzoniero LM, Yu SP, et al. Measurement of intracellular free zinc in living cortical neurons: routes of entry. J Neurosci 1997;17:9554–64.
8. Choi DW, Koh JY. Zinc and brain injury. Annu Rev Neurosci 1998;21:347–75.
9. Capasso M, Jeng JM, Malavolta M, Mocchegiani E, Sensi SL. Zinc dyshomeostasis: a key modulator of neuronal injury. J Alzheimers Dis 2005;8:93–108; discussion 209–15.
10. Pal S, He K, Aizenman E. Nitrosative stress and potassium channel-mediated neuronal apoptosis: is zinc the link? Pflugers Arch 2004;448:296–303.
11. Panayiotidis MI, Bortner CD, Cidlowski JA. On the mechanism of ionic regulation of apoptosis: would the Na$^+$/K$^+$-ATPase please stand up? Acta Physiol 2006;187:205–15.
12. Yu SP. Na(+), K(+)-ATPase: the new face of an old player in pathogenesis and apoptotic/hybrid cell death. Biochem Pharmacol 2003;66:1601–9.
13. Bortner CD, Cidlowski JA. Cell shrinkage and monovalent cation fluxes: role in apoptosis. Arch Biochem Biophys 2007;462:176–88.
14. Yokoyama M, Koh J, Choi DW. Brief exposure to zinc is toxic to cortical neurons. Neurosci Lett 1986;71:351–5.
15. Choi DW, Yokoyama M, Koh J. Zinc neurotoxicity in cortical cell culture. Neuroscience 1988;24:67–79.
16. Weiss JH, Hartley DM, Koh JY, Choi DW. AMPA receptor activation potentiates zinc neurotoxicity. Neuron 1993;10:43–9.
17. Koh JY, Choi DW. Zinc toxicity on cultured cortical neurons: involvement of N-methyl-D-aspartate receptors. Neuroscience 1994;60:1049–57.
18. Jia Y, Jeng JM, Sensi SL, Weiss JH. Zn^{2+} currents are mediated by calcium-permeable AMPA/kainate channels in cultured murine hippocampal neurones. J Physiol 2002;543:35–48.
19. Yin HZ, Sensi SL, Ogoshi F, Weiss JH. Blockade of Ca^{2+}-permeable AMPA/kainate channels decreases oxygen-glucose deprivation-induced Zn^{2+} accumulation and neuronal loss in hippocampal pyramidal neurons. J Neurosci 2002;22:1273–9.
20. Noh KM, Yokota H, Mashiko T, Castillo PE, Zukin RS, Bennett MV. Blockade of calcium-permeable AMPA receptors protects hippocampal neurons against global ischemia-induced death. Proc Natl Acad Sci USA 2005;102:12230–5.
21. Koh JY, Suh SW, Gwag BJ, He YY, Hsu CY, Choi DW. The role of zinc in selective neuronal death after transient global cerebral ischemia. Science 1996;272:1013–16.
22. Frederickson CJ, Rampy BA, Reamy-Rampy S, Howell GA. Distribution of histochemically reactive zinc in the forebrain of the rat. J Chem Neuroanat 1992;5:521–30.
23. Assaf SY, Chung SH. Release of endogenous Zn^{2+} from brain tissue during activity. Nature 1984;308:734–6.
24. Howell GA, Welch MG, Frederickson CJ. Stimulation-induced uptake and release of zinc in hippocampal slices. Nature 1984;308:736–8.
25. Aniksztejn L, Charton G, Ben-Ari Y. Selective release of endogenous zinc from the hippocampal mossy fibers in situ. Brain Res 1987;404:58–64.

26. Frederickson CJ, Hernandez MD, Goik SA, Morton JD, McGinty JF. Loss of zinc staining from hippocampal mossy fibers during kainic acid induced seizures: a histofluorescence study. Brain Res 1988;446:383–6.
27. Vogt K, Mellor J, Tong G, Nicoll R. The actions of synaptically released zinc at hippocampal mossy fiber synapses. Neuron 2000;26:187–96.
28. Suh SW, Garnier P, Aoyama K, Chen Y, Swanson RA. Zinc release contributes to hypoglycemia-induced neuronal death. Neurobiol Dis 2004;16:538–45.
29. Wei G, Hough CJ, Li Y, Sarvey JM. Characterization of extracellular accumulation of Zn^{2+} during ischemia and reperfusion of hippocampus slices in rat. Neuroscience 2004;125:867–77.
30. Frederickson CJ, Giblin LJ, III, Balaji RV, et al. Synaptic release of zinc from brain slices: factors governing release, imaging, and accurate calculation of concentration. J Neurosci Methods 2006;154:19–29.
31. Kay AR. Evidence for chelatable zinc in the extracellular space of the hippocampus, but little evidence for synaptic release of Zn. J Neurosci 2003;23:6847–55.
32. Lavoie N, Peralta MR, III, Chiasson M, et al. Extracellular chelation of zinc does not affect hippocampal excitability and seizure-induced cell death in rats. J Physiol 2007;578:275–89.
33. Sloviter RS. A selective loss of hippocampal mossy fiber Timm stain accompanies granule cell seizure activity induced by perforant path stimulation. Brain Res 1985;330:150–3.
34. Tonder N, Johansen FF, Frederickson CJ, Zimmer J, Diemer NH. Possible role of zinc in the selective degeneration of dentate hilar neurons after cerebral ischemia in the adult rat. Neurosci Lett 1990;109:247–52.
35. Lee JY, Cole TB, Palmiter RD, Koh JY. Accumulation of zinc in degenerating hippocampal neurons of ZnT3-null mice after seizures: evidence against synaptic vesicle origin. J Neurosci 2000;20:RC79.
36. Calderone A, Jover T, Mashiko T, et al. Late calcium EDTA rescues hippocampal CA1 neurons from global ischemia-induced death. J Neurosci 2004;24:9903–13.
37. Choi SM, Choi KO, Lee N, Oh M, Park H. The zinc chelator, N,N,N',N'-tetrakis (2-pyridylmethyl) ethylenediamine, increases the level of nonfunctional HIF-1alpha protein in normoxic cells. Biochem Biophys Res Commun 2006;343:1002–8.
38. Suh SW, Chen JW, Motamedi M, et al. Evidence that synaptically-released zinc contributes to neuronal injury after traumatic brain injury. Brain Res 2000;852:268–73.
39. Mocchegiani E, Bertoni-Freddari C, Marcellini F, Malavolta M. Brain, aging and neurodegeneration: role of zinc ion availability. Prog Neurobiol 2005;75:367–90.
40. Kwak S, Weiss JH. Calcium-permeable AMPA channels in neurodegenerative disease and ischemia. Curr Opin Neurobiol 2006;16:281–7.
41. Berg JM, Shi Y. The galvanization of biology: a growing appreciation for the roles of zinc. Science 1996;271:1081–5.
42. Maret W. Zinc coordination environments in proteins as redox sensors and signal transducers. Antioxid Redox Signal 2006;8:1419–41.
43. Cohen JJ, Duke RC. Glucocorticoid activation of a calcium-dependent endonuclease in thymocyte nuclei leads to cell death. J Immunol 1984;132:38–42.
44. Manev H, Kharlamov E, Uz T, Mason RP, Cagnoli CM. Characterization of zinc-induced neuronal death in primary cultures of rat cerebellar granule cells. Exp Neurol 1997;146:171–8.
45. Sensi SL, Yin HZ, Carriedo SG, Rao SS, Weiss JH. Preferential Zn^{2+} influx through Ca^{2+} permeable AMPA/kainate channels triggers prolonged mitochondrial superoxide production. Proc Natl Acad Sci USA 1999;96:2414–19.
46. Sensi SL, Yin HZ, Weiss JH. AMPA/kainate receptor-triggered Zn^{2+} entry into cortical neurons induces mitochondrial Zn^{2+} uptake and persistent mitochondrial dysfunction. Eur J Neurosci 2000;12:3813–18.
47. Kleiner D. The effect of Zn^{2+} ions on mitochondrial electron transport. Arch Biochem Biophys 1974;165:121–5.
48. Link TA, von Jagow G. Zinc ions inhibit the QP center of bovine heart mitochondrial bc1 complex by blocking a protonatable group. J Biol Chem 1995;270:25001–6.

49. Brown MD, Trounce IA, Jun AS, Allen JC, Wallace DC. Functional analysis of lymphoblast and cybrid mitochondria containing the 3460, 11778, or 14484 Leber's hereditary optic neuropathy mitochondrial DNA mutation. J Biol Chem 2000;275:39831–6.
50. Wudarczyk J, Debska G, Lenartowicz E. Zinc as an inducer of the membrane permeability transition in rat liver mitochondria. Arch Biochem Biophys 1999;363:1–8.
51. Jiang D, Sullivan PG, Sensi SL, Steward O, Weiss JH. Zn(2+) induces permeability transition pore opening and release of pro-apoptotic peptides from neuronal mitochondria. J Biol Chem 2001;276:47524–9.
52. Malaiyandi LM, Honick AS, Rintoul GL, Wang QJ, Reynolds IJ. Zn^{2+} inhibits mitochondrial movement in neurons by phosphatidylinositol 3-kinase activation. J Neurosci 2005;25:9507–14.
53. Bonanni L, Chachar M, Jover-Mengual T, et al. Zinc-dependent multi-conductance channel activity in mitochondria isolated from ischemic brain. J Neurosci 2006;26:6851–62.
54. Sheline CT, Behrens MM, Choi DW. Zinc-induced cortical neuronal death: contribution of energy failure attributable to loss of NAD(+) and inhibition of glycolysis. J Neurosci 2000;20:3139–46.
55. Kim CH, Kim JH, Moon SJ, et al. Pyrithione, a zinc ionophore, inhibits NF-kappaB activation. Biochem Biophys Res Commun 1999;259:505–9.
56. Kim YH, Kim EY, Gwag BJ, Sohn S, Koh JY. Zinc-induced cortical neuronal death with features of apoptosis and necrosis: mediation by free radicals. Neuroscience 1999;89:175–82.
57. McLaughlin B, Pal S, Tran MP, et al. p38 activation is required upstream of potassium current enhancement and caspase cleavage in thiol oxidant-induced neuronal apoptosis. J Neurosci 2001;21:3303–11.
58. Du S, McLaughlin B, Pal S, Aizenman E. In vitro neurotoxicity of methylisothiazolinone, a commonly used industrial and household biocide, proceeds via a zinc and extracellular signal-regulated kinase mitogen-activated protein kinase-dependent pathway. J Neurosci 2002;22:7408–16.
59. Zhang Y, Wang H, Li J, et al. Peroxynitrite-induced neuronal apoptosis is mediated by intracellular zinc release and 12-lipoxygenase activation. J Neurosci 2004;24:10616–27.
60. Bossy-Wetzel E, Talantova MV, Lee WD, et al. Crosstalk between nitric oxide and zinc pathways to neuronal cell death involving mitochondrial dysfunction and p38-activated K^+ channels. Neuron 2004;41:351–65.
61. Zhang CF, Yang P. Zinc-induced aggregation of Abeta (10-21) potentiates its action on voltage-gated potassium channel. Biochem Biophys Res Commun 2006;345:43–9.
62. Atar D, Backx PH, Appel MM, Gao WD, Marban E. Excitation-transcription coupling mediated by zinc influx through voltage-dependent calcium channels. J Biol Chem 1995;270:2473–7.
63. Park JA, Koh JY. Induction of an immediate early gene egr-1 by zinc through extracellular signal-regulated kinase activation in cortical culture: its role in zinc-induced neuronal death. J Neurochem 1999;73:450–6.
64. Csermely P, Szamel M, Resch K, Somogyi J. Zinc can increase the activity of protein kinase C and contributes to its binding to plasma membranes in T lymphocytes. J Biol Chem 1988;263:6487–90.
65. Noh KM, Kim YH, Koh JY. Mediation by membrane protein kinase C of zinc-induced oxidative neuronal injury in mouse cortical cultures. J Neurochem 1999;72:1609–16.
66. Kim YH, Koh JY. The role of NADPH oxidase and neuronal nitric oxide synthase in zinc-induced poly(ADP-ribose) polymerase activation and cell death in cortical culture. Exp Neurol 2002;177:407–18.
67. Noh KM, Koh JY. Induction and activation by zinc of NADPH oxidase in cultured cortical neurons and astrocytes. J Neurosci 2000;20:RC111.
68. Spence MW, Byers DM, Palmer FB, Cook HW. A new Zn^{2+}-stimulated sphingomyelinase in fetal bovine serum. J Biol Chem 1989;264:5358–63.
69. Hannun YA. The sphingomyelin cycle and the second messenger function of ceramide. J Biol Chem 1994;269:3125–8.

70. Brugg B, Michel PP, Agid Y, Ruberg M. Ceramide induces apoptosis in cultured mesencephalic neurons. J Neurochem 1996;66:733–9.
71. Schissel SL, Schuchman EH, Williams KJ, Tabas I. Zn^{2+}-stimulated sphingomyelinase is secreted by many cell types and is a product of the acid sphingomyelinase gene. J Biol Chem 1996;271:18431–6.
72. Park JA, Lee JY, Sato TA, Koh JY. Co-induction of p75NTR and p75NTR-associated death executor in neurons after zinc exposure in cortical culture or transient ischemia in the rat. J Neurosci 2000;20:9096–103.
73. Sheline CT, Wang H, Cai AL, Dawson VL, Choi DW. Involvement of poly ADP ribosyl polymerase-1 in acute but not chronic zinc toxicity. Eur J Neurosci 2003;18:1402–9.
74. Zalewski PD, Forbes IJ, Betts WH. Correlation of apoptosis with change in intracellular labile Zn(II) using zinquin [(2-methyl-8-*p*-toluenesulphonamido-6-quinolyloxy)acetic acid], a new specific fluorescent probe for Zn(II). Biochem J 1993;296 (Part 2):403–8.
75. Fraker PJ, Telford WG. A reappraisal of the role of zinc in life and death decisions of cells. Proc Soc Exp Biol Med 1997;215:229–36.
76. Perry DK, Smyth MJ, Stennicke HR, et al. Zinc is a potent inhibitor of the apoptotic protease, caspase-3. A novel target for zinc in the inhibition of apoptosis. J Biol Chem 1997;272: 18530–3.
77. Chai F, Truong-Tran AQ, Ho LH, Zalewski PD. Regulation of caspase activation and apoptosis by cellular zinc fluxes and zinc deprivation: a review. Immunol Cell Biol 1999;77:272–8.
78. Ho LH, Ratnaike RN, Zalewski PD. Involvement of intracellular labile zinc in suppression of DEVD-caspase activity in human neuroblastoma cells. Biochem Biophys Res Commun 2000;268:148–54.
79. Schrantz N, Auffredou MT, Bourgeade MF, Besnault L, Leca G, Vazquez A. Zinc-mediated regulation of caspases activity: dose-dependent inhibition or activation of caspase-3 in the human Burkitt lymphoma B cells (Ramos). Cell Death Differ 2001;8:152–61.
80. Marini M, Frabetti F, Canaider S, Dini L, Falcieri E, Poirier GG. Modulation of caspase-3 activity by zinc ions and by the cell redox state. Exp Cell Res 2001;266:323–32.
81. Ahn YH, Kim YH, Hong SH, Koh JY. Depletion of intracellular zinc induces protein synthesis-dependent neuronal apoptosis in mouse cortical culture. Exp Neurol 1998;154:47–56.
82. Adler M, Shafer H, Hamilton T, Petrali JP. Cytotoxic actions of the heavy metal chelator TPEN on NG108-15 neuroblastoma-glioma cells. Neurotoxicology 1999;20:571–82.
83. Virag L, Szabo C. Inhibition of poly(ADP-ribose) synthetase (PARS) and protection against peroxynitrite-induced cytotoxicity by zinc chelation. Br J Pharmacol 1999;126:769–77.
84. Bancila V, Nikonenko I, Dunant Y, Bloc A. Zinc inhibits glutamate release via activation of pre-synaptic K channels and reduces ischaemic damage in rat hippocampus. J Neurochem 2004;90:1243–50.
85. Cote A, Chiasson M, Peralta MR, III, Lafortune K, Pellegrini L, Toth K. Cell type-specific action of seizure-induced intracellular zinc accumulation in the rat hippocampus. J Physiol 2005;566:821–37.
86. Hyun HJ, Sohn JH, Ha DW, Ahn YH, Koh JY, Yoon YH. Depletion of intracellular zinc and copper with TPEN results in apoptosis of cultured human retinal pigment epithelial cells. Invest Ophthalmol Vis Sci 2001;42:460–5.
87. Lipton SA, Kater SB. Neurotransmitter regulation of neuronal outgrowth, plasticity and survival. Trends Neurosci 1989;12:265–70.
88. Choi DW. Calcium: still center-stage in hypoxic-ischemic neuronal death. Trends Neurosci 1995;18:58–60.
89. Colvin RA, Fontaine CP, Laskowski M, Thomas D. Zn^{2+} transporters and Zn^{2+} homeostasis in neurons. Eur J Pharmacol 2003;479:171–85.
90. Hershfinkel M, Moran A, Grossman N, Sekler I. A zinc-sensing receptor triggers the release of intracellular Ca^{2+} and regulates ion transport. Proc Natl Acad Sci USA 2001;98: 11749–54.
91. Hershfinkel M, Silverman WF, Sekler I. The zinc sensing receptor, a link between zinc and cell signaling. Mol Med 2007;13:331–6.

92. Sensi SL, Ton-That D, Weiss JH, Rothe A, Gee KR. A new mitochondrial fluorescent zinc sensor. Cell Calcium 2003;34:281–4.
93. McMahon RJ, Cousins RJ. Regulation of the zinc transporter ZnT-1 by dietary zinc. Proc Natl Acad Sci USA 1998;95:4841–6.
94. Liuzzi JP, Cousins RJ. Mammalian zinc transporters. Annu Rev Nutr 2004;24:151–72.
95. Sekler I, Sensi SL, Hershfinkel M, Silverman WF. Mechanism and regulation of cellular zinc transport. Mol Med 2007;13:337–43.
96. Choi DW. Excitotoxic cell death. J Neurobiol 1992;23:1261–76.
97. Stork CJ, Li YV. Intracellular zinc elevation measured with a "calcium-specific" indicator during ischemia and reperfusion in rat hippocampus: a question on calcium overload. J Neurosci 2006;26:10430–7.
98. Berg JM. Zinc fingers and other metal-binding domains. Elements for interactions between macromolecules. J Biol Chem 1990;265:6513–16.
99. Maret W, Krezel A. Cellular zinc and redox buffering capacity of metallothionein/thionein in health and disease. Mol Med 2007;13:371–5.
100. Korichneva I. Zinc dynamics in the myocardial redox signaling network. Antioxid Redox Signal 2006;8:1707–21.
101. Haase H, Rink L. Signal transduction in monocytes: the role of zinc ions. Biometals 2007;20:579–85.
102. Maret W, Vallee BL. Thiolate ligands in metallothionein confer redox activity on zinc clusters. Proc Natl Acad Sci USA 1998;95:3478–82.
103. Jiang LJ, Maret W, Vallee BL. The glutathione redox couple modulates zinc transfer from metallothionein to zinc-depleted sorbitol dehydrogenase. Proc Natl Acad Sci USA 1998;95:3483–8.
104. Aizenman E, Stout AK, Hartnett KA, Dineley KE, McLaughlin B, Reynolds IJ. Induction of neuronal apoptosis by thiol oxidation: putative role of intracellular zinc release. J Neurochem 2000;75:1878–88.
105. Katakai K, Liu J, Nakajima K, Keefer LK, Waalkes MP. Nitric oxide induces metallothionein (MT) gene expression apparently by displacing zinc bound to MT. Toxicol Lett 2001;119:103–8.
106. Frederickson CJ, Cuajungco MP, LaBuda CJ, Suh SW. Nitric oxide causes apparent release of zinc from presynaptic boutons. Neuroscience 2002;115:471–4.
107. Frederickson CJ, Maret W, Cuajungco MP. Zinc and excitotoxic brain injury: a new model. Neuroscientist 2004;10:18–25.
108. Knapp LT, Klann E. Superoxide-induced stimulation of protein kinase C via thiol modification and modulation of zinc content. J Biol Chem 2000;275:24136–45.
109. Cima RR, Dubach JM, Wieland AM, Walsh BM, Soybel DI. Intracellular Ca(2+) and Zn(2+) signals during monochloramine-induced oxidative stress in isolated rat colon crypts. Am J Physiol Gastrointest Liver Physiol 2006;290:G250–G261.
110. Kroncke KD. Cellular stress and intracellular zinc dyshomeostasis. Arch Biochem Biophys 2007;463:183–7.
111. Hao Q, Maret W. Aldehydes release zinc from proteins. A pathway from oxidative stress/lipid peroxidation to cellular functions of zinc. FEBS J 2006;273:4300–10.
112. Maret W. Oxidative metal release from metallothionein via zinc-thiol/disulfide interchange. Proc Natl Acad Sci USA 1994;91:237–41.
113. Maret W. Metallothionein/disulfide interactions, oxidative stress, and the mobilization of cellular zinc. Neurochem Int 1995;27:111–17.
114. Land PW, Aizenman E. Zinc accumulation after target loss: an early event in retrograde degeneration of thalamic neurons. Eur J Neurosci 2005;21:647–57.
115. Lee JY, Hwang JJ, Park MH, Koh JY. Cytosolic labile zinc: a marker for apoptosis in the developing rat brain. Eur J Neurosci 2006;23:435–42.
116. Aizenman E, Sinor JD, Brimecombe JC, Herin GA. Alterations of N-methyl-D-aspartate receptor properties after chemical ischemia. J Pharmacol Exp Ther 2000;295:572–7.
117. Aras MA, Aizenman E. Obligatory role of ASK1 in the apoptotic surge of K⁺ currents. Neurosci Lett 2005;387:136–40.

118. Zhang Y, Aizenman E, DeFranco DB, Rosenberg PA. Intracellular zinc release, 12-lipoxygenase activation and MAPK dependent neuronal and oligodendroglial death. Mol Med 2007;13:350–5.
119. Sensi SL, Jeng JM. Rethinking the excitotoxic ionic milieu: the emerging role of Zn(2+) in ischemic neuronal injury. Curr Mol Med 2004;4:87–111.
120. Redman PT, Jefferson BS, Ziegler CB, et al. A vital role for voltage-dependent potassium channels in dopamine transporter-mediated 6-hydroxydopamine neurotoxicity. Neuroscience 2006;143:1–6.
121. Pal S, Hartnett KA, Nerbonne JM, Levitan ES, Aizenman E. Mediation of neuronal apoptosis by Kv2.1-encoded potassium channels. J Neurosci 2003;23:4798–802.
122. Pal SK, Takimoto K, Aizenman E, Levitan ES. Apoptotic surface delivery of K^+ channels. Cell Death Differ 2006;13:661–7.
123. Redman PT, He K, Hartnett KA, et al. Apoptotic surge of potassium currents is mediated by p38 phosphorylation of Kv2.1. Proc Natl Acad Sci USA 2007;104:3568–73.
124. Obenauer JC, Cantley LC, Yaffe MB. Scansite 2.0: proteome-wide prediction of cell signaling interactions using short sequence motifs. Nucleic Acids Res 2003;31:3635–41.
125. Park KS, Mohapatra DP, Misonou H, Trimmer JS. Graded regulation of the Kv2.1 potassium channel by variable phosphorylation. Science 2006;313:976–9.
126. Du J, Haak LL, Phillips-Tansey E, Russell JT, McBain CJ. Frequency-dependent regulation of rat hippocampal somato-dendritic excitability by the K^+ channel subunit Kv2.1. J Physiol 2000;522 (Part 1):19–31.
127. Misonou H, Mohapatra DP, Park EW, et al. Regulation of ion channel localization and phosphorylation by neuronal activity. Nat Neurosci 2004;7:711–18.
128. Misonou H, Mohapatra DP, Trimmer JS. Kv2.1: a voltage-gated k^+ channel critical to dynamic control of neuronal excitability. Neurotoxicology 2005;26:743–52.
129. Malin SA, Nerbonne JM. Delayed rectifier K^+ currents, IK, are encoded by Kv2 alpha-subunits and regulate tonic firing in mammalian sympathetic neurons. J Neurosci 2002;22: 10094–105.
130. Misonou H, Mohapatra DP, Menegola M, Trimmer JS. Calcium- and metabolic state-dependent modulation of the voltage-dependent Kv2.1 channel regulates neuronal excitability in response to ischemia. J Neurosci 2005;25:11184–93.
131. Mohapatra DP, Park KS, Trimmer JS. Dynamic regulation of the voltage-gated Kv2.1 potassium channel by multisite phosphorylation. Biochem Soc Trans 2007;35:1064–8.
132. Michaelevski I, Chikvashvili D, Tsuk S, et al. Direct interaction of target SNAREs with the Kv2.1 channel. Modal regulation of channel activation and inactivation gating. J Biol Chem 2003;278:34320–30.
133. Lauritzen I, De Weille JR, Lazdunski M. The potassium channel opener (-)-cromakalim prevents glutamate-induced cell death in hippocampal neurons. J Neurochem 1997;69: 1570–9.
134. Abdul M, Hoosein N. Expression and activity of potassium ion channels in human prostate cancer. Cancer Lett 2002;186:99–105.
135. Hegle AP, Marble DD, Wilson GF. A voltage-driven switch for ion-independent signaling by ether-a-go-go K^+ channels. Proc Natl Acad Sci USA 2006;103:2886–91.
136. Yu SP, Yeh CH, Gottron F, Wang X, Grabb MC, Choi DW. Role of the outward delayed rectifier K^+ current in ceramide-induced caspase activation and apoptosis in cultured cortical neurons. J Neurochem 1999;73:933–41.
137. Bortner CD, Cidlowski JA. Caspase independent/dependent regulation of K(+), cell shrinkage, and mitochondrial membrane potential during lymphocyte apoptosis. J Biol Chem 1999; 274:21953–62.
138. Hughes FM, Jr., Cidlowski JA. Potassium is a critical regulator of apoptotic enzymes in vitro and in vivo. Adv Enzyme Regul 1999;39:157–71.
139. Cain K, Langlais C, Sun XM, Brown DG, Cohen GM. Physiological concentrations of K^+ inhibit cytochrome c-dependent formation of the apoptosome. J Biol Chem 2001;276: 41985–90.
140. Brevnova EE, Platoshyn O, Zhang S, Yuan JX. Overexpression of human KCNA5 increases IK V and enhances apoptosis. Am J Physiol Cell Physiol 2004;287:C715–C722.

141. Yu SP, Yeh CH, Sensi SL, et al. Mediation of neuronal apoptosis by enhancement of outward potassium current. Science 1997;278:114–17.
142. Yu SP, Farhangrazi ZS, Ying HS, Yeh CH, Choi DW. Enhancement of outward potassium current may participate in beta-amyloid peptide-induced cortical neuronal death. Neurobiol Dis 1998;5:81–8.
143. Xiao H, Dai X, Mao X. [Inhibition of voltage-activated outward delayed rectifier potassium channel currents in dorsal root ganglion neurons of rats by lead]. Zhonghua Yu Fang Yi Xue Za Zhi 2001;35:108–10.
144. Wang X, Xiao AY, Ichinose T, Yu SP. Effects of tetraethylammonium analogs on apoptosis and membrane currents in cultured cortical neurons. J Pharmacol Exp Ther 2000;295:524–30.
145. Wei L, Xiao AY, Jin C, Yang A, Lu ZY, Yu SP. Effects of chloride and potassium channel blockers on apoptotic cell shrinkage and apoptosis in cortical neurons. Pflugers Arch 2004;448:325–34.
146. Knoch ME, Hartnett KA, Hara H, Kandler K, Aizenman E. Microglia induce neurotoxicity via intraneuronal Zn(2+) release and a K(+) current surge. Glia 2008;56:89–96.
147. Boer K, Spliet WG, van Rijen PC, Redeker S, Troost D, Aronica E. Evidence of activated microglia in focal cortical dysplasia. J Neuroimmunol 2006;173:188–95.
148. De Simoni MG, Perego C, Ravizza T, et al. Inflammatory cytokines and related genes are induced in the rat hippocampus by limbic status epilepticus. Eur J Neurosci 2000;12:2623–33.
149. Ekdahl CT, Claasen JH, Bonde S, Kokaia Z, Lindvall O. Inflammation is detrimental for neurogenesis in adult brain. Proc Natl Acad Sci USA 2003;100:13632–7.
150. Milligan CE, Levitt P, Cunningham TJ. Brain macrophages and microglia respond differently to lesions of the developing and adult visual system. J Comp Neurol 1991;314:136–46.
151. Rizzi M, Perego C, Aliprandi M, et al. Glia activation and cytokine increase in rat hippocampus by kainic acid-induced status epilepticus during postnatal development. Neurobiol Dis 2003;14:494–503.
152. Wang MJ, Lin SZ, Kuo JS, et al. Urocortin modulates inflammatory response and neurotoxicity induced by microglial activation. J Immunol 2007;179:6204–14.
153. Milligan CE, Cunningham TJ, Levitt P. Differential immunochemical markers reveal the normal distribution of brain macrophages and microglia in the developing rat brain. J Comp Neurol 1991;314:125–35.
154. Marin-Teva JL, Dusart I, Colin C, Gervais A, van Rooijen N, Mallat M. Microglia promote the death of developing Purkinje cells. Neuron 2004;41:535–47.
155. Hribar M, Bloc A, Medilanski J, Nusch L, Eder-Colli L. Voltage-gated K$^+$ current: a marker for apoptosis in differentiating neuronal progenitor cells? Eur J Neurosci 2004;20:635–48.
156. Chi XX, Xu ZC. Differential changes of potassium currents in CA1 pyramidal neurons after transient forebrain ischemia. J Neurophysiol 2000;84:2834–43.
157. Deng P, Pang ZP, Zhang Y, Xu ZC. Increase of delayed rectifier potassium currents in large aspiny neurons in the neostriatum following transient forebrain ischemia. Neuroscience 2005; 131:135–46.
158. Underwood DC, Osborn RR, Kotzer CJ, et al. SB 239063, a potent p38 MAP kinase inhibitor, reduces inflammatory cytokine production, airways eosinophil infiltration, and persistence. J Pharmacol Exp Ther 2000;293:281–8.
159. Legos JJ, McLaughlin B, Skaper SD, et al. The selective p38 inhibitor SB-239063 protects primary neurons from mild to moderate excitotoxic injury. Eur J Pharmacol 2002;447: 37–42.
160. Pearce DA, Jotterand N, Carrico IS, Imperiali B. Derivatives of 8-hydroxy-2-methylquinoline are powerful prototypes for zinc sensors in biological systems. J Am Chem Soc 2001;123:5160–1.
161. Murphy TH, Miyamoto M, Sastre A, Schnaar RL, Coyle JT. Glutamate toxicity in a neuronal cell line involves inhibition of cystine transport leading to oxidative stress. Neuron 1989;2:1547–58.
162. Stanciu M, Wang Y, Kentor R, et al. Persistent activation of ERK contributes to glutamate-induced oxidative toxicity in a neuronal cell line and primary cortical neuron cultures. J Biol Chem 2000;275:12200–6.

163. Lee JY, Mook-Jung I, Koh JY. Histochemically reactive zinc in plaques of the Swedish mutant beta-amyloid precursor protein transgenic mice. J Neurosci 1999;19:RC10.
164. Danscher G, Jensen KB, Frederickson CJ, et al. Increased amount of zinc in the hippocampus and amygdala of Alzheimer's diseased brains: a proton-induced X-ray emission spectroscopic analysis of cryostat sections from autopsy material. J Neurosci Methods 1997; 76:53–9.
165. Cherny RA, Legg JT, McLean CA, et al. Aqueous dissolution of Alzheimer's disease Abeta amyloid deposits by biometal depletion. J Biol Chem 1999;274:23223–8.
166. Regland B, Lehmann W, Abedini I, et al. Treatment of Alzheimer's disease with clioquinol. Dement Geriatr Cogn Disord 2001;12:408–14.
167. Ritchie CW, Bush AI, Mackinnon A, et al. Metal-protein attenuation with iodochlorhydroxyquin (clioquinol) targeting Abeta amyloid deposition and toxicity in Alzheimer disease: a pilot phase 2 clinical trial. Arch Neurol 2003;60:1685–91.
168. Hensley K, Hall N, Subramaniam R, et al. Brain regional correspondence between Alzheimer's disease histopathology and biomarkers of protein oxidation. J Neurochem 1995; 65:2146–56.
169. Friedlich AL, Lee JY, van Groen T, et al. Neuronal zinc exchange with the blood vessel wall promotes cerebral amyloid angiopathy in an animal model of Alzheimer's disease. J Neurosci 2004;24:3453–9.
170. Huang X, Atwood CS, Hartshorn MA, et al. The A beta peptide of Alzheimer's disease directly produces hydrogen peroxide through metal ion reduction. Biochemistry 1999;38:7609–16.
171. Yu HB, Li ZB, Zhang HX, Wang XL. Role of potassium channels in Abeta(1-40)-activated apoptotic pathway in cultured cortical neurons. J Neurosci Res 2006;84:1475–84.
172. Pannaccione A, Boscia F, Scorziello A, et al. Up-regulation and increased activity of KV3.4 channels and their accessory subunit MinK-related peptide 2 induced by amyloid peptide are involved in apoptotic neuronal death. Mol Pharmacol 2007;72:665–73.
173. Liss B, Haeckel O, Wildmann J, Miki T, Seino S, Roeper J. K-ATP channels promote the differential degeneration of dopaminergic midbrain neurons. Nat Neurosci 2005;8:1742–51.
174. Dhanasekaran M, Albano CB, Pellet L, et al. Role of lipoamide dehydrogenase and metallothionein on 1-methyl-4-phenyl-1,2,3,6- tetrahydropyridine-induced neurotoxicity. Neurochem Res 2008;33:980–984.

Chapter 7
Mitochondrial Ion Channels in Ischemic Brain

Elizabeth A. Jonas

Abstract Mitochondria produce ATP for many cellular processes including the highly energy dependent events of normal brain function. Mitochondria also exist at the center of a web of signaling within the cell that regulates cell life and death. In many forms of programmed cell death mitochondria play a crucial role in regulating oxidative phosphorylation to produce or conserve precious energy supplies. They also manage cytosolic levels of calcium and zinc, key ions implicated in excitotoxic neuronal death linked to energy deprivation during ischemia, cellular metabolic decline, neurodegeneration, and aging. During the type of programmed cell death known as apoptosis, mitochondria regulate the release of proapoptotic factors such as cytochrome c from the mitochondrial intermembrane space. Cytochrome c and other proapoptotic molecules activate downstream enzymes including caspases that destroy the cell contents, including its nuclear material.

This chapter is divided into three parts. The first part will summarize the current knowledge of the functions of the known channels in the outer membrane including VDAC and BCL-2 family proteins and how they may regulate programmed cell death. The second part will discuss the possible roles of the ion channel activities of the inner membrane in normal and pathological neuronal activity and describe how a complex may form between the ion channel components of the two membranes. The third part will describe the functions of mitochondrial ion channel activity in ischemic brain.

Keywords: Mitochondrial channels; Brain ischemia; Apoptosis; Bcl-2; VDAC; Oxygen free radicals; Zinc, Permeability transition

E.A. Jonas
Department of Internal Medicine (Endocrinology), Yale University School of Medicine,
P.O. Box 208020, 333 Cedar Street, New Haven, CT 06520, USA
e-mail: elizabeth.jonas@yale.edu

7.1 Introduction

Mitochondrial ion channels are intimately involved in numerous cellular processes. Regulated mitochondrial membrane pores release ATP and take up ADP. The influx and efflux of calcium, sodium, potassium, and zinc are tightly regulated by mitochondrial channels. The permeability of the inner membrane to different ions helps determine the membrane potential of the inner membrane, and regulates matrix volume. In addition, the BCL-2 family proteins, which are known to play a role in producing or inhibiting the release of proapoptotic factors from mitochondria, have properties of ion channels. These specialized proteins perform physiological functions, for example, regulating the strength and pattern of synaptic transmission, as well as participating in cell death.

The inner mitochondrial membrane is a barrier similar to the plasma membrane that allows the separation of ion species to produce an electrochemical gradient. The potential energy produced by charge separation can be used for work and for signaling, through the activity of transporters and ion channels. Channels regulate the ATP-producing activity of the mitochondria and the enzymatic activity of the matrix. Matrix enzymes produce ATP through the process of oxidative phosphorylation. Energy is generated by the reduction of NAD(P) to NAD(P)H and FAD to FADH in the tricarboxylic acid cycle. This energy is then used (by oxidation of the reduced species) for the transport of electrons from high- to low-energy electron acceptors while H^+ ions are pumped out across the inner membrane, thereby producing a large electrochemical (voltage) gradient. Movement of H^+ ions back down the gradient across the ATP synthase phosphorylates ADP to produce ATP (phosphorylation).

The gating of outer membrane channels may also regulate the release of metabolites and ions into the cytosol. While little is known about the membrane potential across the outer membrane, it has become clear that outer membrane permeability is tightly controlled, in part by the voltage dependence of outer membrane conductances. Control of outer membrane permeability compartmentalizes the intermembrane space with implications for metabolic regulation and cell fate determination.

7.2 Role of VDAC in Mitochondrial Function

Perhaps the most prevalent ion channel in the mitochondrial outer membrane is VDAC, and no studies of the role of mitochondria in cell death can ignore the biophysical characteristics or physiological behavior of this important molecule (1, 2). The main function of VDAC appears to be its ability to conduct metabolites such as ATP, ADP, NADH, and pyruvate, in addition to other metabolites whose molecular weight can reach almost 1,000. VDAC therefore has different biophysical characteristics than other voltage-gated ion channels (3). Historically, it was thought that VDAC was always open, making the mitochondrial outer membrane like a leaky sieve, but many recent studies have contributed to the present notion that the opening and closing of VDAC are highly regulated (4). VDAC is highly conserved in its tertiary structure from yeast to man (1).

7.3 Biophysical Characteristics of VDAC

VDAC sequences contain an amphiphilic alpha-helix and 12 or more alternating hydrophobic/hydrophilic segments, which form a beta sheet that makes up the wall of a cylinder or barrel (Fig. 7.1) (5). In this configuration, the hydrophilic inner surface through which water and solutes travel faces the pore and the hydrophobic outer surface is buried in the membrane (6).

Recordings of VDAC in artificial lipid membranes demonstrate that all VDACs have a conserved set of biophysical features (1, 7). The VDAC channel has a large single channel conductance of 1–4 nS. The open state of the channel prefers anions over cations 2:1, and strongly prefers metabolically relevant anions such as ATP. The channels prefer to open at voltages close to 0 mV, either in the positive or negative direction, and the channel undergoes rapid transitions to several closed states at potentials more negative than −50 or positive than +50. These *closed* states are permeable to cations, but not to anions such as metabolites.

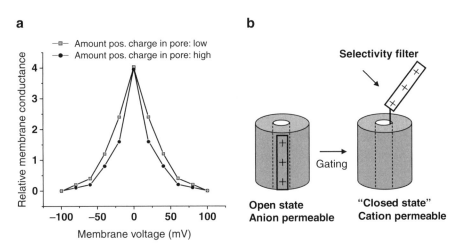

Fig. 7.1 Voltage dependence of membrane conductance of the voltage-dependent anion channel (VDAC). (**a**) The artist's rendition simulates data recorded when the channel protein is reconstituted in artificial membranes. Increasing the number of charged residues in the pore increases the steepness of the voltage dependence. (**b**) Dependence of the open and closed states of VDAC on position of the charged selectivity moiety inside or outside of the pore

The change in selectivity with gating predicts that positive charges may move out of the pore during channel closure and that these charged moieties may comprise the voltage sensor (1). Therefore, increasing the positive charge within the pore increases the steepness of voltage dependence and decreasing the positive charge will decrease the steepness (Fig. 7.1).

Regulators of VDAC include a variety of polyanions, all of which have similar effects (7). When they are added to one side of the membrane, they appear to favor the closure of the channel, possibly by drawing out the positively charged gate region (8). A large (100,000 MW) soluble protein modulator of VDAC, most likely residing within the intermembrane space, increases the voltage dependence of the channel and favors the closed state (9). VDAC may also favor its closed state under normal conditions due to the presence of negative anions and protein modulators in the intermembrane space. Immobile molecules in the intermembrane space may exert osmotic pressure, drawing water through the pore and decreasing hydrostatic pressure within the pore (10). The decline in hydrostatic pressure decreases pore volume and finally favors channel closure (11).

NADH and NADPH significantly affect channel gating. NADH doubles the voltage dependence of VDAC (12). Addition of NADH and NADPH to mitochondria with intact outer membranes results in a sixfold reduction of the permeability of the outer membrane to ADP (13).

7.4 Control of Metabolism by VDAC

Hexokinase is the first enzyme in the glycolytic pathway (14) and it requires ATP. It binds to the outer surface of the outer membrane via its association with VDAC, allowing hexokinase preferential access to the ATP that is leaving the mitochondria and giving mitochondria preferential access to ADP produced by hexokinase. Hexokinase preferentially uses ATP that is made in mitochondria over exogenous ATP. The relationship between VDAC and hexokinase suggests that VDAC is the central regulator of the interaction between glycolysis and mitochondrial respiration (15). Contact sites between the two membranes may contain the colocalized adenine nucleotide transporter (ANT), VDAC, and hexokinase (16). Increasing the proximity of VDAC to inner membrane proteins provides ADP readily to the ANT, thereby increasing the efficiency of production of ATP. An increase in cellular metabolic rate, by activating hexokinase, increases the likelihood that the two membranes will come into close contact (17, 18). The ATP generated by mitochondrial respiration then further stimulates the enzymatic activity of hexokinase.

Extremely high hexokinase levels such as those found in certain tumor cells actually inhibit the conductance of metabolites through VDAC (1, 19). Hexokinase, in this setting, appears to help the cell switch between glycolysis and mitochondrial metabolism. If VDAC closes, presumably no mitochondrial ATP can exit the mitochondria nor can ADP enter, and therefore the cell must switch from mitochondrial metabolism to glycolysis. The homeostatic mechanisms regulating VDAC closure are quite complex, however, since glucose-6-phosphate, the product of the enzymatic activity of hexokinase, antagonizes the binding of hexokinase to VDAC and tends to favor the open state. Therefore, if glycolysis becomes inhibited downstream of glucose-6-phosphate, and glucose-6-phosphate builds up, then VDAC may open again.

7.5 BCL-2 Family Ion Channels: Role in Programmed Cell Death in Neurons

An important mode of cell death occurs when mitochondrial outer membranes are permeabilized by the regulated actions of the BCL-2 family proteins. These proteins exhibit two important features: They produce ion channel activity in intracellular organelle membranes, and they are regulated by interactions with binding partners, including BCL-2 family members and other cellular proteins. As a result, the BCL-2 family proteins control the release of death-inducing mitochondrial factors such as cytochrome *c* into the cytosol. In a strange twist, antiapoptotic proteins can be rapidly converted into proapoptotic molecules by proteolytic cleavage after a cell death stimulus, and this event can dramatically increase the permeability of the mitochondrial outer membrane by increasing the single channel conductance of the pore-forming proteins (20–22). In neurons, the size of the mitochondrial channel conductance and its binding partners determine whether a mitochondrial channel promotes cell survival or death (Fig. 7.4).

Programmed cell death or apoptosis is the genetic predisposition of cells to die (23). In this fashion cells are eliminated developmentally, and throughout the life of an organism, to remove old and damaged cells (24). During pathological brain insults such as ischemia, infection, or trauma, some brain cells die immediately, but others die a delayed death long after the insult, by turning on programmed death pathways (25). BCL-2 family proteins determine the fate of cells exposed to many different pro-death signals including growth factor deprivation, ultraviolet and gamma radiation, viral infections, and free radical formation (23). In addition, BCL-2 family proteins regulate death stimuli that had previously been relegated to the realm of necrosis, such as ischemic cell death in the heart and brain (26).

During the process of programmed cell death, mitochondria appear to receive a signal to undergo permeabilization of their outer membranes (23, 27). The opening of the outer mitochondrial membrane is also sometimes referred to as the activation of the apoptosis channel (28). This event occurs suddenly, leading to the release of the intermembrane space proteins such as cytochrome *c* (27, 29). Release of cytochrome *c* produces two events. On the one hand, loss of cytochrome *c* compromises the ability of mitochondria to produce ATP and eventually to maintain the mitochondrial membrane potential. On the other hand, cytochrome *c* and other factors released from mitochondria activate downstream caspases that chew up hundreds of cellular proteins (30). How exactly mitochondrial membrane permeabilization occurs is still as yet unclear. Nevertheless, it appears that BCL-2 family proteins play a major role in either regulating or directly producing the permeabilization of the outer membrane.

Three categories of BCL-2 proteins appear to be important in determining cell death (Fig. 7.2). These are the antiapoptotic members (such as BCL-x_L, BCL-2, and MCL-1), the proapoptotic members such as Bax and Bak, and a large group of BH-3-only proteins such as BID, BAD, PUMA, and NOXA (31). The antiapoptotic members of the group are similar in structure and sequence to Bax and Bak, and in addition to the BH1, 2, and 3 domains, contain a BH4 domain that is important for the antiapoptotic features of the molecules.

7.6 Actions of BCL-2 Family Proteins Are Regulated by Binding Partners

The mechanisms by which antiapoptotic proteins such as BCL-2 and BCL-x_L prevent cell death are poorly understood. Recent studies have focused on the ability of antiapoptotic proteins such as BCL-x_L to bind to and sequester proapoptotic members of the BCL-2 family (32). BCL-2 and BCL-x_L bind to Bax and to BH3 peptides, preventing the proapoptotic actions of these proteins. The binding partners of the different BCL-2 family proteins within the family are known, but the order of binding events seems important (Fig. 7.3) (31, 32). Under resting conditions, BCL-x_L and BCL-2 are bound to activator BH3 molecules, keeping them in a quiescent state. New data suggest that activation of inactivator BH3-only proteins

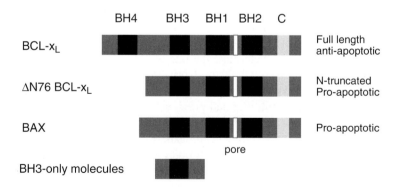

Fig. 7.2 BCL-2 family proteins. Canonical antiapoptotic proteins contain the BH4 domain that is lacking in proapoptotic BCL-2 family members such as cleaved BCL-xL (ΔN BCL-xL) and Bax. BH3-only molecules are traditionally thought to be proapoptotic, but they may also play physiological roles (see text)

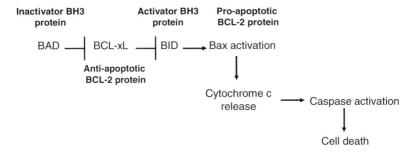

Fig. 7.3 Schematic of an example of one possible hierarchical model of the control of cell death by BCL-2 family proteins. From left to right, after a cell death signal, BAD (the inactivator BH3 protein) sequesters BCL-xL, freeing BID (the activator BH3 protein) to activate Bax. Bax oligomerizes and permeabilizes mitochondrial outer membranes, releasing cytochrome *c* to activate downstream caspase pathways

such as BAD is the first step in initiation of cell death. These inactivator BH3 proteins then bind to the antiapoptotic BCL-2 proteins, BCL-x_L and BCL-2, preventing them from binding to Bax or to activator BH3 peptides such as BID. The activator BH3 peptides can then bind to and activate Bax, or bind to inhibitors of Bak, such as VDAC2 (33). The end result is that Bax or Bak are free to homo-oligomerize and thereby permeabilize outer mitochondrial membranes. This hierarchy model explains the high killing potency of a subset of BH3-only proteins that is known to bind all antiapoptotic proteins (31, 32).

Bax and BCL-x_L share structural homology in the BH1-BH3 domains (34, 35). It is clear from the binding pair studies that both Bax and BCL-x_L interact in complexes with the BH3-only proteins. When other proteins or binding partners compete for their binding, they both may get released from these complexes and can perform other functions.

7.7 BCL-2 Family Proteins Function as Ion Channels

Another important shared feature of Bax and BCL-x_L is that they both produce ion channel activity when inserted into artificial lipid membranes (34, 35). BCL-x_L and Bax are both either localized to the outer mitochondrial membrane, or translocated into the outer membrane upon a death stimulus (36, 37). Despite structural similarity to BCL-x_L, Bax, like other proapoptotic molecules, lacks the BH4 domain that produces antiapoptotic activity (38, 39). At first glance, the ion channel activities of Bax and BCL-x_L seem quite similar (35).

The three-dimensional structure of BCL-x_L is comprised of seven alpha helices (34, 40). Two outer layers of amphipathic helices serve to screen the long hydrophobic alpha helices from the aqueous domain. A long proline-rich loop found between the first and second helices is absent in the proapoptotic members of the family. The loop may be vulnerable to protease digestion, and contains phosphorylation sites. The BH1, 2, and 3 domains of BCL-x_L fold together to give a hydrophobic region involved in homo and heterodimerization, and contain domains important for interaction with the BH3-only proteins. The structure of BCL-x_L is strikingly similar to that of the diphtheria toxin membrane translocation domain and the pore-forming domains of bacterial colicins that kill sensitive cells via the formation of a highly conductive ion channel in the target cell's plasma membrane. The overall organization of the colicin-like channels is that of a hydrophobic region containing the pore, shielded by amphipathic helices that keep molecule soluble in the cytoplasm until insertion into a membrane activates the pore function. Two helices of Bax and BCL-x_L are insufficient to form a pore (34) but their ability to homo and heterodimerize may provide for interactions critical for their pore-forming ability.

It is possible that channel activity observed in artificial lipid membranes is not the same as activities in vivo. Both BCL-x_L and Bax have demonstrated avid pore-forming capability in lipid bilayers if lipid composition of the bilayer is carefully regulated (34, 35, 41, 42). Both Bax and BCL-x_L display similar channel activity with multiple conductances, but BCL-x_L has a linear conductance, whereas Bax appears to be more voltage dependent, is more anion selective than BCL-x_L, and has larger peak conductances (35).

The similar channel activities in these anti and proapoptotic proteins raised several questions, the most important of which was whether the channel activity of the anti and proapoptotic molecules was important at all for their anti and proapoptotic functions, and if so, why were the channel activities so similar and could they prove more dissimilar in vivo?

7.8 Recordings of BCL-2 Family Proteins In Vivo

To examine these issues, the giant synapse of the squid stellate ganglion was used as a model system to study the activity of recombinant BCL-2 proteins in mitochondrial membranes inside a living neuron (22, 43, 44). Employing a concentric electrode arrangement, a clean patch pipette tip of small internal diameter is exposed to mitochondria inside the neuronal presynaptic ending, and channel activity is recorded from a patch of outer membrane on a mitochondrion exposed to the inside of the pipette. When such patches are exposed to recombinant antiapoptotic full length BCL-x_L protein (FL BCL-x_L) multiple conductance activity is observed in mitochondrial patches within 5 min of the start of the recording (Fig. 7.4). Unitary openings correspond to conductances between 100 and 750 pS and the current voltage relationship is slightly outwardly rectifying (44).

Channel activity of recombinant Bax recorded in presynaptic mitochondrial membranes shares some but not all features with that of recombinant full length BCL-x_L. Bax, like BCL-x_L, has been shown to form ion channels in artificial lipid membranes (35), but in order to do so readily in vitro, it must be activated by treatment with detergent, which causes the protein to oligomerize (27, 42, 45).

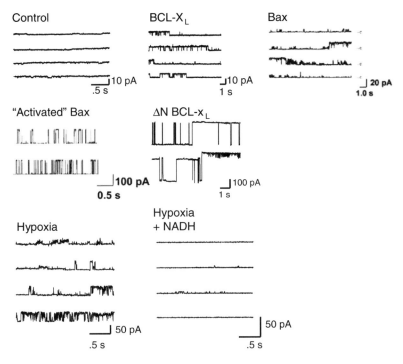

Fig. 7.4 Channel recordings performed on mitochondrial membranes within a living neuronal synapse. Top shows a control recording and recordings of patches exposed to recombinant FL BCL-xL or Bax. The middle panel shows recordings of patches exposed to Bax or ΔN BCL-xL, demonstrating the large conductance activity of the membrane in the presence of these proteins. On the bottom is shown a mitochondrial recording within a hypoxic synapse and a recording of a patch exposed to a VDAC inhibitor (NADH) within a hypoxic synapse

In isolated mitochondria or isolated outer mitochondrial membranes, however, this activation of Bax is apparently mediated by endogenous tBID (N-terminally cleaved/truncated BID) (46) or another BH3-only protein (47). The findings suggest that activation of Bax channel activity requires a mitochondrial membrane component (46).

When purified recombinant Bax protein lacking the C terminus (BaxΔC) or full-length Bax is placed inside the patch pipette used to record from mitochondria in the squid presynaptic terminal, discrete ion channel conductances recorded in the patches range between 100 and 750 pS and demonstrate outward rectification (43).

In contrast to BCL-xL, however, in 5–10% of the channels observed on mitochondria inside the neurons with recombinant Bax, a number of large openings with conductances >750 pS are detected, similar to the large conductances reported for purified Bax in artificial lipid membranes (28). In contrast to the more intermediate conductance openings, the current–voltage relations for the large openings are linear. Thus, it appears that including Bax in the patch pipette induces channel activity with two distinct properties, a large conductance state and a less conductive state that shares some properties with that induced by full-length BCL-x_L.

The findings suggest that the smaller conductance channel activity of Bax could represent the activity of Bax in a form prior to its exposure to a death stimulus or to activator BH3 peptides. However, this *inactive* form of Bax could have additional functions besides waiting to bring on cell death in response to a stimulus. Bax may alter synaptic function in the healthy cell. Bax and Bak can act as prosurvival factors in neurons (48–50). It is possible that their protective effects in such models may result indirectly from their actions on synaptic activity (43, 50).

If the smaller conductance activity represents the activity of Bax in healthy cells, then the infrequently detected large conductance activity could be important for its death-promoting actions. Indeed, proteolytic cleavage of Bax accelerates the onset of its pro-death function (51), and it is conceivable that the spontaneous large activity could be a marker for neurons in which apoptotic signals have been activated by previous damage. As we will see later, BCL-xL also has an alternative (larger) channel conductance that can be activated after a death stimulus, in particular after hypoxia in squid synapse, or ischemic injury in brain.

7.9 Endogenous Death Channels Produced by BAX-Containing Protein Complexes

The first patch clamp recordings of endogenous death channel activity were performed on mitochondria isolated from cultured cells undergoing apoptosis (52). Pavlov et al. were able to detect an ion channel (mitochondrial apoptosis-induced channel, MAC) whose pore diameter was estimated to be of sufficient size (~4 nm) to allow the passage of cytochrome *c* and larger proteins. This channel displays multiple conductances, the largest of which is about 2.5 nS. The channel activity is expressed in mitochondrial outer membranes, inhibited in cells overexpressing BCL-2, and is similar to the activity of pure Bax expressed in artificial lipid membranes. The timing of cytochrome *c* release in apoptotic cells correlates well with the onset of MAC activity and with the translocation of Bax to mitochondrial membranes, further suggesting that channel complexes include Bax protein. Moreover, MAC activity can be immunodepleted from mitochondrial membranes treated with anti-Bax antibodies.

7.10 Interaction of VDAC with BCL-2 Family Proteins

Because of the importance of VDAC in mitochondrial function, a role for VDAC in cell death has been postulated. A description of the studies in which VDAC has been found to interact with BCL-2 family proteins follows.

7.11 VDAC and Apoptosis

The possibility that VDAC is a component of a cell death channel that can, in the absence of other outer membrane proteins, release cytochrome *c* during programmed cell death remains controversial. Although programmed cell death can occur in the absence of VDAC (53) and the functional unit of VDAC is most likely a monomer (54) whose pore size is too small to release proapoptotic factors such as cytochrome *c*, several reports support the tendency of VDAC to self-assemble into dimers, trimers, and tetramers (55, 56), whose pore size could be larger. It has been suggested that VDAC oligomerization could be dependent on previous release of cytochrome *c* (55). VDAC normally only passes certain negatively charged metabolites, most not much larger than ATP. Large, positively charged cytochrome *c* is unlikely to pass through VDAC in its monomeric form since permeation by cations is reduced in the open conformation.

VDAC may also interact with molecular regulators of apoptosis and metabolism such as hexokinase, creatine kinase, and members of the BCL-2 family. The precise mechanism by which association with these molecules regulates the permeability of VDAC to control cell death is unknown (57, 58).

7.12 Interaction of VDAC with BCL-2 Family Members

Although the diameter of VDAC at its largest is not compatible with the release of cytochrome *c*, it is possible that VDAC-interacting proteins could regulate its pore diameter or form multimeric structures with VDAC with a pore large enough to release proapoptotic factors from the mitochondria. One model proposes that Bax interacts with VDAC to provide a large pore through which cytochrome *c* could permeate and that the BH4 domain of antiapoptotic proteins closes VDAC and inhibits cell death (59, 60). Another model suggests that BCL-2 family proteins regulate mitochondrial membrane permeability by closing VDAC during apoptosis. This closure prevents metabolite exchange across the outer membrane resulting in inner membrane swelling and eventual rupture of the outer membrane, thereby releasing cytochrome *c* (Fig. 7.5) (61). This process is prevented by the antiapoptotic protein BCL-xL, which, by interacting with VDAC after a cell death stimulus, maintains metabolic exchange across the outer mitochondrial membrane and prevents cell death (62). Proapoptotic tBID has also been proposed as a VDAC-interacting partner. In a study of VDAC reconstituted into artificial lipid membranes, the activated proapoptotic protein BID induced VDAC channel closure (63).

7.13 Interactions of VDAC with BCL-xL

A recent structural study of BCL-xL and VDAC illuminates potential sites of interaction between the two molecules in an artificial lipid environment. The structural spectrum of recombinantly expressed VDAC (56) reveals a folded pro-

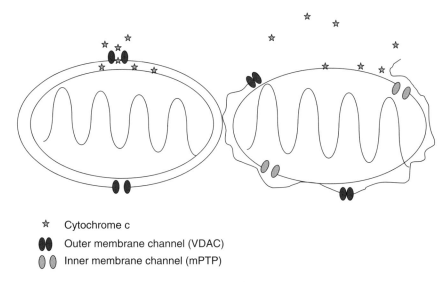

* Cytochrome c
● ● Outer membrane channel (VDAC)
◐ ◐ Inner membrane channel (mPTP)

Fig. 7.5 Two different scenarios explain the role of VDAC in cytochrome *c* release and cell death. On the left, in response to a death stimulus, VDAC opens to a very large conductance, possibly regulated by BCL-2 proteins, and results in the release of cytochrome *c* through the outer membrane, leaving the inner membrane intact. On the right, VDAC closes in response to a death stimulus, resulting in a decrease in metabolite exchange, subsequent swelling of the mitochondrial matrix, rupture of the outer membrane, and release of cytochrome *c*

tein of high beta-sheet content, as described previously. Titration of VDAC with a large excess of metabolites results in chemical shifts indicative of low-affinity binding to VDAC (especially NADH, NAD, and ATP), a finding predicted by the requirement of low-affinity binding for efficient transport of these metabolites. BCL-xL also causes a marked shift of the chemical spectrum of VDAC, indicating significant interaction between VDAC and BCL-xL. In complementary experiments, VDAC in turn causes a significant shift in the chemical spectrum of BCL-xL. VDAC appears to bind to BCL-xL in two areas. The first is the helical hairpin region of BCL-xL near the C-terminus that is the putative site of interaction of BCL-xL with membranes. The other is a region near the N-terminus that may form part of the BH-4 domain, consistent with previous studies suggesting that the BH-4 domain of BCL-xL is involved in VDAC channel modulation (60). The model based on these structural findings predicts that the two membrane proteins are oriented parallel to each other in the membrane and bind to each other along the entire length of the helical hairpin of BCL-xL and one face of the beta barrel of VDAC.

Other structural studies suggest that a complex may form between BCL-xL and VDAC. VDAC usually exists as a trimer in the specific detergents used in the studies, but in the presence of BCL-xL, one VDAC of the trimer may be displaced to form a new trimeric structure with BCL-xL as the third member. The data suggest that BCL-xL may change the structural properties of VDAC and that it could enhance metabolite or cytochrome c conductance in different contexts.

7.14 BCL-xL Interaction with VDAC in Mitochondria of Live Neurons

To prove that an interaction between BCL-xL and VDAC forms in vivo, channel activity of mitochondria was recorded in vivo in the absence and presence of recombinant BCL-xL protein (22). BCL-xL is known to exist in two forms, a full length antiapoptotic form, and an N-truncated version (ΔN BCL-xL) that appears after death stimuli and acts as a cell killer protein (21). In mitochondria of synapses exposed to the death stimulus hypoxia, or mitochondria isolated from ischemic brain, large channel activity appears that has biophysical features of the activity produced by application of ΔN BCL-xL to control mitochondria (21, 22, 26). Both the mitochondrial ion channel produced during hypoxia and the channel activated by the application of ΔN Bcl-xL are strongly inhibited by mM concentrations of NADH (Fig. 7.4). NADH binds with low affinity to VDAC but in large concentrations can block its channel activity (13, 64). Although NADH blocks the activity of ΔN Bcl-xL in mitochondrial membranes, it is ineffective when exposed to ΔN BCL-x_L-induced channel activity in artificial lipid membranes, where VDAC is absent (22, 65, 66). These studies suggest that the large channel activity of mitochondria that have undergone a death stimulus (hypoxia) is produced by a channel made up of a protein complex of VDAC and ΔN Bcl-xL. Whether BCL-xL and VDAC actually interact biochemically or merely biophysically is still in question.

7.15 VDAC2 Inhibits Apoptosis

In mammals, three different *VDAC* genes encoding distinct isoforms have been reported (2). More complex multicellular organisms have all three isoforms, suggesting that the different isoforms have specialized functions. Yeast has two isoforms. Yeast VDAC1 protein forms pores when reconstituted into artificial lipid membranes, but yeast VDAC2 does not. VDAC 1 and 3 are the predominant pore-forming isomers of VDAC in mammalian cells, but a role for VDAC 2 has recently been suggested (33). Cells deficient in VDAC2, but not in VDAC1, are more susceptible to BAK oligomerization and apoptotic cell death, and these events are prevented by overexpression of VDAC2. In the model to explain these findings, the authors suggested that the proapoptotic molecules tBID, BIM, or BAD could displace VDAC2 from BAK enabling homo-oligomerization of BAK and release of cytochrome c through the mitochondrial outer membrane.

7.16 Mitochondrial Inner Membrane Channels

Many studies of inner membrane physiology over the last half century were concerned with the management of calcium fluxes, calcium buffering by the matrix, and catastrophic calcium-induced depolarization of mitochondria (67–69). The next section will attempt to outline the inner membrane ion channel conductances as they relate to the mechanisms of calcium movement between the mitochondrion and the cytosol and the formation of ATP. Ion channels that regulate ion and metabolite fluxes across the inner membrane may form complexes with channels in the outer membrane such that the complex of channels may regulate the flow of ions and metabolites directly across the two membranes between the matrix and the cytosol. Regulation of complex formation could thereby alter metabolism and neuronal function.

7.17 Energy dependence of Mitochondrial Calcium Accumulation

Calcium normally cycles constantly between the mitochondrial matrix and the cytosol. Calcium enters the matrix over its electrochemical gradient via the uniporter, a calcium-selective channel of the inner membrane (70, 71) and can be exchanged with sodium (sodium in, calcium out, Na^+/Ca^{2+} exchanger) linked to a sodium/hydrogen (Na^+/H^+, hydrogen in, sodium out) exchanger and H^+ efflux via respiratory complexes (68). These exchange pathways are used when cytosolic calcium levels are relatively low, and when no net accumulation of calcium occurs in the mitochondrial matrix (Fig. 7.6).

When calcium rises more precipitously in the cytosol, however, mitochondria can accumulate calcium and therefore may serve as major buffers for cytoplasmic calcium. In this setting, H^+ ions can be pumped out in exchange for calcium so that net accumulation of calcium will occur together with net efflux of H^+ ions (68). The net accumulation of calcium leads to a net loss of H^+ ions from the matrix, which results in matrix alkalinization, and increases the difference in pH across the inner membrane. Eventually, no more H^+ can be lost from the matrix without compromising the maximum value of the membrane potential. When this happens, the membrane begins to depolarize toward 0 mV.

More extensive calcium uptake can occur in the presence of anions (such as Pi, acetate, HCO_3^-), which permeate upon cotransport with H^+ ions (Fig 7.6). The cotransport of H^+ and anions is stimulated by the previous alkalinization of the matrix. The resultant reestablishment of the normal concentration of H^+ ions in the matrix then allows the respiratory chain to reestablish the membrane potential by pumping out H^+ ions again, so that there may be only a transient membrane depolarization during calcium uptake. If phosphate is the permeating anion, then a calcium phosphate precipitate forms in the matrix, so that free calcium remains low. Under these conditions there is theoretically possible a massive accumulation

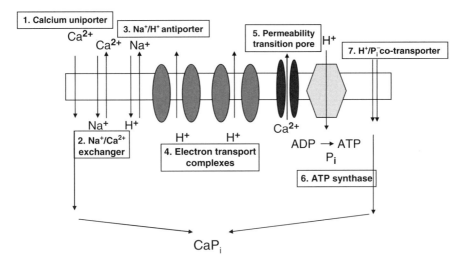

Fig. 7.6 Management of calcium homeostasis in cells by mitochondrial calcium buffering. When cytosolic calcium levels are low, calcium cycles into and out of the matrix using channels and transporters (1–4). When cytosolic calcium levels rise, mitochondria can buffer calcium by the net exchange of calcium for H⁺ ions, compromising the synthesis of ATP (6). The buffering of calcium is aided by formation of a calcium phosphate complex in the matrix (7). Calcium can be rereleased through the transporters (2–4) or through a calcium activated channel that may be related to the mPTP (5)

of precipitated calcium but very little accumulation of free matrix calcium. In the presence of phosphate, unlike with other anions such as acetate that do not form a precipitate, there is no change in osmolarity of the matrix. One H⁺ accumulates for every calcium ion taken up, as the precipitate forms and H⁺ is freed up from hydrogen phosphate. The increase in free H⁺ in the matrix stimulates the respiratory complexes to pump H⁺ back out by accelerating the respiratory chain to maintain the membrane potential, just as if ATP were being made. H⁺ ions can also be pumped out by reversal of the synthase, requiring ATP hydrolysis. All of the earlier described processes take energy, as H⁺ is pumped out but calcium flows in *at the expense of ATP production*. If no phosphate is available, then this reaction is prevented; further alkalinization of the matrix occurs with gradual depletion of the H⁺ pool, finally resulting in depolarization of the membrane potential, preventing further calcium accumulation.

7.18 Voltage-Dependent Inner Membrane Channels: The Calcium Uniporter

Calcium enters the matrix via the uniporter, which appears to be a calcium-selective, voltage-dependent channel (70). By recording in the whole mitoplast configuration, a calcium-selective current can be recorded that is increased by varying cytoplasmic calcium concentrations. The membrane is highly conductive to calcium under these circumstances, suggesting that the permeability pathway constitutes a channel rather than a transporter. Although the molecular identity of the calcium uniporter is still not known, one candidate for calcium uptake in heart is the ryanodine receptor, which is activated by calcium and inhibited by Mg^{2+} and ruthenium red (72). The ryanodine recptor may play an important role in the regulation of calcium uptake during cardiac ischemia.

7.19 Other Inner Membrane Conductances: Mitoplast Recording Technique

Patch clamping of isolated mitochondrial inner membranes in the *mitoplast* configuration has yielded data on several different conductances, but the most well-studied is the large conductance voltage-dependent channel of the inner membrane, variously named in the literature, but most likely the underlying conductance of the mitochondrial permeability transition (73, 74). The channel or pore is termed the mitochondrial permeability transition pore (mPTP).

The first mitoplast recordings were obtained in the 1980s from isolated inner membranes from cuprizone-fed animals. Cuprizone feeding yields giant mitoplasts (5 or more μM in diameter), much larger that those derived from control animals (1–3 μM in diameter), creating an organelle closer to the size of a cell. This type of mitoplast is easier to patch clamp. The first recordings (75, 76) revealed an anion-selective 108-pS channel.

7.20 Channel Activity Correlated with Permeability Transition: The Mitochondrial Permeability Pore (mPTP)

During ischemic cell death, calcium rises in the cytosol, and as we have seen earlier, it is buffered by mitochondria. Mitochondria seem to be able to take up calcium at a certain rate, but if calcium rises too rapidly, the buffering capacity seems to get overwhelmed and mitochondria depolarize suddenly. The cause of the sudden depolarization is most likely the opening of a channel, the mitochondrial permeability transition pore, which plays an important role in cell death, and whose features will now be described.

The permeability transition is characterized by a sudden loss of mitochondrial membrane potential induced by an increase in permeability of the inner membrane to a large number of unrelated solutes (67). Physiologically relevant molecules and those for which there is no known influx pathway enter by diffusion down their concentration gradients, because all energy-dependent processes of the inner membrane are stopped by the permeabilization of the membrane and subsequent dissipation of the potential gradient. Transition is regulated by calcium on the matrix side, and therefore it serves as a calcium efflux pathway from the matrix to the cytosol. Swelling of the matrix occurs after permeability transition because of the rapid influx of ions and small molecules followed by water. Swelling can be measured optically by the decrease in light scattering by the dissolved solutes. Therefore, permeability transition measured optically can be correlated with channel activity as measured by patch clamp recording of mitoplasts.

MCC (Kinnally and Tedeschi, 1994) is a multiconductance, voltage-dependent channel recorded in mitoplast preparations. The activity is more prominent at negative potentials with closings to lower conductances at positive potentials. Channels are inhibited by amiodarone or cyclosporine A. The MCC has biophysical characteristics in common with the permeability transition pore (mPTP) (67) and the mitochondrial megachannel (MMC) (77–79). It is activated by calcium, inhibited by Mg^{2+}, and is voltage dependent (79–82). The similarities between the conductances recorded by different groups also include inhibition by ADP and by alterations in pH.

mPTP is further characterized by its conductance of 1.3 nS in mitoplasts bathed in control solution containing 150 mM KCl (83). The channel is voltage dependent. When the matrix is made positive, the channel is more likely to enter subconductance states. Unlike VDAC, at all negative potentials (of the matrix), the channel remains at a high conductance. The 1.3-nS channel is activated by calcium on the matrix side, and inhibited by competition of calcium with Mg^{2+}, Ba^{2+}, Sr^{2+}, and Mn^{2+} (84). The calcium-activated 1.3-nS conductance is also inhibited by CSA, a known inhibitor of the prolyl isomerase cyclophilin, on the matrix side (77). CSA is also known to inhibit permeability transition measured optically.

The finding that channel activity is inhibited by the binding of CSA to cyclophilin establishes that cyclophilin is in a complex with the channel pore protein, whose molecular nature is still not completely known. Studies of cyclophilin D knockout mice by two independent groups (85, 86) suggest that cyclophilin D, and by inference the mPTP, is involved in ischemic necrosis, because mitochondria isolated from the knockout mice are resistant to calcium-induced and CSA-inhibited permeability transition. Correlated with the resistance to permeability transition is the resistance to necrotic death induced by ROS and calcium overload and to cardiac ischemia/reperfusion injury (86). Tissues from animals overexpressing cyclophilin D have abnormal swollen mitochondria and a propensity toward spontaneous cell death (85). Despite these findings, developmental apoptotic death appears normal, raising suspicions that developmental apoptosis does not require cyclophilin-regulated inner membrane channel activity.

The prevalent hypothesis is that the mPTP contains three components, the ANT, which normally transports ATP out of the inner membrane, VDAC, and cyclophilin D (87). The role of the ANT remains unproven, because in hepatocyte mitochondria isolated from knockout animals lacking both liver-expressed isoforms of ANT, calcium-induced permeability transition still occurs, and is nonresponsive to inhibitors of ANT, such as bongkrekic acid (88).

7.21 Complex of Channels Exists at Contact Points Between Outer and Inner Membranes

It is tempting to speculate that VDAC forms a molecular complex with an inner membrane channel for the purpose of metabolite and ion exchange directly between the matrix and the cytosol. Coimmunoprecipitation studies support this hypothesis suggesting that VDAC, ANT, and cyclophilin D are biochemically linked (87). VDAC clusters at contact sites where mPTP-like activity is present by patch recording (17, 89, 90).

There are estimated to be about 37 contact sites per square micron of mitochondrial membrane, making it likely that there are seven contact sites in a recording of a membrane patch of 0.5 μm. Because of the importance of communication between the matrix and the cytosol, it is possible that a functional mitochondrial channel could require the presence of both inner and outer membrane components at contact sites in series (Fig. 7.7) (91). This would provide for streamlined flow of ions and metabolites from the matrix bypassing the intermembrane space to the cytosol. If this were to be the only type of conduit, however, there would be no way for intermembrane space components to be taken up or released through a channel. For example, cytochrome c, which resides in the intermembrane space, could not get released in the absence of outer membrane rupture (27, 92).

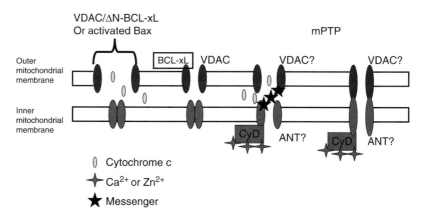

Fig. 7.7 Interaction of inner and outer mitochondrial membrane ion channels in different scenarios (from left to right): (1) the release of cytochrome c from the intermembrane space during cell death through an outer membrane channel formed by Bax or a VDAC/ΔN BCL-xL complex (see text for details). (2) During physiological functioning of mitochondria such as during cell growth, or during inhibition of cell death, VDAC may open to release metabolites, assisted by interaction with BCL-xL. (3) Formation of a two membrane-spanning complex of channel proteins. If the inner membrane channel is activated by the binding of cyclophilin D in the presence of calcium, an outer membrane channel such as VDAC may open in response to activation of a messenger within the intermembrane space, or (4) by formation of a channel complex spanning the two membranes

Opposing the idea that channels of the two membranes exist in a fixed complex is the finding that recordings of MCC (mPTP) in yeast mitoplasts (with contact points) derived from cells lacking VDAC have channel activity that is similar to that of MCC recorded in mitoplasts from wild-type cells (93), and that normal developmental apoptosis (with cytochrome c release) occurs in cyclophilin D knockout animals presumably in the absence of high-conductance inner membrane channel activity (85, 86).

Contact point formation may be regulated by cell metabolic needs. Contact between the membranes may occur depending on the need to release ATP, to permit the entry of ADP into the matrix, or to support the entry of calcium. Contact sites may form when oxidative phosphorylation is taking place (94). In keeping with this idea, ADP, atractyloside, and succinate, a substrate for mitochondrial respiration, appear to induce contacts, whereas glycerol and uncouplers such as DNP or the electron transport inhibitor antimycin A decrease their formation (95, 96). Bonanni et al. have observed a decrease in electron density of contact sites after ischemia, suggesting that channel regulatory proteins may have dispersed from the complex perhaps as a result of damage to the electron transport chain during the ischemic event (26). One possible explanation for the multiconductance state of many mitochondrial channels could be the ability to increase conductance depending on an increasing interaction between the two membranes (Kinnally and Tedeschi, 1994).

7.22 Fundamental Events During Ischemia

We have seen that mitochondrial ion channel activity of VDAC, the calcium uniporter, and mPTP may regulate the functional state of neurons by changing levels of ATP and intracellular calcium. Members of the BCL-2 family also function not only as mitochondrial ion channels, but act within a complex signaling network that regulates cell fate.

Brain ischemia is a pathological condition that serves to exemplify the complexity that underlies decisions of cell life and death in the adult nervous system. Brain ischemia arises as a result of low oxygen tension in the tissues after obstruction of an artery that supplies part of the brain with blood (focal ischemia) or a cardiopulmonary event that significantly lowers systemic blood pressure for an extended period of time (global ischemia) (Zukin et al., 2004) (167). Models of these human disease states have been developed in rodents. Focal ischemia in rodents most closely resembles human stroke where complete blockage of an artery causes rapid necrosis of tissue dependent on that artery for blood supply. Cells in the penumbra, or area that receives only partial supply from the damaged artery, are at risk for delayed cell death that has many of the biochemical and some of the morphological features of programmed cell death. Transient forebrain or global ischemia, in contrast, induces delayed, cell-specific death only of vulnerable hippocampal CA1 pyramidal neurons, whereas the normal function of other brain areas is restored after the ischemic period [Zukin et al., 2004; (25, 97, 98)]. In rodents, this latter experimental paradigm serves as a good model for the study of programmed cell death in the vulnerable neurons.

The events during and after global ischemia can be described as follows: During the ischemic episode, mitochondria in the neurons serve as *oxygen-sensing organelles* and rapidly decrease ATP production (99, 100). Cells throughout the brain exhibit energy depletion and altered energy dynamics (101–104). ATP depletion induces neuronal membrane depolarization and promotes release of synaptic glutamate, a rise in cytosolic Ca^{2+}, reverse operation of glutamate transporters, and swelling of cells [(105–107); Zukin et al., 2004]. After reperfusion, ATP levels and membrane polarization are restored (Zukin et al., 2004). Although histologically detectable neuronal death does not occur until 48 h, apoptotic cascades are activated within 3 h (108, 109). An early event is disruption of the integrity of the mitochondrial outer membrane enabling release of cytochrome *c* from the intermembrane space (99, 110).

7.23 Cellular Events During Ischemia-Excitotoxicity

Several events occur during ischemia and afterward in the phase of reperfusion that greatly influence the behavior of mitochondrial ion channels, and it is clear that regulation of these channels helps determine the fate of a neuron undergoing ischemic injury (74). The first important phenomenon is that cytosolic ATP levels decline. This occurs as the high-energy bonds of ATP and ADP are used to provide energy for cellular kinases and pumps, in particular the Na^+/K^+ ATPase and Ca^{++} ATPases at the plasma membrane and endoplasmic reticulum. The direct effect on mitochondria of low ATP and high phosphate in the setting of a lack of oxygen is complex, but Pi is known to activate the mPTP when free calcium levels rise rapidly in the matrix. Although adenine nucleotides prevent pore opening, continued hypoxia leads eventually to complete loss of adenine nucleotides as they are degraded to the nucleosides and bases (111). Changes in the ATP/ADP ratio also affect the behavior of other mitochondrial channels, in particular the mitoKATPase, as will be described later.

The other important event in ischemia is the rise in cytosolic calcium. Calcium rises as a result of several events. Depletion of intracellular ATP slows the activity of intracellular membrane and plasma membrane calcium ATPases, which are used to extrude calcium from the cell. In addition, depolarization of the plasma membrane occurs because of a lack of ATP for use by the Na^+/K^+ exchanger. This activates voltage-gated calcium channels and the release of glutamate, which further acts on calcium-permeable glutamate receptors postsynaptically to allow the influx of calcium (and zinc) into the cells. Additional glutamate release occurs after reversal of the glutamate uptake transporters upon energy failure in both neurons and glia. High sodium influx through NMDA receptors may prevent neuronal reuptake of glutamate even after reperfusion (112). Intracellular acidification caused by lactic acidosis also contributes to excitotoxicity by leading to increased intracellular

Na$^+$ through Na$^+$/H$^+$ exchange, with resultant impairment of Na$^+$/Ca^{++} exchange. The latter is particularly prominent upon reperfusion, when extracellular acidity is rectified, leading to a high pH gradient across the plasma membrane and cell swelling.

A key element in the activation of calcium influx is uptake of calcium by mitochondria. As described, mitochondria take up calcium via the activity of the uniporter across the inner membrane. The activity of the uniporter is regulated by cytosolic calcium levels such that increasing cytosolic calcium increases its activity. The influx of calcium into the mitochondrial matrix is balanced by the activity of calcium efflux pathways, dependent on the inner membrane Na$^+$/Ca^{++} and Na$^+$/H$^+$ exchangers. As we have seen, in the presence of inorganic phosphate, mitochondria load with large amounts of calcium without altering levels of matrix-free calcium, since calcium and phosphate precipitate. The calcium buffering capacities of mitochondria enable mitochondria to buffer cytosolic calcium whenever the cytoplasmic calcium level rises above the *set point* for the balance of influx and efflux of calcium (113). However, acutely massive elevations in cytosolic calcium, or chronically elevated cytoplasmic calcium above the set point lead to calcium overload in the mitochondria and permeability transition, via the calcium-activated pore (mPTP) in the inner membrane. It is not clear exactly how or if the mPTP always gets activated during ischemia (114), but it is clear that continued calcium accumulation by mitochondria causes a drain on cellular energy, as the mitochondria need to constantly reestablish the proton gradient during long-term calcium uptake. It is possible that the energy requirements in the presence of plasma membrane depolarization after glutamate exposure plus the requirements to reestablish the mitochondrial membrane potential require more energy than the mitochondria can produce during ischemia, and therefore the membrane potential of the mitochondria depolarizes, leading to further activation of the voltage-dependent mPTP.

In addition to calcium and voltage, the mPTP is also inhibited by ATP and ADP, which decline during ischemia. The central importance of mitochondrial calcium uptake in the promotion of cell death from excitotoxicity was demonstrated by showing that prior depolarization of mitochondria slowed calcium uptake by decreasing the electrochemical gradient (115, 116) and protected cells from death. In addition, loss of contractile function in heart on reperfusion of ischemic tissue is related to increased mitochondrial, not cytoplasmic calcium levels (117).

7.24 Role of Oxygen-Free Radicals in Ischemic Neuronal Damage

Free radicals are harmful reactive oxygen species that form when the high-energy-reducing equivalents of the electron transport chain meet cellular oxygen instead of their intended targets in the chain. This occurs normally during running of the electron transport chain for metabolism, but may increase during times when the metabolic rate increases (118). Pathologically, ROS formation occurs during times of cell stress, when the electron transport chain is damaged or inefficient, such as after ischemia, and in genetic diseases where components of the electron transport chain are missing or defective (119, 120).

Elevated oxygen-free radicals have been found to play a role in activating the mPTP. Reperfusion after ischemia leads to overproduction of free radicals, perhaps by a cytosolic reaction catalyzed by xanthine oxidase (121). Elevation of reactive oxygen species may be contributed to by mitochondria after ischemia. Depolarization of the mitochondrial membrane potential during the episode of hypoxia initially may halt free radical formation as mitochondrial metabolism is slowed, but during reperfusion, when the membrane potential is reestablished, components of the electron transport chain that may have become damaged during the ischemic event contribute to free radical formation. The damaged complexes slow electron transport, allowing the escape of electrons out of the complexes, where they meet oxygen in an uncontrolled environment and form radicals (112). Opening of the mPTP may also contribute to electron transport dysfunction upon reperfusion in ways that are poorly understood (81, 122).

7.25 BCL-2 Family Proteins in Ischemic Neuronal Damage

BCL-2 family members are critical to breakdown of the functional integrity of the mitochondrial outer membrane in response to ischemia and other insults (24, 123). BCL-xL is abundantly expressed in neurons of the adult brain (124–126). BCL-xL localizes to the outer mitochondrial membrane (37), and is thought to protect cells from death by regulating export of ATP from mitochondria and/or by blocking activation of proapoptotic BCL-2 family members (61, 65, 127–129). As described, injurious stimuli such as hypoxia promote the N-terminal proteolytic cleavage of BCL-xL, converting it into a BAX-like killer protein ΔN-BCL-xL (20). Proapoptotic BCL-2 family members such as BAX and ΔN-BCL-xL promote the formation of large conductance channels in the mitochondrial outer membrane (22, 43), cytochrome c release (21, 130), and synaptic failure (44).

To record mitochondrial ion channel activity in living neurons undergoing a death stimulus, patch clamp studies took advantage of findings that ischemic insults in neurons can lead to programmed cell death as described earlier (25), and that changes in synaptic efficacy often precede the onset of cell death in hypoxic neurons (26, 44). The rapid onset of the effects of ischemia on synaptic transmission could contribute to determining whether a cell will survive or die, and therefore, channel activity associated with apoptosis may be evoked in the outer membrane of mitochondria within ischemic presynaptic terminals. The presynaptic terminal of the squid giant stellate ganglion is very sensitive to hypoxia, which attenuates synaptic transmission over 10–30 min (131). Thus, the squid synapse provides a good synaptic model in which to study the effects of hypoxia on mitochondrial ion channels. Indeed, patch clamp recordings of channel activity on mitochondria in hypoxic squid presynaptic terminals reveal large multiconductance channel activity (131).

The channel activity produced by hypoxic conditions closely resembles that produced by applying ΔN BCL-xL protein to control mitochondria during patch clamp recordings. Although the possible physiological role of BCL-2 family proteins in regulation of synaptic transmission had not been previously suspected, evidence indicates that hypoxia-induced neuronal death can be attributed, at least in part, to activation of BCL-2 family proteins (20, 51, 99, 108, 109, 132–138). A pro-death role for BCL-x_L during hypoxia could be due to caspase or calpain-like proteolytic cleavage.

Proof of the hypothesis that the ion channel activity of the ischemic synaptic mitochondria is regulated by BCL-xL comes from studies with the small molecule ABT-737, a mimetic of the BH3-only protein BAD that binds to BCL-xL with high affinity within a pocket of the three-dimensional structure that usually binds BH3-only proteins. In confocal time-lapsed microscopy experiments, the binding of ABT-737 displace can be seen to a GFP-tagged BH3-only protein from BCL-xL at mitochondrial surfaces in intact tumor cells. In cancer cell lines, ABT-737 alone effectively induces cell death possibly via its ability, as a BAD mimetic, to displace from BCL-xL the prebound proapoptotic proteins Bax and Bak (139). When injected prior to a hypoxic insult of the synapse, however, ABT-737 prevents the appearance of the large conductance activity of mitochondrial membranes associated with declining synaptic function. Furthermore, ABT-737 attenuates synaptic dysfunction produced by hypoxia or by exogenous injection of ΔN BCL-xL into the synaptic terminal. These results suggest that BCL-xL, in its proapoptotic truncated form, is contributing to mitochondrial ion channel activity in hypoxic neurons.

7.26 Large Channels of Mitochondria from Postischemic Hippocampal CA1 Neurons

In mammalian brain after global ischemia, as described earlier, neurons from the CA1 region of the hippocampus are vulnerable to a type of delayed death with characteristics of programmed cell death, in particular the release of proapoptotic molecules such as cytochrome *c* from mitochondria into the cytosol, the activation of downstream caspases, and the eventual enzymatic destruction of cellular components including DNA (140). Just like in hypoxic squid neurons, the mitochondria isolated from brains of ischemic rats at 50 min after reperfusion demonstrate large channels with multiple levels of conductance in mitochondrial outer membranes, openings at all voltages between −100 and +100 mV, and a reversal potential at about 0 mV. In keeping with the early activation of proteases in ischemic models, global ischemia produces pronounced protease activity in synaptosomes, as assessed by zVAD-FMK fluorescence at 50 min after insult. Control mitochondria from hippocampus exhibit high levels of BCL-xL, with little or no evidence of ΔN-BCL-xL. In contrast, global ischemia induces the appearance of ΔN-BCL-xL by Western blotting, evident at 50 min after insult, suggesting that early caspase or calpain cleavage of BCL-xL to the proapoptotic

ΔN-BCL-xL may cause early changes in mitochondrial conductances after global ischemia. Furthermore, the ischemic channel activity is mimicked by application of recombinant ΔN BCL-xL protein to control mitochondria, and inhibited by a specific antibody against BCL-xL (26).

7.27 Large Channel Activity Associated with VDAC

Because the recordings are performed directly on predominantly intact mitochondria, it is likely that the membrane contacted by the patch pipette is the outer mitochondrial membrane. The conductance of this membrane is known to be reduced by millimolar concentrations of NADH (13). In lipid bilayers, NADH reduces the conductance of VDAC (1) and in squid, NADH reduces the conductance of both ΔN BCL-xL and the hypoxia channel (22, 131). In keeping with these findings, recordings of channel activity in postischemic mammalian brain mitochondria in the presence of NADH (2 mM) applied via the bath perfusate and the patch pipette contain a significantly lower frequency of large conductance activity than recordings of ischemic mitochondria performed in the absence of NADH, suggesting that the ischemia-induced channel requires VDAC.

7.28 Large Zn2+-Activated Channels in Postischemic Mitochondria

Recordings of outer membranes reconstituted into liposomes or of intact mitochondria within living cells reveal that the permeability of the outer membrane is tightly controlled (4, 52, 141, 142). Thus, outer membrane channels may be modulated at the cytosolic or intermembrane face, possibly by ligands and second messengers, or even by contact with proteins of the inner membrane at contact sites between the two membranes (143, 144). Introduction of Ca^{2+} or other divalent ions to the matrix side of the patch in preparations of isolated inner membranes [*mitoplasts*; (83, 145)] and/or application of Zn^{2+} via the bath perfusion (146) activates a nS channel in the inner membrane. This inner membrane channel is blocked by cyclosporin A (77) and the Zn^{2+} chelator TPEN (146) suggesting that the channel is divalent sensitive, as is the mPTP.

Findings suggest that, in ischemic brain, Zn^{2+} localizes to the inside of ischemic mitochondria. Recordings of mitochondria isolated from control brain reveal Zn^{2+}-sensitive channels. In addition, the large conductance activity present in the outer membranes of postischemic mitochondria is exquisitely sensitive to bath application of the specific Zn^{2+} chelator, TPEN, which virtually eliminates the largest conductance state, supporting the role of Zn^{2+} in activation of large channels after ischemia. The presence of zinc in ischemic mitochondria and not in controls is confirmed by fluorescent imaging of isolated mitochondria with the zinc-specific probe RhodZin-3 (26). The findings suggest that Zn^{2+} activates a divalent-sensitive channel on the inner membrane that would in turn activate the VDAC/ΔN BCL-xL complex on the outer membrane.

7.29 Ischemic Tolerance and Mitochondrial Ion Channel Activity

The pro-apoptotic protein BAD is an important regulatory protein during cell death from ischemia and as such plays a central role in ischemic preconditioning or ischemic tolerance. Ischemic tolerance is a well-established phenomenon in which a brief ischemic insult (or preconditioning) protects CA1 neurons against a subsequent more prolonged ischemic challenge (147). We now know that this phenomenon involves mitochondrial ion channels. The protective action of ischemic tolerance involves inhibition of activated caspase-3 (148). Ischemic preconditioning promotes Akt phosphorylation at Ser473 and phosphorylation/inactivation of downstream targets of Akt. PI3K/Akt signaling, which regulates cell growth and survival (149, 150), is required for preconditioning-induced neuroprotection (151). In vitro studies show that phosphorylated BAD normally resides in a complex with the cytosolic chaperone 14-3-3. During ischemia and after other cell death signals, BAD becomes dephosphorylated, dissociates from 14-3-3, and translocates to the outer membrane of the mitochondria, where it binds to BCL-xL and promotes mitochondrial cytochrome c release (152).

Dephosphorylated BAD is the only form of BAD that can bind to BCL-xL or can be found in the mitochondrial membrane (153), suggesting that the binding of dephosphorylated BAD to BCL-xL at mitochondrial membranes is necessary for apoptosis.

BAD also may play an important role in the normal physiology of glucose metabolism. In addition to hypoxia, dephosphorylation of BAD is observed after glucose deprivation and is followed by cell death. Phosphorylated BAD provides for normal glucose homeostasis and the phosphorylation state of BAD may reflect the metabolic state of the cell (154). The pool of BAD that affects glucose homeostasis may be different from the pool that is necessary for apoptosis, because during normal glucose metabolism, unlike in programmed cell death (153), phosphorylated BAD is found in a protein complex at mitochondrial membranes.

The mechanism of the control of apoptosis by BAD translocation to the mitochondrion is still not completely understood. When BAD binds to BCL-xL, it may release BAX, or release activator proapoptotic molecules such as BID (31, 32). BID activates the oligomerization of BAX, which then forms pores in the mitochondrial outer membrane to release proapoptotic factors from the mitochondria (47, 138, 140, 155–157). Evidence that BCL-xL might also contribute directly to ion channel formation after BAD translocation comes from findings that the large channel formation recorded in postischemic mitochondria is correlated with the coprecipitation of BAD with BCL-xL by immunohistochemistry.

Preconditioning of the animals prior to the ischemic episode prevents BAD interaction with BCL-xL, and prevents large channel formation at mitochondrial membranes suggesting that this is one mechanism leading to the preconditioned state. The absence of channel formation is associated with a lack of cytochrome c release, prevention of caspase activation, and a marked attenuation of cell death. In addition, BAD binding to BCL-xL is correlated with the proteolytic cleavage of BCL-xL, to form ΔN BCL-xL, which could be responsible for the change in outer membrane permeability and cytochrome c release from mitochondria. Alternatively, BAD may sequester full length BCL-xL at mitochondrial membranes, allowing BAX or BAK to form the large channel activity (Miyawaki et al., 2008) (168).

7.30 Mito K ATP

The mito KATP channel has properties similar to those of the plasma membrane KATP channel, but with a lower conductance (158). Application of the agonist diazoxide is correlated with protection against injury from ischemia (ischemic preconditioning) in heart and isolated myocytes (159–161). Increased permeability to K⁺ produces tolerance to ischemic injury in heart and in brain (160). The normally small influx of potassium through the ATP-regulated channel may increase during ischemia, and regulates the matrix volume as well as possibly preventing the catastrophic opening of the mPTP. The mechanism of preconditioning by the opening of mitoKATP may also involve signaling pathways, since modulation of mitoKATP is correlated with activation of PKC and tyrosine kinases, and changes in redox state of the mitochondria. Many K⁺ channel openers activate mitoKATP, but those that are selective for mitoKATP are diazoxide, nicorandil, and BMS191095. The only selective inhibitor is hydroxydecanoate (162).

7.31 Mito KCa

A KCa channel has also been detected by patch clamping-isolated inner membrane (mitoplasts). The channel has a conductance of 300 pS and is inhibited by charybdotoxin or iberiotoxin at nM concentrations (163, 164). A KCa opener increases protection against ischemia in rabbit hearts. The increased KCa conductance seems to guard against excessive calcium accumulation after ischemia (by predepolarizing the inner membrane slightly prior to the cytoplasmic calcium rise) and to fine-tune mitochondrial volume. These events appear to produce ischemic preconditioning. Mito KCa may be regulated by PKA, not by PKC, unlike MitoKATP (165). Antibodies to the BK-type KCa channel cross-react with mito KCa channels, and recent reports have demonstrated immunoreactivity in the mitochondrial fraction and by immuno EM in brain (166).

7.32 Conclusions

Mitochondria appear to play a central role in decision making about cell fate after neuronal injury. During ischemia, oxygen and substrate levels decline, positioning the mitochondrion as the first sensor of injury. Metabolic functions of the cell are clearly dependent on the chemical and biophysical properties of mitochondrial inner and outer membranes, and therefore ion channels of these membranes are carefully regulated during physiological and pathophysiological events in the nervous system. It appears that homeostatic regulation of both plasma membrane potential and mitochondrial membrane potential can be overcome by the intensity and rapidity of onset of some death signals such as ischemia, and when neurons are damaged so that they can no longer regulate intracellular ion homeostasis, particularly that of calcium, zinc, and sodium, mitochondrial inner membrane ion channels sense this, become activated, and signal to outer membrane conductances, including those of the BCL-2 family and the VDAC. More work is necessary to understand the complex interrelationships of physiological neuronal function, metabolism, neuronal death, and mitochondrial ion channels.

References

1. Colombini, M., E. Blachly-Dyson, and M. Forte, *VDAC, a channel in the outer mitochondrial membrane*. Ion Channels, 1996. 4: 169–202.
2. Shoshan-Barmatz, V., et al., *The voltage-dependent anion channel (VDAC): function in intracellular signalling, cell life and cell death*. Current Pharmaceutical Design, 2006. 12(18): 2249–70.
3. Hodge, T. and M. Colombini, *Regulation of metabolite flux through voltage-gating of VDAC channels*. Journal of Membrane Biology, 1997. 157(3): 271–9.
4. Jonas, E.A., J. Buchanan, and L.K. Kaczmarek, *Prolonged activation of mitochondrial conductances during synaptic transmission*. Science, 1999. 286(5443): 1347–50.
5. Blachly-Dyson, E., et al., *Selectivity changes in site-directed mutants of the VDAC ion channel: structural implications*. Science, 1990. 247(4947): 1233–6.
6. Blachly-Dyson, E., et al., *Probing the structure of the mitochondrial channel, VDAC, by site-directed mutagenesis: a progress report*. Journal of Bioenergetics and Biomembranes, 1989. 21(4): 471–83.
7. Colombini, M., *Voltage gating in the mitochondrial channel, VDAC*. Journal of Membrane Biology, 1989. 111(2): 103–11.
8. Mangan, P.S. and M. Colombini, *Ultrasteep voltage dependence in a membrane channel*. Proceedings of the National Academy of Sciences of the United States of America, 1987. 84(14): 4896–900.
9. Holden, M.J. and M. Colombini, *The mitochondrial outer membrane channel, VDAC, is modulated by a soluble protein*. FEBS Letters, 1988. 241(1–2): 105–9.
10. Holden, M.J. and M. Colombini, *The outer mitochondrial membrane channel, VDAC, is modulated by a protein localized in the intermembrane space*. Biochimica et Biophysica Acta, 1993. 1144(3): 396–402.
11. Zimmerberg, J. and V.A. Parsegian, *Polymer inaccessible volume changes during opening and closing of a voltage-dependent ionic channel*. Nature, 1986. 323(6083): 36–9.

12. Zizi, M., et al., *NADH regulates the gating of VDAC, the mitochondrial outer membrane channel*. Journal of Biological Chemistry, 1994. 269(3): 1614–16.
13. Lee, A.C., M. Zizi, and M. Colombini, *Beta-NADH decreases the permeability of the mitochondrial outer membrane to ADP by a factor of 6*. Journal of Biological Chemistry, 1994. 269(49): 30974–80.
14. Lemasters, J.J. and E. Holmuhamedov, *Voltage-dependent anion channel (VDAC) as mitochondrial governator – thinking outside the box*. Biochimica et Biophysica Acta, 2006. 1762(2): 181–90.
15. Golshani-Hebroni, S.G. and S.P. Bessman, *Hexokinase binding to mitochondria: a basis for proliferative energy metabolism*. Journal of Bioenergetics and Biomembranes, 1997. 29(4): 331–8.
16. Hashimoto, M. and J.E. Wilson, *Membrane potential-dependent conformational changes in mitochondrially bound hexokinase of brain*. Archives of Biochemistry and Biophysics, 2000. 384(1): 163–73.
17. Brdiczka, D., et al., *Microcompartmentation at the mitochondrial surface: its function in metabolic regulation*. Advances in Experimental Medicine and Biology, 1986. 194: 55–69.
18. Hackenbrock, C.R., *Energy-linked ultrastructural transformations in isolated liver mitochondria and mitoplasts. Preservation of configurations by freeze-cleaving compared to chemical fixation*. Journal of Cell Biology, 1972. 53(2): 450–65.
19. Penso, J. and R. Beitner, *Clotrimazole and bifonazole detach hexokinase from mitochondria of melanoma cells*. European Journal of Pharmacology, 1998. 342(1): 113–17.
20. Cheng, E.H., et al., *Conversion of Bcl-2 to a Bax-like death effector by caspases*. Science, 1997. 278(5345): 1966–8.
21. Clem, R.J., et al., *Modulation of cell death by Bcl-XL through caspase interaction*. Proceedings of the National Academy of Sciences of the United States of America, 1998. 95(2): 554–9.
22. Jonas, E.A., et al., *Proapoptotic N-truncated BCL-xL protein activates endogenous mitochondrial channels in living synaptic terminals*. Proceedings of the National Academy of Sciences of the United States of America, 2004. 101(37): 13590–5.
23. Adams, J.M. and S. Cory, *The Bcl-2 apoptotic switch in cancer development and therapy*. Oncogene, 2007. 26(9): 1324–37.
24. Kroemer, G. and J.C. Reed, *Mitochondrial control of cell death*. Nature Medicine, 2000. 6(5): 513–19.
25. Banasiak, K.J., Y. Xia, and G.G. Haddad, *Mechanisms underlying hypoxia-induced neuronal apoptosis*. Progress in Neurobiology, 2000. 62(3): 215–49.
26. Bonanni, L., et al., *Zinc-dependent multi-conductance channel activity in mitochondria isolated from ischemic brain*. Journal of Neuroscience, 2006. 26(25): 6851–62.
27. Green, D.R. and G. Kroemer, *The pathophysiology of mitochondrial cell death*. Science, 2004. 305(5684): 626–9.
28. Dejean, L.M., et al., *Oligomeric Bax is a component of the putative cytochrome c release channel MAC, mitochondrial apoptosis-induced channel*. Molecular Biology of the Cell, 2005. 16(5): 2424–32.
29. Martinez-Caballero, S., et al., *The role of the mitochondrial apoptosis induced channel MAC in cytochrome c release*. Journal of Bioenergetics and Biomembranes, 2005. 37(3): 155–64.
30. Youle, R.J. and A. Strasser, *The BCL-2 protein family: opposing activities that mediate cell death*. Nature Reviews Molecular Cell Biology, 2008. 9(1): 47–59.
31. Galonek, H.L. and J.M. Hardwick, *Upgrading the BCL-2 network [comment]*. Nature Cell Biology, 2006. 8(12): 1317–19.
32. Kim, H., et al., *Hierarchical regulation of mitochondrion-dependent apoptosis by BCL-2 subfamilies [see comment]*. Nature Cell Biology, 2006. 8(12): 1348–58.
33. Cheng, E.H., et al., *VDAC2 inhibits BAK activation and mitochondrial apoptosis*. Science, 2003. 301(5632): 513–7.
34. Schendel, S.L., M. Montal, and J.C. Reed, *Bcl-2 family proteins as ion-channels*. Cell Death and Differentiation, 1998. 5(5): 372–80.

35. Schlesinger, P.H., et al., *Comparison of the ion channel characteristics of proapoptotic BAX and antiapoptotic BCL-2*. Proceedings of the National Academy of Sciences of the United States of America, 1997. 94(21): 11357–62.
36. Wolter, K.G., et al., *Movement of Bax from the cytosol to mitochondria during apoptosis*. Journal of Cell Biology, 1997. 139(5): 1281–92.
37. Kaufmann, T., et al., *Characterization of the signal that directs Bcl-x(L), but not Bcl-2, to the mitochondrial outer membrane*. Journal of Cell Biology, 2003. 160(1): 53–64.
38. Sugioka, R., et al., *BH4-domain peptide from Bcl-xL exerts anti-apoptotic activity in vivo*. Oncogene, 2003. 22(52): 8432–40.
39. Tsujimoto, Y. and S. Shimizu, *VDAC regulation by the Bcl-2 family of proteins*. Cell Death and Differentiation, 2000. 7(12): 1174–81.
40. Muchmore, S.W., et al., *X-ray and NMR structure of human Bcl-xL, an inhibitor of programmed cell death*. Nature, 1996. 381(6580): 335–41.
41. Minn, A.J., et al., *Bcl-x(L) forms an ion channel in synthetic lipid membranes*. Nature, 1997. 385(6614): 353–7.
42. Antonsson, B., et al., *Bax oligomerization is required for channel-forming activity in liposomes and to trigger cytochrome c release from mitochondria*. Biochemical Journal, 2000. 345 (Part 2): 271–8.
43. Jonas, E.A., J.M. Hardwick, and L.K. Kaczmarek, *Actions of BAX on mitochondrial channel activity and on synaptic transmission*. Antioxidants and Redox Signaling, 2005. 7(9–10): 1092–100.
44. Jonas, E.A., et al., *Modulation of synaptic transmission by the BCL-2 family protein BCL-xL*. Journal of Neuroscience, 2003. 23(23): 8423–31.
45. Hsu, Y.T. and R.J. Youle, *Bax in murine thymus is a soluble monomeric protein that displays differential detergent-induced conformations*. Journal of Biological Chemistry, 1998. 273(17): 10777–83.
46. Roucou, X., et al., *Bax oligomerization in mitochondrial membranes requires tBid (caspase-8-cleaved Bid) and a mitochondrial protein*. Biochemical Journal, 2002. 368 (Part 3): 915–21.
47. Polster, B.M., K.W. Kinnally, and G. Fiskum, *BH3 death domain peptide induces cell type-selective mitochondrial outer membrane permeability*. Journal of Biological Chemistry, 2001. 276(41): 37887–94.
48. Middleton, G. and A.M. Davies, *Populations of NGF-dependent neurones differ in their requirement for BAX to undergo apoptosis in the absence of NGF/TrkA signalling in vivo*. Development, 2001. 128(23): 4715–28.
49. Lewis, J., et al., *Inhibition of virus-induced neuronal apoptosis by Bax*. Nature Medicine, 1999. 5(7): 832–5.
50. Fannjiang, Y., et al., *BAK alters neuronal excitability and can switch from anti- to pro-death function during postnatal development*. Developmental Cell, 2003. 4(4): 575–85.
51. Wood, D.E. and E.W. Newcomb, *Cleavage of Bax enhances its cell death function*. Experimental Cell Research, 2000. 256(2): 375–82.
52. Pavlov, E.V., et al., *A novel, high conductance channel of mitochondria linked to apoptosis in mammalian cells and Bax expression in yeast*. Journal of Cell Biology, 2001. 155(5): 725–31.
53. Baines, C.P., et al., *Voltage-dependent anion channels are dispensable for mitochondrial-dependent cell death [see comment]*. Nature Cell Biology, 2007. 9(5): 550–5.
54. Rostovtseva, T.K., W. Tan, and M. Colombini, *On the role of VDAC in apoptosis: fact and fiction*. Journal of Bioenergetics and Biomembranes, 2005. 37(3): 129–42.
55. Zalk, R., et al., *Oligomeric states of the voltage-dependent anion channel and cytochrome c release from mitochondria*. Biochemical Journal, 2005. 386 (Part 1): 73–83.
56. Malia, T.J. and G. Wagner, *NMR structural investigation of the mitochondrial outer membrane protein VDAC and its interaction with antiapoptotic Bcl-xL*. Biochemistry, 2007. 46(2): 514–25.
57. Wicker, U., et al., *Effect of macromolecules on the structure of the mitochondrial intermembrane space and the regulation of hexokinase*. Biochimica et Biophysica Acta, 1993. 1142(3): 228–39.

58. Brdiczka, D., P. Kaldis, and T. Wallimann, *In vitro complex formation between the octamer of mitochondrial creatine kinase and porin.* Journal of Biological Chemistry, 1994. 269(44): 27640–4.
59. Shimizu, S., M. Narita, and Y. Tsujimoto, *Bcl-2 family proteins regulate the release of apoptogenic cytochrome c by the mitochondrial channel VDAC [see comment] [erratum appears in Nature 2000 Oct 12;407(6805):767].* Nature, 1999. 399(6735): 483–7.
60. Shimizu, S., Y. Shinohara, and Y. Tsujimoto, *Bax and Bcl-xL independently regulate apoptotic changes of yeast mitochondria that require VDAC but not adenine nucleotide translocator.* Oncogene, 2000. 19(38): 4309–18.
61. Vander Heiden, M.G., et al., *Outer mitochondrial membrane permeability can regulate coupled respiration and cell survival.* Proceedings of the National Academy of Sciences of the United States of America, 2000. 97(9): 4666–71.
62. Gottlieb, E., S.M. Armour, and C.B. Thompson, *Mitochondrial respiratory control is lost during growth factor deprivation.* Proceedings of the National Academy of Sciences of the United States of America, 2002. 99(20): 12801–6.
63. Rostovtseva, T.K., et al., *Bid, but not Bax, regulates VDAC channels.* Journal of Biological Chemistry, 2004. 279(14): 13575–83.
64. Wunder, U.R. and M. Colombini, *Patch clamping VDAC in liposomes containing whole mitochondrial membranes.* Journal of Membrane Biology, 1991. 123(1): 83–91.
65. Basanez, G., et al., *Bax-type apoptotic proteins porate pure lipid bilayers through a mechanism sensitive to intrinsic monolayer curvature.* Journal of Biological Chemistry, 2002. 277(51): 49360–5.
66. Basanez, G., et al., *Pro-apoptotic cleavage products of Bcl-xL form cytochrome c-conducting pores in pure lipid membranes.* Journal of Biological Chemistry, 2001. 276(33): 31083–91.
67. Gunter, T.E. and D.R. Pfeiffer, *Mechanisms by which mitochondria transport calcium.* American Journal of Physiology, 1990. 258(5, Part 1): C755–C786.
68. Nicholls, D. and K. Akerman, *Mitochondrial calcium transport.* Biochimica et Biophysica Acta, 1982. 683(1): 57–88.
69. Bernardi, P., *Mitochondrial transport of cations: channels, exchangers, and permeability transition.* Physiological Reviews, 1999. 79(4): 1127–55.
70. Kirichok, Y., G. Krapivinsky, and D.E. Clapham, *The mitochondrial calcium uniporter is a highly selective ion channel.* Nature, 2004. 427(6972): 360–4.
71. Litsky, M.L. and D.R. Pfeiffer, *Regulation of the mitochondrial Ca^{2+} uniporter by external adenine nucleotides: the uniporter behaves like a gated channel which is regulated by nucleotides and divalent cations.* Biochemistry, 1997. 36(23): 7071–80.
72. Beutner, G., et al., *Identification of a ryanodine receptor in rat heart mitochondria.* Journal of Biological Chemistry, 2001. 276(24): 21482–8.
73. Halestrap, A.P., et al., *Mitochondria and cell death.* Biochemical Society Transactions, 2000. 28(2): 170–7.
74. Crompton, M., *The mitochondrial permeability transition pore and its role in cell death.* Biochemical Journal, 1999. 341 (Part 2): 233–49.
75. Sorgato, M.C., B.U. Keller, and W. Stuhmer, *Patch-clamping of the inner mitochondrial membrane reveals a voltage-dependent ion channel.* Nature, 1987. 330(6147): 498–500.
76. Sorgato, M.C., et al., *Further investigation on the high-conductance ion channel of the inner membrane of mitochondria.* Journal of Bioenergetics and Biomembranes, 1989. 21(4): 485–96.
77. Szabo, I. and M. Zoratti, *The giant channel of the inner mitochondrial membrane is inhibited by cyclosporin A.* Journal of Biological Chemistry, 1991. 266(6): 3376–9.
78. Szabo, I. and M. Zoratti, *The mitochondrial megachannel is the permeability transition pore.* Journal of Bioenergetics and Biomembranes, 1992. 24(1): 111–17.
79. Bernardi, P., et al., *Modulation of the mitochondrial permeability transition pore. Effect of protons and divalent cations.* Journal of Biological Chemistry, 1992. 267(5): 2934–9.
80. Scorrano, L., V. Petronilli, and P. Bernardi, *On the voltage dependence of the mitochondrial permeability transition pore. A critical appraisal.* Journal of Biological Chemistry, 1997. 272(19): 12295–9.

81. Bernardi, P., *Modulation of the mitochondrial cyclosporin A-sensitive permeability transition pore by the proton electrochemical gradient. Evidence that the pore can be opened by membrane depolarization.* Journal of Biological Chemistry, 1992. 267(13): 8834–9.
82. Kinnally, K.W., et al., *Calcium modulation of mitochondrial inner membrane channel activity.* Biochemical and Biophysical Research Communications, 1991. 176(3): 1183–8.
83. Petronilli, V., I. Szabo, and M. Zoratti, *The inner mitochondrial membrane contains ion-conducting channels similar to those found in bacteria.* FEBS Letters, 1989. 259(1): 137–43.
84. Szabo, I., P. Bernardi, and M. Zoratti, *Modulation of the mitochondrial megachannel by divalent cations and protons.* Journal of Biological Chemistry, 1992. 267(5): 2940–6.
85. Baines, C.P., et al., *Loss of cyclophilin D reveals a critical role for mitochondrial permeability transition in cell death [see comment].* Nature, 2005. 434(7033): 658–62.
86. Nakagawa, T., et al., *Cyclophilin D-dependent mitochondrial permeability transition regulates some necrotic but not apoptotic cell death [see comment].* Nature, 2005. 434(7033): 652–8.
87. Crompton, M., S. Virji, and J.M. Ward, *Cyclophilin-D binds strongly to complexes of the voltage-dependent anion channel and the adenine nucleotide translocase to form the permeability transition pore.* European Journal of Biochemistry, 1998. 258(2): 729–35.
88. Kokoszka, J.E., et al., *The ADP/ATP translocator is not essential for the mitochondrial permeability transition pore [see comment].* Nature, 2004. 427(6973): 461–5.
89. Sandri, G., M. Siagri, and E. Panfili, *Influence of Ca^{2+} on the isolation from rat brain mitochondria of a fraction enriched of boundary membrane contact sites.* Cell Calcium, 1988. 9(4): 159–65.
90. Moran, O., et al., *Electrophysiological characterization of contact sites in brain mitochondria [erratum appears in J Biol Chem 1990 Jul 5;265(19):11405].* Journal of Biological Chemistry, 1990. 265(2): 908–13.
91. Halestrap, A., *Biochemistry: a pore way to die [comment].* Nature, 2005. 434(7033): 578–9.
92. Brustovetsky, N., et al., *Calcium-induced cytochrome c release from CNS mitochondria is associated with the permeability transition and rupture of the outer membrane.* Journal of Neurochemistry, 2002. 80(2): 207–18.
93. Lohret, T.A. and K.W. Kinnally, *Multiple conductance channel activity of wild-type and voltage-dependent anion-selective channel (VDAC)-less yeast mitochondria.* Biophysical Journal, 1995. 68(6): 2299–309.
94. Knoll, G. and D. Brdiczka, *Changes in freeze-fractured mitochondrial membranes correlated to their energetic state. Dynamic interactions of the boundary membranes.* Biochimica et Biophysica Acta, 1983. 733(1): 102–10.
95. Bucheler, K., V. Adams, and D. Brdiczka, *Localization of the ATP/ADP translocator in the inner membrane and regulation of contact sites between mitochondrial envelope membranes by ADP. A study on freeze-fractured isolated liver mitochondria.* Biochimica et Biophysica Acta, 1991. 1056(3): 233–42.
96. Hackenbrock, C.R., *States of activity and structure in mitochondrial membranes.* Annals of the New York Academy of Sciences, 1972. 195: 492–505.
97. Lee, J.M., G.J. Zipfel, and D.W. Choi, *The changing landscape of ischaemic brain injury mechanisms.* Nature, 1999. 399(6738 Suppl): A7–A14.
98. Lipton, P., *Ischemic cell death in brain neurons.* Physiological Reviews, 1999. 79(4): 1431–568.
99. Sugawara, T., et al., *Mitochondrial release of cytochrome c corresponds to the selective vulnerability of hippocampal CA1 neurons in rats after transient global cerebral ischemia.* Journal of Neuroscience, 1999. 19(22): RC39.
100. Wang, X.Q., et al., *Apoptotic insults impair Na^+, K^+-ATPase activity as a mechanism of neuronal death mediated by concurrent ATP deficiency and oxidant stress.* Journal of Cell Science, 2003. 116 (Part 10): 2099–110.
101. Tian, G.F. and A.J. Baker, *Glycolysis prevents anoxia-induced synaptic transmission damage in rat hippocampal slices.* Journal of Neurophysiology, 2000. 83(4): 1830–9.

102. Howard, E.M., et al., *Electrophysiological changes of CA3 neurons and dentate granule cells following transient forebrain ischemia.* Brain Research, 1998. 798(1–2): 109–18.
103. Fleidervish, I.A., et al., *Enhanced spontaneous transmitter release is the earliest consequence of neocortical hypoxia that can explain the disruption of normal circuit function.* Journal of Neuroscience, 2001. 21(13): 4600–8.
104. Bolay, H., et al., *Persistent defect in transmitter release and synapsin phosphorylation in cerebral cortex after transient moderate ischemic injury [see comment].* Stroke, 2002. 33(5): 1369–75.
105. Choi, D.W., *Calcium and excitotoxic neuronal injury.* Annals of the New York Academy of Sciences, 1994. 747: 162–71.
106. Sattler, R. and M. Tymianski, *Molecular mechanisms of calcium-dependent excitotoxicity.* Journal of Molecular Medicine, 2000. 78(1): 3–13.
107. Nishizawa, Y., *Glutamate release and neuronal damage in ischemia.* Life Sciences, 2001. 69(4): 369–81.
108. Alkayed, N.J., et al., *Estrogen and Bcl-2: gene induction and effect of transgene in experimental stroke.* Journal of Neuroscience, 2001. 21(19): 7543–50.
109. Northington, F.J., et al., *Delayed neurodegeneration in neonatal rat thalamus after hypoxia-ischemia is apoptosis.* Journal of Neuroscience, 2001. 21(6): 1931–8.
110. Ouyang, Y.B., et al., *Survival- and death-promoting events after transient cerebral ischemia: phosphorylation of Akt, release of cytochrome C and Activation of caspase-like proteases.* Journal of Cerebral Blood Flow and Metabolism, 1999. 19(10): 1126–35.
111. Jennings, R.B. and C. Steenbergen, Jr., *Nucleotide metabolism and cellular damage in myocardial ischemia.* Annual Review of Physiology, 1985. 47: 727–49.
112. Nicholls, D.G., *Mitochondrial dysfunction and glutamate excitotoxicity studied in primary neuronal cultures.* Current Molecular Medicine, 2004. 4(2): 149–77.
113. Nicholls, D.G. and S. Chalmers, *The integration of mitochondrial calcium transport and storage.* Journal of Bioenergetics and Biomembranes, 2004. 36(4): 277–81.
114. Reynolds, I.J., *Mitochondrial membrane potential and the permeability transition in excitotoxicity.* Annals of the New York Academy of Sciences, 1999. 893: 33–41.
115. Budd, S.L. and D.G. Nicholls, *Mitochondria, calcium regulation, and acute glutamate excitotoxicity in cultured cerebellar granule cells.* Journal of Neurochemistry, 1996. 67(6): 2282–91.
116. Budd, S.L. and D.G. Nicholls, *A reevaluation of the role of mitochondria in neuronal Ca^{2+} homeostasis.* Journal of Neurochemistry, 1996. 66(1): 403–11.
117. Miyamae, M., et al., *Attenuation of postischemic reperfusion injury is related to prevention of $[Ca^{2+}]m$ overload in rat hearts.* American Journal of Physiology, 1996. 271(5, Part 2): H2145–H2153.
118. Balaban, R.S., S. Nemoto, and T. Finkel, *Mitochondria, oxidants, and aging.* Cell, 2005. 120(4): 483–95.
119. DiMauro, S. and E.A. Schon, *Mitochondrial respiratory-chain diseases [see comment].* New England Journal of Medicine, 2003. 348(26): 2656–68.
120. Kakkar, P. and B.K. Singh, *Mitochondria: a hub of redox activities and cellular distress control.* Molecular and Cellular Biochemistry, 2007. 305(1–2): 235–53.
121. McCord, J.M., *Oxygen-derived free radicals in postischemic tissue injury.* New England Journal of Medicine, 1985. 312(3): 159–63.
122. McEwen, M.L., P.G. Sullivan, and J.E. Springer, *Pretreatment with the cyclosporin derivative, NIM811, improves the function of synaptic mitochondria following spinal cord contusion in rats.* Journal of Neurotrauma, 2007. 24(4): 613–24.
123. Hengartner, M.O., *The biochemistry of apoptosis [see comment].* Nature, 2000. 407(6805): 770–6.
124. Boise, L.H., et al., *bcl-x, a bcl-2-related gene that functions as a dominant regulator of apoptotic cell death.* Cell, 1993. 74(4): 597–608.
125. Gonzalez-Garcia, M., et al., *bcl-x is expressed in embryonic and postnatal neural tissues and functions to prevent neuronal cell death.* Proceedings of the National Academy of Sciences of the United States of America, 1995. 92(10): 4304–8.

126. Krajewski, S., et al., *Immunohistochemical analysis of in vivo patterns of Bcl-X expression.* Cancer Research, 1994. 54(21): 5501–7.
127. Cheng, E.H., et al., *BCL-2, BCL-X(L) sequester BH3 domain-only molecules preventing BAX- and BAK-mediated mitochondrial apoptosis.* Molecular Cell, 2001. 8(3): 705–11.
128. Kluck, R.M., et al., *The release of cytochrome c from mitochondria: a primary site for Bcl-2 regulation of apoptosis [see comment].* Science, 1997. 275(5303): 1132–6.
129. Antonsson, B., et al., *Inhibition of Bax channel-forming activity by Bcl-2.* Science, 1997. 277(5324): 370–2.
130. Fujita, N., et al., *Acceleration of apoptotic cell death after the cleavage of Bcl-XL protein by caspase-3-like proteases.* Oncogene, 1998. 17(10): 1295–304.
131. Jonas, E.A., et al., *Exposure to hypoxia rapidly induces mitochondrial channel activity within a living synapse.* Journal of Biological Chemistry, 2005. 280(6): 4491–7.
132. Martinou, J.C., et al., *Overexpression of BCL-2 in transgenic mice protects neurons from naturally occurring cell death and experimental ischemia.* Neuron, 1994. 13(4): 1017–30.
133. Kitagawa, K., et al., *Amelioration of hippocampal neuronal damage after global ischemia by neuronal overexpression of BCL-2 in transgenic mice [see comment] [comment].* Stroke, 1998. 29(12): 2616–21.
134. Lindsten, T., et al., *The combined functions of proapoptotic Bcl-2 family members bak and bax are essential for normal development of multiple tissues.* Molecular Cell, 2000. 6(6): 1389–99.
135. Condorelli, F., et al., *Caspase cleavage enhances the apoptosis-inducing effects of BAD.* Molecular and Cellular Biology, 2001. 21(9): 3025–36.
136. Kirsch, D.G., et al., *Caspase-3-dependent cleavage of Bcl-2 promotes release of cytochrome c.* Journal of Biological Chemistry, 1999. 274(30): 21155–61.
137. Li, H., et al., *Cleavage of BID by caspase 8 mediates the mitochondrial damage in the Fas pathway of apoptosis.* Cell, 1998. 94(4): 491–501.
138. Luo, X., et al., *Bid, a Bcl2 interacting protein, mediates cytochrome c release from mitochondria in response to activation of cell surface death receptors.* Cell, 1998. 94(4): 481–90.
139. Oltersdorf, T., et al., *An inhibitor of Bcl-2 family proteins induces regression of solid tumours.* Nature, 2005. 435(7042): 677–81.
140. Polster, B.M. and G. Fiskum, *Mitochondrial mechanisms of neural cell apoptosis.* Journal of Neurochemistry, 2004. 90(6): 1281–9.
141. Tedeschi, H. and K.W. Kinnally, *Channels in the mitochondrial outer membrane: evidence from patch clamp studies.* Journal of Bioenergetics and Biomembranes, 1987. 19(4): 321–7.
142. Tedeschi, H., K.W. Kinnally, and C.A. Mannella, *Properties of channels in the mitochondrial outer membrane.* Journal of Bioenergetics and Biomembranes, 1989. 21(4): 451–9.
143. Marzo, I., et al., *Bax and adenine nucleotide translocator cooperate in the mitochondrial control of apoptosis.* Science, 1998. 281(5385): 2027–31.
144. Brenner, C., et al., *Bcl-2 and Bax regulate the channel activity of the mitochondrial adenine nucleotide translocator.* Oncogene, 2000. 19(3): 329–36.
145. Kinnally, K.W., M.L. Campo, and H. Tedeschi, *Mitochondrial channel activity studied by patch-clamping mitoplasts.* Journal of Bioenergetics and Biomembranes, 1989. 21(4): 497–506.
146. Sensi, S.L., et al., *Modulation of mitochondrial function by endogenous Zn^{2+} pools.* Proceedings of the National Academy of Sciences of the United States of America, 2003. 100(10): 6157–62.
147. Gidday, J.M., *Cerebral preconditioning and ischaemic tolerance.* Nature Reviews Neuroscience, 2006. 7(6): 437–48.
148. Tanaka, H., et al., *Ischemic preconditioning: neuronal survival in the face of caspase-3 activation.* Journal of Neuroscience, 2004. 24(11): 2750–9.
149. Plas, D.R., et al., *Akt and Bcl-xL promote growth factor-independent survival through distinct effects on mitochondrial physiology.* Journal of Biological Chemistry, 2001. 276(15): 12041–8.
150. Plas, D.R. and C.B. Thompson, *Akt-dependent transformation: there is more to growth than just surviving.* Oncogene, 2005. 24(50): 7435–42.

151. Yano, S., et al., *Activation of Akt/protein kinase B contributes to induction of ischemic tolerance in the CA1 subfield of gerbil hippocampus*. Journal of Cerebral Blood Flow and Metabolism, 2001. 21(4): 351–60.
152. Chan, P.H., *Mitochondria and neuronal death/survival signaling pathways in cerebral ischemia*. Neurochemical Research, 2004. 29(11): 1943–9.
153. Zha, J., et al., *Serine phosphorylation of death agonist BAD in response to survival factor results in binding to 14-3-3 not BCL-X(L) [see comment]*. Cell, 1996. 87(4): 619–28.
154. Danial, N.N., et al., *BAD and glucokinase reside in a mitochondrial complex that integrates glycolysis and apoptosis [see comment]*. Nature, 2003. 424(6951): 952–6.
155. Shangary, S. and D.E. Johnson, *Peptides derived from BH3 domains of Bcl-2 family members: a comparative analysis of inhibition of Bcl-2, Bcl-x(L) and Bax oligomerization, induction of cytochrome c release, and activation of cell death*. Biochemistry, 2002. 41(30): 9485–95.
156. Plesnila, N., et al., *BID mediates neuronal cell death after oxygen/glucose deprivation and focal cerebral ischemia*. Proceedings of the National Academy of Sciences of the United States of America, 2001. 98(26): 15318–23.
157. Polster, B.M., et al., *Inhibition of Bax-induced cytochrome c release from neural cell and brain mitochondria by dibucaine and propranolol*. Journal of Neuroscience, 2003. 23(7): 2735–43.
158. Inoue, I., et al., *ATP-sensitive K^+ channel in the mitochondrial inner membrane*. Nature, 1991. 352(6332): 244–7.
159. Garlid, K.D., et al., *Cardioprotective effect of diazoxide and its interaction with mitochondrial ATP-sensitive K^+ channels. Possible mechanism of cardioprotection*. Circulation Research, 1997. 81(6): 1072–82.
160. O'Rourke, B., *Evidence for mitochondrial K^+ channels and their role in cardioprotection*. Circulation Research, 2004. 94(4): 420–32.
161. Garlid, K.D. and P. Paucek, *The mitochondrial potassium cycle*. IUBMB Life, 2001. 52(3–5): 153–8.
162. O'Rourke, B., *Mitochondrial ion channels*. Annual Review of Physiology, 2007. 69: 19–49.
163. Siemen, D., et al., *Ca^{2+}-activated K channel of the BK-type in the inner mitochondrial membrane of a human glioma cell line*. Biochemical and Biophysical Research Communications, 1999. 257(2): 549–54.
164. Gu, X.Q., et al., *Hypoxia increases BK channel activity in the inner mitochondrial membrane*. Biochemical and Biophysical Research Communications, 2007. 358(1): 311–16.
165. Sato, T., et al., *Mitochondrial Ca^{2+}-activated K^+ channels in cardiac myocytes: a mechanism of the cardioprotective effect and modulation by protein kinase A*. Circulation, 2005. 111(2): 198–203.
166. Douglas, R.M., et al., *The calcium-sensitive large-conductance potassium channel (BK/MAXI K) is present in the inner mitochondrial membrane of rat brain*. Neuroscience, 2006. 139(4): 1249–61.
167. Zukin RS (2004) in Stroke, eds Mohr J, Choi D, Grotta J, Weir B, Wolf P (Churchill Livingstone, Philadelphia), pp 829–854.
168. Miyawaki T, Mashiko T, Ofengeim D, Flannery RJ, Noh KM, Fujisawa S, Bonanni L, Bennett MV, Zukin RS, Jonas EA (2008) *Ischemic preconditioning blocks BAD translocation, Bcl-xL cleavage, and large channel activity in mitochondria of postischemic hippocampal neurons*. Proceedings of the National Academy of Sciences of the United States of America 105:4892–4897.

Part II
Reactive Oxygen Species, and Gene Expression to Behavior

Chapter 8
Perinatal Panencephalopathy in Premature Infants: Is It Due to Hypoxia-Ischemia?

Hannah C. Kinney and Joseph J. Volpe

Keywords: Cerebral palsy; Cytokines; Periventricular leukomalacia; Oligodendrocytes; Subplate neurons; Thalamus

8.1 Introduction

Despite spectacular advances in the intensive care of the premature infant, 25–50% of survivors still demonstrate an unacceptable burden of cognitive, behavioral, motor, and sensory deficits (1–3). Moreover, it is increasingly clear that the clinical presentation in an individual child born prematurely is extraordinarily complex, without, for example, cerebral palsy in isolation of cognitive deficits (1, 2, 4, 5). It is also increasingly clear from autopsy studies that the underlying basis of these complex clinical abnormalities is (not surprisingly) *not* a single lesion alone, e.g., periventricular leukomalacia (PVL), or even two or three lesions, but rather reflects a combination of multiple gray and white matter lesions, which vary across individuals, thereby accounting for variable clinical presentations (6). Indeed, we believe that the term *perinatal panencephalopathy in premature infants* (PPPI) best captures the full spectrum of the brain pathology seen in premature infants today, as characterized clinically, radiographically, and at autopsy (6). We define this entity at the pathological level as brain injury in the premature infant (<37 weeks at birth) that consists of variable combinations of (a) white matter damage (PVL, diffuse white matter gliosis [DWMG]); (b) gray matter damage (neuronal necrosis, neuronal loss, and/or gliosis) primarily in the thalamus, basal ganglia, cerebellum, and brainstem, with relative sparing of the cerebral cortex; and (c) focal and diffuse axonal injury in the cerebral white matter. The term *perinatal* underscores the presence of this pattern of injury

H.C. Kinney (✉) and J.J. Volpe
Department of Pathology, Children's Hospital Boston, 300 Longwood Avenue, Boston, MA 02115, USA
e-mail: hannah.kinney@childrens.harvard.edu

throughout the perinatal period, as we have observed it in term infants with congenital heart disease (7), as well as in preterm infants (6), who die in the perinatal period.

The key question is *what is the cause of PPPI?* The answer to this question is currently considered uncertain, and multiple possibilities (e.g., hypoxia-ischemia, fetal inflammatory syndrome and cytokines, and thyroid derangements) are intensely debated (5, 8, 9). We are focused upon causation because knowledge of the *primary* cause of a disease still remains the fundamental basis upon which a disease is understood, a clinical diagnosis is made, and treatment is developed (10). Here, we define *cause* as the origin of the disease process that is the generative factor responsible for all subsequent events. The term *pathogenesis*, on the other hand, refers to the sequence of events in the response of cells and tissues to the causative factor, from the initial insult to the full-blown expression of the disease. Historically the concept of *one etiologic agent for one disease* has dominated our thinking about causation in medicine, based primarily upon studies of infections and single-gene disorders (10). Yet, this formulation is no longer tenable, given our increased understanding of the interaction between genetic and environmental factors in disease origins (10), as well as the caustative role of two or more exposure *hits* and of synergy among different generative factors. Indeed, sick premature infants in the intensive care nursery – with different genetic susceptibilities – are bombarded by multiple, simultaneous insults of variable severity, including hypoxia-ischemia, infection, metabolic derangements, and coagulation disturbances, that in turn interfere with an enormous array of interrelated brain developmental programs at a critical period in fetal life. Thus, we expect that the clinical presentation and neuropathology of these infants are complex, variable, and *multifactorial*: clearly, the optimal strategy to prevent PPPI is to prevent preterm birth. For those premature infants who continue to be born, however, the goal remains the development of the most specific therapies possible – a goal dependent upon the identification of the (primary) causative factor and relevant pathogenesis. In this review, we focus upon the role of cerebral hypoxia-ischemia in the cause and pathogenesis of PPPI, i.e., we attempt to answer the question: is PPPI due to hypoxia-ischemia? We begin here with an overview of the human neuropathology of PPPI based primarily upon studies from our laboratory. We then review the evidence for a causative role for hypoxia-ischemia in this entity, based upon clinical, epidemiologic, pathologic, and experimental data. We also consider the role of synergy of hypoxia-ischemia with other factors, specifically infection/inflammation, in the causation of PPPI. We conclude with our current thinking about the role of hypoxia-ischemia and its interactions with multiple other factors in this entity.

8.2 The Neuropathology of PPPI

Historically, the focus of the analysis of brain injury in premature infants has been upon PVL (and germinal matrix hemorrhages), and not gray matter injury, with attribution of the neuroanatomic substrate of cerebral palsy to PVL (11–15). This focus upon white matter injury reflects in part the subtlety of neuronal pathology in the premature brain by conventional histology, and thus the need for specialized tissue markers to facilitate its recognition, e.g., immunomarkers for reactive astrocytes [glial fibrillary aciditic protein [GFAP] (6) and for oxidative and nitrative stress, and Golgi analysis for subcellular (dendritic/spine/axonal) injury 16)]. In the seminal paper of the neuropathology of PVL in 1969, Banker and Larroche, for example, reported only *mild neuronal injury* in the cerebral cortex, hippocampus, subiculum, basis pontis, and cerebellar dentate nucleus and Purkinje cells in the majority of PVL brains (12). Subsequent neuropathologic studies reported only rare *anoxic neuronal injury* in association with PVL (12–15). Occasional reports emphasized that PVL did not occur in isolation, but rather was associated primarily with germinal matrix hemorrhages (13). The paradigm shift in conceptualizing preterm brain injury as combined white *and* gray matter injury reflects increased clinical emphasis upon the predominance of cognitive impairments in survivors of prematurity, with or without cerebral palsy (1, 2, 4, 5), and the search for the neuroanatomic substrate of these impairments that are traditionally considered to involve gray matter (neuronal cell body) function, as opposed to white matter (oligodendrocyte) function (6). Indeed, disabilities in cognition and learning are more common today than cerebral palsy in extremely low birth weight preterm infants, affecting 20–50% of such infants, compared with cerebral palsy, which affects 10–20% (1–3). In addition, modern quantitative volumetric MRI studies of premature infants with PVL exhibit reduced volumes of the cerebral cortex, thalamus, basal ganglia, hippocampus, and cerebellum (17–22). In support of these clinical and radiographic observations, we found neuronal loss almost exclusively in PVL, with gray matter lesions in the thalamus, basal ganglia, and cerebellum in a third or more of PVL cases, in a recent survey of the entire spectrum of lesions in the premature brain at autopsy in the perinatal period (6) (Fig. 8.1). Here, we first describe the key features of the pathology of white matter injury in premature infants, followed by those of gray matter, with the recognition that these types of lesions occur in various combinations with each other in individual brains.

White Matter Injury in PPP1. PVL is a developmental lesion of the cerebral white matter characterized by two key components: (a) focal periventricular necrosis and (b) diffuse reactive gliosis and microglial activation of the surrounding white matter (11, 23). The greatest period of risk for PVL is in the premature infant during mid- to late gestation (24–32 weeks), although it also occurs in full-term neonates, particularly those with congenital cardiac or pulmonary disease (7). In the recent survey of the neuropathology of 89 premature infants who died in the perinatal period at our hospital between 1997 and 1999, 41% of the cases had PVL, and 41% had DWMG (diffuse astrogliosis without focal necrosis) (6). Of note, 82% of the PVL cases had necrotic foci that were only detected microscopically, and were <1 mm in diameter; macroscopically evident periventricular cysts (<5 mm in diameter) were noted in only one PVL case, and chalky-white necrotic foci (2–3 mm) (*white spots*) were visible in two cases. These autopsy data underline the modern radiographic data suggesting that cystic PVL is no longer the major expression of white matter disease in premature infants, and beg the question if the MRI findings of hyperintensity reflect the combination of microcysts and diffuse gliosis of the white matter (1). In the neuropathology survey, all cerebral lobes

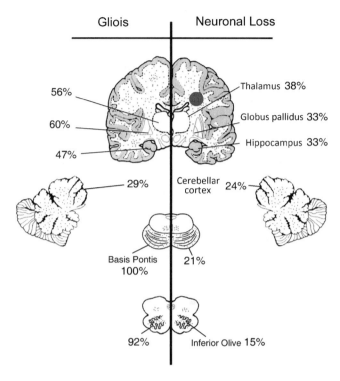

Fig. 8.1 Summary diagram of the neuropathology of PPPI. The incidence of gray matter damage (any degree of neuronal loss) (*left* panel) is compared with the incidence of gliosis (any degree) (*right* panel). Gliosis of the cerebral and cerebellar white matter, basis pontis, brainstem tegmentum, and inferior olives is depicted by small red dots, and focal, periventricular necrosis in the cerebral white matter (PVL) is denoted by large, solid, black circle (*See Color Plates*)

demonstrated a similar incidence of PVL (23–28%), except for the temporal lobe (12%). The incidence of PVL tended to increase with age, but only significantly so in the frontal lobe. In contrast to PVL, DWMG significantly increased in almost all sites with increasing gestational and postnatal ages, and by 37+ weeks, it was present in 100% of the cases in the fronto-parieto-temporal lobes, 90% of the cases in the cerebellum, and 80–86% of the cases in the internal capsule and corpus callosum. The presence of hypertrophic astrocytes in the cerebral white matter of premature infants without focal necrosis, as reported in this series in the DWMG group, has been recognized for decades, but its clinical and pathologic significance remains unknown. Focal necrosis and diffuse hypertrophic astrocytes that are associated with *globules* and *acutely damaged glia* have been considered histological manifestations of the same disorder of immature cerebral white matter for which the term acquired perinatal telencephalopathy (PTL) has been coined (24). Yet, hypoxic-ischemic white matter injury may follow a continuum of damage, from

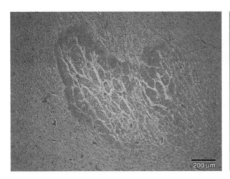 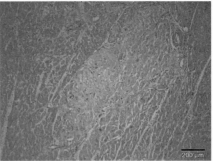

Fig. 8.2 Coagulative necrosis in acute PVL (**a**) is morphologically similar to that in acute myocardial infarction (**b**), with preservation of cellular outlines, nuclear changes and loss, and eosionophilic hypereosinophilia or pallor. Coagulative necrosis, irrespective of the organ, is considered to be due to ischemia (see text) (*See Color Plates*)

mild (gliosis [hypertrophic astrocytes] alone) to severe (periventricular necrosis combined with gliosis) (11, 23). Yet, astrocytes may also normally undergo hypertrophy in the late fetal and perinatal white matter as an obligatory developmental change, potentially due to the *physiological oxidative stress* of active myelin sheath synthesis, and thus may not be a marker of pathology at all (25). The so-called *myelination glia* are immature oligodendrocytes (OLs) that express markers such as O4, and are *morphologically* similar to GFAP-positive reactive astrocytes; coimmunolabelling studies by us, however, show no overlap in the expression of O4 and GFAP (6). Further studies are needed to examine the significance of astrocytic hypertrophy in developmental pathology.

The necrotic component of PVL consists of sharply circumscribed foci of necrosis in the periventricular regions of the lateral ventricles. These necrotic foci measure 2–6 mm in diameter and are within 15 mm of the ventricular wall. They have a specific temporal evolution, beginning with *coagulation necrosis* within the first 24 h after the insult: this consists of nuclear pyknosis, tissue vacuolation (indicative of edema), staining changes, typically increased or reduced eosinophilia, and acutely necrotic, swollen axons (spheroids) (Fig. 8.2). Necrosis affects all of the cellular elements in these periventricular foci, i.e., oligodendrocyte precursors, astrocytes, blood vessels, and axons. Within a few days, marginal astrocytic proliferation and capillary hyperplasia develop, and within 3–7 days, microglia proliferate and lipid-laden cells (macrophages) accumulate. The swollen axons may eventually mineralize, staining for calcium and/or iron. In the chronic lesions, cavitation occurs within a few weeks; alternatively, the focal necrosis heals with a dense glial scar. The healed lesions stain palely for myelin (e.g., with Luxol-fast-blue or immunocytologically with antibodies against myelin basic protein [MBP]) as myelin production is reduced within them (Fig. 8.3) (26). When white matter damage is severe or longstanding, there may be an overall reduction in cerebral white matter volume, ventriculomegaly (hydrocephalus ex vacuo), and thinning of the corpus callosum (15). In the white matter surrounding the necrotic foci, i.e., the so-called

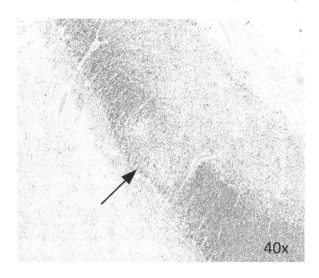

Fig. 8.3 Reduced MBP-immunostaining in a healed necrotic focus (*arrow*) in the optic radiation of a case of PVL at 40 postconceptional weeks

diffuse component of PVL, there is variable astrogliosis (11, 23, 26). In addition, immunohistochemical analysis of the diffuse component with CD68, a microglial/macrophagocytic marker, indicates that there is a previously underappreciated, diffuse activation of microglia (26).

During the period when infants are at greatest risk for PVL, i.e., 24–32 postconceptional weeks, premyelinating late OL progenitors (O4$^+$ and O1$^+$) (pre-OLs) predominate in the human cerebral white matter (27, 28). Lethal injury to pre-OLs in the immature (nonmyelinated) cerebral white matter is postulated to be the key feature of PVL (5, 11, 23, 26, 29, 30), resulting in hypomyelination, as documented by neuroimaging studies (31). Two neuropathologic studies specifically suggest a loss of pre-OLs in PVL (26, 32), yet the overall number of PVL cases studied in these two reports is limited. Although the study by Iida et al. reported a loss of OLs in a larger PVL dataset, this result was determined using a nonspecific OL marker, ferritin (33). Indeed, the analysis of pre-OLs in human PVL has been hampered by the lack of immunological markers that can be applied to formalin-fixed, paraffin-embedded tissue in the archives of pathology departments. Recently, we reported the use of the Olig2 immunomarker in white matter analysis in PVL. Olig2 is a basic helix-loop-helix transcription factor, essential for OL development (34, 35). Expression of Olig2 in the vertebrate brain appears to be the earliest indicator of OL differentiation, and it continues to be expressed in many OLs of the adult brain (34, 35). Thus, Olig2 serves as a marker of the OLs throughout their lineage, including mature (MBP-positive) myelinating OLs, notably in archival tissue (26, 34). We found that Olig2 cell density is not reduced in the focal and/or diffuse components of PVL ($n = 18$) compared with that in controls ($n = 18$) adjusted for age (34). Moreover, there is no difference in Olig2 cell density among acute, subacute, and/or chronic stages of PVL, suggesting that Olig2 cells do not progressively die over the temporal evolution of the white matter damage. Yet, we also found a specific increase in Olig2 cell density in areas within and immediately adjacent to the necrotic foci (regardless of histopathologic age) compared with areas 4–6 mm distant from them (34). There was a trend for increased density of immunolabeled cells with Ki67, a marker of cell proliferation in the focal and diffuse components. These data raise the possibility that OL progenitor proliferation is triggered to generate new OLs, thereby compensating for a *masked* OL cell loss in PVL that occurs hyperacutely, prior to the time-periods *captured* in our histological sections. It also suggests the possibility that the maintenance of Olig2 cell density overall, and the increased cell density at the necrotic focus in particular, is due to OL migration from the site of OL progenitors, i.e., the subventricular zone of the lateral ventricle. Despite the *maintenance* of total OL density in the 18 PVL cases in our series, we found abnormalities of O4 and MBP immunostaining, with an increase and persistence of OLs with perikaryal (as opposed to distal cytoplasmic process) MBP immunostaining in the diffusely damaged, premyelinated white matter, and abortive MBP expression in the periventricular necrotic foci. Our data suggest that the basis of the myelin deficits in survivors of PVL includes (a) inadequate tissue repair due to suboptimal proliferation and migration of OL progenitors to the necrotic core, (b) an arrest in maturation of the OL progenitor or pre-OL to the mature phenotype, (c) an inability of the mature OL to produce or traffic MPB to produce sufficient myelin sheaths, and/or (d) primary axonal injury with defective axonal signaling for myelin initiation and production by OLs (34).

Gray Matter Injury in PPP1: Neurons. In our recent study of the neuropathology of prematurity, we analyzed gray matter injury in cases with PVL ($n = 17$) compared with cases without PVL ($n = 24$) who died within the perinatal period (6). Neuronal loss and/or gliosis were considered markers of subacute and chronic injury, and were found in a third or more of PVL cases (Fig. 8.1). Indeed, neuronal loss was found almost exclusively in PVL cases than in non-PVL cases, with significantly increased incidence and severity in the thalamus (38%), globus pallidus (33%), and cerebellar dentate nucleus (29%) (Fig. 8.1). The incidence of gliosis was significantly increased in PVL cases compared with that in the non-PVL cases in the deep gray nuclei (thalamus/basal ganglia) (50–60% of PVL cases), and basis pontis (100% of PVL cases) (Figs. 8.1 and 8.4). Thalamic and basal ganglionic lesions occurred almost exclusively in infants with PVL; remarkably, not a single infant with DWMG, for example, exhibited neuronal loss in the cerebral cortex, hippocampus, and deep gray nuclei (6). Similarly, gliosis was very unusual in these gray matter regions in the infants with DWMG. We also found that acute neuronal necrosis, considered a marker of terminal injury, was common in all cases in this series, and that 66% of PVL cases and 59% of non-PVL cases had two or more gray matter sites with acute neuronal necrosis. Moreover, the incidence of acute neuronal necrosis was not significantly different at any gray matter site between these two groups when adjusted for gestational and postnatal age. Overall, we found that gray matter lesions in subacute and chronic stages occur in a third or more of PVL cases, indicating that white matter injury generally does not typically occur in isolation, and more than one gray matter lesion is involved in each case (Fig. 8.5).

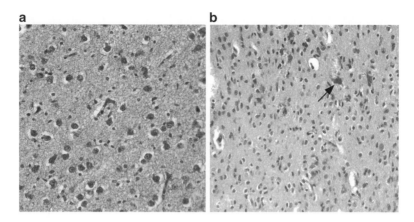

Fig. 8.4 There is severe neuronal loss in the thalamus in premature infants with PVL, demonstrated in a PVL case (**b**) compared with an age-matched control (**a**). A remaining neuron in the PVL case is mineralized (**b**) (*arrow*). Hematoxylin-and-eosin/Luxol-fast-blue, x60

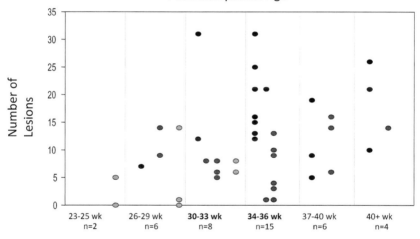

Fig. 8.5 The number of lesions, i.e., neuronal loss and gliosis, in all gray matter sites is presented according to diagnosis and divided into intervals of postconceptional age. The peak incidence period of PVL is in bold. PVL cases are in *black*; DWMG cases are in *red*, and control cases [no focal necrosis and/or diffuse white matter gliosis (DWMG)] are in *green* (*See Color Plates*)

This autopsy study suggests that neuronal loss and/or gliosis in the perinatal period in gray matter sites critical to cognition, memory, and learning, i.e., thalamus, basal ganglia, hippocampus, and cerebellum (36–39), play a role in cognitive defects in long-term survivors of prematurity. The neuroanatomic structures involved with neuronal loss and/or gliosis in this survey of the neuropathology of prematurity

(6) correlate well with the neuroimaging data, which shows volumetric deficits in the thalamus and basal ganglia, cerebellum, hippocampus, and cerebral cortex in survivors of prematurity (with or without PVL) (17–22). The autopsy data also suggest that at least some of the common, less severe motor deficits in the survivors of prematurity are due to gray (thalamic/basal ganglionic, ponto-olivary) damage, compared with the spastic motor deficits, i.e., cerebral palsy, that are attributed to axonal damage in the necrotic foci in PVL, which focally destroy the descending corticospinal tracts. Of note, subplate neurons embedded within the subcortical white matter play a critical role in the development of the cerebral cortex: when these neurons are ablated, for example, afferent thalamocortical axons fail to reach the cortex and abnormal cortical lamination results (40, 41). While subplate neuronal loss has been demonstrated in animal models of hypoxic-ischemic injury (41, 42), it has yet to be reported in PPPI.

Gray Matter Injury in PPP1: Axons. The classic descriptions of PVL include the major observation that axonal swellings or spheroids, indicative of acute axonal transection, are present in the periventricular necrotic foci, i.e., the *focal* component of PVL, and that axonal mineralization typically occurs in the healed foci (11, 23, 43–46). In a study of axonal injury in PVL involving 85 PVL and 30 non-PVL cases ranging in age from 22 to 41 gestational weeks, for example, Deguchi et al. divided white matter *necrosis* into *focal*, *widespread*, and *diffuse* according to its distribution (43). In this schema, *diffuse* referred to *necrosis* extending from the periventricular to subcortical areas. They found immunopositivity to β-amyloid precursor protein (β-APP) (a marker of axonal damage) in the axons around the immediate periphery of foci of all types of necrosis in 76% of PVL cases. Yet the presence of widespread axonal injury in the *diffuse* component has only recently been addressed (46). In the brains of 13 PVL and 17 control cases, we applied immunomarkers of axonal pathology (β-APP and the apoptotic marker fractin), and axonal regeneration (growth-associated protein [GAP]-43). Axonal spheroids were confirmed with β-APP in necrotic foci in the acute and subacute stages, as previously described by other investigators. We found that GAP-43 expression was also present in these spheroids in the necrotic foci, suggesting possible attempts at axonal regeneration. Diffuse axonal damage, on the other hand, was detected by fractin in white matter sites surrounding and distant from acute and organizing foci of necrosis. Fractin, an antibody to the 32-kD product of caspase-cleaved actin, is a marker of apoptosis-related events (46, 47). It has been demonstrated in apoptotic cell bodies in pontosubicular neuronal necrosis due to perinatal hypoxic-ischemic injury, degenerating cell bodies, axons, and dendrites in Alzheimer disease, and animal models of medial forebrain bundle axotomy (46, 47). In the white matter surrounding focal cavitation or scar formation, immunostaining for fractin and β-APP was negative, suggesting chronic axonal loss and absence. Thirty-one percent of these PVL cases had thalamic damage, and 15%, neuronal injury in the cerebral cortex overlying PVL. In essence, the major finding of this study is that diffuse axonal injury occurs in the gliotic cerebral white matter in the acute and organizing stages of PVL, and that axonal injury is not restricted solely to necrotic foci or their immediate periphery, but rather, it occurs in the diffusely gliotic component, in many instances far from the focal necrosis. This finding suggests that widespread and diffuse axonal injury (surrounding and extending beyond the necrotic foci) could cause, at least in part, the reduced white matter volume and secondary ventriculomegaly seen in long-term survivors with PVL.

The diffuse axonal injury that occurs in PVL (i.e., PPPI) likely results from one or more of the following basic mechanisms: (a) primary ischemia to the axon (48), (b) impaired differentiation secondary to OL cell loss and abnormal OL–axonal interactions that are critical for axonal maturation and function (49, 50), (c) distant Wallerian degeneration secondary to axonal transaction in the periventricular necrotic foci, and/or (d) degeneration of afferent axons from the thalamus to the cortex and vice versa, secondary to primary injury to the projecting neurons in these gray matter sites. The onset of myelination involves initial contact between the OL and axon via signals directly on the axon itself (49), as well as release of promyelinating factors from neurons and astrocytes (50). If axons in PVL are damaged, proper axon–OL interactions will likely be hindered, resulting in failed or reduced myelination. The observation by us of diffuse axonal injury without reduction in OLs in PVL suggests the possibility that primary injury to the axon in PVL results in a decrease in axonal signals to the proper complement of OLs necessary for proper and complete myelination. Studies of premature infants by advanced MRI techniques have shown abnormalities consistent with axonal abnormalities. Thus, diffusion tensor imaging of VLBW infants, as early as term equivalent age (51, 52) and then later in childhood (53, 54), has shown diminished relative anisotropy, an MRI measure of preferred directionality of diffusion, especially in infants with cerebral white matter injury. This impairment in anisotropic diffusion could be caused by either a disturbance of axonal number, size, packing, or of axonal membranes or intracellular constituents (55), or alternatively, a disturbance in ensheathment of axons by pre-OLs. A definitive correlation between pathologic findings concerning axonal injury and the diffusion tensor MRI data cannot be made without MRI data obtained at or close to the time of autopsy; such information has yet to be available.

Other Types of Brain Injury Associated with PPPI. Brain injury in premature infants at autopsy may be complicated by multiple derangements, including viral, bacterial, or fungal infection, hemorrhages, hyperbilirubinemia (kernicterus), and hypoglycemia. The pathologic changes that result from these disorders can exacerbate and/or obliterate those of PPPI. Of note, the neuronal loss and gliosis that we describe in PPPI (see earlier) reflect the *end stage* of virtually all metabolic derangements in gray matter, including hypoxia-ischemia, hyperbilirubinemia, and hypoglycemia. A specific type of neuronal insult may result in a specific topography of gray matter injury, based upon region-specific vulnerabilities. Nevertheless, multiple insults share the same lethal pathway, as witnessed by the same *morphological* "end stage" of necrosis and apoptosis in ischemia, hyperbilirubinemia, and hypoglycemia – insults that differentially mediate cell injury via excitotoxicity (23). Hemorrhages of various kinds are frequently seen in premature infants in association with injury to gray and/or white matter structures (6). In our autopsy survey of the neuropathology of prematurity, for example, germinal matrix hemorrhage with or without intraventricular hemorrhage involved 50% of the cases (unpublished data). Finally, given that multiple genetic disorders are associated with prematurity, perinatal encephalopathy is not infrequently superimposed upon brain malformations, e.g., Potter's syndrome (6).

8.3 Strategy Toward Establishing the Causative Role for Hypoxia-Ischemia in PPPI

We turn now to the consideration of a causative role for hypoxia-ischemia in PPPI. Historically, hypoxia-ischemia has been considered the leading candidate for the primary cause of PVL, the defining feature of the white matter injury in PPPI, at least as early as the seminal pathologic description by Banker and Larroche almost four decades ago (12). The fundamental abnormality in cerebral hypoxia-ischemia is a deficit of brain oxygen that occurs via two mechanisms: hypoxemia, defined as a diminished amount of oxygen in the blood supply; and ischemia, a diminished amount of blood perfusing the brain (5, 11). In premature infants, hypoxemia and/or ischemia most commonly result from pulmonary immaturity and respiratory distress/failure, typically complicated by systemic hypotension (5, 11). In PVL, cerebral ischemia/reperfusion is postulated to trigger excessive intracellular calcium resulting from nonexcitotoxic (pump failure-mediated) and excitotoxic (glutamate [AMPA/kainate] receptor-mediated) toxicity (56–72). Excitotoxicity to pre-OLs is critically mediated by inotrophic glutamate receptors that are located on them (64, 66–69, 71), as well as by the glutamate transporter GLT1 (62). The nonreceptor-mediated mechanism involves glutamate competition for the cystine transporter and promotion of cystine efflux under conditions of high extracellular levels of glutamate (56, 73). The result is depletion of intracellular glutathione (which requires cysteine for biosynthesis) and cell death by oxidative stress. However, the substantial levels (millimolar) of glutamate required for this effect suggest that this mechanism may not operate in vivo under most pathologic conditions. The receptor-mediated mechanism, which requires micromolar levels of glutamate, is more likely to occur in vivo, as shown directly in animal models (65, 70, 72). The principal sources of elevated extracellular glutamate in cerebral white matter with hypoxia-ischemia are glutamate transporters (62). Failure of glutamate uptake and actual reversal of transport occur in the setting of energy failure because of the failure of the Na^+/K^+ ion pump, as these transporters are high-affinity, sodium-dependent systems. Activation of the calcium permeable non-NMDA (AMPA/kainate) receptors by glutamate triggers mitochondrial dysfunction and intracellular free radical accumulation (66, 70). When the generation of reactive oxygen species (ROS) and reactive nitrogen species (RNS) exceeds the rate at which endogenous defenses scavenge them, proteins, lipids, DNA, and other molecules become targets for oxidative and nitrative modification, which leads in turn to cell death either by necrosis or apoptosis (30, 74). Recent studies in experimental systems have shown that a second type of glutamate receptor, the NMDA receptor, is present on the processes of pre-OLs, and that excessive activation of these receptors, as occurs in models of ischemia, leads to excessive calcium influx, generation of ROS/RNS, and loss of processes (67, 68). Thus, the data indicate that excessive glutamate in human cerebral white matter during the peak period of PVL could lead both to death of developing OLs by activation of AMPA receptors on the cell soma and to loss of oligodendroglial processes by activation of NMDA receptors on cell processes (34). Both events could result in impaired myelination and axonal function or structure or both. The tissue injury results in activation of astrocytes and microglia with the production and release of cytokines, including interferon-γ (IFN-γ), which is cytotoxic to OLs (75–78). This inflammatory response, if excessive and/or prolonged, may propagate free radical injury. Indeed, reactive microglia/macrophages in the inflammatory response are major sources of the ROS, nitric oxide (NO), and peroxynitrite (79, 80). On balance, astrocytes are protective against RNS and ROS toxicity to other cells, but recent data suggest that they may also contribute to cytotoxicity by the upregulation of iNOS and peroxynitrite injury (80).

The chain of cellular events (pathogenesis) in the white matter due to hypoxia-ischemia is essentially the same in the gray matter, with the key difference that the targeted cell in the gray matter is the neuron, as opposed to the developing OL in the white matter (23). Thus, we propose that white and gray matter injury triggered by cerebral ischemia occurs simultaneously in the sick premature infant with respiratory compromise and systemic hypotension, leading to glutamate, free radical, and cytokine toxicity to developing OLs and neurons, with differential topographic patterns of injury based upon the developmental (and genetic) susceptibilities of each of these cells. A reasonable strategy to establish the causative role of hypoxia-ischemia involves fulfilling the following criteria: (a) determine that PPPI or its subcomponents, e.g., PVL, precisely mimic the known *morphological* pathology of hypoxia-ischemia in humans and animal models; (b) determine that PPPI or its subcomponents are associated with functional markers of hypoxia-ischemia, e.g., decreased cerebral blood flow (CBF), in premature infants; (c) determine that PPPI or its subcomponents are associated with epidemiological risk factors for hypoxia-ischemia; (d) determine underlying vulnerabilities that make developing oligodendrocytes and neurons, the major cellular targets in PPPI, susceptible to hypoxia-ischemia; and (e) prevent PPPI or its subcomponents in animal models via intervention in the molecular and cellular pathways critical to the pathogenesis of glutamate, free radical, and/or cytokine toxicity in premature infants or perinatal animal models. These criteria and the supporting evidence for them in PVL are summarized in Tables 8.1 and 8.2. In the following discussion, we focus specifically upon PVL, the major white matter subcomponent of PPPI, because more cellular information in particular is available about it in the human brain compared with the gray matter component. The same reasoning process, however, could be applied to considering the causative role of hypoxia-ischemia in the gray matter component with equally compelling supportive evidence.

Table 8.1 Summary of major evidence for a causative role of hypoxia-ischemia (HI) in PVL

Criteria	Precedent	Human PVL
Tissue morphology in PVL mimics known HI lesions in humans	Coagulative necrosis in acute infarcts in systemic organs and brain in humans	Coagulative necrosis in acute necrotic periventricular foci
Morphology of downstream effects of HI mimics PVL morphology	Animal and cell culture models with markers of oxidative and nitrative stress, inflammatory cytokines	Presence of cellular markers of oxidative and nitrative stress, and inflammatory response with cytokines
Brain lesions located in arterial end zones	Perinatal and adult HI in arterial end zones	Periventricular focal necrosis in arterial end zones
Neuronal morphology in PVL mimics known HI neuronal lesions in animal models	Neuronal necrosis, loss, and gliosis in perinatal brain damage due to HI in animal models	Acute neuronal necrosis, loss, and gliosis in thalamus and other gray matter regions in PVL
Clinical correlation with risk factors for HI	Human neuroimaging and pathologic studies correlated with clinical and epidemiologic risk factors	HI risk factors for PVL include hypovolemia, oliguria, abrupt hypotension, patent ductus arteriosus, congenital heart disease, and mechanical ventilation
Prevention due to HI intervention	Reduction in infarct size in animal models with agents that interfere with excitotoxicity, oxidative and nitrative stress, and cytokine toxicity	Unknown

Table 8.2 Factors contributing to the vulnerability of cerebral white matter to hypoxia-ischemia in the human fetus

Predominance of pre-OLs (O1, O4), known to be vulnerable to ROS and RNS injury compared with mature (MBP) OLs
Developmental delay in the maturation of the superoxide dismutases
Transient overexpression of the glutamate transporter GLT1
Transient overexpression of AMPA receptors, with lack of the relative expression of GluR2
Expression of IFN-γ receptors on pre-OLs
Transient elevation in the density of ameboid microglia
Immaturity of axonal projections and cellular structure (e.g., neurofilaments)
Immaturity of cerebral arterial vascularization

8.4 Evidence for the Causative Role for Hypoxia-Ischemia in PVL Based upon Human Clinical Studies

Premature infants appear to have a propensity for the development of cerebral ischemia and as a consequence, injury to cerebral white matter. This propensity relates to two key factors: (a) periventricular vascular anatomic and physiological characteristics and (b) an impairment in the regulation of CBF.

Periventricular Vascular Anatomical and Physiological Factors. Although the presence of arterial border zones is controversial (15, 81–86), PVL in our experience typically occurs in deep white matter regions that represent arterial border or end zones (Fig. 8.6) (26). The blood vessels penetrating the brain from the pial surface, the long and short penetrators, are derived from the middle cerebral artery, and to a lesser extent, from the posterior and anterior cerebral arteries; the putative blood vessels supplying the immediate periventricular area are derived from the lenticulostriate arteries and other basal penetrating and choroidal arteries (84). Ventriculofugal arteries in the human brain are believed to run from the lateral ventricle to terminate in deep white matter (81, 82, 84–86). A putative vascular border zone between ventriculopetal arteries of long medullary arteries and ventriculofugal arteries may thereby explain how hypoperfusion of the fetal brain produces periventricular lesions: the arterial border zones represent distal fields, which are the most susceptible to a fall in perfusion pressure and CBF. Independent studies have demonstrated the existence of ventriculofugal arteries in fetal and neonatal brains (82, 85, 86). A physiological correlate of these vascular anatomical factors appears to be the extremely low blood flow to cerebral white matter in the human premature newborn, first shown by positron emission tomographic studies of CBF (5, 87). The likelihood of extremely low white matter flows was suggested initially by the results of xenon clearance studies that documented mean global CBF values in ventilated human premature infants of only approximately 10–12 mL/100 g/min. Subsequent xenon studies confirmed these very low mean global values (5, 88, 89). Our studies of *regional* CBF by PET confirmed the low global values, but more importantly showed that values in cerebral white matter in surviving preterm infants with normal or near normal neurological outcome ranged from only 1.6 to 3.0 mL/100 g/min (5, 87). These remarkably low values in white matter were approximately 25% of those in cortical gray matter, a regional difference confirmed in a study utilizing single photon emission tomography (89). Our blood flow values of less than 5.0 mL/100 g/min in normal or near-normal cerebral white matter in the preterm infant are markedly less than the threshold value for *viability* in adult human brain of 10 mL/100 g/min (*normal* CBF in the adult is approximately 50 mL/100 g/min) (90). The very low values of volemic flow in cerebral white matter in the human premature infant suggest that there is a minimal margin of safety for blood flow to cerebral white matter in such infants.

Cerebral Ischemia – Impaired Cerebrovascular Autoregulation – Pressure-Passive Cerebral Circulation. The vascular end zones and border zones just described would

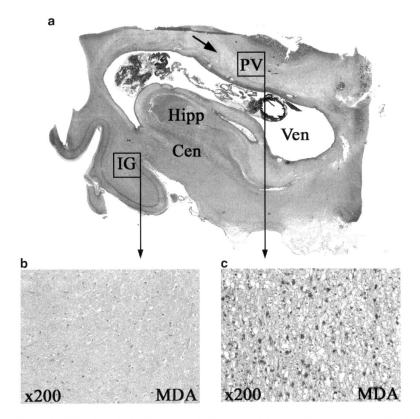

Fig. 8.6 A. In PVL, there is focal necrosis in the periventricular white matter, associated with DWMG in the surrounding deep white matter and relative sparing in the intragyral white matter. The damage in the diffuse white matter is characterized by pallor (*arrow*). The damage is located in the end arterial zones. There is a heterogeneous distribution of reactive glial cells with positive immunostaining for selected markers of oxidative injury (malondialdehyde-protein adducts [MDA] shown) in the different components of the cerebral white matter, as demonstrated in a typical PVL case at 35 postconceptional weeks in the parietal white matter at the level of the atrium of the lateral ventricle. (**b**) The intragyral white matter distant from the periventricular focus of necrosis lacks MDA-immunopositive cells; the central white matter is likewise not involved (CEN). (**c**) In contrast, the MDA-immunopositive glial cells are heavily concentrated in the periventricular white matter surrounding the focal necrosis. *VENT* ventricle, *PV* periventricular region, *HIPP* hippocampus, *CEN* central white matter, *IG* intragyral white matter. Hematoxylin-and-eosin/Luxol-fast-blue (*See Color Plates*)

render the premature infant's brain particularly vulnerable to injury in the presence of cerebral ischemia. Of particular importance in the genesis of impaired CBF, and thereby cerebral ischemia is an apparent impairment of cerebrovascular regulation in sick premature infants (5). Thus, earlier seminal studies employing the technique of radioactive xenon clearance showed that certain premature infants, mechanically ventilated and often clinically unstable, appear to exhibit a pressure-passive cerebral circulation, an observation confirmed multiple times with less invasive methods (5, 91–95). Thus, in such sick premature infants with a pressure-passive cerebral circulation, it would be expected that when blood pressure falls, as occurs commonly in such infants, so would CBF, with the consequence being ischemia in the distribution of the arterial end zones and border zones in cerebral white matter. Moreover, this particular danger is compounded by the demonstration that blood flow to cerebral white matter of the infant is very low (see earlier) and that thereby a minimal margin of safety exists. The proportion of infants with a pressure-passive cerebral circulation and the duration of the abnormality are substantial. In one serial study of 32 mechanically ventilated premature infants from the first hours, near-infrared spectroscopy demonstrated a pressure-passive cerebral circulation in 53% (94). In a later more detailed study of a larger number of infants ($n = 90$), pressure-passive periods were identified at some point(s) early in the neonatal course in 95% of the infants, and the overall mean proportion of the pressure-passive time was 20%; some infants were pressure-passive more than 50% of the time (93). The likelihood of a pressure-passive state increased with decreasing gestational age and periods of hypotension, although the large majority of pressure-passive periods was not accompanied by hypotension. Indeed the nadirs of blood pressure often are not markedly low, and thus could be readily overlooked with routine monitoring. Nevertheless, the cumulative effects of repeated modest declines in CBF, including potentiation of excitotoxicity and free radical accumulation, could be considerable. This concern is accentuated by the repeated finding from neuropathological and imaging studies that the incidence of PVL increases with advancing postnatal age (see earlier). In the study of Tsuji et al., nearly all the cases of PVL (and severe intraventricular hemorrhage) were in the pressure-passive group (94). Although the numbers were small, the observations suggested that (a) infants with impaired cerebrovascular autoregulation and a pressure-passive cerebral circulation could be identified by near-infrared spectroscopy prior to the occurrence of white matter injury, (b) the circulatory abnormality is related to the occurrence of such injury, and (c) if the pressure-passive state could be corrected, perhaps the white matter injury could be prevented.

The propensity to a pressure-passive abnormality in premature infants may relate in part to an absent muscularis around penetrating cerebral arteries and arterioles in the third trimester in the human brain (83, 96). Additional potential reasons for a pressure-passive cerebral circulation in such infants include hypercarbia or hypoxemia related to respiratory disease, the mechanical trauma of labor and/or vaginal delivery to the easily deformed cranium of the premature infant, and the occurrence of normal blood pressures that are dangerously close to the downslope of a normal *immature* autoregulatory curve (5, 92, 94, 97). A combination of these factors or still-to-be-defined perturbants, e.g., cytokine vascular effects, is likely operative. Yet, even in the presence of intact cerebrovascular autoregulation, marked cerebral vasoconstriction or severe systemic hypotension could lead to sufficiently impaired CBF to the cerebral vascular end zones and border zones to result in ischemic cerebral white matter injury. This explanation may account for the demonstrated relationships between marked hypocarbia or hypotension and PVL (5, 98). A recent study of 905 infants (birth weight, 501–1,249 g) evaluated a *cumulative* index of hypocarbia over the first 7 days of life, and found a strong association with the occurrence of PVL (identified as echolucencies on cranial ultrasonography) (99). Infants with the highest quartile of cumulative index of exposure to hypocarbia had more than a fivefold increased risk of PVL when compared with those infants in the lowest quartile (99). Additional evidence supportive of a relation between impaired CBF, i.e., ischemia, and the occurrence of PVL includes the association of the lesion with markers of hypoxic-ischemic events (e.g., neonatal acidosis, elevations of plasma uric acid on the first day of life), episodes of mean arterial blood pressure less than 30 mm Hg, hypovolemia, oliguria, abrupt decreases in blood pressure in chronically hypertensive premature newborns (in whom cerebrovascular autoregulation might be present but the curve shifted to the right), patent ductus arteriosus, congenital heart disease, and severely ill infants treated with extracorporeal membrane oxygenation or cardiac surgery (100, 101).

8.5 Evidence for the Causative Role for Hypoxia-Ischemia in PVL Based upon Human Pathologic Studies

In essence, we conceptualize the focal component of periventricular necrosis as the central *core* of tissue damage, i.e., an infarct, resulting from the most severe ischemic insult in the arterial end zones, and the diffuse component as the *penumbra* in the surrounding tissue in which the reduction in blood flow is less severe, and the capacity to recover is present if perfusion is restored (102). This premise is based upon the major morphological features that are shared between PVL and established infarcts in human brain and other tissues (Table 8.1). In this regard, we place great emphasis upon the finding of *coagulative necrosis in PVL*, the earliest morphological stage of the focal component, because this abnormal morphology is indicative of hypoxia-ischemia, irrespective of the age or organ (e.g., heart, kidney, and brain) (Fig. 8.2), and is exemplary of the fundamental tenet of anatomic pathology that specific morphologic patterns of tissue injury indicate specific types of causative insults. In coagulative necrosis, the general tissue architecture and basic cellular outlines are preserved, but the individual cells demonstrate increased eosinophilia, or, alternatively, marked pallor, in hematoxylin-and-eosin-stained sections (Fig. 8.2) (10). The hypereosinophilia is attributed in part to loss of the normal basophilia imparted by RNA in the cytoplasm and in part to increased binding of eosin to denatured intracytoplasmic proteins (10). Nuclear changes, e.g., pyknosis and karyorrhexis, result from nonspecific breakdown of DNA (10). Coagulative necrosis is the earliest tissue reaction seen in cerebral infarcts due to demonstrable vascular occlusion in the immature and adult brain, and is histologically identical to systemic infarcts, such as observed in acute myocardial infarction (Fig. 8.2) (Table 8.1). The presence of coagulative necrosis in *early* PVL (prior to the masking effects of secondary inflammation) provides strong, but typically overlooked, evidence that the pathogenesis of the white matter lesions involves an ischemic trigger. We also place great emphasis upon the topographic distribution of the periventricular necrosis because it is present in the vascular end zones, as historically recognized (12, 81, 82, 84). But of less appreciated importance, the surrounding gliosis is typically within the deep white matter as well, with relative sparing of the more distant central and intragyral white matter (Fig. 8.6) (26). Oligodendrocyte progenitors are depleted in the subventricular zone within hours of hypoxic-ischemic injury in the perinatal rat model (103), suggesting that their outward migration results in subventricular depletion. Our observation that Olig2 density is preserved in the diffuse components of PVL may reflect, at least in part, a less severe reduction in blood flood in the penumbra compared with that in the central core (Fig. 8.6) (34). Our additional observation of increased Olig2 cell density directly within the necrotic focus in PVL (see earlier) suggests the possibility that OL progenitors migrate toward these sites to attempt repair, an idea supported by neonatal and adult rodent models of hypoxia-ischemia (34, 103, 104).

Markers of Free Radical Injury in PVL. Of direct relevance to PVL, data from perinatal rat models and rodent cell culture indicate that developing OLs are preferentially vulnerable to ROS in comparison to mature OLs that express MBP (5, 23, 26, 29, 30, 32, 73, 74, 79). Autopsy studies by others and by us provide compelling evidence of free radical injury to developing oligodendrocytes (26, 32), as well as to cerebral cortical neurons (103a), in preterm brains in the perinatal period. In our study of 17 PVL cases compared with 28 controls, in which immunocytochemical markers for oxidative (hydroxynonenal) and nitrative (nitrotyrosine) attack were utilized, abundant staining was documented in the diffuse lesion, notably in O1 and O4 OLs (Fig. 8.7) (26). In a study of 33 cases with early periventricular white matter injury and controls, Back et al. also demonstrated that F2-isoprostane, a lipid peroxidation marker, is increased in the early stages of the white matter damage, in conjunction with a $71\% \pm 8\%$ depletion of OL progenitors (32). Studies of cerebrospinal fluid (CSF) in living premature infants also support toxicity by free radical attack, specifically ROS in PVL. In a longitudinal study of premature infants, those with MRI evidence for PVL at term exhibited earlier in the neonatal period sharply elevated CSF levels of oxidative products, detected as protein carbonyls, when compared with cerebrospinal fluid levels in premature infants without MRI evidence for PVL (105). Of relevance to RNS, our group has preliminary evidence of a significant increase in the number of iNOS-positive glia in the diffuse component of PVL (unpublished data). Moreover, inducible nitric oxide synthase (iNOS) is expressed strongly in the reactive astrocytes in the diffuse lesion, but, interestingly, and not prominently in the activated microglia. These data suggest that a key source of nitric oxide in the human lesion, potentially leading to a major portion of the nitrative stress identified previously in diffuse PVL, is the reactive astrocyte. It is likely that the superoxide anion necessary for combination with nitric oxide and thereby generation of the most injurious RNS, peroxynitrite, is derived from both the abundant activated microglia in diffuse PVL and the pre-OL itself. The finding of diffuse axonal injury in PVL with the antibody to fractin suggests primary injury due to

8 Perinatal Panencephalopathy in Premature Infants

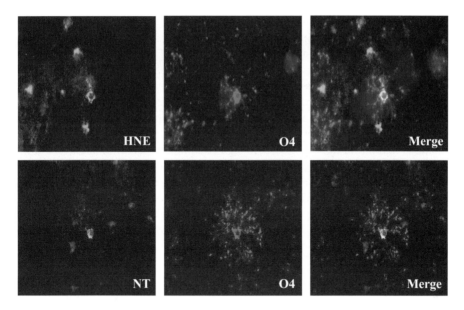

Fig. 8.7 There is oxidative and nitrative injury to developing OLs. Nitrotyrosine (NT) colocalizes (*yellow* signal) with the marker for premyelinating OLs (O4, *red*). Hydroxynonenal (HNE) adducts colocalize (*yellow* signal) with markers for premyelinating OLs (O4, *red*). Diffuse component of PVL in infant at 44 postconceptional weeks (*See Color Plates*)

lesser degrees of ischemia/reperfusion to axons diffusely coursing throughout the cerebral white matter in the penumbra. Of note, NO and peroxynitrite are directly toxic to axons in experimental models (106).

Markers of Excitotoxicity in PVL. In a recent study, we analyzed the expression of EAAT2 (the human counterpart of GLT1) in cerebral white matter from PVL and control cases (107). We found a significantly higher percentage of EAAT2-immunopositive astrocytes in PVL (51.8% ± 5.6%) compared with that in control white matter (21.4% ± 5.6%) ($p = 0.004$). Macrophages in the necrotic foci in PVL also expressed EAAT2 (107). Premyelinating oligodendrocytes in both PVL and control cases expressed EAAT2, without qualitative difference in expression. Our immunocytochemical data suggest that the overall increase in EAAT2 expression in PVL reflects a combination of factors, including an increase in the overall density and number of reactive astrocytes expressing EAAT2, an increase in expression of EAAT2 by individual astrocytes, and an increase in EAAT2 by macrophages that infiltrate the foci of periventricular necrosis (107). The upregulation of EAAT2 in reactive astrocytes and its presence in macrophages in PVL may reflect a protective mechanism against excitotoxicity in vulnerable white matter, and is interpreted by us as a marker of excitotoxic mechanisms in PVL.

Clinicopathologic Correlations of PVL/PPPI. A major issue in our autopsy studies of PVL/PPPI is the failure to correlate significantly all or some of the brain lesions with markers of hypoxia-ischemia based upon a review of the clinical neonatal chart. In the recent survey of the neuropathology of premature term infants dying at our hospital, the incidence of various clinical variables, e.g., chorioamnionitis, maternal fever at delivery, history of cesarean section, and congenital anomalies, was not significantly different among infants with PVL and those without PVL; the mean Apgar scores, for example, were less than 7 at 1 and 5 min in all groups (6). Cardiorespiratory disorders, requirement for and duration of mechanical ventilation, and incidence of cardiorespiratory resuscitation were also not significantly different between the PVL and non-PVL cases. Moreover, infectious and inflammatory disorders e.g., pneumonia, necrotizing enterocolitis, and sepsis, occurred in all groups, and did not differ significantly among them. It is very likely that there are specific clinical factors at work that we do not yet know, or were not analyzed in this study, e.g., lowest oxygen levels, alterations in acid–base status, and dysfunction in cerebral autoregulation. It is clear that clinicopathologic correlations will require more meaningful ways of analysis than complicated and incomplete clinical neonatal charts.

8.6 Evidence for the Causative Role for Hypoxia-Ischemia in PVL Based upon Pathologic Studies in Animal Models

The importance of cerebral ischemia in pathogenesis of PVL is illustrated in many models in developing animals (primarily sheep, piglet, rat) in which ischemia alone has been shown to lead to selective or predominant cerebral white matter injury (72, 108–122) (Table 8.1). The principal cellular target has been identified as the developing OL (72, 108, 109, 115–120, 122). Indeed in one careful study in fetal sheep, under conditions of ischemia, the regional distribution of pre-OLs correlated closely with the regional distribution of the cerebral white matter injury (122). In several ischemic models, other cellular elements in cerebral white matter, e.g., axons, subplate neurons, or adjacent neuronal structures, e.g., basal ganglia, cerebral cortex, have been variably affected, but overall the principal cellular target has been the pre-OL. The repeated demonstrations of maturation-dependent, selective or predominant white matter injury in immature animal models subjected to hypoxia-ischemia represent especially strong evidence for the pathogenetic role of hypoxia-ischemia (Table 8.1). Indeed, because it is unclear whether the immature animal models exhibit the periventricular anatomic and physiologic factors or the disturbances in regulation of CBF described earlier for human premature infants, it would not be surprising if animal models failed to replicate the role for hypoxia-ischemia. Thus, the positive experimental observations are particularly compelling in this context. The demonstration of protection against hypoxic-ischemic injury by drugs that act along the lethal pathways triggered by hypoxia-ischemia, e.g., erythropoietin (antiapoptosis, angiogenesis) and AMPA antagonists, is also compelling evidence of these pathways in the pathogenesis of the brain damage (72, 123, 124) (Table 8.1).

8.7 Intrinsic Vulnerability of the Cerebral White Matter of the Premature Newborn to Hypoxia-Ischemia

A growing body of evidence shows that the cerebral white matter developing OLs, the principal target in the diffuse component of PVL in the premature brain, exhibits a particular maturation-dependent vulnerability to both free radical attack and excitotoxicity, as summarized next. Clearly there is no single one factor, but rather, a convergence of multiple factors that account for the maturational vulnerability of the developing white matter to hypoxia-ischemia (Table 8.1).

Vulnerability to Free Radical Attack. The pre-OL is the predominant phase of the OL lineage observed in human cerebral white matter in the second and third trimesters during the peak period of vulnerability to PVL (32, 33) (Table 8.2). The mechanisms underlying the maturation-dependent vulnerability of developing OLs to ROS attack have been addressed in both human brain and in experimental studies (26, 29, 30, 32, 73, 74). The findings in human brain indicate a delay in development of enzymes at the superoxide dismutase (SOD) step, both MnSOD and Cu/ZnSOD, and catalase (125). Additionally, the possibility that hydrogen peroxide accumulates and is converted to the hydroxyl radical by the Fenton reaction is suggested by the observations of others of the early appearance of iron in developing human white matter, and the acquisition of iron by developing OLs for differentiation (126). Supportive of a relationship between iron and PVL is the observation that for many weeks after human intraventricular hemorrhage, a disorder that sharply increases the risk of PVL, cerebrospinal fluid levels of nonprotein-bound iron are markedly increased (127). Taken together, the findings indicate a maturation-dependent window of vulnerability to oxidative attack during OL development.

Vulnerability to Excitotoxicity. As noted earlier, the critical initiators of excitotoxic cell death are glutamate receptors, and developing OLs express AMPA/kainate type glutamate receptors whose activation results in cell death. Data in animal models show an overexpression of AMPA receptors on the cell somata of pre-OLs vs. mature oligodendroglia (128). Moreover, these pre-OL receptors are deficient in the GluR2 subunit, a feature which renders them calcium-permeable and thereby capable of toxicity. The overexpression of the calcium-permeable receptors in immature rat cerebral white matter correlates with the period of vulnerability to hypoxic-ischemic white matter injury (128). The relevance of these observations in rats is underscored by the recent demonstration in human cerebral white matter pre-OLs of a similar overexpression of AMPA receptors that also are GluR2-deficient and presumably calcium-permeable during the peak period of occurrence of PVL (128) (Table 8.2). In addition, there is a maturation-dependent overexpression of GLT1/EAAT2 in the human cerebral white matter during the peak period of PVL (129); the principal locus for this transporter is the pre-OL (129) (Table 8.2).

Maturation-Dependent Changes in Microglia. There is a transient elevation in microglial cell density in cerebral white matter in the last few weeks of gestation as compared with postnatal ages, i.e., in the peak period of PVL (130). This transient abundance during mid-to-late gestation raises the prospect that these cells participate in developmental processes in the ongoing organization of the white matter (130). These processes include vascularization, involution of the germinal matrix, programmed cell death, axonal development, and myelination (130). The particular involvement of microglia in the pathogenesis of cerebral white matter disease in the premature infant may significantly relate to this developmental overexpression, as it suggests a *priming* of the white matter at the peak period such that microglia are concentrated in the white matter at the right time and in the right place to contribute to its injury when activated, and to thereby contribute to the vulnerability of fetal white matter to hypoxia-ischemia (Table 8.2).

Maturation-Dependent Changes in Axons. Similarly, axons are in a state of very active development in cerebral white matter of the human premature infant (131). The confluence of developing ascending/descending projection fibers, commissural fibers, and corticocortical association fibers in the peritrigonal region is particularly prominent at this time and, notably, is a particular site of predilection for PVL. We have characterized the developmental profile of axons in the cerebral white matter, including with an immunomarker to GAP-43, a synapse-related molecule that is a neuronal membrane phosphoprotein with possible linkage to the submembrane actin cytoskeleton (131). During development, the highest levels of GAP-43 appear along the entire length of the axon as it is elongating, then in preterminal branches and their growth cones in the period in which end arbors are being elaborated (131). After the establishment of stable synapses, most neurons cease expressing GAP-43 at high levels. In the parietal white matter, GAP-43 expression peaks between 30 postconceptional weeks and term (37–40 postconceptional weeks), and corresponds to the onset of expression of phosphorylated neurofilament, a crucial cytoskeletal protein in the axon (131). This critical period for axonal development in the human cerebral white matter coincides with the *window of vulnerability* to PVL (Table 8.2).

8.8 The Causative Role of Synergistic Factors in PVL

An important series of clinical, epidemiological, neuropathologic, and experimental studies indicates that maternal intrauterine infection and fetal systemic inflammation are involved in the pathogenesis of a proportion of cases of PVL, and are the basis for the proposition that maternal infection/fetal inflammation are the *primary* cause of PVL, *exclusive* of hypoxia-ischemia (8). Yet, we review here the mounting evidence that infection/inflammation act synergistically with ischemia to produce white matter injury.

Clinical and Epidemiological Observations. These observations suggest a link between maternal/fetal infection, inflammation, and PVL (132–141), with the possibility that maternal intrauterine infection causes a fetal systemic inflammatory response (in part detected by proinflammatory cytokines in the blood), which in turn results in injury to cerebral white matter (133, 142). A relationship to PVL and/or associated cerebral palsy in premature infants is suggested by the finding of increased incidence of these outcomes in the presence of (a) evidence for maternal/fetal infection/inflammation, e.g., chorioamnionitis, funisitis, premature rupture of membranes (132, 134, 136); (b) elevated levels of proinflammatory cytokines, especially IL-6 and IL-1β, in amniotic fluid or umbilical cord blood (133, 134); and (c) evidence for intrauterine T-cell activation (137). It is noteworthy, however, that nearly all of these studies utilized the ultrasonographic finding of echolucencies, i.e., apparent *cystic* PVL, for the diagnosis of PVS. As noted earlier, cystic PVL now is recognized to account for less than 10% of white matter injury (1), and thus the likely presence of noncystic PVL in the *controls* makes the interpretation of these data difficult. Moreover, two recent studies of former preterm infants show no significant relationship between levels of proinflammatory cytokines in early neonatal blood and the later diagnosis of cerebral palsy (138, 139), and a recent study of 126 infants showed no correlation between clinical chorioamnionitis and ultrasonographically demonstrated white matter injury (140). Moreover, in a careful study of 100 consecutive premature infants studied by MRI at term, there was no significant relationship between the occurrence of chorioamnionitis or prolonged rupture of membranes and moderate to severe white matter injury (55). Of note, an increased rate of cerebral palsy and other neurodevelopmental disabilities has been reported in preterm infants after neonatal sepsis (143). This relationship is particularly apparent among extremely low birth weight infants (401–1,000 g birth weight) of whom 65% exhibit at least one infection during the postnatal period (143). Moreover, because as many as 25% of all infants <1,500 g experience neonatal sepsis (143), the contributory role of postnatal infection/inflammation could be considerable; more data are needed.

Human Neuropathologic Observations. A potential role for inflammation and specifically, cytokines in the pathogenesis of PVL is supported by the neuropathologic finding of proinflammatory cytokines in the human lesion (78, 142, 144). Cytokines comprise a heterogeneous group of polypeptide mediators that activate the immune response and inflammatory responses (5, 11, 23, 77, 78). In the CNS, microglia induce reactive astrocytosis via release of TNF, IL-1, IL-6, and IFN-γ. In vitro studies indicate that TNF-α and IFN-γ have toxic effects upon OLs (77). In an immunohistochemical study of human perinatal brain tissues, we found immunolocalization of IFN-γ in cells with the morphologic features of oligodendroglia, astrocytes and, in the necrotic foci of PVL, macrophages (78). Given the direct toxicity of IFN-γ to OLs (77, 145) and its capability to induce NOS activity in astrocytes (146), as well as to induce proliferation in astrocytes and microglia (145), this cytokine may be important in the pathogenesis of PVL. Developing OLs in the human perinatal brain express the IFN-γ receptor, reinforcing their vulnerability to this cytokine (78) (Table 8.2). In the neuropathologic series of Kadhim et al., 19 cases with PVL and cytokines in the lesion had *asphyxia* in the background (141). Thus, the possibility that hypoxia-ischemia could have led to the inflammatory response in these cases of human PVL is important to consider. This possibility is underscored by the observation in animal models and to some extent in human adult stroke that ischemia/reperfusion is associated with a brisk inflammatory response, characterized by activation of microglia, secretion of cytokines, mobilization, adhesion and migration of macrophages and inflammatory cells, and reactive astrocytosis (75, 76, 147–149). An important interacting role for systemic infection/inflammation with hypoxia-ischemia in a portion of the 19 cases of PVL well-studied by Kadhim et al. is suggested by the finding that 8 of the 19 cases with *asphyxia* were complicated by either fetal inflammation (chorioamnionitis) or neonatal infection (141). Importantly, brain cytokine immunoreactivity (TNF-α, IL-1β) in the PVL cases with infection was at least double the immunoreactivity in the PVL cases without infection. Thus, the possibility of a *potentiating interaction between a hypoxic-ischemic insult and systemic infection/inflammation in the inflammatory response (including activation of microglia) in brain is suggested.* In this regard, activated microglia secrete toxic diffusible products, e.g., ROS and RNS, that are likely more important than cytokines in the genesis of white matter injury. Moreover, the means of activation of microglia by the molecular products of infection is crucial to the cascade to injury. Few data are available for human brain, but involvement of the innate immune system in brain via toll-like receptors on microglia and the results of the activation of these receptors are likely to be relevant (see later). Although the mechanisms by which maternal/fetal infection and inflammation initiate deleterious molecular and structural events in the brain are not fully understood, it is clear that microglia play a central role. It is important here to reemphasize the finding in our study of human PVL of the marked microglial activation discovered in the diffuse component of the lesion (26). Moreover, a resident transient population of microglia (presumably capable of the same sensitizing relationship between the two insults shown in the experimental models) is concentrated in human cerebral white matter at the correct time and in the correct distribution to be activated and become injurious (see earlier) (130).

Infection and Hypoxia-Ischemia-Potentiating Insults. Deleterious interactions between maternal/fetal and postnatal infection/inflammation and hypoxia-ischemia likely occur at several steps in the cascade to white matter injury (149–152). Indeed the relative inconsistency of data supporting a role for infection *alone* in pathogenesis of PVL suggests that a second factor, perhaps hypoxia-ischemia, is necessary. Notably, abundant experimental studies in immature animals show that prior exposure to the molecular products of infection can cause *subthreshold* hypoxic-ischemic insults to be markedly injurious (149–152). Additional support for a potentiating relationship for infection/inflammation and hypoxia/ischemia vis-á-vis PVL is suggested by a recent study of 61 consecutively born preterm infants (<32 gestational weeks): the risk of white matter injury, identified by MRI and ultrasonography, was enhanced with a history of histologic chorioamnionitis only in the presence of concurrent placental vascular disturbances consistent with a placental perfusion defect (153).

The combination of infection/inflammation and hypoxia-ischemia could result in deleterious circulatory effects, impacting ultimately the cerebral circulation. Thus, in several experimental studies deleterious systemic hemodynamic effects have been shown to accompany fetal or neonatal infection and/or exposure to lipopolysaccharide (LPS) or cytokines (149–152). That a similar interaction occurs in the premature infant is suggested from important studies by Yanowitz et al. (153–155). Thus, in one report of 55 premature infants (25–32 gestational weeks), chorioamnionitis was evident in 22 placentae and was associated both with elevated blood IL-6 concentrations and decreased mean arterial blood pressures in the first hours of life (153). In a subsequent report, decreased blood pressures were shown to persist for the entire first week of life (154). Moreover, because several cytokines stimulated by infection, especially TNF-α and IL-1β, exhibit vasoactive properties, it is reasonable to postulate that these inflammatory molecules contribute to impaired cerebrovascular autoregulation, likely important in pathogenesis of PVL (see earlier) (156). Under such circumstances, modest decreases in arterial blood pressure could result in decreases in CBF. Thus, the possibility of amplifying or at least additive effects on the systemic and cerebral circulations is considerable in the presence of both infection/inflammation and hypoxia-ischemia.

8.9 The Potential Role of Cumulative Hypoxic-Ischemic Insults in PPPI

Our studies with near-infrared spectroscopy of cerebral hemodynamics in premature infants show frequent, but often not marked, declines in CBF that are associated with an increased risk for PVL (93). These observations suggest that PVL may develop as a result of an *accumulation of hypoxic-ischemic insults,* a possibility that is supported by the demonstration by neuropathologic studies and recent MRI studies that white matter injury in premature infants increases markedly in frequency with increasing duration of survival (6, 34, 157). Moreover, in our neuropathologic studies of PVL, lesions of multiple ages, i.e., acute, subacute, and chronic, typically are present in association with each other (6, 7, 26, 34, 46, 71). Does the lesion become more frequent during the neonatal period because of (a) an additive effect of many small decreases in the CBF, (b) alterations in excitatory amino acid receptor subunit composition caused by the initial insult(s), thereby rendering the receptors more sensitive to subsequent insults, as shown in cultured pre-OLs and in an animal model of PVL (66), and/or (c) a sensitizing effect of maternal/fetal infection, not severe enough to produce injury alone, on subsequent or concomitant hypoxic-ischemic insults, perhaps also not severe enough to produce injury alone? *These possibilities of course are not mutually exclusive.*

8.10 Conclusions

At the outset of this review, we asked the question: Is PPPI due to hypoxia-ischemia? We believe the answer is *yes*, at least for white matter injury, given the interrelated confirmatory evidence from multiple disciplines, e.g., neonatology, human neuropathology, neuroscience, epidemiology, and neuroradiology (Tables 8.1 and 8.2). Yet, the search for a *single* or even primary cause of PPPI seems outmoded, in light of the current appreciation of the complexity of causation (see earlier), and the evidence for multiple insults occurring in sick premature infants in multiple organ systems at simultaneous and multiple time points. In addition, a growing body of multidisciplinary evidence suggests that infection/inflammation and hypoxia-ischemia potentiate each other to produce PVL. Clearly, the mechanisms involved in the deleterious interaction of hypoxic/ischemia and infection/inflammation in PPPI are a major direction for developing the means to ameliorate white and gray matter brain injury in the premature infant. Taken together, these data indicate that more than one cause triggers the final common pathways to cell death, i.e., glutamate, cytokine, and free radical toxicity. Consequently, targeting these shared pathways is an important strategy in preventing the death of developing OLs and neurons. Moreover, the replacement of OL progenitors by stem cell strategies may represent the key means to override OL and neuronal cell death overall, irrespective of the cause.

Acknowledgments We are grateful for the many contributions of our dedicated collaborators to the work in perinatal brain injury from our institution. We also appreciate the help of Mr. Richard A. Belliveau and Ms. Irene Miller in manuscript preparation. This work was supported by grants from the NINDS (PO1-NS38475) and NICHD (Children's Hospital Developmental Disabilities Research Center) (P30-HD18655).

References

1. Volpe JJ. Cerebral white matter injury of the premature infant – more common than you think. Pediatrics 2003; 112: 176–180.
2. Woodward LJ, Edgin JO, Thompson D, Inder TE. Object working memory deficits predicted by early brain injury and development in the preterm infant. Brain 2005; 128: 2578–2587.
3. Ancel PY, Livinec F, Larroque B, et al. Cerebral palsy among very preterm children in relation to gestational age and neonatal ultrasound abnormalities: The EPIPAGE cohort study. Pediatrics 2006; 117: 828–835.
4. Nosarti C, Giouroukou E, Micall N, Rifkin L, Morris RG, Murray RM. Impaired executive functioning in young adults born very preterm. J Int Neuropsychol Soc 2007; 13: 571–581.
5. Volpe JJ. Neurology of the Newborn. 4th ed. Philadelphia, PA: WB Saunders, 2001.
6. Pierson CR, Folkerth RD, Billiards SS, Trachtenberg FL, Drinkwater ME, Volpe JJ, Kinney HC. Gray matter injury associated with periventricular leukomalacia in the premature infant. Acta Neuropathol 2007; 114: 619–631.
7. Kinney HC, Panigrahy A, Newburger JW, Jonas RA, Sleeper LA. Hypoxic-ischemic brain injury in infants with congenital heart disease dying after cardiac surgery. Acta Neuropathol 2005; 110: 563–578.
8. Dammann O, Kuban KCK, Leviton A. Perinatal infection, fetal inflammatory response, white matter damage, and cognitive limitations in children born preterm. Ment Retard Dev Disabil Res Rev 2002; 8: 46–50.
9. Nelson KB. Is it HIE? Any why that matters. Acta Paediatr 2007; 96: 1113–1114.
10. Kumar V, Abbas AK, Fausto N. Robbins and Cotran pathologic basis of disease. Philadelphia, PA: Elsevier Saunders, 2005, pp. 21–25.

11. Kinney HC, Haynes RL, Folkerth RD. White matter disorders in the perinatal period. In: Golden JA, Harding BN (eds). Pathology and Genetics: Acquired and Inherited Diseases of the Developing Nervous System. Basel: ISN Neuropathology Press, 2004, pp. 156–170.
12. Banker BQ, Larroche JC. Periventricular leukomalacia of infancy. A form of neonatal anoxic encephalopathy. Arch Neurol 1962; 7: 386–410.
13. Armstrong DL, Sauls CD, Goddard-Finegold J. Neuropathologic findings in short-term survivors of intraventricular hemorrhage. Am J Dis Child 1987; 141: 617–621.
14. Armstrong D, Norman MG. Periventricular leucomalacia in neonates. Complications and sequelae. Arch Dis Child 1974; 49: 367–375.
15. DeReuck J, Chattha AS, Richardson E Jr. Pathogenesis and evolution of periventricular leukomalacia in infancy. Arch Neurol 1972; 27: 229–236.
16. Marin Padilla M. Developmental neuropathology and impact of perinatal brain damage. III. Gray matter lesions of the neocortex. J Neuropathol Exp Neurol 1999; 58: 407–429.
17. Peterson BS, Vohr B, Staib LH, Cannistraci CJ, Dolberg A, Schneider KC, Katz KH, Westerveld M, Sparrow S, Anderson AW, Duncan CC, Makuch RW, Gore JC, Ment LR. Regional brain volume abnormalities and long-term cognitive outcome in preterm infants. JAMA 2000; 284: 1939–1947.
18. Inder TE, Warfield SK, Wang H, Huppi PS, Volpe JJ. Abnormal cerebral structure is present at term in premature infants. Pediatrics 2005; 115: 286–294.
19. Lin Y, Okumura A, Hayakawa F, Kato K, Kuno T, Watanabe K. Quantitative evaluation of thalami and basal ganglia in infants with periventricular leukomalacia. Dev Med Child Neurol 2001; 43: 481–485.
20. Ricci D, Anker S, Cowan F, Pane M, Gallini F, Luciano R, Donvito V, Baranello G, Cesarini L, Bianco F, Rutherford M, Romagnoli C, Atkinson J, Braddick O, Guzzetta F, Mercuri E. Thalamic atrophy in infants with PVL and cerebral visual impairment. Early Hum Dev 2006; 82: 591–595.
21. Isaacs EB, Lucas A, Chong WK, Wood SJ, Johnson CL, Marshall C, Vargha-Khadem F, Gadian DG. Hippocampal volume and everyday memory in children of very low birth weight. Pediatr Res 2000; 47: 713–720.
22. Limperopoulos C, Soul JS, Haidar H, Huppi PS, Bassan H, Warfield SK, Robertson RL, Moore M, Akins P, Volpe JJ, du Plessis AJ. Impaired trophic interactions between the cerebellum and the cerebrum among preterm infants. Pediatrics 2005; 116: 844–850.
23. Kinney HC, Armstrong DL. Perinatal neuropathology. In: Graham DI, Lantos PE (eds) Greenfield's Neuropathology. London: Arnold, 2002, pp. 557–559.
24. Leviton A, Gilles FH. Acquired perinatal leukoencephalopathy. Ann Neurol 1984; 16: 1–8.
25. Haynes RL, Folkerth RF, Szweda LI, Volpe JJ, Kinney HC. Lipid peroxidation during human cerebral myelination. J Neuropathol Exp Neurol 2006; 65: 894–904.
26. Haynes RL, Folkerth RD, Keefe RJ, Sung I, Swzeda LI, Rosenberg PA, Volpe JJ, Kinney HC. Nitrosative and oxidative injury to premyelinating oligodendrocytes is accompanied by microglial activation in periventricular leukomalacia in the human premature infant. J Neuropathol Exp Neurol 2003; 62: 441–450.
27. Back SA, Luo NL, Borenstein NS, Levine JM, Volpe JJ, Kinney HC. Late oligodendrocyte progenitors coincide with the developmental window of vulnerability for human perinatal white matter injury. J Neurosci 2001; 21: 1302–1312.
28. Back SA, Luo NL, Borenstein NS, Volpe JJ, Kinney HC. Arrested oligodendrocyte lineage progression during human cerebral white matter development: dissociation between the timing of progenitor differentiation and myelinogenesis. J Neuropathol Exp Neurol 2002; 61: 197–211.
29. Back SA, Gan X, Li Y, Rosenberg PR, et al. Maturation-dependent vulnerability of oligodendrocytes to oxidative stress-induced death caused by glutathione depletion. J Neurosci 1998; 18: 6241–6253.
30. Haynes RL, Baud O, Li J, et al. Oxidative and nitrative injury in periventricular leukomalacia: a review. Brain Pathol 2005; 15: 225–233.

31. Counsell SJ, Allsop JM, Harrison MC, et al. Diffusion-weighted imaging of the brain in preterm infants with focal and diffuse white matter abnormality. Pediatrics 2003; 112 (Part 1): 1–7.
32. Back SA, Luo NL, Mallinson RA, O'Malley JP, et al. Selective vulnerability of preterm white matter to oxidative damage defined by F(2)-isoprostanes. Ann Neurol 2005; 58: 108–120.
33. Iida K, Takashima S, Ueda K. Immunohistochemical study of myelination and oligodendrocyte in infants with periventricular leukomalacia. Pediatr Neurol 1995; 13: 296–304.
34. Billiards SS, Haynes RL, Folkerth RD, Borenstein NS, Trachtenberg FL, Rowitch DH, Ligon KL, Volpe JJ, Kinney HC. Myelin abnormalities despite total oligodendrocyte cell preservation in periventricular leukomalacia. Brain Pathol 2008; 18: 156–163.
35. Ligon KL, Kesari S, Kitada M, et al. Development of NG2 neural progenitor cells requires Olig gene function. Proc Natl Acad Sci USA 2006; 103: 7853–7858.
36. Sur M, Rubenstein JL. Patterning and plasticity of the cerebral cortex. Science 2005; 310: 805–810.
37. Monchi O, Petrides M, Strafella AP, Worsley KJ, Doyon J. Functional role of the basal ganglia in the planning and execution of actions. Ann Neurol 2006; 59: 257–264.
38. Leutgeb S, Leutgeb JK, Moser MB, Moser EI. Place cells, spatial maps and the population code for memory. Curr Opin Neurobiol 2005; 15: 738–746.
39. Schmahmann JD, Caplan D. Cognition, emotion and the cerebellum. Brain 2006; 129: 290–292.
40. Ghosh A, Shatz CJ. A role for subplate neurons in the patterning of connections from thalamus to neocortex. Development 1993; 117: 1031–1047.
41. McQuillen PS, Ferriero DM. Perinatal subplate neuron injury: implications for cortical development and plasticity. Brain Pathol 2005; 15: 250–260.
42. McQuillen PS, Sheldon RA, Shatz CJ, Ferriero DM. Selective vulnerability of subplate neurons after early neonatal hypoxia-ischemia. J Neurosci 2003; 23: 3308–3315.
43. Deguchi K, Oguchi K, Takashima S. Characteristic neuropathology of leukomalacia in extremely low birth weight infants. Pediatr Neurol 1997; 16: 296–300.
44. Meng SZ, Arai Y, Deguchi K, Takashima S. Early detection of axonal and neuronal lesions in prenatal-onset periventricular leukomalacia. Brain Dev 1997; 19: 480–484.
45. Arai Y, Deguchi K, Mizuguchi M, et al. Expression of beta-amyloid precursor protein in axons of periventricular leukomalacia brains. Pediatr Neurol 1995; 13: 161–163.
46. Haynes RL, Billiards SS, Borenstein NS, Volpe JJ, Kinney HC. Diffuse axonal injury in periventricular leukomalacia. Pediatr Res 2008; 63: 656–661.
47. Suurmeijer AJ, van der Wijk J, van Veldhuisen DJ, Yang F, Cole GM. Fractin immunostaining for the detection of apoptotic cells and apoptotic bodies in formalin-fixed and paraffin-embedded tissue. Lab Invest 1999; 79: 619–620.
48. Underhill SM, Goldberg MP. Hypoxic injury of isolated axons is independent of ionotropic glutamate receptors. Neurobiol Dis 2007; 25: 284–290.
49. Taveggia C, Zanazzi G, Petrylak A, Yano H, Rosenbluth J, Einheber S, Xu X, Esper RM, Loeb JA, Shrager P, Chao MV, Falls DL, Role L, Salzer JL. Neuregulin-1 type III determines the ensheathment fate of axons. Neuron 2005; 47: 681–694.
50. Stevens B, Porta S, Haak LL, Gallo V, Fields RD. Adenosine: a neuron-glial transmitter promoting myelination in the CNS in response to action potentials. Neuron 2002; 36: 855–868.
51. Huppi PS, Murphy B, Maier SE, Zientara GP, Inder TE, Barnes PD, Kikinis R, Jolesz FA, Volpe JJ. Microstructural brain development after perinatal cerebral white matter injury assessed by diffusion tensor magnetic resonance imaging. Pediatrics 2001; 107: 455–460.
52. Counsell SJ, Dyet LE, Larkman DJ, Nunes RG, Boardman JP, Allsop JM, Fitzpatrick J, Srinivasan L, Cowan FM, Hajnal JV, Rutherford MA, Edwards AD. Thalamo-cortical connectivity in children born preterm mapped using probabilistic magnetic resonance tractography. Neuroimage 2007; 34: 896–904.
53. Vangberg TR, Skranes J, Dale AM, Martinussen M, Brubakk AM, Haraldseth O. Changes in white matter diffusion anisotropy in adolescents born prematurely. Neuroimage 2006; 32: 1538–1548.
54. Thomas B, Eyssen M, Peeters R, Molenaers G, Van Hecke P, De Cock P, Sunaert S. Quantitative diffusion tensor imaging in cerebral palsy due to periventricular white matter injury. Brain 2005; 128: 2562–2577.

Color Plates

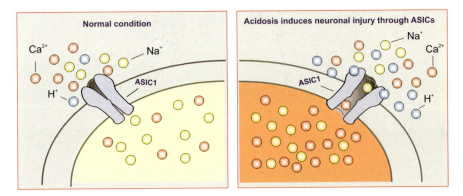

Fig. 2.1 Role of ASIC activation in acidosis-induced neuronal injury. (See complete caption on page 34)

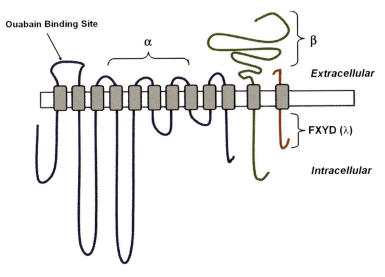

Fig. 4.1 Molecular structure of the Na+/K+-ATPase. The diagraph illustrates the α, β, and λ subunits in the Na+/K+-ATPase protein

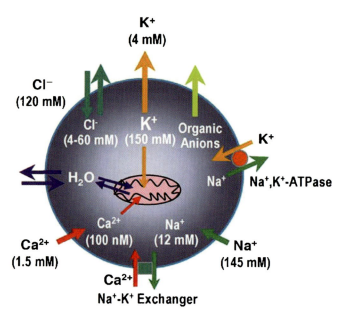

Fig. 4.2 Ionic distributions across the membranes and regulation pathways for K+ homeostasis. (See complete caption on page. 62)

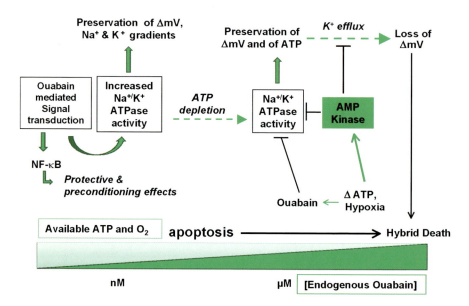

Fig. 4.3 Strategies for modulation of Na+/K+-ATPase under pathological conditions.

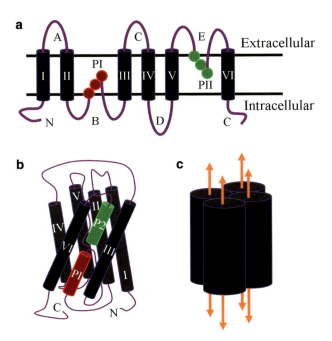

Fig. 5.1 Structural model of AQP water channel. (See complete caption on page 81)

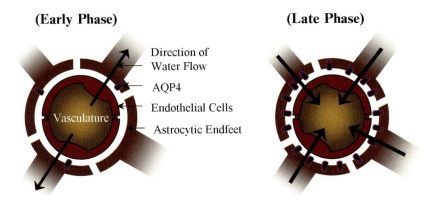

Fig. 5.2 Proposed model of AQP4-mediated resolution of brain edema. (See complete caption on page 85)

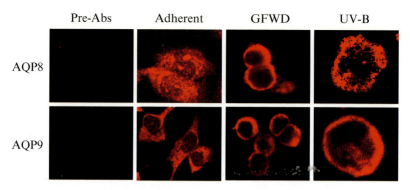

Fig. 5.3 Aquaporin 8 and 9 translocation in RUCA-1 cells (106) during growth factor withdrawal (GFWD) and ultraviolet-B (UV-B)-induced apoptosis. (See complete caption on page 87)

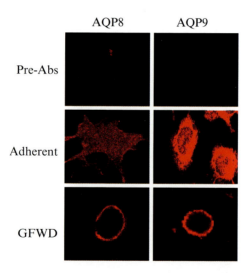

Fig. 5.4 Aquaporin translocation during apoptosis in ovarian granulosa cells. (See complete caption on page 88)

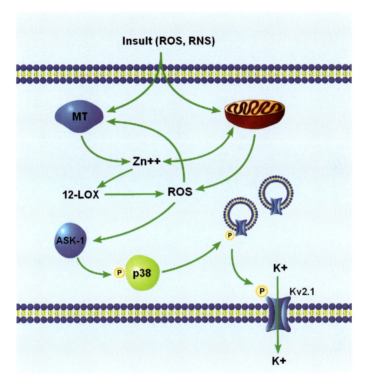

Fig. 6.1 Schematic of proposed pathway linking ROS/RNS-mediated liberation of intracellular Zn^{2+} and a subsequent enhancement of Kv2.1-mediated K^+ currents. (See complete caption on page 101)

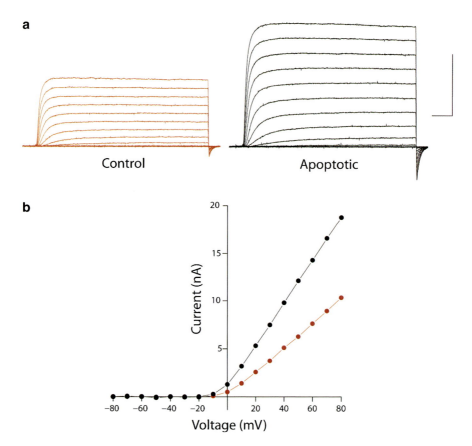

Fig. 6.2 DTDP induces characteristic apoptotic K$^+$ current surge. (See complete caption on page 102)

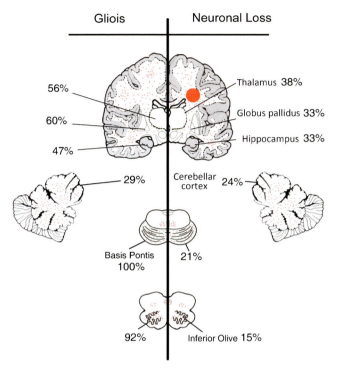

Fig. 8.1 Summary diagram of the neuropathology of PPPI. (See complete caption on page 156)

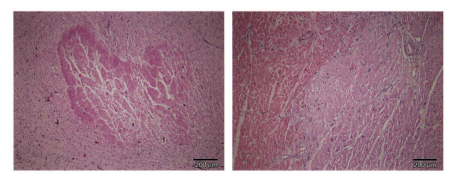

Fig. 8.2 Coagulative necrosis in acute PVL (**a**) is morphologically similar to that in acute myocardial infarction (**b**), with preservation of cellular outlines, nuclear changes and loss, and eosionophilic hypereosinophilia or pallor. (See complete caption on page 157)

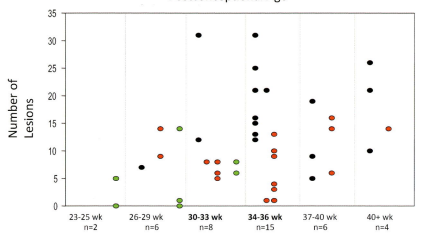

Fig. 8.5 The number of lesions, i.e., neuronal loss and gliosis, in all gray matter sites is presented according to diagnosis and divided into intervals of postconceptional age. (See complete caption on page 160)

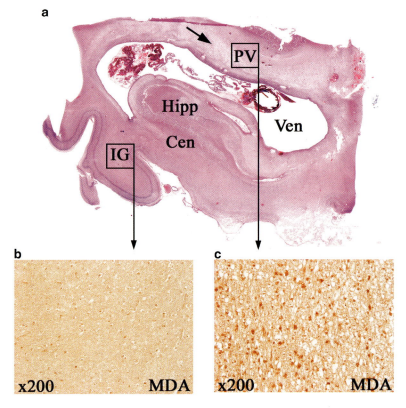

Fig. 8.6 A. In PVL, there is focal necrosis in the periventricular white matter, associated with DWMG in the surrounding deep white matter and relative sparing in the intragyral white matter. (See complete caption on page 167)

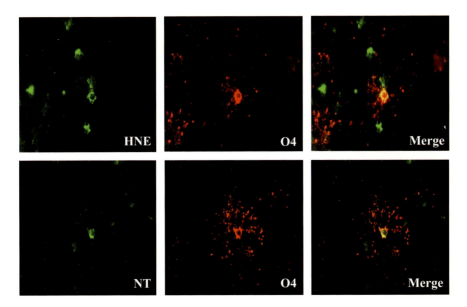

Fig. 8.7 There is oxidative and nitrative injury to developing OLs. Nitrotyrosine (NT) colocalizes (*yellow* signal) with the marker for premyelinating OLs (O4, *red*). (See complete caption on page 171)

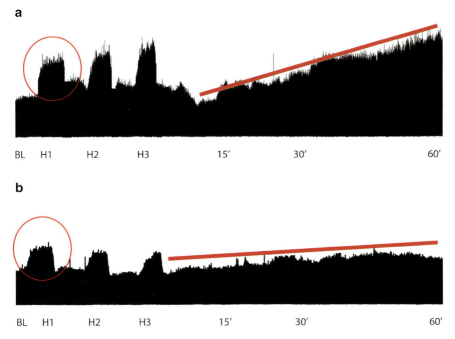

Fig. 9.1 Phrenic nerve motor neurogram recordings of adult male rats exposed to normoxia (**a**) or perinatal hypoxia (**b**). (See complete caption on page 200)

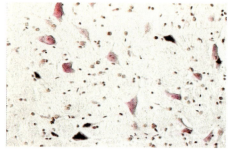

Fig. 10.3 *Top*: Histological section of the brain stem at the level of plate 61 (Paxinos and Watson) showing TUNEL-positive neurons in the RVLM, counterstained with H&E, from a rat 4 days after cardiac arrest and resuscitation. *Middle* and *Bottom*: Dots indicating locations of TUNEL-positive neuronal cell bodies at the indicated brain stem levels

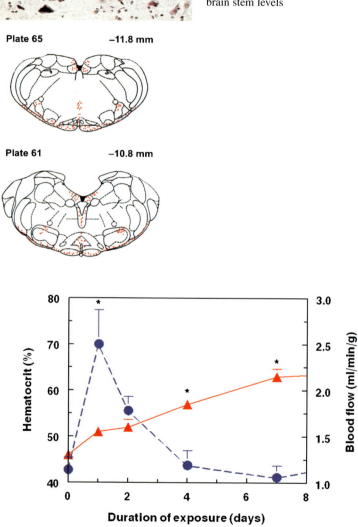

Fig. 10.5 Hematocrit change and blood flow change in brain stem during hypoxia. The blood flow in frontal cortex (circles, dashed line) was increased significantly on the first and second day and returned to the baseline by the fourth day of hypoxic exposure, whereas by day 4 of exposure, the hematocrit (triangles, solid line) was significantly elevated and remained elevated thereafter (* indicates values significantly different from the normoxic baseline, $p < 0.05$)

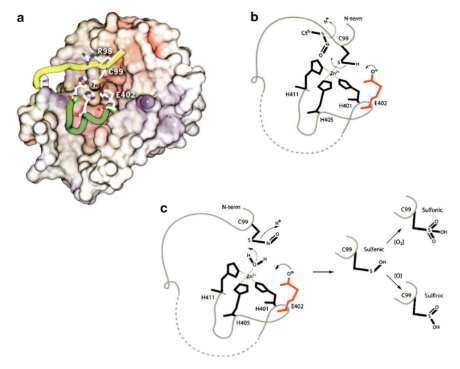

Fig. 11.1 Model of MMP-9 activation by S-nitrosylation and subsequent oxidation. (See complete caption on page 227)

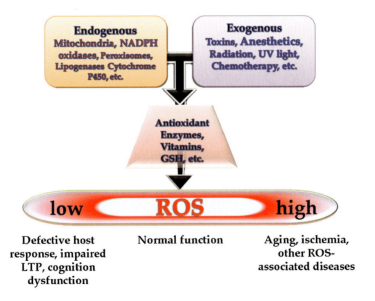

Scheme 12.1 A critical balance of ROS levels is needed for normal nervous system function, with pathological conditions, including hypoxia-ischemia, associated with elevated ROS at one end of the spectrum, and inadequate ROS levels leading to decreased synaptic activity, impaired long-term potentiation (LTP), and cognitive dysfunction at the other (based on Finkel and Holbrook 2000)

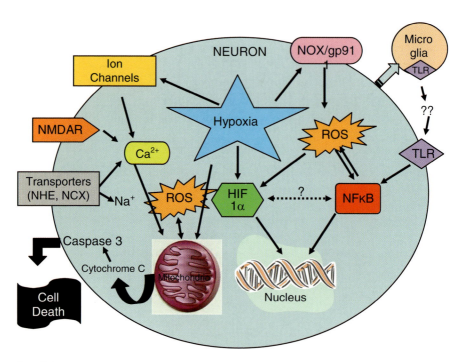

Fig. 13.1 Multiple pathways are activated or altered by exposure to chronic constant hypoxia during the early postnatal period. (See complete caption on page 267)

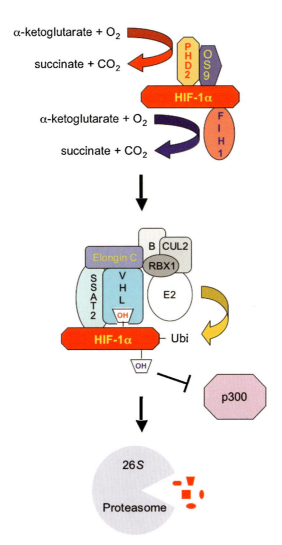

Fig. 14.1 HIF-1 activity is regulated by O_2-dependent hydroxylation. Under normoxic conditions, HIF-1α is synthesized, but is rapidly subjected to prolyl hydroxylation (*red*) by PHD2. OS9 promotes hydroxylation by binding to both PHD2 and HIF-1α. FIH-1 mediates asparaginyl hydroxylation (*blue*), which blocks the binding of the coactivator p300 (or the related protein CBP). Prolyl hydroxylation of HIF-1α is required for binding of VHL, which together with SSAT2 recruits elongin C, which in turn recruits a ubiquitin-protein ligase complex containing elongin B (B), ring box protein 1 (RBX1), cullin 2 (CUL2), and an E2 ubiquitin-conjugating enzyme. Ubiquitination of HIF-1α targets the protein for degradation by the 26*S* proteasome

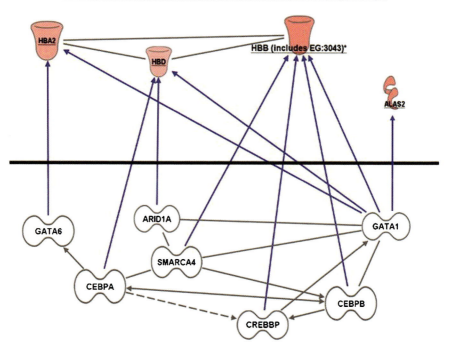

Fig. 15.3 Graphical representation of transcription factor network that regulates the expression of genes related to oxygen transport in CCH-treated mice. The relationship of transcription factors and the target genes was summarized by Network analysis using Ingenuity knowledge database (Ingenuity System, http://www.ingenuity.com). The network indicates that upregulation of oxygen delivery system requires coordination of multiple transcription factors

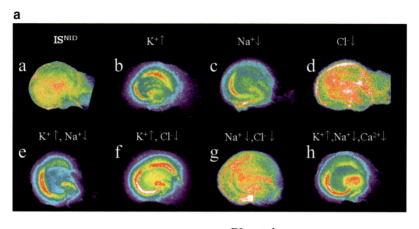
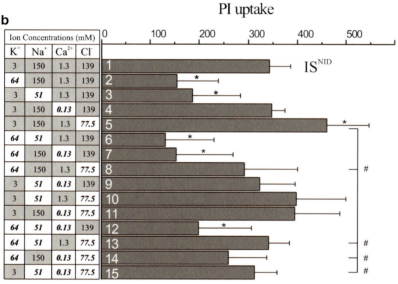

Fig. 16.1 Effects of ion shifts on cell death in astrocyte-rich area in ischemic solution (IS)-treated hippocampal slices. (**a**) pseudocolor images of propidium iodide (PI) fluorescence acquired from eight different hippocampal slices treated with nonion disturbed IS (IS^{NID}) (i), and IS^{NID} with different combinations of ion changes as shown above each slice (ii–viii). Pseudocolor represents the PI fluorescence intensity with purple indicating the least and white the most intense level. Suffix '↑' or '↓' of each labeling indicates that the concentration of that ion in modified IS^{NID} was adjusted to the one used in IS. (**b**) Bar graph showing PI uptake in astrocyte-rich area under different experimental conditions. Different combinations of ion changes are shown in a table to the left of the bar graph. Each row in the table presents ion concentrations of major species used in the IS^{NID} and the gray bars right next to the row show the PI uptake under each experimental condition. $^*p < 0.05$, vs. IS^{NID} group (bar 1), $^\#p < 0.05$, vs. IS^{NID} with low Cl$^-$ group (bar 5), $n = 12$ for each group, Student's t test

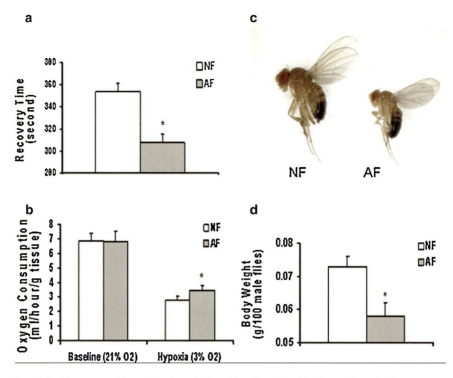

Fig. 17.1 Phenotype of hypoxia-selected flies. (See Complete Caption on Page 327)

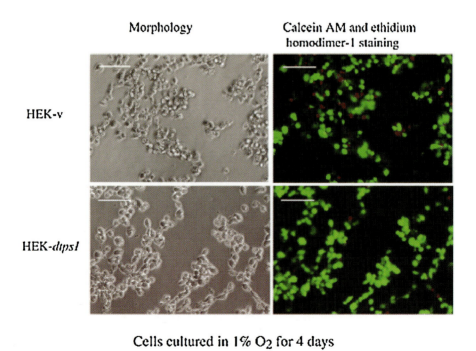

Fig. 17.3 Morphology, calcein, and PI of HEK cells in culture. (See Complete Caption on Page 331)

55. Prayer D, Barkovich AJ, Kirschner DA, Prayer LM, Roberts TP, Kucharczyk J, Moseley ME. Visualization of nonstructural changes in early white matter development on diffusion-weighted MR images: evidence supporting premyelination anisotropy. Am J Neuroradiol 2001;22: 1572–1576.
56. Oka A, Belliveau MJ, Rosenberg PA, Volpe JJ. Vulnerability of oligodendroglia to glutamate: pharmacology, mechanisms and prevention. J Neurosci 1993; 13: 1441–1453.
57. Liu Y, Silverstein FS, Skoff R, Barks JDE. Hypoxic-ischemic oligodendroglial injury in neonatal rat brain. Pediatr Res 2002; 51: 25–33.
58. Yoshioka A, Backskai B, Pleasure D. Pathophysiology of oligodendroglial excitotoxicity. J Neurosci Res 1996; 46: 427–438.
59. Jelinski SE, Yager JY, Juurlink BHJ. Preferential injury of oligodendroblasts by a short hypoxic-ischemic insult. Brain Res 1999; 815: 150–153.
60. Rosenberg PA, Dai W, Gan XD, Ali S, et al. Mature myelin basic protein expressing oligodendrocytes are insensitive to kainate toxicity. J Neurosci Res 2003; 71: 237–245.
61. McDonald JW, Althomsons SP, Hyrc KL, Choi DW, et al. Oligodendrocytes from forebrain are highly vulnerable to AMPA/kainate receptor-mediated excitotoxicity. Nat Med 1998; 4: 291–297.
62. Rao VLR, Bowen KK, Dempsey RJ. Transient focal cerebral ischemia down-regulates glutamate transporters GLT-1 and EAAC1 expression in rat brain. Neurochem Res 2001; 26: 497–502.
63. Ness JK, Romanko MJ, Rothstein RP, Wood TL, Levison SW. Perinatal hypoxia-ischemia induces apoptotic and excitotoxic death of periventricular white matter oligodendrocyte progenitors. Dev Neurosci 2001; 23: 203–208.
64. Sanchez-Gomez MV, Matute C. AMPA and kainate receptors each mediate excitotoxicity in oligodendroglial cultures. Neurobiol Dis 1999; 6: 475–485.
65. Back SA, Craig AS, Kayton R, Luo NL, et al. Hypoxia-ischemia preferentially triggers glutamate depletion from oligodendroglia and axons in perinatal cerebral white matter. J Cereb Blood Flow Metab 2007; 27: 334–347.
66. Deng W, Rosenberg PA, Volpe JJ, Jensen FE. Calcium-permeable AMPA/kainate receptors mediate toxicity and preconditioning by oxygen-glucose deprivation in oligodendrocyte precursors. Proc Natl Acad Sci USA 2003; 100: 6801–6806.
67. Micu I, Jiang Q, Coderre E, et al. NMDA receptors mediate calcium accumulation in myelin during chemical ischaemia. Nature 2006; 439: 988–992.
68. Salter MG, Fern R. NMDA receptors are expressed in developing oligodendrocyte processes and mediate injury. Nature 2005; 438: 1167–1171.
69. Deng W, Wang H, Rosenberg PA, Volpe JJ, et al. Role of metabotropic glutamate receptors in oligodendrocyte excitotoxicity and oxidative stress. Proc Natl Acad Sci USA 2004; 101: 7751–7756.
70. Deng W, Yue Q, Rosenberg PA, Volpe JJ, et al. Oligodendrocyte excitotoxicity determined by local glutamate accumulation and mitochondrial function. J Neurochem 2006; 96: 213–222.
71. Sanchez-Gomez MV, Alberdi E, Ibarretxe G, Torre I, et al. Caspase-dependent and caspase-independent oligodendrocyte death mediated by AMPA and kainate receptors. J Neurosci 2003; 23; 9519–9528.
72. Follett P, Rosenberg P, Volpe JJ, Jensen FE. NBQX attenuates excitotoxic injury in developing white matter. J Neurosci 2000; 20: 9235–9241.
73. Yonezawa M, Back SA, Gan X, Rosenberg PA, et al. Cystine deprivation induces oligodendroglial death: rescue by free radical scavengers and by a diffusible glial factor. J Neurochem 1996; 67: 566–573.
74. Baud O, Li J, Zhang Y, Neve RL, et al. Nitric oxide-induced cell death in developing oligodendrocytes is associated with mitochondrial dysfunction and apoptosis-inducing factor translocation. Eur J Neurosci 2004; 20: 1713–1726.
75. Okuma Y, Uehara T, Miyazaki H, Miyasaka T, et al. The involvement of cytokines, chemokines and inducible nitric oxide synthase (iNOS) induced by a transient ischemia in neuronal survival/death in rat brain. Folia Pharmacol 1998; 111: 37–44.

76. Bona E, Andersson A-L, Blomgren K, Gilland E, et al. Chemokine and inflammatory cell response to hypoxia-ischemia in immature rats. Pediatr Res 1999; 45: 500–509.
77. Andrews T, Zhang P, Bhat NR. TNF-α potentiates IFN gamma-induced cell death in oligodendrocyte progenitors. J Neurosci Res 1998; 54: 574–583.
78. Folkerth RD, Keefe RJ, Haynes RL, Trachtenberg FL, Volpe JJ, Kinney HC. Interferon-gamma expression in periventricular leukomalacia in the human brain. Brain Pathol 2004; 14: 265–274.
79. Li J, Baud O, Vartanian T, Volpe JJ, et al. Peroxynitrite generated by inducible nitric oxide synthase and NADPH oxidase mediates microglial toxicity to oligodendrocytes. Proc Natl Acad Sci USA 2005; 102: 9936–9941.
80. Brown GC. Mechanisms of inflammatory neurodegeneration: iNOS and NADPH oxidase. Biochem Soc Trans 2007; 35: 1119–1121.
81. Nakamura Y, Okudera T, et al. Vascular architecture in white matter of neonates: its relationship to periventricular leukomalacia. J Neuropathol Exp Neurol 1994; 53: 582–589.
82. Takashima S, Tanaka K. Development of cerebrovascular architecture and its relationship to periventricular leukomalacia. Arch Neurol 1978; 35: 11–16.
83. Kuban KC, Gilles FH. Human telencephalic angiogenesis. Ann Neurol 1985; 17: 539–548.
84. Rorke LB. Anatomical features of the developing brain implicated in pathogenesis of hypoxic–ischemic injury. Brain Pathol 1992; 2: 211–221.
85. Van den Bergh R. Centrifugal elements in the vascular pattern of the deep intracerebral blood supply. Angiology 1960; 20: 88–94.
86. Wigglesworth JS, Pape KE. An integrated model for haemorrhagic and ischaemic lesions in the newborn brain. Early Hum Dev 1978; 2: 179–199.
87. Altman DI, Powers WJ, Perlman JM, Herscovitch P, et al. Cerebral blood flow requirement for brain viability in newborn infants is lower than in adults. Ann Neurol 1988; 24: 218–226.
88. Greisen G, Pryds O. Low CBF, discontinuous EEG activity, and periventricular brain injury in ill, preterm neonates. Brain Dev 1989; 11: 164–168.
89. Borch K, Greisen G. Blood flow distribution in the normal human preterm brain. Pediatr Res 1998; 41: 28–33.
90. Powers WJ, Grubb RL, Darriet D, Raichle ME. Cerebral blood flow and cerebral metabolic rate of oxygen requirements for cerebral function and viability in humans. J Cereb Blood Flow Metab 1985; 5: 600–608.
91. Muller AM, Morales C, Briner J, Baenziger O, et al. Loss of CO_2 reactivity of cerebral blood flow is associated with severe brain damage in mechanically ventilated very low birth weight infants. Eur J Paediatr Neurol 1997; 5: 157–163.
92. Pryds O, Edwards AD. Cerebral blood flow in the newborn infant. Arch Dis Child 1996; 74: F63–F69.
93. Soul JS, Hammer PE, Tsuji M, Saul P, et al. Fluctuating pressure-passivity is common in the cerebral circulation of sick premature infants. Pediatr Res 2007; 61: 467–473.
94. Tsuji M, Saul JP, du Plessis A, Eichenwald E, et al. Cerebral intravascular oxygenation correlates with mean arterial pressure in critically ill premature infants. Pediatrics 2000; 106: 625–632.
95. Lemmers PM, Toet M, van Schelven LJ, van Bel F. Cerebral oxygenation and cerebral oxygen extraction in the preterm infant: the impact of respiratory distress syndrome. Exp Brain Res 2006; 173: 458–467.
96. Nelson MD, Gonzalez-Gomez I, Gilles FH. The search for human telencephalic ventriculofugal arteries. Am J Neuroradiol 1991; 12: 215–222.
97. von Siebenthal K, Beran J, Wolf M, Keel M, et al. Cyclical fluctuations in blood pressure, heart rate and cerebral blood volume in preterm infants. Brain Dev 1999; 21: 529–534.
98. Wiswell TE, Graziani LJ, Kornhauser MS, Stanley C, et al. Effects of hypocarbia on the development of cystic periventricular leukomalacia in premature infants treated with high-frequency jet ventilation. Pediatrics 1996; 98: 918–924.
99. Shankaran S, Langer JC, Kazzi SN, Laptook AR, et al. Cumulative index of exposure to hypocarbia and hyperoxia as risk factors for periventricular leukomalacia in low birth weight infants. Pediatrics 2006; 118: 1654–1659.

100. Graziani LJ, Baumgart S, Desai S, Stanley C, et al. Clinical antecedents of neurologic and audiologic abnormalities in survivors of neonatal extracorporeal membrane oxygenation. J Child Neurol 1997; 12: 415–422.
101. Shortland DB, Gibson NA, Levene MI, Archer LN, et al. Patent ductus arteriosus and cerebral circulation in preterm infants. Dev Med Child Neurol 1990; 32: 386–393.
102. Bandera E, Botteri M, Minelli C, Sutton A, Abrams KR, Latronico N. Cerebral blood flow threshold of ischemic penumbra and infarct core in acute ischemic stroke: a systematic review. Stroke 2006; 37; 1334–1339.
103. Levison SW, Rothstein RP, Romanko MJ, Snyder MJ, et al. Hypoxia-ischemia depletes the rat perinatal subventricular zone of oligodendrocyte progenitors and neural stem cells. Dev Neurosci 2001; 23: 234–247.
103a. Folkerth RD, Trachtenberg FL, Haynes RL. Oxidative injury in the cerebral cortex and subplate neurons in periventricular leukomalacia. J Neuropathol Exp Neurol 2008; 67: 677–686.
104. Zaidi AU, Bessert DA, Ong JE, Xu H, et al. New oligodendrocytes are generated after neonatal hypoxic-ischemic brain injury in rodents. Glia 2004; 46: 380–390.
105. Inder TE, Mocatta T, Darlow B, Spencer C, et al. Elevated free radical products in the cerebrospinal fluid of VLBW infants with cerebral white matter injury. Pediatr Res 2002; 52: 213–218.
106. Touil T, Deloire-Grassin MS, Vital C, Petry KG, Broachet B. In vivo damage of CNS myelin and axons induced by peroxynitrite. Neuroreport 2001; 12: 3637–3644.
107. DeSilva TM, Billiards SS, Borenstein NS, Trachenberg FL, Volpe JJ, Kinney HC, Rosenberg PA. Glutamate transporter EAAT2 expression is upregulated in reactive astrocytes in human periventricular leukomalacia. J Comp Neurol 2008; 508: 238–248.
108. Baud O, Daire JL, Dalmaz Y, Fontaine RH, et al. Gestational hypoxia induces white matter damage in neonatal rats: a new model of periventricular leukomalacia. Brain Pathol 2004; 14: 1–10.
109. Cai ZW, Pang Y, Xiao F, Rhodes PG. Chronic ischemia preferentially causes white matter injury in the neonatal rat brain. Brain Res 2001; 898: 126–135.
110. Duncan JR, Cock ML, Harding R, Rees SM. Relation between damage to the placenta and the fetal brain after late-gestation placental embolization and fetal growth restriction in sheep. Am J Obstet Gynecol 2000; 183: 1013–1022.
111. Ikeda T, Murata Y, Quilligan EJ, Choi BH, et al. Physiologic and histologic changes in near-term fetal lambs exposed to asphyxia by partial umbilical cord occlusion. Am J Obstet Gynecol 1998; 178: 24–32.
112. Ivacko JA, Sun R, Silverstein FS. Hypoxic-ischemic brain injury induces an acute microglial reaction in perinatal rats. Pediatr Res 1995; 39: 39–47.
113. Lou HC, Lassen NA, Tweed WA, Johnson G, et al. Pressure passive cerebral blood flow and breakdown of the blood-brain barrier in experimental fetal asphyxia. Acta Paediatr Scand 1970; 68: 57–63.
114. Mallard EC, Rees S, Stringer M, Cock MI, et al. Effects of chronic placental insufficiency on brain development in fetal sheep. Pediatr Res 1998; 43: 262–270.
115. Matsuda T, Okuyama K, Cho K, Hoshi N, et al. Induction of antenatal periventricular leukomalacia by hemorrhagic hypotension in the chronically instrumented fetal sheep. Am J Obstet Gynecol 1999; 181: 725–730.
116. Meng S, Qiao M, Scobie K, Tomanek B, et al. Evolution of magnetic resonance imaging changes associated with cerebral hypoxia-ischemia and a relatively selective white matter injury in neonatal rats. Pediatr Res 2006; 59: 554–559.
117. Olivier P, Baud O, Evrard P, Gressens P, et al. Prenatal ischemia and white matter damage in rats. J Neuropathol Exp Neurol 2005; 64: 998–1006.
118. Uehara H, Yoshioka H, Kawase S, Nagai H, et al. A new model of white matter injury in neonatal rats with bilateral carotid artery occlusion. Brain Res 1999; 837: 213–220.
119. Petersson KH, Pinar H, Stopa EG, Faris RA, et al. White matter injury after cerebral ischemia in ovine fetuses. Pediatr Res 2002; 51: 768–776.

120. Skoff RP, Bessert DA, Barks JDE, Song DK, et al. Hypoxic-ischemic injury results in acute disruption of myelin gene expression and death of oligodendroglial precursors in neonatal mice. Int J Dev Neurosci 2001; 19: 197–208.
121. Rees S, Stringer M, Just Y, Hooper SB, et al. The vulnerability of the fetal sheep brain to hypoxemia at mid-gestation. Brain Res Dev Brain Res 1997; 103: 103–118.
122. Riddle A, Luo NL, Manese M, Beardsley DJ, et al. Spatial heterogeneity in oligodendrocyte lineage maturation and not cerebral blood flow predicts fetal ovine periventricular white matter injury. J Neurosci 2006; 26: 3045–3055.
123. Li Y, Zhongyang L, Keogh CL, Yu SP, Wei L. Erythropoientin-induced neurovascular protection, angiogenesis, and cerebral blood flow restoration after focal ischemia in mice. J Cereb Blood Flow Metab 2007; 27: 1043–1054.
124. Matuez E, Moricz K, Gigler G, Benedek A, Barkoczy J, Levay G, Harsing LG, Szenasi G. Therapeutic time window of neuroprotection by non-competitive AMPA antagonists in transient and permanent focal cerebral ischemia in rats. Brain Res 2006; 1123: 60–67.
125. Folkerth RD, Haynes RL, Borenstein N, Belliveau RA, Trachtenberg FL, Rosenberg PA, Volpe JJ, Kinney HC. Developmental lag of superoxide dismutases relative to other antioxidant enzymes in premyelinated human telencephalic white matter. J Neuropathol Exp Neurol 2004; 14: 265–274.
126. Connor JR, Menzies SL. Relationship of iron to oligodendrocytes and myelination. Glia 1996; 17: 83–93.
127. Savman K, Nilsson UA, Blennow M, Kjellmer I, et al. Non-protein-bound iron is elevated in cerebrospinal fluid from preterm infants with posthemorrhagic ventricular dilation. Pediatr Res 2001; 49: 208–212.
128. Talos DM, Follett PL, Folkerth RD, Fishman RE, et al. Developmental regulation of alpha-amino-3-hydroxy-5-methyl-4-isoxazole-propionic acid receptor subunit expression in forebrain and relationship to regional susceptibility to hypoxic/ischemic injury. II. Human cerebral white matter and cortex. J Comp Neurol 2006; 497: 61–77.
129. DeSilva TM, Kinney HC, Borenstein NS, Trachtenberg FL, Irwin N, Volpe JJ, Rosenberg PA. The glutamate transporter EAAT2 is transiently expressed in developing human cerebral white matter. J Comp Neurol 2007; 501: 879–890.
130. Billiards SS, Haynes RL, Folkerth RD, Trachtenberg FL, Liu L, Volpe JJ, Kinney HC. Development of microglia in the cerebral white matter of the human fetus and infant. J Comp Neurol 2006; 497: 1990–2008.
131. Haynes RL, Borenstein NS, DeSilva TM, Folkerth RD, Liu LG, Volpe JJ, Kinney HC. Axonal development in the cerebral white matter of the human fetus and infant. J Comp Neurol 2005; 484: 156–167.
132. De Felice C, Toti P, Laurini RN, Stumpo M, et al. Early neonatal brain injury in histologic chorioamnionitis. J Pediatr 2001; 138: 101–104.
133. Yoon BH, Jun JK, Romero R, Park KH, et al. Amniotic fluid inflammatory cytokines (interleukin-6, interleukin-1β, and tumor necrosis factor-α), neonatal brain white matter lesions, and cerebral palsy. Am J Obstet Gynecol 1997; 177: 19–26.
134. Yoon BH, Romero R, Park JS, Kim CJ, et al. Fetal exposure to an intra-amniotic inflammation and the development of cerebral palsy at the age of three years. Am J Obstet Gynecol 2000; 182: 675–681.
135. Yoon BH, Romero R, Yang SH, Jun JK, et al. Interleukin-6 concentrations in umbilical cord plasma are elevated in neonates with white matter lesions associated with periventricular leukomalacia. Am J Obstet Gynecol 1996; 174: 1433–1440.
136. Perlman JM, Risser R, Broyles RS. Bilateral cystic periventricular leukomalacia in the premature infant: associated risk factors. Pediatrics 1996; 97: 822–827.
137. Duggan PJ, Maalouf EF, Watts TL, Sullivan MHF, et al. Intrauterine T-cell activation and increased cytokine concentrations in preterm infants with cerebral lesions. Lancet 2001; 358: 1699–1700.
138. Nelson KB, Grether JK, Dambrosia JM, Walsh E, et al. Neonatal cytokines and cerebral palsy in very preterm infants. Pediatr Res 2003; 53: 600–607.

139. Kaukola T, Saryaraj E, Patel DD, Tchernev V, et al. Cerebral palsy is characterized by protein mediators in cord serum. Ann Neurol 2004; 55: 186–194.
140. Locatelli A, Vergani P, Ghidini A, Assi F, et al. Duration of labor and risk of cerebral white-matter damage in very preterm infants who are delivered with intrauterine infection. Am J Obstet Gynecol 2005; 193: 928–932.
141. Kadhim HJ, Tabarki B, Verellen G, De Prez C, et al. Inflammatory cytokines in the pathogenesis of periventricular leukomalacia. Neurology 2001; 56: 1278–1284.
142. Deguchi K, Mizuguchi M, Takashima S. Immunohistochemical expression of tumor necrosis factor a in neonatal leukomalacia. Pediatr Neurol 1996; 14: 13–16.
143. Stoll BJ, Hansen NI, AdamsChapman I, Fanaroff AA, et al. Neurodevelopmental and growth impairment among extremely low-birth-weight infants with neonatal infection. JAMA 2004; 292: 2357–2365.
144. Kadhim HJ, Tabarki B, De Prez C, Rona A-M, et al. Interleukin-2 in the pathogenesis of perinatal white matter damage. Neurology 2002; 58: 1125–1128.
145. Vartanian T, Li Y, Zhao M, Stefansson K. Interferon-gamma reduced oligodendrocyte cell death: implications for the pathogenesis of multiple sclerosis. Mol Med 1995; 1: 732–743.
146. Lee SC, Dickerson DW, Liu W, Brojann CF. Induction of nitric oxide synthase activity in human astrocytes by interleukin-1 beta and interferon-gamma. J Neuroimmunol 1993; 46; 19–24.
147. Sairanen T, Carpen O, Karjalainen-Lindsberg ML, Paetau A, et al. Evolution of cerebral tumor necrosis factor-alpha production during human ischemic stroke. Stroke 2001; 32: 1750–1757.
148. Zhang Z, Chopp M, Powers C. Temporal profile of microglial response following transient (2h) middle cerebral artery occlusion. Brain Res 1997; 744: 189–198.
149. Wang X, Hagberg H, Nie C, Zhu C, et al. Dual role of intrauterine immune challenge on neonatal and adult brain vulnerability to hypoxia-ischemia. J Neuropathol Exp Neurol 2007; 66: 552–561.
150. Ikeda T, Mishima K, Aoo N, Egashira N, et al. Combination treatment of neonatal rats with hypoxia-ischemia and endotoxin induces long-lasting memory and learning impairment that is associated with extended cerebral damage. Am J Obstet Gynecol 2004; 191: 2132–2141.
151. Larouche A, Roy M, Kadhim H, Tsanaclis AM, et al. Neuronal injuries induced by perinatal hypoxic-ischemic insults are potentiated by prenatal exposure to lipopolysaccharide: animal model for perinatally acquired encephalopathy. Dev Neurosci 2005; 27: 134–142.
152. Garnier Y, Coumans ABC, Berger R, Jensen A, et al. Endotoxemia severely affects circulation during normoxia and asphyxia in immature fetal sheep. J Soc Gynecol Invest 2003; 8: 134–142.
153. Yanowitz TD, Jordan JA, Gilmour CH, Towbin R, et al. Hemodynamic disturbances in premature infants born after chorioamnionitis: association with cord blood cytokine concentrations. Pediatr Res 2002; 51: 310–316.
154. Yanowitz TD, Baker RW, Roberts JM, Brozanski BS. Low blood pressure among very-low-birth-weight infants with fetal vessel inflammation. J Perinatol 2004; 24: 299–304.
155. Yanowitz TD, Potter DM, Bowen A, Baker RW, et al. Variability in cerebral oxygen delivery is reduced in premature neonates exposed to chorioamnionitis. Pediatr Res 2006; 59: 299–304.
156. Cheranov SY, Jaggar JH. TNF-alpha dilates cerebral arteries via NAD(P)H oxidase-dependent Ca^{2+} spark activation. Am J Physiol Cell Physiol 2005; 290: C964–C971.
157. Maalouf EF, Duggan PJ, Counsell S, Rutherford MA, et al. Comparison of findings on cranial ultrasound and magnetic resonance imaging in preterm infants. Pediatrics 2001; 107: 719–727.

Chapter 9
Effects of Intermittent Hypoxia on Neurological Function

David Gozal

Abstract Sleep-disordered breathing is a frequent condition across the age spectrum in the clinical setting, and is characterized by recurring episodes of hypoxia during sleep, as well as by disruption of sleep integrity. It has become clear that this highly prevalent condition leads to substantial morbidities primarily affecting the central nervous and cardiovascular systems and is also associated with marked alterations in respiratory patterning. Substantial advances have occurred in the last decade on our understanding of the pathophysiological role played intermittent hypoxia in these altered phenotypes. In this chapter, the current conceptual frameworks on the mechanisms underlying the consequences of episodic hypoxemia during sleep will be reviewed, with particular attention to cognitive deficits and altered neural control of breathing. When appropriate, differences in the response patterns as dictated by developmental stages at which intermittent hypoxia occurs will also be addressed.

Keywords: sleep apnea; long-term facilitation; neurobehavioral deficits; oxidative stress; inflammation

9.1 Intermittent Hypoxia

Chronic episodic or intermittent hypoxia (IH) is one of the primary consequences associated with several prevalent clinical conditions, such as sleep apnea, apnea of prematurity, chronic obstructive pulmonary disease, pulmonary fibrosis, and nocturnal asthma. All of these diseases can lead to development of significant morbidities, primarily affecting the CNS and cardiovascular systems, and cognitive and psychological dysfunctions have emerged as recognized consequences of the occurrence of IH during sleep for long periods of time spanning over months to years (1). Although the disorders associated with IH have heterogeneous pathophysiological processes

---

D. Gozal
Department of Pediatrics, Comer Children's Hospital – The University of Chicago,
5721 S. Maryland Avenue, MC 8000, Suite K-160, Chicago, IL 60637
email: dgozal@peds.bsd.uchicago.edu

that contribute to their unique clinical phenotypes, recent clinical and experimental evidence suggests that exposures to prolonged periods of IH may play a critical role in the onset and progression of the neurobehavioral and cognitive morbidity that will ultimately develop in the context of these disorders. For many years, attention to hypoxia revolved around the study of the effects elicited by prolonged sustained hypoxia, a situation that is only infrequently identified in clinical settings, and that is more traditionally associated with high-altitude climbing activities. In the last decade or so, we have witnessed a major shift from the study of adaptive mechanisms to sustained hypoxia to the study of animal and human models aiming to unravel the potential adaptive and injury-inducing mechanisms that are operational in the context of IH. Thus, it is imperative that we understand better the time domains and pathways involved in the global effects of IH on the brain. In this chapter, I will therefore focus on our current understanding of the effects of IH on cognitive function. In addition, I will review the differences in the effects of IH on different brain regions.

Intermittent hypoxia (IH) is traditionally defined as the recurring occurrence of low blood oxygen levels, followed by subsequent reoxygenation episodes. Obviously, this definition encompasses a wide variety of situations that would essentially fit such rather liberal criteria, such that it would be unfeasible and unrealistic to attempt and compare how all the possible permutations allowed by definition of IH may interfere with CNS homeostatic functioning. Nevertheless, a number of critical factors need to be carefully delineated and defined such as to enable critical comparisons across different laboratories. These factors include the frequency and duration of the hypoxia-reoxygenation cycles, the severity of the hypoxic stimulus, the kinetics of time to nadir and time to reoxygenation, the magnitude of reoxygenation (less severe hypoxia vs. normoxia vs. hyperoxia), as well as the period of exposure. Indeed and as a corollary of these properties, different IH paradigms may be associated with dramatically different consequences, for example, increased (preconditioning) or decreased tissue resistance to subsequent injury, plasticity of neural mechanisms that regulate cardiorespiratory functions, altered regulation of function, and ultimately neuronal, glial, or other CNS cell-type injury and death (2–4). Thus, it becomes apparent how difficult it would be to generalize across conditions, and for this reason, I will primarily review recent clinical and experimental findings from studies of patients with obstructive sleep apnea (OSA), and animal models of the hypoxia-reoxygenation events that are most frequently encountered among such patients.

9.2 OSA and Cognition

OSA adversely impacts multiple CNS regions, ultimately leading to substantial neurocognitive and behavioral deficits (5–7). In adults, the major neurocognitive manifestations of OSA include excessive daytime sleepiness, mood disturbances, particularly depression and reduced self-esteem, and impaired cognition (5, 8, 9). Clinical and epidemiological data from studies of patients with OSA have led to the overarching consensus that the repeated episodes of upper airway obstruction during sleep and the resultant IH episodes from such events indeed play a significant role in the cognitive deficits observed in these patients. Clinical imaging studies have recently implicated neurodegenerative changes within brain regions underlying the neural processing of memory and executive function, such as the hippocampus and prefrontal cortex, which in turn constitute a direct consequence of the exposure to the episodic hypoxia associated with OSA (10–16). For example, neuroimaging studies have reported that adult OSA patients are at increased risk for developing regional gray and white matter losses, and for displaying alterations in markers of neuronal integrity, as well as for showing changes in prefrontal lobe vascular perfusion (17–26). Although not all studies have replicated these findings (27), a recent review of the literature concluded that neurodegenerative processes are indeed operational in OSA (28). More recently, Morrell and Twigg have confirmed and extended their findings that extensive loss of gray matter occurs in the hippocampal and parahippocampal brain regions of OSA patients (29), lending further credence to the hypothesis positing that disruption of neuroanatomical integrity is a consequence and underlying factor in OSA-associated neurcognitive morbidity. Indeed, more recent data further show that patients with OSA appear to be at increased risk for developing silent brain infarcts in the context of altered platelet function and aggregability (30).

Since OSA is a disease that affects the entire age spectrum of patients, it is worthwhile mentioning that children with OSA will manifest a rather unique behavioral phenotype in comparison to adults. For example, in addition to diminished learning capabilities (31, 32), children with OSA are reported to have neurobehavioral disturbances, as well as an increased incidence of hyperactivity and retarded growth (33). The similarity in cognitive and behavioral deficits observed in children who suffer from OSA and children with ADHD has led to the suggestion that OSA and hyperactive/impulsive subtype of ADHD, although distinct clinical entities, may also share common pathophysiological mechanisms (34). A review of the clinical evidence strongly supports the concept that even mild chronic and intermittent hypoxia in childhood may lead to alterations in behavior and cognitive performance (35). Furthermore, children with symptoms of breathing disturbance during sleep are more likely to be academically underachieving before treatment (36, 37), and that such academic problems may persist even after their sleep-disordered breathing has resolved, if treatment was not implemented in a timely fashion (37). These findings along with a recent report on the presence of altered neurotransmitter and neuronal integrity markers in prefrontal and hippocampal regions in children afflicted with OSA (38) suggest that exposure to the pathophysiological components of OSA, namely IH and sleep fragmentation (episodic arousals), may exert unique and long-lasting changes on neural function and development.

9.3 Cognitive and Behavioral Effects of IH

Growing awareness of the potential neurocognitive consequences of episodic hypoxia has led to development and application of animal models that attempt to reproduce the cardiorespiratory and sleep patterns of patients with sleep apnea, such as to allow for more thorough investigation of the roles played by each of these perturbations on cognitive function. For example, Gozal et al. reasoned that if rats were exposed to IH during sleep, and if such exposures generated the oxyhemoglobin saturation patterns that characterize moderate to severe patients with sleep apnea, albeit without intrusion on sleep architecture, then inferences may be possible on the mechanisms involved in the anatomical and functional pathologies that such IH exposures would yield (39). Using this paradigm, we demonstrated that exposures lasting 14 days and consisting of an intermittent hypoxia profile characterized by alternating epochs of 90 s of hypoxia (10% O_2) and 90 s of reoxygenation to normoxic conditions during the habitual sleep times of adult male rats led to major functional impairments on the acquisition and retention of a hippocampal-dependent learning task, i.e., the spatial reference version of the Morris water maze (39). In contrast, we found no deficits in IH-exposed animals on a simple cued version water maze task, indicating that the impairments in spatial learning were not due to alterations in sensorimotor function. Since these initial findings, subsequent studies by our laboratory and other laboratories have shown that cognitive deficits consistent with impaired functioning of the hippocampus and/or prefrontal cortex will develop in both rats and mice following exposures to IH (39–42). Furthermore, IH was associated with reduced probability to induce long-term potentiation of field-evoked potentials following a tetanic pulse train stimulation of Shaffer collaterals in the CA1 region of the hippocampus (43), a finding that is compatible with the reported alterations in sodium and potassium conductances, decreased energy neurotransmitter substrates, as well as sodium bicarbonate exchangers identified in hippocampal neurons of mice exposed to IH (44–47). More recently, male rats tested on a delayed matching to place task in the water maze following a similar IH exposure were found to have significant impairments in a working memory task, lending further support that prefrontal structures are adversely affected following IH (48). Furthermore, cholinergic neurons appear to be particularly vulnerable to IH (48), a finding that has been corroborated for neurons within the hippocampal formation, whereby the potential for alterations in G protein-mediated signal transduction seems to contribute to IH-related hippocampal dysfunction (49).

Of note, longer lasting chronic IH exposures have also been implicated in the emergence of hypersomnolence, a rather frequent and sometimes prominent symptom in patients with sleep apnea. Exposures to IH during the rodent sleep cycle for 8 weeks resulted in impaired maintenance of wakefulness in mice along with reduced neuronal cell counts within regions that promote wakefulness, thereby implicating IH as a major underlying contributor to the excessive daytime sleepiness observed in patients with OSA (50). Recent additional studies have not only confirmed earlier findings but further extended such findings to other brain regions such as the cerebellum, while implicating oxidative stress-related mechanisms (51, 52). Collectively, the reported findings illustrate that exposures to IH are capable of replicating some of cardinal neurocognitive morbidities associated with OSA in adults, and provide further support for the hypothesis that chronic exposures to IH constitute an underlying cause of cognitive and behavioral dysfunction.

To further explore the viability of this conceptual framework, it was necessary to further examine whether the unique behavioral features of OSA in children could be replicated in the developing rodent through exposures to IH during sleep. Exposures to IH from postnatal day 10–25, an age period that ontogenetically corresponds to the age range at which peak incidence of OSA occurs in children (31, 33), were associated with rather prominent neurobehavioral deficits in the spatial reference version of the Morris water maze (41). Strikingly, when IH was applied during this postnatal period, major increases in locomotion activity in the open field emerged among male juvenile rats (tested at 30 days of age), and a higher propensity for these animals to jump from the platform during early training trials in the water maze were also documented (41). Such observations are markedly reminiscent of the male preponderance reported among children with ADHD, and provide a potential pathogenetic link between the increased prevalence of ADHD-like symptoms and OSA in children. Of note, close scrutiny of the behaviors exhibited by the juvenile rats exposed to IH revealed short duration of social interactions and reduced exploratory activity when exposed to novel objects (Gozal, unpublished observations). Furthermore, similar enhancements in locomotor activity were also observed in rat pups exposed to intermittent hypoxia at 7–11 days of postnatal age when tested at 35 days of age, although no effects of gender were consistently detected in this study (41, 53). As I alluded to in the Introduction, the differences in phenotypic responses to IH could be related to the varying degrees and duration of the intermittent hypoxia profiles used in each of the studies, as well as could be accounted for by the age at which the animals were exposed and tested. Furthermore, the enhanced locomotor activity observed in juvenile rats is unique to this age group, since no patterns of altered locomotor activity emerged among adult rats exposed to similar IH profiles. Interestingly, the hyperactivity patterns typically resolve over time as the animals reach adulthood (Row and Gozal 2005, unpublished observations).

One of the important questions on IH-induced cognitive deficits involves the reversibility of such deficits upon discontinuation of IH. Unfortunately, we currently have only limited information in this regard, and clearly more studies will be needed to elucidate critical exposure durations that may curtail the functional recovery from IH. Based on the available information, exposures to IH appear to impose long-term behavioral consequences in both adult and developing animals. For example, adult male rats exposed to IH for 2 weeks will recover only partially from their spatial learning deficits after 2 weeks of recovery under normoxic conditions (39). It is unclear whether the recovery trajectory is slower than the temporal characteristics of the induction of spatial learning deficits in the rat. Furthermore, we still do not know whether longer exposures will translate into completely irreversible deficits. For example, deficiencies in spatial working memory learning performance were found in male, but not in female rats, after they were exposed to IH from postnatal day 10 to 25, and then allowed to recover till age 4 months, before being tested (54). Anatomical studies of such animals revealed the presence of residual alterations in the prefrontal cortical dendritic arborization patterns of neurons, along with the presence of long-lasting changes in the tissue concentrations of monoaminergic neurotransmitters (54). Furthermore, juvenile rats exposed to IH for only 4 days at the age of 7 days displayed working memory impairments when tested at 65 days of age (55). Exposures to intermittent hypobaric hypoxia from birth until the age of 19 days impaired spatial learning as much as 7 months after cessation of IH (56, 57). Taken together, these studies would suggest that the behavioral consequences of IH may be long-lasting and only partially reversible. As mentioned, however, it remains unclear whether these changes are directly due to intermittent hypoxia-induced residual damage to selected brain regions, or instead represent compensatory processes brought about by the dysregulation of subcortical structures (see later).

9.4 Pathophysiology of IH-Induced Cognitive Deficits

There is no doubt that hypoxia is a frequent occurrence in neonatal, pediatric, and adult clinical settings, and it is now well established that hypoxia constitutes a major pathological factor associated with impairment of brain function and neuronal cell injury, the latter being ultimately associated with neurodegeneration and cell death (56–58). Although our current knowledge on the effects of mild intermittent hypoxia on neuronal cell function and survival is far from complete, a number of recent studies have begun to shed some light on some of the pathways accounting for the relationships between the behavioral and neuronal consequences of such IH exposures. Experimental studies have shown that exposures to IH lead to increased neuronal cell apoptosis, and that the alteration further leads to the emergence of cytoarchitectural disorganization in brain regions involved in learning and memory, such as the hippocampal CA1 region and the frontoparietal cortex (39). In the rat model, the apoptotic increases induced by IH are time-dependent, with progressive increases over the initial 24–48h after initiation of IH, peaking at 48–72h, and slowly decreasing thereafter, albeit remaining higher than normative apoptotic rates found in normoxic rats (39). In mice, neurons within the hippocampal formation appear to be more resistant, since a more severe degree of hypoxia is required to elicit similar findings (Gozal, unpublished observations), and the latency to peak apoptosis is also prolonged, with increases in apoptotic markers such as cleaved caspase 3 or TUNEL peaking at 7 days of exposure. Regional disparities are also apparent in the CNS of rats exposed to IH. Indeed, IH-induced apoptosis was extensively present in the CA1 region of the hippocampus, and yet the CA3 region and dorsocaudal brainstem were virtually unaffected, indicating that IH selectively affects specific brain regions, such that the study of the mechanisms accounting for regional susceptibility differences may ultimately reveal potential contributing factors underlying the mechanisms through which IH elicits neuronal cell losses (59–61). In addition, enhanced glial proliferation has also been shown to develop following IH exposures (39), and may represent the pathological activation of microglia, such as has been associated with various neurodegenerative diseases (62, 63). Reductions in *N*-acetyl aspartate/creatine (NAA/Cr) ratios, a noninvasive marker of neuronal integrity, were reported in mice exposed to an intermittent hypoxia profile consisting of alternations between 21 and 11% oxygen every 4 min for 4 weeks (64). Further, irreversible decreases in myelination of the corpus callosum have also been shown in a rodent model, a finding that may explain compromised connectivity between cortical regions (65). We should also point out that IH-induced effects on neuronal survival greatly differ from those of a sustained hypoxic exposure of similar magnitude (66), and that such differences are primarily due to the disproportionate recruitment of caspase mechanisms by IH to elicit increased cellular death (66).

The structural changes described heretofore as induced by IH are paralleled by alterations in the cellular and molecular substrates of synaptic plasticity within the hippocampus and cortex. For example, IH exposures will modify the excitability of hippocampal CA1 neurons (44–46, 67). It is very likely that such changes in excitability account for the reduced ability of hippocampal neurons to enable and sustain long-term potentiation (43). It is now conclusively demonstrated that long-term potentiation within the hippocampus is a NMDA receptor-dependent mechanism, and therefore it is not surprising that IH exposures will lead to reductions in the density of NMDA R1 receptor expressing cells in the CA1 field of hippocampus (39), and will also diminish the number of NMDA receptor binding sites within surviving neurons (68). Based on such findings, it has been postulated that IH will trigger the activation of NMDA-mediated excitotoxic processes, thereby accounting for the unique vulnerability of NMDA R1 expressing cells to IH (39, 69).

At the cellular level, IH has been demonstrated to induce significant decreases in the Ser-133 phosphorylation of the cAMP-response element binding protein (CREB), a transcription factor that mediates critical components of neuronal survival and memory consolidation in mammals within the hippocampal CA1 region. Decreases in CREB phosphorylation, without changes in total CREB expression reached their nadir between 6 h and 3 days of IH exposure, and gradually returned toward baseline levels by 14–30 days of IH exposures (69, 70). The changes in CREB phosphorylation paralleled concomitant changes in neuronal apoptosis induced by IH.

IH was also found to induce similar time-dependent biphasic changes in protein kinase B (AKT)-dependent survival pathways within the CA1 region (71), as well as differentially recruit endoplasmic reticulum proteins such as valosin-associated protein that are involved in cell survival (72). These changes were temporally correlated with the initial increases and subsequent decreases in neuronal apoptosis (71). Collectively, these studies indicate that IH impacts neuronal function at the anatomical, cellular, and molecular levels.

Although hippocampal and cortical regions appear to be especially vulnerable to IH, recent evidence suggests that additional brain areas may also be adversely affected by IH exposures. Exposure to 8 weeks of IH results in oxidative injuries in sleep–wake regions of the brain, such as the basal forebrain, and such changes are manifested by increased nitrosylation and carbonylation within these regions that in turn mediated increases in somnolence and increased susceptibility to short-term sleep loss as observed in mice exposed to IH (50, 73–75). In addition, 1-month exposures to IH led to increased glutathione reductase activity and thiobarbituric acid reactive substance (TBARS) levels in the pons and cerebellum of rats (76), and basal forebrain (Gozal 2006, unpublished observations). The latter findings have important implications for cognitive function, since the basal forebrain also provides cholinergic innervation of hippocampal and cortical structures involved in learning and memory, suggesting that such changes may also contribute to learning and memory impairments. Consistent with this hypothesis, animals exposed to 2 weeks of IH were found to have reduced expression of acetyl choline esterase-labeled neurons in the medial septal nucleus, in both the vertical and horizontal limbs of the diagonal band, and the substantia inominata after 14 days of IH exposure, suggesting that a loss of cholinergic neuronal phenotype within the basal forebrain may play a role in the cognitive impairments associated with IH exposures (44).

Experimental studies in rats have delineated a developmental period from P10 to P25 during which IH exposures result in a dramatic increases in number of cortical and hippocampal neurons undergoing apoptosis in comparison to both neonatal and adult rats (77). These findings are consistent with previous work demonstrating that the brain is particularly vulnerable to hypoxia during periods of maturation and development. Hypoxic episodes occurring during these critical periods of development have a serious effect on brain maturation with anatomical consequences ranging from cell death to hampered differentiation of dendrites and axons, as well as compromised neurite outgrowth and synapse formation (55–57). Severe perinatal and postnatal forms of hypoxia/ischemia or prolonged anoxia are associated with cognitive and motor impairments and in some cases death (78–80). In addition, epidemiological studies indicate that milder forms of perinatal and postnatal hypoxia may be associated with increased risk for behavioral disorders such as ADHD (81–83), suggesting that increased developmental susceptibility may account for the alteration in behavioral phenotype seen after early postnatal IH exposures (41). Similar developmental windows of susceptibility have been demonstrated in rodent models that employed significantly more severe hypoxia or anoxia (84, 85), suggesting that the effects of hypoxia in the immature mammal may involve additive or synergistic effects between ongoing developmental apoptosis and IH-induced apoptosis. Hypoxic injury during early periods of vulnerability may result in disruption of brain growth and maturation and ultimately lead to long-term neurocognitive morbidity (39, 40, 47, 53, 76). Although the degree to which IH-induced changes may be reversible is still a matter of debate, and could be related to the markers and techniques employed (64), the available evidence suggests that IH exposure is capable of inducing long-term neuroanatomical consequences. Studies have shown that early postnatal exposure to IH is associated with disorganization of the processes of astroglial and oligodendroglial cells in the cortex and hippocampus (55), as well as with decreased dendritic branching and disruption of monoaminergic pathways in the prefrontal cortex in 5-month old male, but not female rats, exposed to IH from P10 to P25 (53). These findings, when taken in conjunction with behavioral studies, indicate that IH exerts long-term consequences on the cellular networks underlying spatial reference memory, such as the hippocampus proper, and most particularly the CA1 region, as well as the prefrontal and cortical substrates that mediate spatial working memory and long-term memory retrieval (86, 87).

Of note, an additional period of increased vulnerability to IH is manifest during aging. Indeed, aging rats exposed to room air or IH display more severe spatial learning impairments compared with similarly exposed young rats; furthermore, the decrements in performance coincide with the magnitude of IH-induced decreases in CREB phosphorylation (88), as well as with decreases in proteasomal activity. Indeed, while reduced activity of the 26S proteasome occurred in both young and aging rats exposed to IH, the changes were substantially greater in the aging animals, suggesting that formation of protein aggregates was more likely in aging rats, and that such increases in denatured proteins within the cell would promote neuronal cell dysfunction, and increased cell losses. Indeed, neuronal apoptosis, as shown by cleaved caspase 3 expression, was particularly increased in the brain of aging rats exposed to IH (88). Taken together, such findings point to a unique window of vulnerability among aging rodents exposed to IH, which is reflected at least in part, by the more prominent decreases in CREB phosphorylation and a marked inability of the ubiquitin-proteasomal pathway to adequately clear and process degraded proteins (88).

Patients suffering from OSA have increased plasma levels of oxidative stress markers, suggesting that oxidative injury and inflammation may be an underlying mechanism of OSA-associated morbidities (89). In support of such assumption, increased expression of oxidative stress markers also occurs in the brains of rodents exposed to IH (42, 75, 90), and such effects are also detectable in other non-CNS tissues and plasma, with the magnitude of oxidative stress being dose-dependent, i.e., the more severe the hypoxia or the longer lasting the IH (91). Increased inflammation and oxidative stress have been implicated as a primary factor in the adverse neuronal and behavioral consequences of exposures to IH, since antioxidant therapies, or overexpression of antioxidant genes (e.g., *MnSOD*), or even targeted disruption of prooxidant genes (e.g., NADPH oxidase), will attenuate IH-induced learning deficits and also reduce IH-induced hypersomnolence in rodents (42, 74, 92). Prostaglandin E_2 tissue concentrations, a marker for the expression and activity of the inflammatory protein cyclooxygenase 2 (COX-2), are increased in hippocampal and cortical regions of the rat brain after exposures to IH (93). In addition, lipid peroxidation of polyunsaturated fatty acids, a well-documented mechanism of cellular injury, is substantially increased after IH exposures, as evidenced by the increased presence of malondialdehyde (MDA), a commonly used indicator of oxidative stress in cells and tissues (42, 46). In addition, increased carbonylation- and nitrosylation-induced oxidative injury has been found in hypoxia sensitive following 8 weeks of IH exposure in mice (46, 73, 75). These processes may occur, at least in part, due to activation of astroglia and subsequent loss of buffering functions that ultimately contribute to pathological processes, such as increased glial proliferation and activation (39, 75). Astroglial and microglial cells are thought to play the central roles in inflammatory processes in the brain (94). Activated astroglial and microglial cells express high levels of the inducible isoform of nitric oxide synthase (iNOS) and COX-2, with both enzymes initiating pathways ultimately leading to production of a variety of reactive oxygen species. Inhibition of both iNOS and COX-2 is very efficient in limiting experimental ischemia/reperfusion injury (95). Furthermore, cytokine release facilitates production of ROS superoxide and hydrogen peroxide as well as iNOS and COX-2 by microglia, thus perpetuating inflammation and aggravating ongoing oxidative stress (96). Consistent with these findings, we showed that both pharmacological and genetic inhibition of iNOS conferred increased resistance against IH-induced learning deficits (97), and other investigators have shown that such interventions are also effective against IH-mediated hypersomnolence (74). Similarly, pharmacological inhibition of COX-2 results in marked attenuations of IH-induced spatial learning deficits in rats (97). Also worthy of mention is the potentially adverse role of excessive platelet-activating factor receptor (PAFR) activity as triggered by IH, since genetic ablation of PAFR in mice was accompanied by marked improvements in the overall levels COX-2-derived activity products as well as reductions in iNOS activity and oxidative stress in brain cortex (98).

There is now compelling support for the role played by the pro- and antioxidant cellular systems in neuronal injury associated with IH. In addition to aforementioned considerations, we have reported that transgenic mice overexpressing Cu,Zn-superoxide dismutase exposed to CIH conditions had a lower level of steady-state ROS production and reduced neuronal apoptosis in brain cortex compared with that of normal control mice (90), and equivalent findings occur for MnSOD transgenic mice as well (92). Furthermore, NADPH oxidase-dependent production of the superoxide radical (O_2^-) has been identified as a major contributor to oxidative injury in the brain and other target organs under conditions of inflammation and severe hypoxia (99–102). Long-term IH increases the expression of NADPH oxidase, which in turn underlies the increased neuronal inflammation and oxidative stress observed in animal models of OSA (7).

Finally, we have recently demonstrated that reduced stability of lipid membranes as brought about by genetic ablation of apolipoprotein E increases markedly the susceptibility to IH, via exaggerated production of lipid peroxidation and inflammatory products (103).

Alterations in neurotransmitter systems may also play a role in the neurobehavioral disturbances seen after IH exposures. Experimental and clinical studies implicate the disruption of norepinephrine and dopaminergic pathways in the development of hyperactivity and working memory dysfunction characteristic of disorders of minimal brain dysfunction, such as ADHD (104–107). Dopamine D1 receptors are operationally important in working memory mechanisms (103), whereas polymorphisms in the dopamine D4 receptor gene in children modulate behavioral temperament profiles characterized by features of *novelty seeking*, which included impulsive and exploratory behavior (106, 108). Furthermore, D4 receptors appear to be essential for hyperactivity and impaired behavioral inhibition in rodents (109, 110). Expression of dopamine and modulation of its receptors are subject to significant developmental alterations (111), which can be readily influenced during critical periods of development by factors such as gender and external stress (112, 113). The available evidence from our experimental models suggests that dysregulation of dopaminergic function is of significance in the context of IH-induced neurobehavioral deficits (114).

Several studies have shown dysregulation of dopaminergic and catecholaminergic systems following IH exposure. Li et al. (115) exposed rats to 8 h daily of varying fractional concentrations of inspired oxygen and carbon dioxide for 35 days, consisting of brief (3–6 s) episodic (twice every minute) eucapnic (3.5% FiO_2 and 10% $FiCO_2$) or hypocapnic (3.5% FiO_2 and 0% $FiCO_2$) challenges with hypoxia or room air (21% FiO_2 and 0.03% $FiCO_2$). In this study, episodic hypocapnic hypoxia produced a decrease in dopamine turnover and eucapnic hypoxia increased norepinephrine levels in the hypothalamus, suggesting that episodic eucapnic and hypocapnic hypoxia may affect metabolism of these neurotransmitters in the CNS (115). Decker et al. found enhanced expression of vesicular monoamine transporter (VMAT) and of D1 dopamine receptors in the striatum of posthypoxic (exposed from days 7–11) rats at 80 days of age (40). Furthermore, reduced levels of extracellular striatal dopamine were also found in these posthypoxic rats during both the dark and light phases of the circadian cycle (53). Kheirandish et al. reported alterations in frontocortical catecholaminergic pathways resulting from exposure to IH during a critical period of postnatal developmental (54). Taken together, these findings suggest that alterations and disruption of tyrosine hydroxylase-related pathways may underlie, or at least contribute to neurobehavioral vulnerability.

9.5 Environmental and Lifestyle Modulation of End-Organ Susceptibility

Diseases associated with intermittent hypoxia, such as OSA, are typically complex disorders that are unlikely to be determined by a single gene, and most likely represent the interactions of several genetic and/or environmental risk factors. Indeed, obesity, craniofacial genetic determinants, and hormonal regulation of upper airway musculature are all contributors to a certain extent, and the magnitude of such contribution varies from patient to patient (116, 117). Although no environmental factors have yet been identified to cause OSA, environmental exposure to alcohol, sedatives, smoking, and/or fragmented sleep exacerbates OSA severity (118, 119). Genetic factors, such as apolipoprotein E (ApoE) allelic variants, which among other things operate as risk factor modifiers for Alzheimer's disease and atherosclerosis, have also been associated with increased incidence and severity of OSA (120–123). However, to date the impact of genes and environmental factors on the cognitive consequences of OSA remains relatively unknown. Given the now well-accepted concept that both genes and environment play a critical role in maintaining neural function under both pathological and normal conditions, a better understanding of the potential contributions of both genetic and environmental factors to the neurocognitive deficits associated with OSA is therefore essential for our understanding of the pathogenesis of the disease (124–130).

Recent evidence from animal models suggests that both genetic and environmental factors are capable of modulating susceptibility to IH. For example, and as mentioned earlier, apolipoprotein E knockout mice show increased susceptibility to IH-induced spatial learning deficits in the water maze, as well as increased brain levels of the inflammatory mediator prostaglandin E2 and elevated concentrations of the oxidative stress marker, malondialdehyde (114). Similar enhancements of cognitive vulnerability to IH exposures occurred within the hippocampal CA1 region along with substantial decreases in CREB phosphorylation in rats fed on an obesity-inducing (high in both fat and refined carbohydrate content) diet (131). Conversely, environmental enrichment through spatial task training attenuated the extent and magnitude of IH-induced cognitive deficits (132). In addition, the potential to recruit a response of de novo neurogenesis that becomes functionally relevant needs to be mentioned as well (133). These studies clearly illustrate that both intrinsic internal factors (i.e., genetic) as well as external environmental factors must be taken into account in our understanding of the large phenotypic variance that characteristically occurs in patients with OSA.

9.6 Effects of Sustained and Intermittent Hypoxia on Respiratory Control

Thus far, this chapter has focused on more rostral structures such as the hippocampus and frontal cortex and the potential effects of IH. However, in addition to such functional and structural changes induced by hypoxia, it is worthwhile considering the fact that long-term hypoxic conditions may also lead to marked alterations in respiratory patterning. I will therefore succinctly delineate some of the more salient modifications in breathing induced by long-term hypoxia, and thus illustrate the diversity and wide range of effects that conditions such as hypoxia may elicit in different brain regions.

9.6.1 Sustained Hypoxia and Control of Breathing

The effect of sustained hypoxia (SH) on respiratory control is undoubtedly one of the most extensively investigated areas of respiratory physiology. It has long been recognized that exposures to environmental hypoxia, such as may occur with high-altitude climbing, will lead to time-dependent increases in ventilation. However, great differences emerge between mature and developing mammals when exposed to the same hypoxic paradigm. In adults, exposure to sustained hypoxia enhances the response to subsequent acute hypoxic challenges (134, 135). The enhanced hypoxic ventilatory response (HVR) is mediated by the combination of increased sensitivity of the peripheral chemoreceptors and centrally mediated adaptations within the excitatory networks underlying HVR (136–138); however, these effects will gradually decrease over time after either humans or animals are removed from the hypoxic environment. In sharp contrast, perinatal sustained hypoxic exposures will elicit a progressive decrease in hypoxic ventilatory sensitivity and a relative blunting of HVR (139–141).

This reduction in HVR may be accounted in part by reduced O_2 sensitivity of the carotid bodies (142–144). Intriguingly, the blunted sensitivity to hypoxia will persist into adulthood, suggesting that perinatal SH induces developmental (meta)plasticity of the HVR, even if some degree of functional recovery of carotid body chemosensitivity may spontaneously occur (139). Consequently, the mechanisms underlying this form of developmentally regulated neural plasticity are currently unknown and still subject to debate. Recent evidence has demonstrated that hypoxic phrenic responses are intact following perinatal hypoxia, supporting the assumption that peripheral chemoreceptor function as well as related central pathway functions are preserved in adult rats following perinatal hypoxia (145). Instead, Bavis et al. proposed that the site of long-term plasticity must reside downstream to the active phrenic motor neurons, and therefore be related to neuromuscular transmission, function of respiratory muscles, respiratory mechanics, or feedback control. Furthermore, these investigators also proposed that gender influences are also of importance in perinatal hypoxia-induced respiratory plasticity, since female gender displays diminished susceptibility for long-term alterations in respiratory output (145).

In addition to its influence on the abrupt dynamic changes associated with acute hypoxia, i.e., HVR, sustained exposures to environmental hypoxia have also been associated with plasticity of the resting normoxic ventilatory output. In general, SH leads to a progressive time-dependent (days to weeks) increase in resting minute ventilation, which has been termed ventilatory acclimatization to hypoxia (VAH) (146), a phenomenon that has been well documented in a variety of mammalian species including humans (147–150). The process of adaptation to hypoxia that results in ventilatory acclimatization appears to involve both ends of the synaptic pathways underlying the ventilatory response, such that development of increased sensitivity of the peripheral chemoreceptors occurs in parallel with enhanced/more efficient integration of afferent inputs and amplification of centrally generated efferent output (134). Soon after return to normoxia, this process is gradually reduced and resting minute ventilation usually reverts to preexposure levels, all this being conditional on the exposure taking place after the critical window of development. In fact, the duration of the phenomenon termed VAH will depend, at least to a certain extent, on the preceding duration of the sustained hypoxic exposure (151). In contrast with adult mammals, experimental evidence has also revealed that perinatal SH leads to increases in resting minute ventilation, with most studies reporting that the enhanced normoxic ventilation (after long-term perinatal SH has ceased) will persist into adulthood (145, 152, 153).

9.6.2 *Effects of Intermittent Hypoxia on Respiratory Control*

Despite the wealth of studies focusing on the contribution of SH to ventilatory plasticity relatively little is known about the effects of IH on respiratory control. A growing body of evidence suggests that IH is indeed a very different stimulus from SH, and may therefore lead to vastly different consequences. As discussed earlier, the severity of the hypoxic stimulus, its duration within the cycle, the ramp or square pattern of stimulus presentation, the number of iterations applied, and the overall duration of exposure within the circadian or ultradian period are all important elements of the IH stimulus, such that radically different responses may occur (154–157). In addition, the developmental stage at which hypoxia occurs is also of paramount importance. In the following paragraphs, I will report on two discrete categories of IH, namely acute IH (AIH), i.e., lasting from minutes to hours, and CIH, i.e., lasting from days to weeks.

9.7 Effects of Acute Intermittent Hypoxia on Respiratory Plasticity

Adult anesthetized, vagotomized, paralyzed, mechanically ventilated, isocapnic rats exposed to an acute IH stimulus consisting of 5 min of hypoxia ($F_{IO_2} = 11\%$) separated by 5 min of hyperoxia ($F_{IO_2} = 50\%$) exhibit an enhancement of phrenic motor output, a phenomenon that has been termed long-term facilitation (LTF; Fig. 9.1) (158). Furthermore, LTF can not be elicited by equivalent exposures to

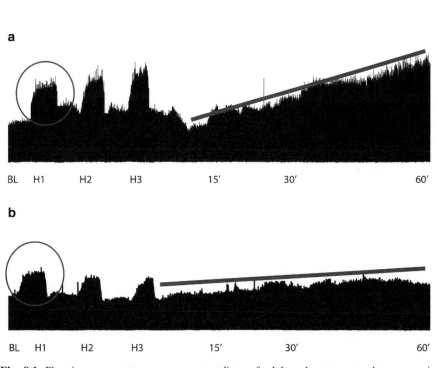

Fig. 9.1 Phrenic nerve motor neurogram recordings of adult male rats exposed to normoxia (**a**) or perinatal hypoxia (**b**). During acute hypoxia (circle), the hypoxic phrenic nerve output response was reduced compared with the normoxic controls, and similarly phrenic nerve long-term facilitation was markedly attenuated after perinatal hypoxia (line) (*See Color Plates*)

noncyclic hypoxia, which apparently is due to SH-induced activation of okadaic-sensitive phosphatases (159, 160). In recent years, Mitchell and colleagues have extensively reported on the involvement of serotoninergic pathways as the critical mediator for this form of neural plasticity (161). According to the proposed model, activation of postsynaptic 5-HT receptors, principally 5-HT$_{2A}$ subtypes, will lead to downstream signaling cascades involving PKC. PKC could then both act directly or indirectly by modulation of other kinases, and potentiate the activation of glutamate receptor channels such as NMDA receptors, thus leading to enhanced phrenic output. Furthermore, the neurotrophic factor BDNF was also critically implicated in the initiation and consolidation of LTF, and is operational either presynaptically or postsynaptically, through binding to the cognate receptor, TrkB (162). In addition, adenosine receptor subtypes and the critical role of reactive oxygen species in phrenic LTF have been recently unraveled (163, 164), and the potential need for apnea has also been postulated (165). Other recent studies have pointed to profound genetic as well as gender-related influences in this model (166, 167). However, while our understanding in the adult animal model has considerably advanced, little is known about the effects of IH on phrenic LTF (pLTF) in the early postnatal period, this being largely due to the formidable challenges presented by the surgical preparation. Despite such obstacles, there is some initial evidence suggestive that LTF exhibits age-dependent expression. Indeed, in a study comparing phrenic and hypoglossal LTF in young (3–4-month old) and aged (13-month old) male Sprague-Dawley rats, significant decreases in LTF emerged in the aged rats possibly related to age-associated changes in serotonin and its receptors (166). However, the opposite, i.e., increased vLTF at later ages, appears to occur in female rats, suggesting complex interactions between the respiratory control and estrogen–progesterone systems in the context of AIH-induced phrenic nerve respiratory plasticity (167). Furthermore, the characteristics of phrenic LTF in rats younger than 1 month of age and the interactions between development and prepubertal gender-related hormones are currently unknown. Of note, antecedent exposures to long-term IH during postnatal life will ultimately lead to marked blunting of AIH-induced LTF (Fig. 9.1), while at the same time induce increased neurokinin 1 receptor expressing neurons in regions such as the pre-Bötzinger complex (168, 169).

Other models of acute IH-induced plasticity have assessed ventilation in freely behaving animals. Ventilatory long-term facilitation (vLTF) represents the enhancement of respiratory output during normoxia following acute IH exposures (170). Thus, resting V_E as measured by whole body flow-through plethysmography displayed considerable enhancements of respiratory output following an acute isocapnic IH exposure using a similar stimulation paradigm to that previously used for phrenic nerve LTF. The augmented V_E was primarily accounted for by elevated respiratory frequency without significant contributions from tidal volume (V_T). Interestingly, Olson et al. were unable to demonstrate these ventilatory enhancements after a comparable exposure to continuous hypoxia (170). The excitatory effects of AIH on ventilation have been also confirmed by Ling et al. in rats and have implicated serotoninergic mechanisms (171). Furthermore, altered characteristics of respiratory pattern in the absence of global ventilatory output changes and of chemosensitivity were identified following a short number of brief hypoxic exposures in waking humans (172, 173). The presence of LTF was further documented in sleeping men during nonrapid eye movement sleep (174). Subsequent studies have demonstrated that the magnitude of the IH-induced ventilatory increases is amplified by pretreatment with chronic IH (175, 176). Recent studies have indicated that vLTF is greater in 1-month- vs. 2-month-old rats indicating that younger animals may have an increased capacity for IH-induced ventilatory plasticity (177). Taken together these studies would suggest that the developmental stage at which the IH paradigm is applied is a major determinant of the overall magnitude and characteristics of the ensuing vLTF.

Studies conducted in the immediate postnatal period, using somewhat different IH paradigms in rabbits [e.g., 10 min 10% O_2 followed by 10 min 21% O_2, for 5 cycles (178)] or in rats [e.g., 21% O_2 alternating with 5% O_2, 9 cycles over 16 h, initiated 6 h after birth (179)] demonstrated increased normoxic V_E, i.e., vLTF, mediated primarily through increases in V_T (179). Under these circumstances, AIH also induced increases in HVR that correlated with increased sensory output from ex vivo carotid bodies harvested from similarly exposed pups (179). Using a different AIH protocol [5 min 10% O_2 followed by 10 min 21% O_2, for 8 cycles (155)], Gozal and Gozal were unable to elicit enhancements of the early phase of HVR; however, this AIH protocol revealed attenuations of HVD that correlated with increased expression of neuronal nitric oxide synthase (nNOS) within the caudal brainstem. Thus, pharmacological inhibition of nNOS resulted in normalization of HVD (155).

In summary, incremental evidence suggests that AIH elicits an excitatory form of neuroplasticity that is particularly efficient in the developing animal.

9.8 Effects of Chronic Intermittent Hypoxia on Respiratory Plasticity

The technical requirements associated with prolonged IH exposures have somewhat precluded extensive exploration of the effects of CIH. The recent availability of computer-driven servo-controlled environmental systems has allowed for more extensive exploration of this issue.

Indeed, evidence from CIH studies conducted in rats exposed prenatally using 90-s alternations of 10% O_2 and 21% O_2 supports the presence of marked developmental respiratory plasticity (180). Furthermore, it is important to note that when CIH is applied during the prenatal period it leads to lifelong alterations in ventilatory control (180). Waters et al. found that a CIH profile consisting of 30 min a day for 6 days attenuates HVR as demonstrated by diaphragmatic EMG expressed as a function of normoxic baseline activity or as ventilatory output measured using a pneumotachograph in piglets (181, 182). Additionally, these investigators reported significant increases in the substance P within the nTS, possibly as a compensatory mechanism for changes in NK1 receptor expression, since the latter appear to be critical mediators of HVR (183).

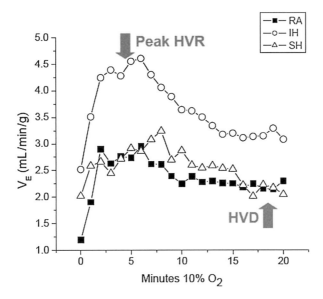

Fig. 9.2 Individual ventilatory measurements in rats undergoing acute hypoxic challenges (20 min, 10% O_2) following either 30 days of early postnatal room air (RA), intermittent hypoxia (IH), or sustained hypoxia (SH)

In more recent work from our laboratory, we have illustrated differences in respiratory patterns in both adult (184) and developing rats (185) following prolonged periods of IH, with prominent consequences on not only resting ventilation but also on HVR characteristics (Fig. 9.2).

Additional laboratories have also recently undertaken the exploration of brainstem regions other than the phrenic motor nerve nucleus that may contribute to respiratory long-term facilitation, such that we should anticipate an explosive expansion of the conceptual framework and network pathways underlying the adaptive strategies of respiratory function in the context of episodic hypoxia (186–193).

9.9 Summary

This chapter reviewed the currently available evidence on IH and the CNS. The cumulative evidence supports the concept that long-term exposures to IH are detrimental to the rostral CNS regions, particularly with regard to cognitive and behavioral functioning. More specifically, functional aspects of hippocampal, prefrontal cortical, as well as related subcortical structures are affected by prolonged IH, and such effects involve interdependent and cooperative interactions between inflammatory and oxidative stress pathways.

However, in parallel with such IH-induced effects, evidence derived from studies on brainstem-related respiratory regions would support very opposite conclusions, namely that neuronal pools within these regions undergo changes following IH that aim to promote survival and improved functionality.

Additionally, unique periods of susceptibility during the lifespan are present and dictate the response patterns to IH, and further presented are initial findings suggestive of interactions between disease severity, genetically determined susceptibility, and environmental and lifestyle conditions that through their joint interactions will ultimately determine the magnitude of end-organ morbidity associated with diseases characterized by IH during sleep. We further contend that utilization of IH paradigms that mimic the oxygenation patterns of a specific sleep disorder may be useful in the elucidation of specific mechanisms underlying particular aspects of OSA-associated manifestations.

Acknowledgments DG is supported by National Institutes of Health grants HL69932, HL65270, and 2P50HL60296, and by The Commonwealth of Kentucky Research Challenge for Excellence Trust Fund, and the Children's Foundation Endowment for Sleep Research.

References

1. Prabhakar NR. Oxygen sensing during intermittent hypoxia: cellular and molecular mechanisms. J Appl Physiol 2001; 90:1986–1994.
2. McGuire M, Zhang Y, White DP, Ling L. Effect of hypoxic episode number and severity on ventilatory long-term facilitation in awake rats. J Appl Physiol 2002; 93:2155–2161.
3. Prabhakar NR, Kline DD. Ventilatory changes during intermittent hypoxia: importance of pattern and duration. High Alt Med Biol 2002; 3:195–204.
4. Waters KA, Gozal D. Responses to hypoxia during early development. Respir Physiol Neurobiol 2003; 136:115–129.
5. Decary A, Rouleau I, Montplaisir J. Cognitive deficits associated with sleep apnea syndrome: a proposed neuropsychological test battery. Sleep 2000; 23:369–381.
6. Punjabi NM, Polotsky VY. Disorders of glucose metabolism in sleep apnea. J Appl Physiol 2005; 99:1998–2007.
7. Vgontzas AN, Bixler EO, Chrousos GP. Sleep apnea is a manifestation of the metabolic syndrome. Sleep Med Rev 2005; 9:211–224.
8. Kales A, Caldwell AB, Cadieux RJ, et al. Severe obstructive sleep apnea. II. Associated psychopathology and psychosocial consequences. J Chronic Dis 1985; 38:427–434.
9. Roehrs T, Merrion M, Pedrosi B, et al. Neuropsychological function in obstructive sleep apnea syndrome (OSAS) compared to chronic obstructive pulmonary disease (COPD). Sleep 1995; 18:382–388.
10. Beebe DW, Gozal D. Obstructive sleep apnea and the prefrontal cortex: towards a comprehensive model linking nocturnal upper airway obstruction to daytime cognitive and behavioral deficits. J Sleep Res 2002; 11:1–16.
11. Beebe DW, Groesz L, Wells C, et al. The neuropsychological effects of obstructive sleep apnea: a meta-analysis of norm-referenced and case-controlled data. Sleep 2003; 26:298–307.
12. Ayalon L, Peterson S. Functional central nervous system imaging in the investigation of obstructive sleep apnea. Curr Opin Pulmon Med 2007; 13:479–483.
13. Robbins J, Redline S, Ervin A, Walsleben JA, Ding J, Nieto FJ. Associations of sleep-disordered breathing and cerebral changes on MRI. J Clin Sleep Med 2005; 1:159–165.
14. Zimmerman ME, Aloia MS. A review of neuroimaging in obstructive sleep apnea. J Clin Sleep Med 2006; 2:461–471.

15. Tonon C, Vetrugno R, Lodi R, Gallassi R, Provini F, Iotti S, Plazzi G, Montagna P, Lugaresi E, Barbiroli B. Proton magnetic resonance spectroscopy study of brain metabolism in obstructive sleep apnoea syndrome before and after continuous positive airway pressure treatment. Sleep 2007; 30:305–311.
16. Halbower AC, Degaonkar M, Barker PB, Earley CJ, Marcus CL, Smith PL, Prahme MC, Mahone EM. Childhood obstructive sleep apnea associates with neuropsychological deficits and neuronal brain injury. PLoS Med 2006; 3:e301.
17. Ficker JH, Feistel H, Moller C, et al. Changes in regional CNS perfusion in obstructive sleep apnea syndrome: initial SPECT studies with injected nocturnal 99mTc-HMPAO. Pneumologie 1997; 51(9):926–930.
18. Kamba M, Suto Y, Ohta Y, et al. Cerebral metabolism in sleep apnea. Evaluation by magnetic resonance spectroscopy. Am J Respir Crit Care Med 1997; 156(1):296–298.
19. Kamba M, Inoue Y, Higami S, et al. Cerebral metabolic impairment in patients with obstructive sleep apnoea: an independent association of obstructive sleep apnoea with white matter change. J Neurol Neurosurg Psychiatry 2001; 71(3):334–339.
20. Macey PM, Henderson LA, Macey KE. Brain morphology associated with obstructive sleep apnea. Am J Respir Crit Care Med 2002; 166(10):1382–1387.
21. Alchanatis M, Deligiorgis N, Zias N, et al. Frontal brain lobe impairment in obstructive sleep apnoea: a proton MR spectroscopy study. Eur Respir J 2004; 24(6):980–986.
22. Bartlett DJ, Rae C, Thompson CH. Hippocampal area metabolites relate to severity and cognitive function in obstructive sleep apnea. Sleep Med 2004; 5(6):593–596.
23. Thomas RJ, Rosen BR, Stern CD, et al. Functional imaging of working memory in obstructive sleep-disordered breathing. J Appl Physiol 2005; 98(6):2226–2234.
24. Ayalon L, Ancoli-Israel S, Klemfuss Z, et al. Increased brain activation during verbal learning in obstructive sleep apnea. Neuroimage 2006; 31(4):1817–1825.
25. Davies CW, Crosby JH, Mullins RL, et al. Case control study of cerebrovascular damage defined by magnetic resonance imaging in patients with OSA and normal matched control subjects. Sleep 2001; 24(6):715–720.
26. Gilman S, Chervin RD, Koeppe RA, et al. Obstructive sleep apnea is related to a thalamic cholinergic deficit in MSA. Neurology 2003; 61(1):35–39.
27. O'Donoghue FJ, Briellmann RS, Rochford PD, et al. Cerebral structural changes in severe obstructive sleep apnea. Am J Respir Crit Care Med 2005; 171(10):1185–1190.
28. Nowak M, Kornhuber J, Meyrer R. Daytime impairment and neurodegeneration in OSAS. Sleep 2006; 29(12):1521–1530.
29. Morrell MJ, Twigg G. Neural consequences of sleep disordered breathing: the role of intermittent hypoxia. Adv Exp Med Biol 2006; 588:75–88.
30. Minoguchi K, Yokoe T, Tazaki T, et al. Silent brain infarction and platelet activation in obstructive sleep apnea. Am J Respir Crit Care Med 2007; 175(6):612–617.
31. Gozal D. Sleep-disordered breathing and school performance in children. Pediatrics 1998; 102:616–620.
32. Beebe DW. Neurobehavioral morbidity associated with disordered breathing during sleep in children: a comprehensive review. Sleep 2006; 29(9):1115–1134.
33. Gozal D. Morbidity of obstructive sleep apnea in children: facts and theory. Sleep Breath 2001; 5(1):35–42.
34. O'Brien LM, Mervis CB, Holbrook CR, et al. Neurobehavioral correlates of sleep-disordered breathing in children. J Sleep Res 2004; 13(2):165–172.
35. Bass JL, Corwin M, Gozal D, et al. The effect of chronic or intermittent hypoxia on cognition in childhood: a review of the evidence. Pediatrics 2004; 114(3):805–816.
36. Urschitz MS, Wolff J, Sokollik C, et al. Nocturnal arterial oxygen saturation and academic performance in a community sample of children. Pediatrics 2005; 115(2):e204–e209.
37. Gozal D, Pope DW Jr. Snoring during early childhood and academic performance at ages thirteen to fourteen years. Pediatrics 2001; 107(6):1394–1399.
38. Halbower AC, Degaonkar M, Barker PB, et al. Childhood obstructive sleep apnea associates with neuropsychological deficits and neuronal brain injury. PLoS Med 2006; 3(8):e301.

39. Gozal D, Daniel JM, Dohanich GP. Behavioral and anatomical correlates of chronic episodic hypoxia during sleep in the rat. J Neurosci 2001; 21(7):2442–2450.
40. Decker MJ, Hue GE, Caudle WM, et al. Episodic neonatal hypoxia evokes executive dysfunction and regionally specific alterations in markers of dopamine signaling. Neuroscience 2003; 117(2):417–425.
41. Row BW, Kheirandish L, Neville JJ, et al. Impaired spatial learning and hyperactivity in developing rats exposed to intermittent hypoxia. Pediatr Res 2002; 52(3):449–453.
42. Row BW, Liu R, Xu W, et al. Intermittent hypoxia is associated with oxidative stress and spatial learning deficits in the rat. Am J Respir Crit Care Med 2003; 167(11):1548–1553.
43. Payne RS, Goldbart A, Gozal D, et al. Effect of intermittent hypoxia on long-term potentiation in rat hippocampal slices. Brain Res 2004; 1029(2):195–199.
44. Zhao P, Xue J, Gu XQ, Haddad GG, Xia Y. Intermittent hypoxia modulates Na+ channel expression in developing mouse brain. Int J Dev Neurosci 2005; 23(4):327–333.
45. Douglas RM, Miyasaka N, Takahashi K, Latuszek-Barrantes A, Haddad GG, Hetherington HP. Chronic intermittent but not constant hypoxia decreases NAA/Cr ratios in neonatal mouse hippocampus and thalamus. Am J Physiol Regul Integr Comp Physiol 2007; 292:R1254–R1259.
46. Gu XQ, Haddad GG. Maturation of neuronal excitability in hippocampal neurons of mice chronically exposed to cyclic hypoxia. Am J Physiol Cell Physiol 2003; 284:C1156–C1163.
47. Douglas RM, Xue J, Chen JY, Haddad CG, Alper SL, Haddad GG. Chronic intermittent hypoxia decreases the expression of Na/H exchangers and HCO3-dependent transporters in mouse CNS. J Appl Physiol 2003; 95:292–299.
48. Row BW, Kheirandish L, Cheng Y, et al. Impaired spatial working memory and altered choline acetyltransferase (CHAT) immunoreactivity and nicotinic receptor binding in rats exposed to intermittent hypoxia during sleep. Behav Brain Res 2007; 177(2):308–314.
49. Hambrecht VS, Vlisides PE, Row BW, et al. Cholinergic and opioid activation of G proteins in rat hippocampus: modulation by hypoxia. Hippocampus 2007; 17(10):934–942.
50. Veasey SC, Davis CW, Fenik P, et al. Long-term intermittent hypoxia in mice: protracted hypersomnolence with oxidative injury to sleep-wake brain regions. Sleep 2004; 27(2):194–201.
51. Pae EK, Chien P, Harper RM. Intermittent hypoxia damages cerebellar cortex and deep nuclei. Neurosci Lett 2005; 375:123–128.
52. Ramanathan L, Gozal D, Siegel JM. Antioxidant responses to chronic hypoxia in the rat cerebellum and pons. J Neurochem 2005; 93(1):47–52.
53. Decker MJ, Jones KA, Solomon IG, et al. Reduced extracellular dopamine and increased responsiveness to novelty: neurochemical and behavioral sequelae of intermittent hypoxia. Sleep 2005; 28(2):169–176.
54. Kheirandish L, Gozal D, Pequignot JM, et al. Intermittent hypoxia during development induces long-term alterations in spatial working memory, monoamines, and dendritic branching in rat frontal cortex. Pediatr Res 2005; 58(3):594–599.
55. Simonova Z, Sterbova K, Brozek G, et al. Postnatal hypobaric hypoxia in rats impairs water maze learning and the morphology of neurones and macroglia in cortex and hippocampus. Behav Brain Res 2003; 141(2):195–205.
56. Nyakas C, Buwalda B, Kramers RJ, et al. Postnatal development of hippocampal and neocortical cholinergic and serotonergic innervation in rat: effects of nitrite-induced prenatal hypoxia and nimodipine treatment. Neuroscience 1994; 59(3):541–559.
57. Nyakas C, Buwalda B, Luiten PG. Hypoxia and brain development. Prog Neurobiol 1996; 49(1):1–51.
58. Nyakas C, Markel E, Schuurman T, et al. Impaired learning and abnormal open-field behaviours of rats after early postnatal anoxia and the beneficial effect of the calcium antagonist nimodipine. Eur J Neurosci 1991; 3(2):168–174.
59. Gozal E, Gozal D, Pierce WM, et al. Proteomic analysis of CA1 and CA3 regions of rat hippocampus and differential susceptibility to intermittent hypoxia. J Neurochem 2002; 83(2):331–345.
60. Klein JB, Gozal D, Pierce WM, Thongboonkerd V, Scherzer JA, Sachleben LR, Guo SZ, Cai J, Gozal E. Proteomic identification of a novel protein regulated in CA1 and CA3 hippocampal regions during intermittent hypoxia. Respir Physiol Neurobiol 2003; 136(2–3):91–103.

61. Klein JB, Barati MT, Wu R, et al. Akt-mediated valosin-containing protein 97 phosphorylation regulates its association with ubiquitinated proteins. J Biol Chem 2005; 280(36): 31870–31881.
62. Lue LF, Walker DG, Brachova L, et al. Involvement of microglial receptor for advanced glycation endproducts (RAGE) in Alzheimer's disease: identification of a cellular activation mechanism. Exp Neurol 2001; 171(1):29–45.
63. Vlassara H. The AGE-receptor in the pathogenesis of diabetic complications. Diabetes Metab Res Rev 2001; 17(6):436–443.
64. Douglas RM, Miyasaka N, Takahashi K, et al. Chronic intermittent but not constant hypoxia decreases NAA/Cr ratios in neonatal mouse hippocampus and thalamus. Am J Physiol Regul Integr Comp Physiol 2007; 292(3):R1254–R1259.
65. Kanaan A, Farahani R, Douglas RM, et al. Effect of chronic continuous or intermittent hypoxia and reoxygenation on cerebral capillary density and myelination. Am J Physiol Regul Integr Comp Physiol 2006; 290(4):R1105–R1114.
66. Gozal E, Sachleben LR Jr, Rane MJ, et al. Mild sustained and intermittent hypoxia induce apoptosis in PC-12 cells via different mechanisms. Am J Physiol Cell Physiol 2005; 288(3): C535–C542.
67. Gu XQ, Haddad GG. Decreased neuronal excitability in hippocampal neurons of mice exposed to cyclic hypoxia. J Appl Physiol 2001; 91(3):1245–1250.
68. Pichiule P, Chavez JC, Boero J, et al. Chronic hypoxia induces modification of the N-methyl-D-aspartate receptor in rat brain. Neurosci Lett 1996; 218(2):83–86.
69. Albin RL, Greenamyre JT. Alternative excitotoxic hypotheses. Neurology 1992; 42(4): 733–738.
70. Goldbart A, Row BW, Kheirandish L, et al. Intermittent hypoxic exposure during light phase induces changes in cAMP response element binding protein activity in the rat CA1 hippocampal region: water maze performance correlates. Neuroscience 2003; 122(3): 585–590.
71. Goldbart A, Cheng ZJ, Brittian KR, et al. Intermittent hypoxia induces time-dependent changes in the protein kinase B signaling pathway in the hippocampal CA1 region of the rat. Neurobiol Dis 2003; 14(3):440–446.
72. Klein JB, Barati MT, Wu R, Gozal D, Sachleben LR Jr, Kausar H, Trent JO, Gozal E, Rane MJ. Akt-mediated valosin-containing protein 97 phosphorylation regulates its association with ubiquitinated proteins. J Biol Chem 2005; 280(36):31870–31881.
73. Sanfilippo-Cohn B, Lai S, Zhan G, et al. Sex differences in susceptibility to oxidative injury and sleepiness from intermittent hypoxia. Sleep 2006; 29(2):152–159.
74. Zhan G, Fenik P, Pratico D, et al. Inducible nitric oxide synthase in long-term intermittent hypoxia: hypersomnolence and brain injury. Am J Respir Crit Care Med 2005; 171(12): 1414–1420.
75. Zhan G, Serrano F, Fenik P, et al. NADPH oxidase mediates hypersomnolence and brain oxidative injury in a murine model of sleep apnea. Am J Respir Crit Care Med 2005; 172(7): 921–929.
76. Ramanathan L, Gozal D, Siegel JM. Antioxidant responses to chronic hypoxia in the rat cerebellum and pons. J Neurochem 2005; 93(1):47–52.
77. Gozal E, Row BW, Schurr A, et al. Developmental differences in cortical and hippocampal vulnerability to intermittent hypoxia in the rat. Neurosci Lett 2001; 305(3):197–201.
78. Gray PH, Tudehope DI, Masel JP, et al. Perinatal hypoxic-ischaemic brain injury: prediction of outcome. Dev Med Child Neurol 1993; 35(11):965–973.
79. du Plessis AJ, Johnston MV. Hypoxic-ischemic brain injury in the newborn. Cellular mechanisms and potential strategies for neuroprotection. Clin Perinatol 1997; 24(3):627–654.
80. Tuor UI, Del Bigio MR, Chumas PD. Brain damage due to cerebral hypoxia/ischemia in the neonate: pathology and pharmacological modification. Cerebrovasc Brain Metab Rev 1996; 8(2):159–193.
81. Zappitelli M, Pinto T, Grizenko N. Pre-, peri-, and postnatal trauma in subjects with attention-deficit hyperactivity disorder. Can J Psychiatry 2001; 46(6):542–548.

82. Lou HC. Etiology and pathogenesis of attention-deficit hyperactivity disorder (ADHD): significance of prematurity and perinatal hypoxic-haemodynamic encephalopathy. Acta Paediatr 1996; 85(11):1266–1271.
83. Krageloh-Mann I, Toft P, Lunding J, et al. Brain lesions in preterms: origin, consequences and compensation. Acta Paediatr 1999; 88(8):897–908.
84. Towfighi J, Mauger D, Vannucci RC, et al. Influence of age on the cerebral lesions in an immature rat model of cerebral hypoxia-ischemia: a light microscopic study. Brain Res Dev Brain Res 1997; 100(2):149–160.
85. Vannucci RC, Vannucci SJ. A model of perinatal hypoxic-ischemic brain damage. Ann N Y Acad Sci 1997; 835:234–249.
86. Miyashita Y. Cognitive memory: cellular and network machineries and their top–down control. Science 2004; 306:435–440.
87. Dalley JW, Cardinal RN, Robbins TW. Prefrontal executive and cognitive functions in rodents: neural and neurochemical substrates. Neurosci Biobehav Rev 2004; 28(7):771–784.
88. Gozal D, Row BW, Kheirandish L, et al. Increased susceptibility to intermittent hypoxia in aging rats: changes in proteasomal activity, neuronal apoptosis and spatial function. J Neurochem 2003; 86(6):545–552.
89. Lavie L. Obstructive sleep apnoea syndrome – an oxidative stress disorder. Sleep Med Rev 2003; 7(1):35–51.
90. Xu W, Chi L, Row BW, et al. Increased oxidative stress is associated with chronic intermittent hypoxia-mediated brain cortical neuronal cell apoptosis in a mouse model of sleep apnea. Neuroscience 2004; 126(2):313–323.
91. Li J, Savransky V, Nanayakkara A, et al. Hyperlipidemia and lipid peroxidation are dependent on the severity of chronic intermittent hypoxia. J Appl Physiol 2007; 102(2):557–563.
92. Shan X, Chi L, Ke Y, Luo C, Qian SY, St Clair D, Gozal D, Liu R. Manganese superoxide dismutase protects mouse cortical neurons from chronic intermittent hypoxia-mediated oxidative damage. Neurobiol Dis 2007; 28(2):206–215.
93. Li RC, Row BW, Gozal E, et al. Cyclooxygenase 2 and intermittent hypoxia-induced spatial deficits in the rat. Am J Respir Crit Care Med 2003; 168(4):469–475.
94. Heales SJ, Lam AA, Duncan AJ, et al. Neurodegeneration or neuroprotection: the pivotal role of astrocytes. Neurochem Res 2004; 29(3):513–519.
95. Duncan AJ, Heales SJ. Nitric oxide and neurological disorders. Mol Aspects Med 2005; 26(1–2):67–96.
96. Halliwell B. Phagocyte-derived reactive species: salvation or suicide? Trends Biochem Sci 2006; 31(9):509–515.
97. Li RC, Row BW, Kheirandish L, et al. Nitric oxide synthase and intermittent hypoxia-induced spatial learning deficits in the rat. Neurobiol Dis 2004; 17(1):44–53.
98. Row BW, Kheirandish L, Li RC, et al. Platelet-activating factor receptor-deficient mice are protected from experimental sleep apnea-induced learning deficits. J Neurochem 2004; 89(1):189–196.
99. Ozaki M, Haga S, Zhang HQ, et al. Inhibition of hypoxia/reoxygenation-induced oxidative stress in HGF-stimulated antiapoptotic signaling: role of PI3-K and Akt kinase upon rac1. Cell Death Differ 2003; 10(5):508–515.
100. Li JM, Shah AM. Endothelial cell superoxide generation: regulation and relevance for cardiovascular pathophysiology. Am J Physiol Regul Integr Comp Physiol 2004; 287(5): R1014–R1030.
101. Adibhatla RM, Hatcher JF. Phospholipase A2, reactive oxygen species, and lipid peroxidation in cerebral ischemia. Free Radic Biol Med 2006; 40(3):376–387.
102. Wang Q, Tompkins KD, Simonyi A, et al. Apocynin protects against global cerebral ischemia-reperfusion-induced oxidative stress and injury in the gerbil hippocampus. Brain Res 2006; 1090(1):182–189.
103. Kheirandish L, Row BW, Li RC, et al. Apolipoprotein E-deficient mice exhibit increased vulnerability to intermittent hypoxia-induced spatial learning deficits. Sleep 2005; 28(11): 1412–1417.

104. Williams GV, Castner SA. Under the curve: critical issues for elucidating D1 receptor function in working memory. Neuroscience 2006; 139(1):263–276.
105. Mill J, Caspi A, Williams BS, et al. Prediction of heterogeneity in intelligence and adult prognosis by genetic polymorphisms in the dopamine system among children with attention-deficit/hyperactivity disorder: evidence from 2 birth cohorts. Arch Gen Psychiatry 2006; 63(4):462–469.
106. Mill J, Fisher N, Curran S, et al. Polymorphisms in the dopamine D4 receptor gene and attention-deficit hyperactivity disorder. Neuroreport 2003; 14(11):1463–1466.
107. Viggiano D, Ruocco LA, Arcieri S, et al. Involvement of norepinephrine in the control of activity and attentive processes in animal models of attention deficit hyperactivity disorder. Neural Plast 2004; 11(1–2):133–149.
108. Zhang K, Tarazi FI, Baldessarini RJ. Role of dopamine D(4) receptors in motor hyperactivity induced by neonatal 6-hydroxydopamine lesions in rats. Neuropsychopharmacology 2001; 25(5):624–632.
109. Viggiano D, Vallone D, Sadile A. Dysfunctions in dopamine systems and ADHD: evidence from animals and modeling. Neural Plast 2004; 11(1–2):97–114.
110. Avale ME, Falzone TL, Gelman DM, et al. The dopamine D4 receptor is essential for hyperactivity and impaired behavioral inhibition in a mouse model of attention deficit/hyperactivity disorder. Mol Psychiatry 2004; 9(7):718–726.
111. Herlenius E, Lagercrantz H. Development of neurotransmitter systems during critical periods. Exp Neurol 2004; 190 (Suppl 1):S8–S21.
112. Becker JB. Gender differences in dopaminergic function in striatum and nucleus accumbens. Pharmacol Biochem Behav 1999; 64(4):803–812.
113. Andersen SL, Teicher MH. Sex differences in dopamine receptors and their relevance to ADHD. Neurosci Biobehav Rev 2000; 24(1):137–141.
114. Row BW. Intermittent hypoxia and behavior: is dopamine to blame? Sleep 2005; 28(2):165–167.
115. Li R, Bao G, el-Mallakh RS, et al. Effects of chronic episodic hypoxia on monoamine metabolism and motor activity. Physiol Behav 1996; 60(4):1071–1076.
116. Taheri S, Mignot E. The genetics of sleep disorders. Lancet Neurol 2002; 1(4):242–250.
117. Palmer LJ, Redline S. Genomic approaches to understanding obstructive sleep apnea. Respir Physiol Neurobiol 2003; 135(2–3):187–205.
118. Partinen M, Kaprio J, Koskenvuo M, et al. Genetic and environmental determination of human sleep. Sleep 1983; 6(3):179–185.
119. Partinen M, Telakivi T. Epidemiology of obstructive sleep apnea syndrome. Sleep 1992; 15(6 Suppl):S1–S4.
120. Kadotani H, Kadotani T, Young T, et al. Association between apolipoprotein E epsilon4 and sleep-disordered breathing in adults. JAMA 2001; 285(22):2888–2890.
121. O'Hara R, Schroder CM, Kraemer HC, et al. Nocturnal sleep apnea/hypopnea is associated with lower memory performance in APOE epsilon4 carriers. Neurology 2005; 65(4):642–644.
122. Gozal D, Sans Capdevila O, Kheirandish-Gozal L, et al. Apolipoprotein E ε4 allele, neurocognitive dysfunction, and obstructive sleep apnea in school-aged children. Neurology 2007; 69(3):243–249.
123. Bliwise DL. Sleep apnea, APOE4 and Alzheimer's disease 20 years and counting? J Psychosom Res 2002; 53(1):539–546.
124. Mohammed AH, Henriksson BG, Soderstrom S, et al. Environmental influences on the central nervous system and their implications for the aging rat. Behav Brain Res 1993; 57(2):183–191.
125. Snowdon DA, Kemper SJ, Mortimer JA, et al. Linguistic ability in early life and cognitive function and Alzheimer's disease in late life. Findings from the Nun Study. JAMA 1996; 275(7):528–532.
126. Torasdotter M, Metsis M, Henriksson BG, et al. Environmental enrichment results in higher levels of nerve growth factor mRNA in the rat visual cortex and hippocampus. Behav Brain Res 1998; 93(1–2):83–90.

127. Young D, Lawlor PA, Leone P, et al. Environmental enrichment inhibits spontaneous apoptosis, prevents seizures and is neuroprotective. Nat Med 1999; 5(4):448–453.
128. Kempermann G, Gast D, Gage FH. Neuroplasticity in old age: sustained fivefold induction of hippocampal neurogenesis by long-term environmental enrichment. Ann Neurol 2002; 52(2):135–143.
129. Kheirandish L, Gozal D. Neurocognitive dysfunction in children with sleep disorders. Dev Sci 2006; 9:388–399.
130. Gozal D, Kheirandish L. Oxidant stress and inflammation in the snoring child: confluent pathways to upper airway pathogenesis and end-organ morbidity. Sleep Med Rev 2006; 10(2):83–96.
131. Goldbart AD, Row BW, Kheirandish-Gozal L, et al. High fat/refined carbohydrate diet enhances the susceptibility to spatial learning deficits in rats exposed to intermittent hypoxia. Brain Res 2006; 1090(1):190–196.
132. Row BW, Goldbart A, Gozal E, et al. Spatial pre-training attenuates hippocampal impairments in rats exposed to intermittent hypoxia. Neurosci Lett 2003; 339(1):67–71.
133. Gozal D, Row BW, Gozal E, et al. Temporal aspects of spatial task performance during intermittent hypoxia in the rat: evidence for neurogenesis. Eur J Neurosci 2003; 18(8): 2335–2342.
134. Dwinell MR, Powell FL. Chronic hypoxia enhances the phrenic nerve response to arterial chemoreceptor stimulation in anesthetized rats. J Appl Physiol 1999; 87(2):817–823.
135. Aaron EA, Powell FL. Effect of chronic hypoxia on hypoxic ventilatory response in awake rats. J Appl Physiol 1993; 74(4):1635–1640.
136. Gamboa A, Leon-Velarde F, Rivera-Ch M, et al. Selected contribution: acute and sustained ventilatory responses to hypoxia in high-altitude natives living at sea level. J Appl Physiol 2003; 94(3):1255–1262.
137. Leon-Velarde F, Gamboa A, Rivera-Ch M, Palacios JA, Robbins PA. Selected contribution: peripheral chemoreflex function in high-altitude natives and patients with chronic mountain sickness. J Appl Physiol 2003; 94(3):1269–1278.
138. Alea OA, Czapla MA, Lasky JA, Simakajornboon N, Gozal E, Gozal D. PDGF-beta receptor expression and ventilatory acclimatization to hypoxia in the rat. Am J Physiol Regul Integr Comp Physiol 2000; 279(5):R1625–R1633.
139. Eden GJ, Hanson MA. Effects of chronic hypoxia from birth on the ventilatory response to acute hypoxia in the newborn rat. J Physiol 1987; 392:11–19.
140. Frappell PB, Mortola JP. Hamsters vs. rats: metabolic and ventilatory response to development in chronic hypoxia. J Appl Physiol 1994; 77(6):2748–2752.
141. Mortola JP, Morgan CA, Virgona V. Respiratory adaptation to chronic hypoxia in newborn rats. J Appl Physiol 1986; 61(4):1329–1336.
142. Sladek M, Parker RA, Grogaard JB, Sundell HW. Long-lasting effect of prolonged hypoxemia after birth on the immediate ventilatory response to changes in arterial partial pressure of oxygen in young lambs. Pediatr Res 1993; 34(6):821–828.
143. Sterni LM, Bamford OS, Wasicko MJ, Carroll JL. Chronic hypoxia abolished the postnatal increase in carotid body type I cell sensitivity to hypoxia. Am J Physiol 1999; 277(3, Part 1): L645–L652.
144. Jackson A, Nurse C. Plasticity in cultured carotid body chemoreceptors: environmental modulation of GAP-43 and neurofilament. J Neurobiol 1995; 26(4):485–496.
145. Bavis RW, Olson EB Jr, Vidruk EH, Fuller DD, Mitchell GS. Developmental plasticity of the hypoxic ventilatory response in rats induced by neonatal hypoxia. J Physiol 2004; 557(Part 2):645–660.
146. Bisgard GE. The role of arterial chemoreceptors in ventilatory acclimatization to hypoxia. Adv Exp Med Biol 1994; 360:109–122.
147. Olson EB Jr, Dempsey JA. Rat as a model for humanlike ventilatory adaptation to chronic hypoxia. J Appl Physiol 1978; 44(5):763–769.
148. Dempsey JA, Forster HV, doPico GA. Ventilatory acclimatization to moderate hypoxemia in man. The role of spinal fluid (H$^+$). J Clin Invest 1974; 53(4):1091–1100.

149. Fatemian M, Kim DY, Poulin MJ, Robbins PA. Very mild exposure to hypoxia for 8 h can induce ventilatory acclimatization in humans. Pflugers Arch 2001; 441(6):840–843.
150. Donoghue S, Fatemian M, Balanos GM, Crosby A, Liu C, O'Connor D, Talbot NP, Robbins PA. Ventilatory acclimatization in response to very small changes in PO_2 in humans. J Appl Physiol 2005; 98(5):1587–1591.
151. Bisgard GE. Increase in carotid body sensitivity during sustained hypoxia. Biol Signals 1995; 4(5):292–297.
152. Okubo S, Mortola JP. Long-term respiratory effects of neonatal hypoxia in the rat. J Appl Physiol 1988; 64(3):952–958.
153. Okubo S, Mortola JP. Control of ventilation in adult rats hypoxic in the neonatal period. Am J Physiol 1990; 259(4, Part 2):R836–R841.
154. Waters KA, Gozal D. Responses to hypoxia during early development. Respir Physiol Neurobiol 2003; 136(2–3):115–129.
155. Gozal D, Gozal E. Episodic hypoxia enhances late hypoxic ventilation in developing rat: putative role of neuronal NO synthase. Am J Physiol 1999; 276(1, Part 2):R17–R22.
156. Peng YJ, Prabhakar NR. Effect of two paradigms of chronic intermittent hypoxia on carotid body sensory activity. J Appl Physiol 2004; 96(3):1236–1242.
157. McGuire M, Zhang Y, White DP, Ling L. Effect of hypoxic episode number and severity on ventilatory long-term facilitation in awake rats. J Appl Physiol 2002; 93(6): 2155–2161.
158. Fuller DD, Bach KB, Baker TL, Kinkead R, Mitchell GS. Long term facilitation of phrenic motor output. Respir Physiol 2000; 121(2–3):135–146.
159. Baker TL, Fuller DD, Zabka AG, Mitchell GS. Respiratory plasticity: differential actions of continuous and episodic hypoxia and hypercapnia. Respir Physiol 2001; 129(1–2):25–35.
160. Wilkerson JE, Satriotomo I, Baker-Herman TL, Watters JJ, Mitchell GS. Okadaic acid-sensitive protein phosphatases constrain phrenic long-term facilitation after sustained hypoxia. J Neurosci 2008; 28(11):2949–2958.
161. Fuller DD, Bach KB, Baker TL, Kinkead R, Mitchell GS. Long term facilitation of phrenic motor output. Respir Physiol 2000; 121(2–3):135–146.
162. Baker-Herman TL, Fuller DD, Bavis RW, et al. BDNF is necessary and sufficient for spinal respiratory plasticity following intermittent hypoxia. Nat Neurosci 2004; 7(1):48–55.
163. Golder FJ, Ranganathan L, Satriotomo I, Hoffman M, Lovett-Barr MR, Watters JJ, Baker-Herman TL, Mitchell GS. Spinal adenosine A2a receptor activation elicits long-lasting phrenic motor facilitation. J Neurosci 2008; 28:2033–2042.
164. Mahamed S, Mitchell GS. Simulated apneas induce serotonin dependent respiratory long term facilitation in rats. J Physiol 2008;586(8): 2171–2181.
165. Macfarlane PM, Mitchell GS. Respiratory long-term facilitation following intermittent hypoxia requires reactive oxygen species formation. Neuroscience 2008;152(1): 189–197.
166. Zabka AG, Behan M, Mitchell GS. Selected contribution: time-dependent hypoxic respiratory responses in female rats are influenced by age and by the estrus cycle. J Appl Physiol 2001; 91(6):2831–2838.
167. Zabka AG, Behan M, Mitchell GS. Long term facilitation of respiratory motor output decreases with age in male rats. J Physiol 2001; 531 (Part 2):509–514.
168. Reeves SR, Mitchell GS, Gozal D. Early postnatal chronic intermittent hypoxia modifies hypoxic respiratory responses and long-term phrenic facilitation in adult rats. Am J Physiol Regul Integr Comp Physiol 2006; 290(6):R1664–R1671.
169. Reeves SR, Guo SZ, Brittian KR, Row BW, Gozal D. Anatomical changes in selected cardio-respiratory brainstem nuclei following early postnatal chronic intermittent hypoxia. Neurosci Lett 2006; 402:233–237.
170. Olson EB Jr, Bohne CJ, Dwinell MR, et al. Ventilatory long-term facilitation in unanesthetized rats. J Appl Physiol 2001; 91(2):709–716.
171. Ling L, Fuller DD, Bach KB, Kinkead R, Olson EB Jr, Mitchell GS. Chronic intermittent hypoxia elicits serotonin-dependent plasticity in the central neural control of breathing. J Neurosci 2001; 21(14):5381–5388.

172. Morris KF, Gozal D. Persistent respiratory changes following intermittent hypoxic stimulation in cats and human beings. Respir Physiol Neurobiol 2004; 140(1):1–8.
173. Babcock M, Shkoukani M, Aboubakr SE, Badr MS. Determinants of long-term facilitation in humans during NREM sleep. J Appl Physiol 2003; 94:53–59.
174. Babcock MA, Badr MS. Long-term facilitation of ventilation in humans during NREM sleep. Sleep 1998; 21:709–716.
175. Zabka AG, Mitchell GS, Olson EB Jr, Behan M. Selected contribution: chronic intermittent hypoxia enhances respiratory long-term facilitation in geriatric female rats. J Appl Physiol 2003; 95(6):2614–2623.
176. McGuire M, Zhang Y, White DP, Ling L. Chronic intermittent hypoxia enhances ventilatory long-term facilitation in awake rats. J Appl Physiol 2003; 95(4):1499–1508.
177. McGuire M, Ling L. Ventilatory long-term facilitation is greater in 1-month- versus 2-month-old awake rats. J Appl Physiol 2005; 98(4):1195–1201.
178. Trippenbach T. Ventilatory and metabolic effects of repeated hypoxia in conscious newborn rabbits. Am J Physiol 1994; 266(5, Part 2):R1584–R1590.
179. Peng YJ, Rennison J, Prabhakar NR. Intermittent hypoxia augments carotid body and ventilatory response to hypoxia in neonatal rat pups. J Appl Physiol 2004; 97(5):2020–2025.
180. Gozal D, Reeves SR, Row BW, Neville JJ, Guo SZ, Lipton AJ. Respiratory effects of gestational intermittent hypoxia in the developing rat. Am J Respir Crit Care Med 2003; 167(11):1540–1547.
181. Waters KA, Laferriere A, Paquette J, Goodyer C, Moss IR. Curtailed respiration by repeated vs. isolated hypoxia in maturing piglets is unrelated to NTS ME or SP levels. J Appl Physiol 1997; 83(2):522–529.
182. Waters KA, Tinworth KD. Depression of ventilatory responses after daily, cyclic hypercapnic hypoxia in piglets. J Appl Physiol 2001; 90(3):1065–1073.
183. Wickstrom HR, Berner J, Holgert H, Hokfelt T, Lagercrant H. Hypoxic response in newborn rat is attenuated by neurokinin-1 receptor blockade. Respir Physiol Neurobiol 2004; 140:19–31.
184. Reeves SR, Gozal E, Guo SZ, Sachleben LR Jr, Lipton AJ, Gozal D. Effect of long-term intermittent and sustained hypoxia on hypoxic ventilatory and metabolic responses in the adult rat. J Appl Physiol 2003; 95:1767–1774.
185. Reeves SR, Gozal D. Respiratory and metabolic responses to early postnatal chronic intermittent hypoxia and sustained hypoxia in the developing rat. Pediatr Res 2006; 60:680–686.
186. Kline DD, Ramirez-Navarro A, Kunze DL. Adaptive depression in synaptic transmission in the nucleus of the solitary tract after in vivo chronic intermittent hypoxia: evidence for homeostatic plasticity. J Neurosci 2007; 27:4663–4673.
187. Neverova NV, Saywell SA, Nashold LJ, Mitchell GS, Feldman JL. Episodic stimulation of alpha1-adrenoreceptors induces protein kinase C-dependent persistent changes in motoneuronal excitability. J Neurosci 2007; 27:4435–4442.
188. de Paula PM, Tolstykh G, Mifflin S. Chronic intermittent hypoxia alters NMDA and AMPA-evoked currents in NTS neurons receiving carotid body chemoreceptor inputs. Am J Physiol Regul Integr Comp Physiol 2007; 292(6):R2259–R2265.
189. Lovett-Barr MR, Mitchell GS, Satriotomo I, Johnson SM. Serotonin-induced in vitro long-term facilitation exhibits differential pattern sensitivity in cervical and thoracic inspiratory motor output. Neuroscience 2006; 142(3):885–892.
190. Peña F, Ramirez JM. Hypoxia-induced changes in neuronal network properties. Mol Neurobiol 2005; 32(3):251–283.
191. McGuire M, Zhang Y, White DP, Ling L. Phrenic long-term facilitation requires NMDA receptors in the phrenic motonucleus in rats. J Physiol 2005; 567 (Part 2):599–611.
192. Reeves SR, Gozal D. Protein kinase C activity in the nucleus tractus solitarii is critically involved in the acute hypoxic ventilatory response, but is not required for intermittent hypoxia-induced phrenic long-term facilitation in adult rats. Exp Physiol 2007; 92(6):1057–1066.
193. Reeves SR, Gozal D. Platelet-activating factor receptor modulates respiratory adaptation to long-term intermittent hypoxia in mice. Am J Physiol Regul Integr Comp Physiol 2004; 287(2):R369–R374.

Chapter 10
Brainstem Sensitivity to Hypoxia and Ischemia

Joseph C. LaManna, Paola Pichiule, Kui Xu, and Juan Carlos Chávez

Abstract The response of the brain to hypoxic challenge identifies brain stem regions that participate in the physiological adaptation process and are also the most sensitive to hypoxic/ischemic stress. Adaptation to prolonged mild hypoxia includes increased ventilation through carotid body and nucleus tractus solitarius pathways, which are sensitive to arterial pO_2. The neurons of the rostral ventral lateral medulla (RVLM) drive cerebrovasodilator circuits in response to arterial oxygen content. Cerebral cortical capillary density is increased in response to signals generated by local tissue pO_2 deficiency.

Global cerebral ischemia, such as occurs after cardiac arrest and resuscitation, is accompanied by an increased delayed mortality due to cardiorespiratory failure. The same brain stem nuclei, such as the nucleus paragigantocellularis, that are involved in physiological adaptation to hypoxia appear to be the very same neuronal populations that exhibit postreperfusion programmed cell death.

Keywords: Angiogenesis; Hypoxia-inducible factors; Vascular endothelial growth factor (VEGF); Angiopoietins; Cerebral blood flow (CBF); Cardiac arrest and resuscitation

10.1 Introduction

The mammalian brain is responsible for a significant fraction of the total body energy consumption. For example, the human body generates about 100 W of energy at rest and the brain is responsible for about 15 W, the majority of which is produced through oxidative phosphorylation dependent on glucose and oxygen. There is very little possibility for storage of either glucose or oxygen within the brain tissue and, thus, it is no surprise that the mammalian brain depends on the continuous delivery

---

J.C. LaManna (✉), P. Pichiule, K. Xu, and J.C. Chávez
Department of Neurology (BRB), Case Western Reserve University School of Medicine, 10900 Euclid Avenue, Cleveland OH 44106-4938, USA
e-mail: jcl4@po.cwru.edu

of the appropriate amounts of glucose and oxygen to maintain its function. From this reasoning we can draw several informative conclusions. The first is that the necessity of maintaining sufficient delivery of glucose and oxygen in the face of dynamic changes in neuronal activity and, thus, energy demand requires acute and chronic mechanisms that adjust the rate of delivery. A corollary of this conclusion is that disruption of delivery by interference with blood flow such as that in ischemia, or in content of glucose and/or oxygen in the blood, which can occur, for example, through starvation or asphyxia, will quickly lead to brain dysfunction.

The primary mechanism for responding to increased demand for glucose and oxygen is through increased blood flow (1). This phenomenon is best illustrated by consideration of the response to focal brain activation as revealed by functional magnetic resonance imaging studies (fMRI). These studies show not only an increase in blood delivered to the focally activated brain region, but also that the increased blood is primarily arterial and, therefore, oxygen rich. Indeed, it has been shown that the predominant effect of neuronal activation on oxygen availability is a large increase in tissue oxygen pressure. Glucose delivery would also be increased, of course, and the evidence suggests that although both oxygen consumption and glucose consumption are increased under these activating conditions, the increase in glucose consumption is much greater and out of proportion to the increase in oxygen consumption. As expected then, the acute response to hypoxic exposure is increased blood flow. Hypoglycemia will also result in increased blood flow, but since there is normally a twofold greater glucose influx across the blood brain barrier than used metabolically, there is more margin for maintaining delivery for small decreases in blood glucose.

Another consideration is that the availability of mechanisms that can regulate glucose and oxygen delivery makes possible a wider range of environmental niches for mammalian species and specifically allows habitation at higher altitudes despite the decreased ambient oxygen pressures encountered in those locations. This acclimatization depends on the ability of the organism to make long-term adaptations to chronic hypoxic environments. These adaptations can and do occur at many levels as indicated in Table 10.1 (2, 3).

10.2 The Acute Response to Hypoxia

10.2.1 Cerebral Vasodilation and Blood Flow

Acute effects of hypoxia on CNS include capillary hemoglobin disoxygenation, decreased tissue pO_2, increased cerebral blood volume (vasodilation), faster capillary mean transit time, and increased cerebral blood flow (2).

The most studied response to hypoxia in the CNS is an increase in vasodilation and blood flow. If the hypoxia is mild, then the compensation allows continued consciousness and function. Hypoxia can be subcategorized as *mild, moderate,*

Table 10.1 Structural and functional sites of adaptation to hypoxia

Lung
    Ventilation
        Increased frequency
        Increased tidal volume
    O_2 diffusion – lung to blood
        Increased surface area (size)
O_2 transport in blood
    Increased HCT
    Hb affinity (right shift)
    Increased flow
O_2 diffusion to tissue
    Increased capillary density
O_2 utilization in tissue
    Decreased metabolic demand (decreased activity)
    Increased metabolic efficiency (decreased body temperature)

or *severe*, but all are accompanied by increased blood flow (3). The mechanisms that are responsible for this hypoxic response have been studied for many decades. Indeed, the studies of Roy and Sherrington in the late nineteenth century (4) could be considered as the first quantitative studies of the cerebrovascular responses to the challenge of low oxygen availability.

Certainly, there is a local signaling component that contributes to increased local blood flow. It is fairly well known that the local mechanism involves vasodilation through nitric oxide (5). Other potential local signals include potassium ions, protons, and adenosine. And these all play a part, which varies depending on the severity of the hypoxia.

10.2.2 Brain Stem

More interesting, perhaps, is that there is a significant proportion of the blood flow response to mild and moderate hypoxia that is generated through neuronal circuits, which originate in brainstem centers (6). Quantitative studies over the past 20 years have led to the conclusion that as much as half of the increased cortical flow response can be eliminated by removing the influence of these brainstem centers. The brainstem is especially sensitive to hypoxic exposure and even mild to moderate hypoxia, which can be tolerated for long periods of exposure, and are accompanied by significant metabolic alterations in many brain stem nuclei (7).

The role of the brain stem in the hypoxic response is complex, but perhaps, at the risk of oversimplifying, it can be subdivided into two categories, as depicted in the scheme shown in Fig. 10.1. The first would include the pathways associated with the signals from the oxygen-sensing cells of the carotid body. This pathway is through the nucleus tractus solitarius, interacting with many

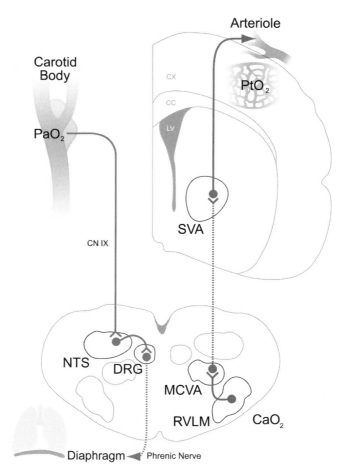

Fig. 10.1 This is a simplified scheme depicting the major brainstem and cortical components of the central nervous system response to hypoxic exposure. *NTS* nucleus tractus solitarius, *RVLM* rostral ventral lateral medulla, *MCVA* medullary cerebrovasodilator area, *DRG* dorsal respiratory group, *CN IX* glossopharyngeal, *SVA* subthalamic cerebrovascular area, *BF* basal forebrain, *CX* cerebral cortex

brain stem and spinal centers and ultimately acting on the diaphragm through increased phrenic nerve activity. In this case the hypoxic signal is related to the arterial pO_2 (8).

The second category incorporates the oxygen-sensing neurons intrinsic to the brain stem and located in the rostral ventral lateral medulla (RVLM). These neurons project to the medullary cerebrovasodilator area (MCVA) and thence through the subthalamic cerebrovasodilator area (SVA) to the diffuse cortical projections emanating from the basal forebrain and the midline thalamic centromedian complex neurons to increase cerebral blood flow (9). It is estimated that 50% or more of the increased cerebral blood flow in response to hypoxia is mediated by these brain stem pathways.

10.3 Global Ischemia

High secondary mortality rates are associated with cardiac arrest after initially successful resuscitation (10, 11). This mortality may be due to the secondary failure of cardiovascular and respiratory brainstem centers. Neuronal death in the brainstem could be due to necrosis following edema-induced ischemia, but also might be due to neuronal apoptosis, such as that occuring in the hippocampus after global ischemia and reperfusion (11).

The brainstem centers most important for cortical blood flow are also the most sensitive to brainstem ischemia (12) as that occurs, for example, after cardiac arrest and resuscitation. In fact, after 12 min of cardiac arrest followed by resuscitation induced in conscious, adult male Wistar rats there were clear signs of neuronal apoptosis and programmed cell death in the brain stem. Bax and activated caspase 3 were detected for a few days after resuscitation, perhaps indicative of reperfusion injury (Fig. 10.2a, b). TUNEL-positive cells in the sensitive regions such as the RVLM (Fig. 10.3 top) can be detected after 2–4 days. These are distributed throughout the brainstem nuclei with cardiovascular and respiratory significance (Fig. 10.3 middle and bottom), for example, the nucleus paragintocellularis, a source of sympathetic neuronal outflow, which is known to be activated during ischemia (12).

10.4 Chronic Hypoxia

10.4.1 Adaptation to Prolonged Mild Hypoxia

The main systemic long-term changes that contribute to chronic hypoxic adaptation are increased ventilation, fall in core body temperature, right-shifted hemoglobin saturation curve, bicarbonate ion excretion, increased packed red cell volume, and loss of body mass (2). These structural and metabolic changes are phased in over the first few weeks of exposure. For the central nervous system, the important variables are transiently increased cerebral blood flow, increased hematocrit, and then increased capillary density as a result of angiogenesis (1).

10.4.2 The Ventilatory Response

One brainstem region that obviously participates in the hypoxic response is the nucleus tractus solitarius. Exposure of unanesthetized rats to acute and sustained hypoxic stress of 1–4 h is known to cause activation of the *c-fos* gene within the neurons of the nucleus tractus solitarius, resulting in expression of fos-like immunoreactive protein (Fos) (13). As shown in Fig. 10.4 (Haxhiu, LaManna et al., unpublished observations), following 7 days of hypoxic exposure, the number of

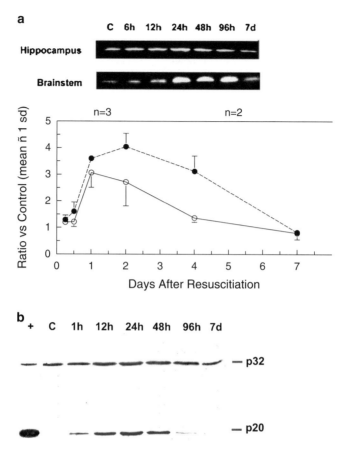

Fig. 10.2 (a) *Top*: Western blot analysis of Bax in samples from the hippocampus and brainstem of rats subjected to cardiac arrest and resuscitation. *Bottom*: Quantitative densitometry of Bax expression in the hippocampus ($n = 3$, open circles) and brain stem ($n = 2$, closed circles) after cardiac arrest and resuscitation. (b) Caspase activity in the brain stem after cardiac arrest and resuscitation. Activated caspase-3 is indicated by the presence of the p20 fragment. The leftmost column is a positive control (+)

NTS cells expressing Fos (24 ± 4 mean $\pm$ sem) was significantly higher than in control rats breathing room air (7 ± 2), but lower than in rats exposed to hypoxia for 4 h (55 ± 8). In two other groups, rats were exposed to 7 days of hypoxia followed by 1-h exposure to 8% O_2 in N_2.

10.4.3 Cerebral Blood Flow

The initial increase in cerebral blood flow persists for a few days, but then returns toward the preexposure levels despite the continued exposure to a low oxygen

10   Brainstem Sensitivity to Hypoxia and Ischemia

Fig. 10.3 *Top*: Histological section of the brain stem showing TUNEL-positive neurons in the RVLM, counterstained with H&E, from a rat 4 days after cardiac arrest and resuscitation. *Middle* and *Bottom*: Dots indicating locations of TUNEL-positive neuronal cell bodies at the indicated brain stem levels (*See Color Plates*)

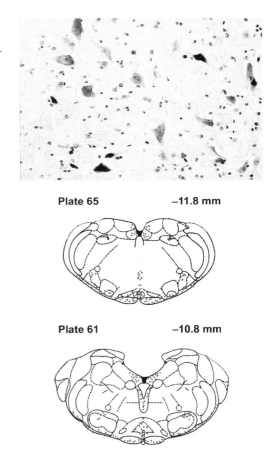

Fig. 10.4 Fos protein expression in the nucleus tractus solitarius in rats exposed to hypoxia. Control rats were kept at normal inspired oxygen and carbon dioxide partial pressure. Rats were exposed to 10% oxygen ($FiO_2 = 0.10$) for 4 h or 7 days. After 7 days of exposure to 10% O_2, rats were exposed to 8% O_2 for 1 h (+hypoxia).

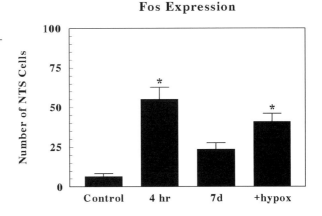

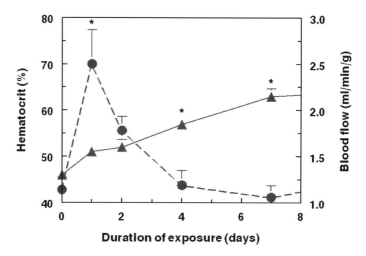

Fig. 10.5 Hematocrit change and blood flow change in brain stem during hypoxia. The blood flow in frontal cortex (circles, dashed line) was increased significantly on the first and second day and returned to the baseline by the fourth day of hypoxic exposure, whereas by day 4 of exposure, the hematocrit (triangles, solid line) was significantly elevated and remained elevated thereafter (* indicates values significantly different from the normoxic baseline, $p < 0.05$) (*See Color Plates*)

environment and consequent lower brain tissue pO_2. The transient flow increase appears to occur generally throughout the brain regions. Figure 10.5, which is a plot of data reported previously (14), shows the response of the rat brain stem blood flow to prolonged mild hypoxia over the time span of a week. The renormalization of brain stem blood flow parallels the increasing hematocrit, which strongly suggests that the stimulus controlling the blood flow response is more likely associated with the normalized arterial oxygen content rather than the continued low arterial pO_2 (15). The low arterial pO_2 is the more important signal that drives the ventilatory response, while the continued lower cerebral cortical tissue pO_2 drives the angiogenesis response (1, 16).

10.4.4 Angiogenesis

Figure 10.6 demonstrates the diffuse tissue hypoxia in the rat brain cortex in response to 10% O_2. The image depicts the tissue areas that are in a more reduced state using the EF-5 method (17, 18). The hypoxic stimulus persists in gray and white for at least 2 weeks, until the newly sprouted capillaries become functional. Tissue pO_2 remains low despite the renormalization of the arterial oxygen content because the delivery of oxygen from the capillary to the tissue is governed by the Fick principle, and so the driving force for the diffusion of oxygen into the tissue is the concentration gradient and this still remains low as long as the ambient oxygen

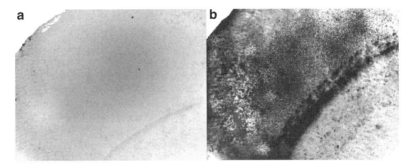

Fig. 10.6 EF-5 stain (negative fluorescent image) of control, normoxic rat cerebral cortex (**a**) and cerebral cortex from a rat exposed for 4h to 10% O_2, (**b**) Diffuse neuronal hypoxia (indicated by dark areas positive for EF-5) in the cortical layers and extensive and severe hypoxia in the subcortical white matter are apparent

pressure is low. The increase in capillary density results in a decrease in the intercapillary distance and, therefore, decreases the diffusion distance.

The process of angiogenesis is controlled by several linked metabolic pathways and signaling systems that are just now beginning to be understood (19). The increase in capillary density is regulated by hypoxia-inducible factor-1. This transcription factor is responsible for upregulation of vascular endothelial growth factor, which mediates endothelial cell growth and cell division. Functional angiogenesis also requires the upregulation of cyclooxygenase and prostaglandin E2. The increased PGE2 results in increased angiopoietin-2. Angiopoietin-2 interferes with the mechanical stability of the capillary, which is normally mediated by the constitutive angiopoietin-1.

10.5 Conclusions

In this brief review it has only been possible to scratch the surface of what is known concerning brain adaptation to hypoxic environments. Still, there are several conclusions that can be drawn and a few new concepts that might be suggested. We can conclude that brainstem metabolism appears to be sensitive to even mild hypoxia. Evidence exists that brain stem regions responsible for cardiorespiratory regulation and cerebrovasodilation, especially in response to metabolic hypoxic/ischemic challenges, participate actively in the adaptation process. Finally, we might conclude that these brain stem nuclei that are activated during hypoxic adaptation undergo neurodegeneration after reversible ischemia, and this loss of function might be responsible for delayed mortality after cardiac arrest and resuscitation.

From a conceptual point of view, we can propose that long-term physiological adaptation to hypoxic environments includes initial transient, brain stem-mediated increases in blood flow, but long-term locally driven increases in capillary density. Overall, the brain demonstrates surprising plasticity in metabolic and structural components in physiological acclimatization and adaptation to environmental variations, and it appears that these same adaptive mechanisms and pathways may be the site of vulnerability to pathological stresses such as ischemia and may be a target for new therapeutic strategies.

Acknowledgment We would like to thank Bernadette Erokwu, Michelle Puchowicz, and Constantinos Tsipis for their help in these studies. These studies have been funded over the past 15 years by support from the AHA and the NINDS. This paper is dedicated to the memory of our friend and colleague Musa Haxhiu.

References

1. Xu K, LaManna JC. Chronic hypoxia and the cerebral circulation. J Appl Physiol 2006; 100(2):725–730.
2. LaManna JC, Chavez JC, Pichiule P. Structural and functional adaptation to hypoxia in the rat brain. J Exp Biol 2004; 207 (Part 18):3163–3169.
3. LaManna JC. Hypoxia in the central nervous system. Essays Biochem 2007; 43:139–152.
4. Roy CS, Sherrington CS. On the regulation of the blood-supply of the brain. J Physiol 1890; 11:85–108.
5. Iadecola C, Pelligrino DA, Moskowitz MA, Lassen NA. Nitric oxide synthase inhibition and cerebrovascular regulation. J Cereb Blood Flow Metab 1994; 14(2):175–192.
6. Underwood MD, Iadecola C, Reis DJ. Lesions of the rostral ventrolateral medulla reduce the cerebrovascular response to hypoxia. Brain Res 1994; 635:217–233.
7. LaManna JC, Haxhiu MA, Kutina-Nelson KL et al. Decreased energy metabolism in brainstem during central respiratory depression in response to hypoxia. J Appl Physiol 1996; 81(4):1772–1777.
8. Guyenet PG. Neural structures that mediate sympathoexcitation during hypoxia. Respir Physiol 2000; 121(2–3):147–162.
9. Golanov EV, Christensen JR, Reis DJ. Neurons of a limited subthalamic area mediate elevations in cortical cerebral blood flow evoked by hypoxia and excitation of neurons of the rostral ventrolateral medulla. J Neurosci 2001; 21(11):4032–4041.
10. Hoxworth JM, Xu K, Zhou Y, Lust WD, LaManna JC. Cerebral metabolic profile, selective neuronal loss, and survival of acute and chronic hyperglycemic rats following cardiac arrest and resuscitation. Brain Res 1999; 821(2):467–479.
11. Crumrine RC, LaManna JC. Regional cerebral metabolites, blood flow, plasma volume and mean transit time in total cerebral ischemia in the rat. J Cereb Blood Flow Metab 1991; 11:272–282.
12. Guyenet PG, Brown DL. Unit activity in nucleus paragigantocelluaris lateralis during cerebral ischemia in the rat. Brain Res 1986; 364:301–314.
13. Haxhiu MA, Strohl KP, Cherniack NS. The *N*-methyl-D-aspartate receptor pathway is involved in hypoxia-induced c-Fos protein expression in the rat nucleus of the solitary tract. J Auton Nerv Syst 1995; 55:65–68.
14. Xu K, Puchowicz MA, LaManna JC. Renormalization of regional brain blood flow during prolonged mild hypoxic exposure in rats. Brain Res 2004; 1027:188–191.
15. Brown MM, Wade JPH, Marshall J. Fundamental importance of arterial oxygen content in the regulation of cerebral blood flow in man. Brain 1985; 108:81–93.
16. Chávez JC, Agani F, Pichiule P, LaManna JC. Expression of hypoxic inducible factor 1α in the brain of rats during chronic hypoxia. J Appl Physiol 2000; 89:1937–1942.
17. Chavez JC, LaManna JC. Activation of hypoxia inducible factor-1 in the rat cerebral cortex after transient global ischemia: potential role of insulin like growth factor-1. J Neurosci 2002; 22(20):8922–8931.

18. Evans SM, Joiner B, Jenkins WT, Laughlin KM, Lord EM, Koch CJ. Identification of hypoxia in cells and tissues of epigastric 9L rat glioma using EF5 [2-(2-nitro-1H-imidazol-1-yl)-*N*-(2,2,3,3,3-pentafluoropropyl) acetamide]. Br J Cancer 1995; 72(4):875–882.
19. Dore-Duffy P, LaManna JC. Physiologic angiodynamics in the brain. Antioxid Redox Signal 2007; 9(9):1363–1371.

Chapter 11
Matrix Metalloproteinases in Cerebral Hypoxia-Ischemia

Zezong Gu, Jiankun Cui, and Stuart A. Lipton

Abstract Matrix metalloproteinase enzymes (MMPs) have been implicated in the pathogenesis of cerebral ischemia (stroke) and neurodegenerative diseases. Until recently, however, the mechanism of activation of MMPs in these disorders remained unclear. Following a hypoxic-ischemic insult, we identified a novel posttranslational modification (PTM) of MMPs involving S-nitrosylation (transfer of an NO group to a critical cysteine thiol) that is responsible for their activation (1). During cerebral ischemia/reperfusion in vivo, MMP-9 colocalizes with neuronal nitric oxide synthase (nNOS). In vitro, we showed that NO generated by nNOS activates the proform of MMP-9, forming S-nitrosylated MMP-9 (or SNO–MMP-9), which induces neuronal apoptosis. Both in vitro and in vivo mass spectrometry identified the activated, stable derivative of MMP-9 as a sulfinic or sulfonic acid, whose formation was triggered by S-nitrosylation following nitrosative and oxidative stress. These findings suggest the existence of an extracellular proteolysis pathway to neuronal cell death in which S-nitrosylation leads to activation of MMPs, and further oxidation results in stable PTM to a sulfinic or sulfonic acid derivative with pathological activity. Our group and others have shown that aberrant proteolytic activity of MMPs, especially MMP-9, contributes directly to neuronal injury, apoptosis, and brain damage after acute cerebral ischemia (1–4). We further investigated potential mechanisms for this toxic effect. We found that activated MMP-9 degrades the extracellular matrix (ECM) component laminin and that this degradation induces neuronal apoptosis in a model of transient focal cerebral ischemia in mice. We determined that the highly specific, mechanism-based, MMP-2/9 inhibitor, SB-3CT, blocks MMP-9 activity, including downstream

Z. Gu (✉), J. Cui
Department of and Anatomical Sciences, University of Missouri
School of Medicine, One Hospital Drive, M263 Medical Sciences Building, Columbia,
MO 65212, USA,
e-mail: guze@health.missouri.edu

S.A. Lipton
Center for Neuroscience, Aging and Stem Cell Research,
Burnham Institute for Medical Research,
La Jolla, CA 92037, USA.

laminin cleavage, thus rescuing neurons from apoptosis (5). We thus concluded that MMP-9 is a potential drug target in acute cerebral ischemia and that SB-3CT derivatives may have significant therapeutic potential in acute stroke patients. Importantly, however, as Eng Lo and colleagues (6) have recently shown, long-term activity of MMPs may be necessary for recovery from ischemic damage, so it appears that only short-term therapy with an MMP antagonist should be entertained in future clinical trials.

Keywords: Blood–brain barrier; Extracellular matrix; Matrix metalloproteinase; Mechanism-based inhibitor; Neurovascular unit; Nitrosative and oxidative stress; Proteolysis; Stroke

11.1 Introduction

Matrix metalloproteinases (MMPs) constitute a family of extracellular soluble or membrane-bound endopeptidases that are prominently involved in maintaining extracellular matrix (ECM) integrity and modulating interactions between cells during development and tissue remodeling (7, 8). Under normal physiological conditions, the activity of MMPs is precisely regulated at several levels – transcription of MMP message, activation of MMP proforms, and inhibition of MMPs by endogenous tissue inhibitors (TIMPs). Imbalance of MMP activity is thought to contribute to many neurological disorders, including stroke, Alzheimer's disease, HIV-associated dementia, and multiple sclerosis (8, 9), as well as other inflammatory and malignant processes (7, 8). MMP-9 in particular is significantly elevated in humans after stroke (10). Mice deficient in MMP-9 manifest a reduction in cerebral infarct size; in addition, treatment with MMP inhibitors or antibodies also reduces infarct size (10–13). In contrast, MMP-2 levels are acutely increased in the brains of baboons after stroke (14). Members of the MMP family share several structural features including propeptide, catalytic, and hemopexin domains (except MMP-7). One cysteine residue in the conserved autoinhibitory region of the propeptide domain coordinates a catalytic zinc ion lying at the catalytic center of the enzyme. This cysteine replaces a Zn^{2+}-bound water molecule representing the nucleophile in peptide bond hydrolysis by MMPs, thus inhibiting activity of the latent form of the enzyme. Disruption of the Zn^{2+}–cysteine interaction exposes Zn^{2+} in the active site allowing H_2O to bind, and consequently activates the MMP zymogen by a mechanism known as the *cysteine switch* (15, 16). Posttranslational modifications (PTMs) of MMPs, such as glycosylation, covalent dimerization, and noncovalent interactions, have been described previously (17). In the present review, we focus on the signaling cascade of MMPs induced predominately by redox-based PTM on the critical protein thiol in the cysteine switch. This redox reaction, induced by NO, is known as S-nitroyslation. This chemical reaction, resulting in the formation of SNO-MMP, contributes to the pathogenesis of neurodegenerative disorders, in particular brain ischemia.

11 Matrix Metalloproteinases in Cerebral Hypoxia-Ischemia

11.2 Crystal Structure Model of S-Nitrosylation of MMPs

Nitric oxide (NO) is a signaling molecule implicated in regulation of many biological processes in the nervous system, including neurotransmitter release, plasticity, and apoptosis (18–20). The chemical reactions of NO are largely dictated by its redox state (21). NO can modulate the biological activity of many proteins by reacting with cysteine thiol

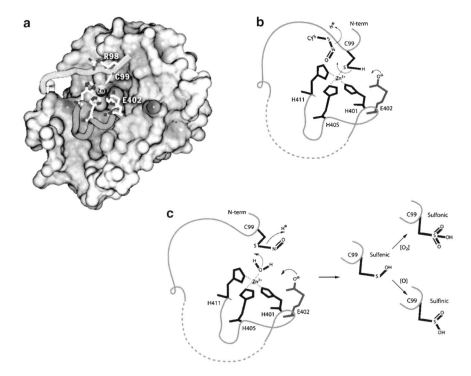

Fig. 11.1 Model of MMP-9 activation by S-nitrosylation and subsequent oxidation. (**a**) Molecular surface of partial sequence of human MMP-9 (from [97]Pro to [411]His without the fibronectin repeats found between [216]Val and [391]Gln) based on the known crystal structure (15). Color coded by charge with positive charge in *violet*, negative charge in *red*, propeptide domain ([97]Pro to [106]Arg) designated by a *yellow* ribbon, and catalytic domain ([401]His to [411]His) in *green*. In proMMP-9, Zn^{2+} is coordinated by a cysteine and three histidine residues. R98, C99, and E402 fit the proposed consensus motif for S-nitrosylation (21). *R* Arg, *C* Cys, *E* Glu, *H* His. (**b, c**) Proposed structure-based chemistry of NO-induced MMP-9 activation. Reactivity of the catalytic cysteine sulfur of MMP-9 appears to be enhanced by increased nucleophilicity of [402]Glu (shown in *red*) to S-nitrosylating agents (SNOC = Cys-SNO, for example). The sulfur bound at the zinc site appears to be highly nucleophilic, which may give high initial reactivity to NO from its endogenous donors. The SNO-MMP-9 propeptide domain appears to be more easily broken up in this highly polar environment and replaced by a nucleophilic water molecule. Reaction with H_2O of the S-nitrosothiol group forms sulfenic acid (–SOH), as observed in glutathione reductase (35). The reversible sulfenic acid can serve as an intermediate leading to subsequent more persistent oxidation steps via ROS to sulfinic (–SO_2H) and sulfonic (–SO_3H) acids [reproduced with permission from (1)] (*See Color Plates*)

to form an S-nitrosylated derivative. Such reactions regulate the activity of many circulating, membrane-bound, cytosolic, and nuclear proteins, including hemoglobin, NMDA receptors, caspases, and NF-κB (22–25). Cerebral ischemia and reperfusion result in nitrosative and oxidative stress, and hence the production of reactive nitrogen/oxygen species (RNS/ROS) (26, 27). The regulation of protein function by S-nitrosylation has led to the proposal that nitrosothiols function as PTMs analogous to phosphorylation or methylation. Although the factors governing cysteine reactivity toward nitrosylating agents are not completely understood, critical features include basic and acidic residues flanking the reactive cysteine, either in linear sequence or as a consequence of the three-dimensional organization of the protein, which catalyze the nitrosylation and denitrosylation steps (28). We noted such a motif in MMP-9 when we made an atomic model of its structure using the related MMP-2 crystal structure (PDB code 1CK7) (Fig. 11.1) (15). In proMMPs, a glutamate (E402 in MMP-9) essential for enzyme catalytic activity is located ~2.8 Å from the cysteine sulfur, and may act as a base to remove the sulfhydryl proton (in a similar fashion, in the activated enzyme, this glutamate acts as a base to activate the Zn^{2+}-bound water). The reactivity of the cysteine sulfur is likely to be further enhanced by its binding to Zn^{2+}, which increases its nucleophilicity. S-nitrosylation of this cysteine would be expected to reduce the nucleophilicity of the cysteine sulfur, weakening the bond to the zinc ion, and thus activating the enzyme. Therefore, we asked whether S-nitrosylation could mechanistically trigger the cysteine switch to activate MMPs under pathophysiologically relevant conditions. Here, we demonstrate a novel extracellular proteolytic cascade in which S-nitrosylation leads to oxidation and hence activation of MMP-9 with consequent neuronal apoptosis. Moreover, we show that this MMP cascading pathway occurs during stroke.

11.3 Neuronal NOS-Associated Activation of MMP During Cerebral Hypoxia/Ischemia

In the initial experiments, we examined the association of MMP-9 and NO immediately after focal cerebral ischemia and reperfusion in rodents. Gelatin zymography revealed an increase in both the level of proMMP-9 and MMP-9 activity in the ischemic hemisphere compared with the contralateral control hemisphere (1). Immunoblotting with an anti-MMP-9 antibody also showed increased MMP-9 levels in the ischemic hemisphere. The slight decrease in actin in the damaged hemisphere may reflect incipient cell loss at this early time point. Under our conditions, MMP-2 was not activated. Similar changes in MMP-9 have recently been reported after human embolic stroke (10). In situ zymography and immunocytochemistry were used to examine the cellular localization of MMP-9 enzymatic activity. We found that MMP activity was significantly elevated in ischemic brain parenchyma after ischemia and reperfusion (1). Moreover, to further demonstrate the pathophysiological relevance of these findings, the activation of MMP was abrogated after stroke in nNOS knockout mice or in wild-type animals that were treated with the relatively specific nNOS inhibitor 3-bromo-7-nitroindazole

(3br7NI). Previously, neuroprotection had been demonstrated under conditions of NOS inhibition (29). In wild-type animals not treated with NOS inhibitors, immunocytochemistry revealed that many cells positive for the neuronal marker NeuN manifested MMP activity (1).

Accumulating evidence has shown that brain damage after middle cerebral artery (MCA) occlusion involves ischemia/reperfusion-induced production of NO (27, 29). We therefore examined the immunoreactivity of nNOS and MMP-9 in the ischemic cortex by double immunofluorescent staining. We observed substantial colocalization of MMP-9 and nNOS (1), indicating the coincident production of NO and MMP-9 activity following ischemia and reperfusion.

11.4 S-Nitrosylation of Recombinant MMP Leads to Its Activation

Next, we tested whether MMP-9 could be S-nitrosylated and thus activated by NO in vitro. To eliminate the effects of TIMP-1 binding to the hemopexin domain, which might interfere with catalysis and activation of MMP-9, we used recombinant proMMP-9 encoding the propeptide and catalytic domains of MMP-9 but lacking the hemopexin domain (R-proMMP-9). R-proMMP-9, purified from conditioned medium of stably transfected human embryonic kidney 293 (HEK293) cells (30), was incubated with the physiological NO donor S-nitrosocysteine (SNOC). The generation of S-nitrosothiol was detected by measurement of the fluorescent compound 2,3-naphthyltriazole (NAT). NAT is stoichiometrically converted from 2,3-diaminonaphthalene (DAN) by NO released from S-nitrosylated proteins and thus provides a quantitative measure of S-nitrosothiol formation (31). R-proMMP-9 after exposure to SNOC resulted in significant S-nitrosothiol formation. To insure that the S-nitrosothiol generated under these conditions represented S-nitroso-MMP-9 (SNO–MMP-9) rather than residual SNOC, we examined the stability of these S-nitrosothiols at different incubation times. We found that the S-nitrosylation product of SNOC-treated R-proMMP-9 was much more stable than SNOC alone; within 15 min of incubation, over 95% of the SNOC had decayed while over 80% of SNO–MMP-9 remained. This temporal separation allowed us to distinguish SNOC from SNO–MMP-9 in the fluorescent S-nitrosothiol assay. Previously, it had been reported that MMPs were not activated by NO donors but only by peroxynitrite (ONOO$^-$) (32, 33); however, those studies did not monitor S-nitrosothiol formation immediately after exposure to specific endogenous nitrosylating agents, as performed here.

To determine if S-nitrosylation of R-pro-MMP-9 resulted in its activation, we compared the effects of the known exogenous MMP-9 activator, p-aminophenylmercuric acetate (APMA) with those of SNOC and another nitrosylating agent, acidified sodium nitrite. Gelatin zymography revealed that incubation with APMA, SNOC, or acidified sodium nitrite led to a partial conversion of the 53.5 kD

R-proMMP-9 into the 41.2 kD activated form of MMP-9; the respective masses were confirmed by mass spectrometry. The activation was inhibited in the presence of the MMP-specific hydroxamate inhibitor GM6001 (Gu and Lipton, unpublished observations). We then compared the activity of R-proMMP-9 incubated with APMA or SNOC by assaying the ability to cleave a synthetic peptide substrate. The initial velocity of R-proMMP-9 activation was 4.80 µM h^{-1} by APMA compared with 0.88 µM h^{-1} by SNOC. One potential caveat in these findings, however, is that the R-pro-MMP-9 lacked the hemopexin domain, which may be necessary for full activity of MMP-9. Therefore, we repeated these experiments with full-length recombinant proMMP-9 and found that S-nitrosylation led to similar activation of the full-length form (1). Taken together, these findings demonstrate that MMP-9 can undergo S-nitrosylation, and furthermore show that NO can directly promote activation of MMP-9.

11.5 MMP Proteolysis-Mediated Neuronal Apoptosis in Cerebrocortical Cultures

Next, we evaluated the effects of NO-activated MMP-9 on neuronal cell apoptosis in cerebrocortical cultures. MMP activity was assessed by in situ zymography; neurons were identified by immunoreactivity for MAP-2, and nuclear morphology was monitored with Hoechst 33342. We found that the percentage of neurons exhibiting MMP activity significantly increased after exposure to R-proMMP-9 preactivated with SNOC compared with R-proMMP-9 alone. SNOC from which NO was dissipated did not activate R-proMMP-9 and did not increase the percentage of neurons exhibiting MMP activity. Additionally, 18 h after exposure to SNOC-activated R-proMMP-9, we assessed apoptotic neurons by staining with anti-MAP-2 and terminal-deoxynucleotidyl-transferase-mediated deoxyuridine triphosphate nick-end labeling (TUNEL) in conjunction with condensed nuclear morphology assessed with Hoechst 33342. For these experiments, R-proMMP-9 was preexposed and hence preactivated by SNOC; NO had already been released from SNOC by the time the cultures were incubated with the activated MMP, as evidenced by measurement with an NO-sensitive electrode (19). Hence, direct release of NO from SNOC or the formation of peroxynitrite (ONOO$^-$) due to the release of NO from SNOC and subsequent reaction with superoxide anion (O$_2^-$) could not have accounted for the observed neuronal apoptosis (19). Extracellular proteolytic cascades triggered by MMPs can disrupt cell–matrix interactions, contribute to cell disconnection, and lead to a form of apoptotic cell death known as anoikis (34). NO-activated MMP-9 resulted in significantly increased neuronal apoptosis, whereas treatment with the MMP inhibitor GM6001 blocked the neuronal cell death. We also observed many neurons coming up off the dish after exposure to NO-activated MMP-9. These results strongly suggested that even high levels of inactivated proMMP-9 protein do not have a deleterious effect on neurons. However, NO-triggered activation converts MMP-9 into a neurotoxin.

11.6 Oxidative Modification Following S-Nitrosylation of MMP In Vitro and In Vivo

From the experiments using DAN to NAT conversion to show nitrosothiol generation, we knew that SNO–MMP-9 formation was associated with MMP-9 activation. However, nitrosothiols can be relatively short-lived and their reaction can be reversed by chemical reducing agents (35). Alternatively, S-nitrosothiol formation could potentially lead to further oxidative reactions that would result in more prolonged activation of MMPs. It has been suggested that S-nitrosylation of proteins via NO may generally represent a signal transduction cascade, whereas subsequent oxidation via ROS can lead to more persistent modifications under some conditions (35). In fact, our observation that SNO–MMP-9 was not stable over time and decayed within 1½ h while MMP-9 activation continued over the ensuing day, suggested that a more long-lasting derivative of activated MMP-9 might be produced, especially in the presence of an oxidative insult. To assess the possibility of these additional oxidative products and further identify the chemical nature of the NO-triggered-modification of MMP-9 responsible for activation, we conducted peptide mass fingerprinting as described previously (36). Mass spectra were obtained after in-gel digestion of human R-proMMP-9 by trypsin using matrix-assisted laser desorption/ionization time-of-flight (MALDI-TOF) mass spectrometry. To perform chemical reduction and in-gel digests without disrupting the peptide fragments of interest in the MALDI-TOF analysis, free cysteines had to be first protected to avoid cleavage followed by uncontrolled disulfide formation. Therefore, we initially protected cysteine by iodoacetamide alkylation in the absence of SNOC exposure. Among eleven mass peaks, we observed seven signature masses of human MMP-9 fragments that were virtually identical (<0.1% variation) to those predicted from theoretical tryptic fragments of MMP-9 deduced from the published amino acid sequences (1). One of these peaks represented the region responsible for the cysteine switch in the propeptide domain of proMMP-9 (CGVPDLGR, 816 Da) that had been alkylated with iodoacetamide (57 Da), to yield a molecular mass of 873.4 Da (acet-CGVPDLGR). In contrast, in vitro exposure of R-proMMP-9 to SNOC prior to attempted iodoacetamide alkylation yielded a tryptic fragment on MALDI-TOF analysis at 848.3 Da, indicating the addition of a stable 32 Da adduct to the propeptide domain CGVPDLGR (816 Da) instead of alkylation. Masking thiol groups by prior alkylation with iodoacetamide before exposure to SNOC (yielding a mass of 873.8 Da) blocked the addition of the 32 Da adduct. In this chemical context, therefore, the 32 Da adduct represented addition of two oxygen molecules to the cysteine residue to form a sulfinic acid derivative (SO$_2$H-CGVPDLGR at 848 Da) (29). In another approach, which empirically allowed us to avoid the mandatory step of initial protection of the critical cysteine residue in the propeptide domain by alkylation, we digested R-proMMP-9 in solution under native conditions rather than in acrylamide gel slices. Using this method, we found a mass peak at 816.7 Da, representing the propeptide domain fragment CGVPDLGR. We then observed a 48 Da shift rather than a 32 Da shift in the mass spectrum of the 816 Da fragment after SNOC exposure, yielding a peak at 864.8 Da, consistent with further oxidation to the sulfonic acid derivative (SO$_3$H-CGVPDLGR).

Our results and those of others show that MMP-9 is activated during focal cerebral ischemia and reperfusion (stroke) and that inhibition of MMP activity ameliorates parenchymal brain damage (11–13). Furthermore, production of NO and ROS is known to occur in the early phase of reperfusion after focal ischemia (26, 27). Thus, we next asked if the oxidation products of MMP-9 that we encountered in vitro after S-nitrosylation were also presented in vivo during focal ischemia and reperfusion. We examined mass spectra of tryptic fragments from affinity-precipitated MMP-9 obtained from rat brain after a 2-h MCA occlusion plus 15-min reperfusion injury or from the contralateral (control) side of the brain. For these experiments, we performed in-gel digestion with trypsin because gel separation offered better protein resolution. MALDI-TOF analysis of specimens obtained from the control side of the brain revealed that after reduction and alkylation by iodoacetamide (57 Da), the rat propeptide domain fragment (CGVPDVGK, mass 774 Da) yielded a peak at 830.3 Da, representing the alkylated fragment (acet-CGVPDVGK). In contrast, on the side of the brain with the stroke, the propeptide domain was not as susceptible to reduction and alkylation as evidenced by the appearance of an additional peak indicating a propeptide tryptic fragment at 821.8 Da; this peak represented the addition of a 48 Da adduct in accord with sulfonic acid derivatization of the thiol group (SO$_3$H-CGVPDVGK), and was similar to that found in vitro after NO activation of human MMP-9. Additionally, MALDI-TOF mass fingerprinting analysis revealed that of the 19 cysteine residues present in MMP-9, only the cysteine in the propeptide domain that coordinates Zn^{2+} in the active site was more persistently modified to a sulfinic (-SO$_2$H) or sulfonic (-SO$_3$H) acid in these experiments. Our findings indicate that S-nitrosylation of this cysteine residue in the prodomain followed by further oxidation to a sulfinic or sulfonic acid derivative leads to activation of MMP-9. Unlike S-nitrosylation, these latter oxidative reactions can be more persistent under certain conditions and therefore could contribute to the pathophysiological activation of MMP-9, as found during cerebral ischemia and reperfusion. Interestingly, activation of the enzyme can occur prior to cleavage, but after sulfinic or sulfonic acid modification, since we were able to observe these derivatives in our peptide analysis of proMMP-9. One of the pathways proposed for oxidation of the nitrosylated cysteine is via hydrolysis to form a sulfenic acid: E-S-N = O + H$_2$O → E-S-OH + HNO (35). The sulfenic acid is labile and susceptible to facile oxidation to the stable sulfinic or sulfonic acid derivatives that we observed during MALDI-TOF peptide fingerprinting. The MMP is set up to carry out hydrolysis of a peptide bond using an activated water molecule, and it is likely that the same machinery can be used to hydrolyze nitrosocysteine. To confirm the pathophysiological relevance of these findings, we performed the same ischemia and reperfusion experiments after nNOS inhibition with 3br7NI, which is known to be neuroprotective and decrease stroke size. Under these conditions with NO formation blocked, the sulfinic and sulfonic acid oxidation products of activated MMP-9 were not observed in our MALDI-TOF analysis. One caveat with these findings is that nNOS deletion or NOS inhibition diminishes stroke damage, and hence one could argue that other stroke-related processes responsible for MMP activation would be reduced. Nonetheless, taken together with the data demonstrating S-nitrosylation of MMPs and the fact that we found MMPs activated in this fashion cause neuronal apoptosis in vitro makes it likely that NO activation of MMPs participates in neuronal injury in vivo.

11.7 Increased MMP Gelatinolytic Activity Spatially Associated with Neuronal Laminin in the Ischemic Brain

We had thus shown that activated MMP-9 directly induces neuronal apoptosis both in vitro and in vivo after focal cerebral ischemia/reperfusion (1). Additional work had reported that MMP-9 contributes to delayed neuronal cell death in the hippocampus after transient global ischemia (37). Although several reports (3, 4, 38–40) had suggested that basement membrane proteins are involved in an MMP-9 proteolytic pathway, it was unclear how MMP substrates contribute to neuronal cell death. To identify potential targets of MMP-9 proteolysis in the ischemic cortex, we examined basal membrane protein laminin by immunohistochemistry. Pan-laminin (pan-Ln) antibody staining followed by deconvolution microscopy detected high immunoreactivity in the proximity of NeuN-positive cells and microvascular structures. We confirmed such neuronal laminin immunoreactivity by immunostaining with a neuron-specific antibody α5 subunit of Ln-10 (41). Gelatinolytic activity largely colocalized with Ln-positive neurons in the ischemic cortex within 2 h of reperfusion, suggesting colocalization of MMP-9 activity with neuronal laminin in the early stages of brain damage after transient ischemia. To corroborate these observations, we asked if purified, activated MMP-9 can cleave laminin. We incubated activated MMP-9 with brain lysates and followed laminin cleavage by analyzing digested samples by SDS-PAGE and Western blotting. We found that MMP-9 cleaved laminin subunits in a dose-dependent manner to generate a 51 kDa fragment. As controls, latent MMP-9 (proMMP-9) or catalytically active MT1-MMP did not degrade laminin. A broad-spectrum MMP inhibitor, GM6001, unlike a cocktail of non-MMP protease inhibitors, inhibited MMP-9 proteolysis of neuronal laminin. Taken together, the data indicated that MMP-9 can cleave laminin on the neuronal surface.

11.8 Inhibition of MMP Proteolysis Prevents Laminin Degradation and Rescues Neurons from Ischemia

As discussed earlier, we had found that S-nitrosylation of MMP-9 leads to its activation and that increased MMP-9 activity in the ischemic cortex does not occur in nNOS KO mice or in wild-type mice after specific inhibition of nNOS activity. We then further demonstrated that laminin degradation was inhibited in the ischemic cortex during stroke in nNOS KO mice and in wild-type mice treated with the specific nNOS antagonist – 3br-7NI. Deconvolution microscopy revealed that laminin immunoreactivity was significantly reduced in the ischemic cortex of wild-type mice compared to the contralateral hemisphere control determined 24 h after reperfusion. Laminin degradation in the ischemic cortex was also attenuated in nNOS KO mice. These results are consistent with a link between MMP-9 activation by S-nitrosylation and laminin cleavage. Next, we examined the time course of laminin degradation and apoptotic cell death in the ischemic cortex. In situ zymography 24 h after reperfusion revealed that increased gelatinolytic activity was associated with apoptotic cells (detected by TUNEL plus condensed nuclei). Moreover, we found decreased laminin immunoreactivity within 3 h of reperfusion compared to the control contralateral cortex. Remaining laminin was mostly associated with elongated microvascular structures rather than neurons. Modest neuronal cell death occurred at 3 h but more massive death occurred at 24 h, as evidenced by condensed nuclei. These results indicate that increased MMP gelatinolytic activity-induced laminin degradation in the ischemic cortex occurs prior to cell death after MMP activation. Recently, we applied a newly developed mechanistically based thiirane inhibitor of MMPs, designated SB-3CT, and examined its effects on MMP-mediated proteolysis of laminin in the mouse stroke model. SB-3CT is the first mechanism-based MMP inhibitor that is selective for the gelatinases MMP-2 and MMP-9 (42). SB-3CT coordinates the catalytic zinc ion, contributing to both slow binding and mechanism-based inhibition. This type of inhibition is unique among MMP inhibitors developed to date (42, 43). Western blots of brain extracts demonstrated partial degradation of laminin (especially of the 360 and 170 kDa subunits) to a 51 kDa fragment in the ischemic hemisphere after transient MCA occlusion. Degradation of laminin and production of the 51 kDa laminin fragment were reduced in the ischemic hemisphere of mice injected with SB-3CT, indicating that specific MMP-9 inhibition significantly protected laminin from proteolysis. Moreover, we found that SB-3CT offered significant protection of the brain in the face of acute hypoxic-ischemic insult. Taken together, our findings suggest that SNO-MMP leads to laminin degradation and neuronal damage during cerebral hypoxia-ischemia, and that specific inhibition of MMP-9 activity produces neuroprotection, as summarized in Fig. 11.2.

Similarly, Lo and colleagues reported that BB-94, a broad-spectrum MMP inhibitor, reduced delayed neuronal cell death in the hippocampus 3 days after transient global ischemia (37). Moreover, Asahi et al. (3) demonstrated that lack of MMP-9 stabilized the blood–brain barrier (BBB) by preventing degradation of BBB-associated proteins. These MMP inhibitors also ameliorate BBB disruption and reperfusion injury experimentally induced by tissue plasminogen activator (tPA) during an ischemic stroke, which increases the risk of bleeding (44). Importantly, however, MMPs may take on a beneficial role as the days progress after the onset of a stroke. At these later time points, MMPs appear to help mediate repair, and thus long-term inhibition of MMP activity may cause detrimental effects. As Lo and colleagues (6) recently have shown, inhibition of

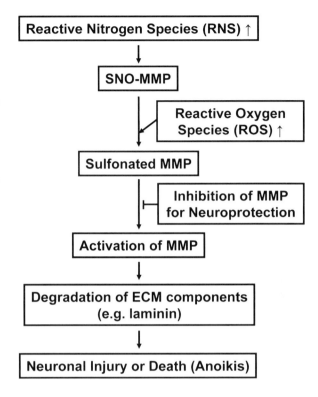

Fig. 11.2 Proposed reaction scheme of SNO-MMP producing abnormal proteolysis of ECM components, e.g., laminin, and therefore contributing to neuronal cell injury and death during cerebral hypoxia-ischemia. Note that S-nitrosylation of MMP, to form SNO-MMP, stimulates MMP enzyme activity, but further oxidation of the MMP to sulfonated forms may produce more persistent activation

MMPs, particularly MMP-9, during this later time window hinders brain repair in mice and may even increase the risk for hemorrhagic transformation following hypoxia-ischemia.

11.9 Summary

S-Nitrosylation and subsequent oxidation of protein thiol in the prodomain of MMP-9 can lead to enzyme activation. It is likely that other homologous MMPs are activated in a similar manner. This series of reactions confers responsiveness of the ECM to nitrosative and oxidative stress. Such stress is relevant to a number of pathophysiological conditions, including cerebral hypoxia-ischemia and several neurodegenerative disorders. Additionally, our results support a model in which extracellular proteolytic cascades triggered by activation of MMPs via NO (1) can disrupt ECM components such as laminin, contribute to cell detachment, and lead to a form of apoptotic cell death known as anoikis (35). In fact, the effect of MMP-9 is functionally similar to that of tPA, which is known to contribute to laminin degradation (41, 45, 46) and can also activate MMPs (6). This point underscores the fact that tPA, which is frequently used as an antithrombolytic clot buster and hence as a therapy for stroke, can also contribute to neuronal cell death (46).

MMP-9 may be produced by various cell types, including neutrophils and macrophages, which are known to migrate into the brain after damage to the BBB because of ischemia/reperfusion injury (8). Specific inhibition of MMP-9 can prevent the dire consequences of excessive activation by either NO or tPA. We show that SB-3CT, a mechanism-based and selective gelatinase inhibitor, provides significant protection against brain damage in an experimental model of focal cerebral ischemia, consistent with a role for MMP-9 in laminin degradation and consequent neuronal cell death after stroke. We conclude that targeting MMP-9 in acute stroke patients is a highly promising therapeutic approach that deserves additional exploration. The use of novel MMP inhibitors such as SB-3CT or derivatives, which exhibit selective inhibition of MMP-9/MMP-2 gelatinases, should minimize the undesired side effects caused by broad spectrum MMP inhibitors in previous clinical trials. Our studies are the first to demonstrate a significant pharmacological benefit of this new type of mechanism-based MMP inhibitor in an animal model of stroke. These results hold promise for the successful application of SB-3CT derivatives to stroke patients.

Acknowledgments Z.G. and S.A.L. were supported in part by NIH grants and by the American Heart Association.

References

1. Gu Z, Kaul M, Yan B, et al. S-Nitrosylation of matrix metalloproteinases: signaling pathway to neuronal cell death. Science 2002;297:1186–90.
2. Rosenberg GA, Navratil M, Barone F, and Feuerstein G. Proteolytic cascade enzymes increase in focal cerebral ischemia in rat. J Cereb Blood Flow Metab 1996;1006:16:360–6.
3. Asahi M, Wang X, Mori T, et al. Effects of matrix metalloproteinase-9 gene knock-out on the proteolysis of blood–brain barrier and white matter components after cerebral ischemia. J Neurosci 2001;19:7724–32.
4. Horstmann S, Kalb P, Koziol J, Gardner H, and Wagner S. Profiles of matrix metalloproteinases, their inhibitors, and laminin in stroke patients: influence of different therapies. Stroke 2003;34:2165–70.
5. Gu Z, Cui J, Brown S, et al. A highly specific inhibitor of matrix metalloproteinase-9 rescues laminin from proteolysis and neurons from apoptosis in transient focal cerebral ischemia. J Neurosci 2005;25:6401–8.
6. Zhao BQ, Wang S, Kim HY, et al. Role of matrix metalloproteinases in delayed cortical responses after stroke. Nat Med 2006;12:441–5.
7. Sternlicht MD and Werb Z. How matrix metalloproteinases regulate cell behavior. Ann Rev Cell Dev Biol 2001;17:463–516.
8. Yong VW. Metalloproteinases: mediators of pathology and regeneration in the CNS. Nat Rev Neurosci 2005;6:931–44.
9. Campbell IL and Pagenstecher A. Matrix metalloproteinases and their inhibitors in the nervous system: the good, the bad and the enigmatic. Trends Neurosci 1999;22:285–7.
10. Montaner J, Alvarez-Sabin J, Molina C, et al. Matrix metalloproteinase expression is related to hemorrhagic transformation after cardioembolic stroke. Stroke 2001;32:1759–66.

11. Asahi M, Asahi K, Jung JC, et al. Role for matrix metalloproteinase-9 after focal cerebral ischemia: effects of gene knockout and enzyme inhibition with BB-94. J Cereb Blood Flow Metab 2000;20:1681–9.
12. Gasche Y, Fujimura M, Morita-Fujimura Y, et al. J Cereb Blood Flow Metab 1999;19:1020–8.
13. Romanic AM, White RF, Arleth AJ, Ohlstein EH, and Barone FC. Matrix metalloproteinase expression increases after cerebral focal ischemia in rats: inhibition of matrix metalloproteinase-9 reduces infarct size. Stroke 1998;29:1020–30.
14. Heo JH, Lucero J, Abumiya T, et al. Matrix metalloproteinases increase very early during experimental focal cerebral ischemia. J Cereb Blood Flow Metab 1999;19:624–33.
15. Morgunova E, Tuuttila A, Bergmann U, et al. Structure of human pro-matrix metalloproteinase-2: activation mechanism revealed. Science 1999;284:1667–70.
16. Van Wart HE and Birkedal-Hansen H. The cysteine switch: a principle of regulation of metalloproteinase activity with potential applicability to the entire matrix metalloproteinase gene family. Proc Natl Acad Sci USA 1999;87:5578–82.
17. Van den Steen PE, Dubois B, Nelissen I, et al. Biochemistry and molecular biology of gelatinase B or matrix metalloproteinase-9 (MMP-9). Crit Rev Biochem Mol Biol 2002;37:375–536.
18. Dawson TM and Snyder SH. Gases as biological messengers: nitric oxide and carbon monoxide in the brain. J Neurosci 1994;14:5147–59.
19. Lipton SA, Choi YB, Pan ZH, et al. A redox-based mechanism for the neuroprotective and neurodestructive effects of nitric oxide and related nitroso-compounds. Nature 1993;364:626–32.
20. Melino G, Bernassola F, Knight RA, et al. S-Nitrosylation regulates apoptosis. Nature 1997;388:432–3.
21. Stamler JS. Redox signaling: nitrosylation and related target interactions of nitric oxide. Cell 1994;78:931–6.
22. Choi YB, Tenneti L, Le DA, et al. Molecular basis of NMDA receptor-coupled ion channel modulation by S-nitrosylation. Nat Neurosci 2000;3:15–21.
23. Jaffrey SR, Erdjument-Bromage H, Ferris CD, Tempst P, and Snyder SH. Protein S-nitrosylation: a physiological signal for neuronal nitric oxide. Nat Cell Biol 2001;3:193–7.
24. Jia L, Bonaventura C, Bonaventura J, and Stamler JS. S-nitrosohaemoglobin: a dynamic activity of blood involved in vascular control. Nature 1996;380(6571):221–6.
25. Matthews JR, Botting CH, Panico M, Morris HR, and Hay RT. Inhibition of NF-kappaB DNA binding by nitric oxide. Nucleic Acids Res 1996;24:2236–42.
26. Kumura E, Kosaka H, Shiga T, Yoshimine T, and Hayakawa T. Elevation of plasma nitric oxide end products during focal cerebral ischemia and reperfusion in the rat. J Cereb Blood Flow Metab 1994;14:487–91.
27. Sato S, Tominaga T, Ohnishi T, and Ohnishi ST. Electron paramagnetic resonance study on nitric oxide production during brain focal ischemia and reperfusion in the rat. Brain Res 1994;647(1):91–6.
28. Stamler JS, Toone EJ, Lipton SA, and Sucher NJ. (S)NO signals: translocation, regulation, and a consensus motif. Neuron 1997;18(5):691–6.
29. Huang Z, Huang PL, Panahian N, et al. Effects of cerebral ischemia in mice deficient in neuronal nitric oxide synthase. Science 1994;265(5180):1883–5.
30. Kridel SJ, Chen E, Kotra LP, et al. Substrate hydrolysis by matrix metalloproteinase-9. J Biol Chem 2001;276(23):20572–8.
31. Wink DA, Kim S, Coffin D, et al. Detection of S-nitrosothiols by fluorometric and colorimetric methods. Methods Enzymol 1999;301:201–11.
32. Eberhardt W, Beeg T, Beck KF, et al. Nitric oxide modulates expression of matrix metalloproteinase-9 in rat mesangial cells. Kidney Int 2000;57:59.
33. Okamoto T, Akaike T, Nagano T, et al. Activation of human neutrophil procollagenase by nitrogen dioxide and peroxynitrite: a novel mechanism for procollagenase activation involving nitric oxide. Arch Biochem Biophys 1997;342:261.
34. Cardone MH, Salvesen GS, Widmann C, et al. The regulation of anoikis: MEKK-1 activation requires cleavage by caspases. Cell 1997;90:315–23.

35. Stamler JS and Hausladen A. Oxidative modifications in nitrosative stress. Nat Struct Biol 1998;5(4):247–9.
36. Yan B and Smith JW. A redox site involved in integrin activation. J Biol Chem 2000;275(51): 39964–72.
37. Lee SR, Tsuji K, and Lo EH. Role of matrix metalloproteinases in delayed neuronal damage after transient global cerebral ischemia. J Neurosci 2004;24:671–8.
38. Castellanos M, Leira R, Serena J, et al. Plasma metalloproteinase-9 concentration predicts hemorrhagic transformation in acute ischemic stroke. Stroke 2003;34:40–6.
39. Chen ZL, Indyk JA, and Strickland S. The hippocampal laminin matrix is dynamic and critical for neuronal survival. Mol Biol Cell 2003;14:2665–76.
40. Hamann GF. Unriddling the role of matrix metalloproteinases in human cerebral stroke. Stroke 2003;34(1):40–6.
41. Indyk JA, Chen ZL, Tsirka SE, et al. Laminin chain expression suggests that laminin-10 is a major isoform in the mouse hippocampus and is degraded by the tissue plasminogen activator/plasmin protease cascade during excitotoxic injury. Neuroscience 2003;116:359–71.
42. Brown S, Bernardo MM, Li ZH, et al. Potent and selective mechanism-based inhibition of gelatinases. J Am Chem Soc 2000;122:6799–800.
43. Kleifeld O, Kotra LP, Gervasi DC, et al. X-ray absorption studies of human matrixmetalloproteinase-2 (MMP-2) bound to a highly selective mechanism based inhibitor. Comparison with the latent and active forms of the enzyme. J Biol Chem 2001;276:17125–31.
44. Pfefferkon T and Rosenberg GA. Closure of the blood–brain barrier by matrix metalloproteinase inhibition reduces rtPA-mediated mortality in cerebral ischemia with delayed reperfusion. Stroke 2003;34:2025–30.
45. Chen ZL and Strickland S. Neuronal death in the hippocampus is promoted by plasmin-catalyzed degradation of laminin. Cell 1997;91(7):917–25.
46. Wang YF, Tsirka SE, Strickland S, et al. Tissue plasminogen activator (tPA) increases neuronal damage after focal cerebral ischemia in wild-type and tPA-deficient mice. Nat Med 1998;4:228–31.

Chapter 12
Oxidative Stress in Hypoxic-Ischemic Brain Injury

Laura L. Dugan, M. Margarita Behrens, and Sameh S. Ali

Abstract Hypoxic-ischemic brain injury (stroke) continues to be the third leading cause of death in the USA. Hypoxia in the absence of ischemia, such as that occurs in individuals with sleep apnea, chronic obstructive pulmonary disease, and a number of other conditions, also represents a major and growing health issue. A great deal has been learned about the processes that contribute to short-term damage and long-term dysfunction of the nervous system after hypoxia or ischemia, and the mechanisms that underlie vulnerability of the brain to hypoxic-ischemic injury. Oxidative stress, due to pathological changes in the homeostasis of reactive oxygen species (ROS), reflecting abnormal production and/or impaired clearance of ROS, has been implicated as a key mechanism that contributes to tissue damage and functional deficits in hypoxic-ischemic brain injury. This chapter will briefly review prior literature on the sources and molecular targets of ROS in hypoxia-ischemia, and will focus on newer studies implicating inflammatory signaling and redox dysregulation in hypoxic-ischemic brain injury.

Keywords: hypoxia; ischemia; brain injury, stroke; sleep apnea; reactive oxygen species; oxidative stress; mitochondria; NADPH oxidase; inflammation

12.1 Reactive Oxygen Species: Overview

Under normal physiologic circumstances, living organisms maintain certain levels of reactive oxygen species (ROS) in cells and tissues. An array of cellular factors contribute to maintaining this *redox* homeostasis through sophisticated and synchronized signaling pathways and regulatory responses. The main elements dictating the equilibrium of redox status in cells and tissues include oxygenation (tissue pO_2), rates of production of ROS, reactive nitrogen species (RNS), and reactive thiol/sulfur species (RSS); levels of antioxidants such as glutathione and vitamins; and

---

L.L. Dugan (✉), M.M. Behrens, and S.S. Ali
Division of Geriatric Medicine, Department of Medicine, University of California, San Diego, CA 92093, USA
e-mail: ladugan@ucsd.edu

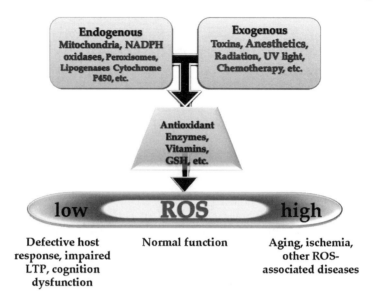

Scheme 12.1 A critical balance of ROS levels is needed for normal nervous system function, with pathological conditions, including hypoxia-ischemia, associated with elevated ROS at one end of the spectrum, and inadequate ROS levels leading to decreased synaptic activity, impaired long-term potentiation (LTP), and cognitive dysfunction at the other (1) (*See Color Plates*)

the abundance and activity of the enzymatic arsenal that metabolize ROS, e.g., superoxide dismutase(s), glutathione peroxidase(s), catalase, peridoxin, and many others. Partial or complete failure of any of these elements, or of their interactive communication, will shift normal homeostasis toward pathology. Hypoxic-ischemic nervous system injury is among more than 100 disease conditions linked to ROS and loss of redox homeostasis (1) (Scheme 12.1).

Reactive oxygen species generally refer to the oxygen free radicals superoxide anion and hydroxyl radical (oxygen molecules carrying an unpaired electron), and the nonradical molecule, hydrogen peroxide (H_2O_2), which can be rapidly converted to hydroxyl radical via the Fenton reaction. In the presence of transition metals, superoxide, H_2O_2, and hydroxyl radicals can also be interconverted by the Haber–Weiss reaction (2). Of note, molecular oxygen (O_2) actually qualifies as a biradical because of its two unpaired electrons, but its reactivity with nonradical compounds is limited due to electron spin restrictions. Under specific conditions, however, O_2 can react with other radicals, accepting one electron to form the reactive superoxide radical $O_2^{\cdot-}$ (2). Thus, superoxide is the single-electron reduction product of O_2 and is the primary ROS produced by a broad array of enzyme complexes, including the mitochondrial electron transport chain, the NADPH oxidase family, xanthine oxidase, and many P450 enzymes.

In keeping with the high metabolic rate of adult neural tissue, the brain has significant concentrations of both enzymatic and small molecule antioxidant defenses. Three members of the superoxide dismutase (SOD) family of enzymes are expressed in brain: Sod1 (Cu,Zn-SOD), a highly abundant cytosolic protein, Sod2 (MnSOD), the mitochondrial form, and extracellular (EC-SOD), a glycosylated extracellular form with homology to Sod1. All three convert $O_2^{\cdot-}$ to H_2O_2, which is then further metabolized to water and oxygen by catalase (cytosol only) or glutathione peroxidase (cytosol, mitochondria). Several other enzymes unrelated to SOD1, for example, the *atx1-HAH* gene product, are also capable of acting as SODs. It remains to be determined to what extent EC-SOD and these alternative dismutases can act as antioxidant systems in humans.

Catalase and glutathione peroxidase provide two important cellular systems for eliminating H_2O_2. Catalase is more abundant in astrocytes than in neurons and in white matter than in gray matter, but it can be induced in neurons by neurotrophins. Catalase can be irreversibly inactivated by oxidation and demonstrates decreased activity after ischemia-reperfusion.

Four members of the glutathione peroxidase family of selenoproteins have been identified. Classical glutathione peroxidase (GPx1), plasma GPx (GPx3), GPx2, and a fourth enzyme, phospholipid hydroperoxide glutathione peroxidase (PHGPx). All use glutathione (GSH) as a cofactor to convert H_2O_2 to water. GPx1, PHGPx, and required components of the glutathione peroxidase system, GSH, glutathione reductase, and NADPH, are present in brain in both cytoplasm and mitochondria. In addition, small-molecule antioxidants including GSH, ascorbic acid (vitamin C), vitamin E, flavonoids, as well as a number of thiol-containing proteins, including thioredoxin and metallothionine, also contribute to the cellular antioxidant pool.

While most free radical species are thought of as highly unstable, many species can persist for seconds to minutes. Free-radical reactions are in fact intrinsic to a majority of the metabolic and synthetic reactions carried out by eukaryotic cells and, as such, are required for life. ATP production by the mitochondrial electron chain, for example, uses a controlled set of oxygen radical reactions to couple the reduction of free-radical electrons with the movement of protons across the mitochondrial membrane. Generally, such reactions are very efficient, but under pathological conditions, leak of radical electrons can produce ROS including $O_2^{\cdot-}$, and as discussed further later, this inefficiency is enhanced by a variety of events that occur during hypoxia-ischemia.

One major consequence of increased production of ROS is damage to cellular macromolecules, resulting in fragmentation or structural changes in lipids, proteins, and nucleic acids. The brain has a number of characteristics that make it especially susceptible to this free radical-mediated damage. Brain lipids contain high concentrations of polyunsaturated fatty acids (PUFAs), which are susceptible to lipid peroxidation. Many brain regions also contain relatively high concentrations of iron. Both of these factors increase the susceptibility of brain cell membranes to lipid peroxidation. Because the brain is almost completely dependent on aerobic metabolism, mitochondrial respiration is higher than in most other tissues, increasing the risk of free radical production by mitochondria, and suggesting, in turn, that free radical-mediated damage to mitochondria and resultant impaired metabolism might be poorly tolerated by the brain due to lack of significant nonmitochondrial sources of energy production.

12.2 Sources of ROS in Hypoxia-Ischemia

The pathology of hypoxia-ischemia has both spatial and temporal components. Injury severity progresses from the central core, with rapid and presumably irreversible injury, to the penumbra as an at-risk area surrounding the core, to remote regions not directly affected by ischemia, but potentially vulnerable to changes in synaptic activity, growth factor secretion, ion shifts, and production of inflammatory mediators triggered by the ischemic insult. Injury also progresses temporally from the initial acute phases of ischemia, in which ionic shifts and energy failure initiate rapid necrosis and apoptosis in the ischemic core. When reperfusion occurs, additional inflammatory and stress-response pathways are activated, and finally, compensatory remodeling, structural, and synaptic changes occur. Likewise, as discussed further later, ROS generation during ischemia-reperfusion has both spatial and temporal aspects. Mitochondria appear to be an important and perhaps dominant source of free radicals in brain during the acute phase of hypoxia-ischemia. Reperfusion, which is associated with a dramatic increase in ROS, recruits additional sources of ROS/RNS, including NOS isoforms, NADPH oxidase(s), and xanthine oxidase, and this process differs between cell types and region.

12.2.1 Mitochondria

The most compelling data supporting the importance of mitochondrial ROS to ischemic brain injury comes from an extensive series of studies using mice with genetic manipulation of intrinsic antioxidant systems, reviewed extensively by Sugawara et al. (3, 4). These studies show that Sod2(–/+) mice, which are partially deficient in mitochondrial superoxide dismutase activity, have enhanced infarct volume and elevated superoxide production after transient ischemia. Sod2(–/+) mice also exhibit increased neuronal apoptosis following ischemia-reperfusion. Complementing these studies, treatment of rats during ischemia with rotenone, which inhibits mitochondrial Complex I, blocked increased ROS production during reperfusion (5), again implicating the mitochondrial electron transport chain (ETC) as a major source of ROS during hypoxia-reperfusion

Mitochondrial ROS production during the acute phase of hypoxia-ischemia may reflect calcium-dependent dysregulation of mitochondria due to NMDA N-methyl-D-aspartate receptor-mediated excitotoxicity. Extracellular glutamate levels, which rise rapidly during ischemia, can overactivate NMDA receptors, increasing intracellular calcium through NMDA receptors, a process that has been shown to rapidly enhance ROS production by mitochondria (6–8). NMDA receptor antagonists have been shown to be both neuroprotective and to reduce mitochondrial ROS production in both cell culture and animal models of ischemia.

Research spanning four decades indicates that the effects of hypoxia-ischemia on mitochondrial function and ROS production progress over time. Studies on mitochondria isolated from ischemic brain (9–14), and from in vivo metabolic imaging studies (15), suggest that ischemia-reperfusion causes both short-term and long-term alterations in mitochondrial function. Mitochondria isolated from brain after brief focal ischemia show reversible defects in State 3 (ADP-dependent), and State 4 (ADP-independent) respiration. These deficits appear to resolve after a brief period of reperfusion. Baseline ATP production by mitochondria isolated from ischemic brain is decreased relative to control mitochondria, but can be stimulated to the same degree as control mitochondria, an observation that is believed to reflect decreased demands for ATP production after ischemia, instead of damage to mitochondria per se (9). Longer periods of ischemia lead to prolonged impairment in State 3 respiration, a finding attributed to loss/metabolism of adenine nucleotides during ischemia, and the need to resynthesize ADP for full mitochondrial function to return (9). In addition, although mitochondrial metabolism appears to return to *normal* soon after ischemia-reperfusion, many studies have found a secondary decline minutes to hours later (12, 16, 17). The cause of this delayed decline is believed to be due to oxidative damage to mitochondria during reperfusion (12, 18). This may reflect increased ROS production by mitochondria, themselves. Furthermore, fatty acids released during ischemia may also contribute to progressive impairment of mitochondrial function (11, 19).

Metabolic imaging studies have suggested that mitochondrial function in postischemic brain may actually remain impaired for hours to days after the ischemic insult (15, 20–24). Although the mechanisms for this impairment have not been fully defined, oxidative loss of pyruvate dehydrogenase (PDH) activity due to oxidation of critical thiols on PDH after ischemic-reperfusion (25, 26) may be one process. Decreased activity of PDH, the *gatekeeper* for entry of glycolytic products into the Kreb's cycle, will substantially impair mitochondrial energy metabolism, especially in brain, where there is little use of alternative substrates, such as fatty acids.

Oxidative damage to mitochondria and resulting metabolic impairment may have several consequences for mitochondrial function, including not only lower ATP synthesis, but decreased mitochondrial membrane potential (ψ_m) and release of proapoptotic factors. Mitochondria are important regulators of intracellular Ca^{2+} (8, 27, 28), and Ca^{2+} uptake is regulated by ψ_m. Mitochondria also integrate signals involved in apoptotic cell death (29–31), and loss of mitochondrial membrane potential (ψ_m) can result in release of cytochrome *c* from the inner mitochondrial membrane; the free cytochrome *c* may then trigger caspase-dependent apoptosis. Multiple studies (32–34) report release of cytochrome *c* and mitochondrial caspase 9 during CNS ischemia, consistent with a decrease in ψ_m in at least some mitochondrial populations. In addition to its ability to trigger apoptosis, release of cytochrome *c* from mitochondria may enhance ROS production by the ETC by limiting the efficient transfer of electrons through the cytochrome aa_3 in cytochrome oxidase (Complex III), enhancing upstream superoxide radical leak at ubiquinone (35). Evidence that this occurs in vivo was reported in piglet brain during ischemia (36). That study followed the oxidation state of the aa_3 cytochrome of Complex III in brain, and found progressive reduction of the cytochrome during ischemia, as expected if electron transfer from Complex III to cytochrome *c* was blocked. The reduced status of the cytochrome was associated with loss of

ATP production (36). Similar results have been reported in brain gerbils (37), and rats (38) subjected to hypoxia. These results suggest that the activity and efficiency of electron transport chain (ETC) components is progressively impaired following hypoxia-ischemia. Continued loss of ETC function will result in a decrease in mitochondrial ψ. Although decreased ψ_m under certain circumstances has been reported to lower mitochondrial ROS production (39, 40), critical loss of cytochrome c and oxidative damage to mitochondrial electron transport chain complexes might enhance ROS production acutely.

In summary, mitochondria appear to be an important early source of ROS production in brain following hypoxia-ischemia, with enhanced oxidant generation reflecting increased Ca^+, and possibly reduced efficiency of electron flux through the ETC. Subsequently, increased ROS production results in oxidative damage to PDH and ETC components and release of cytochrome c accompanied by enhanced mitochondrial uncoupling and loss of ψ_m. These secondary changes might lessen mitochondrial ROS generation, but would also metabolically compromise mitochondria. Compensatory changes in gene expression for mitochondrial ETC proteins, such as decreased expression of Complex I and increased expression of Complex IV may also act to counteract increased ROS production. Finally, long-term changes in mitochondrial function and ROS production due to oxidative damage to both mitochondrial and nuclear DNA, and activation of various inflammatory signaling cascades, may also contribute to late changes in mitochondrial metabolic function and oxidant production.

12.2.2 The NADPH Oxidase (Nox) Family of Superoxide-Generating Enzymes in Hypoxia-Ischemia

The first NADPH oxidase described, $gp91^{phox}$ or Nox2, was originally identified in neutrophils as the enzyme complex responsible for the respiratory burst, an essential host response to microbial invasion. It is now clear that Nox2 and other recently-identified members of the Nox family (Nox1, 3–5, Duox1, 2), which are encoded by separate genes and utilize different activating protein subunits, are expressed in a diverse array of tissues and cells types, including neurons, astrocytes, and microglia in brain (41–46). These isoforms appear to produce less superoxide, suggesting an involvement in intra- and intercellular redox signaling (47). Expression of Nox2 and Nox4 was reported in the CNS (42, 43, 48, 49), and specifically, activation of Nox2-containing NADPH oxidase has been related to brain injury after ischemia-reperfusion (50).

At least six proteins form part of the Nox2-complex: a membrane flavocytochrome b588 core complex consisting of two subunits (Nox2 and $p22^{phox}$), and three cytoplasmic components ($p47^{phox}$, $p67^{phox}$, and Rac1 or Rac2). The enzyme is inactive in resting cells, but becomes activated to generate superoxide upon assembly of the core with the cytoplasmic components through the interaction of the Src homology-3 (SH3) domains on $p47^{phox}$ and $p67^{phox}$ with the proline-rich motifs on p22^{phox}. The heme protein core of Nox2 is the catalytic subunit of the oxidase. The C-terminal half of Nox2 contains sequences for FAD and NADPH binding, which are conserved within the other members of the Nox family. p47^{phox} is a basic protein containing two SH3 domains involved in protein/protein interactions. Under resting conditions, these SH3 domains are masked by autoinhibitory domains. Upon stimulation, p47^{phox} becomes highly phosphorylated and binds to p22^{phox} carrying with it p67^{phox}. p67^{phox} contains many motifs for protein binding such as two SH3 domains (central and C-terminal), two proline-rich regions, and an N-terminal TRP (tetratricopeptide repeat) for binding to Rac 1 or 2, which are crucial for activation of the complex. Rac exists in the GDP-bound state in resting cells and shifts to the GTP-bound state when cells are stimulated. GTP-bound Rac binds solely to p67^{phox}. Defects in any of the first four genes mentioned earlier result in chronic granulomatous disease (CGD).

Nox4 is a 578 amino acid protein that exhibits 39% identity to gp91phox, and was initially identified as the predominant Nox (renox) in kidney cells. However, it was later identified in many cell types (43) as primarily intracellular and able to form an active oxidase upon binding to p22^{phox} without the need for other subunits (50). All components of the NADPH oxidase complex have been detected in various brain regions in mice, including cortex, hippocampus, amygdala, striatum, and thalamus (41).

Several groups have studied the activation of Nox2 in brain microvasculature following ischemia-reperfusion, and from these studies, a major role for superoxide production by this enzyme has emerged. Initial studies showed that *gp91^{phox}(−/−)* mice had smaller infarct volumes following transient ischemia (52), and follow-up studies suggested that Nox2 expression by both microvascular endothelial cells and by neutrophils or monocytes/macrophages contributed to ischemic injury. Nox2 is induced by ischemia in the core region of the infarct (53) in cells that express CD11b and MHC class II markers, suggesting that they are activated macrophages invading the infarct. Recent studies using the Nox inhibitor, apocynin, have confirmed the importance of Nox(s) to hypoxia (54) and ischemia (55, 56), and as discussed later, may be especially important to loss of blood–brain barrier (BBB) integrity (50). An increase in NADPH-oxidase derived superoxide production was observed in the endothelial cells of brain microvasculature after experimental stroke in mice, which was absent in the gp91phox(−/−) strain (50). This increase in superoxide production was shown to increase the permeability of the blood–brain barrier, thus allowing the activation of glial cells by blood-borne inflammatory mediators that can further damage the blood–brain barrier, thus worsening ischemic injury to brain parenchyma [see (57) for a recent extensive review]. The mechanism by which Nox2-dependent NADPH oxidase is activated in microvasculature cells is currently incompletely understood, but may be triggered by the sudden increase in glucose seen with reperfusion, which has been shown to activate the oxidase (58). Inflammatory signaling through NFκB may also induce expression of Nox2 (59); also see for a recent review ROS and inflammatory pathways in cerebral ischemia (60).

Although early activation of Nox2 during ischemia-reperfusion seems to occur mainly within the cerebral microvasculature, this can subsequently lead to substantial activation of astrocytes and microglia in the brain parenchyma, with the consequent release of large amount of proinflammatory cytokines that may induce the NADPH oxidase(s) in not only glial cells, but neurons (46). As described earlier, Nox2 is expressed by neurons in several brain regions, and thus it would be expected that its activation during ischemia-reperfusion would worsen the injury process. Indeed, we have recently shown that an abrupt increase in excitatory transmission, secondary to pharmacological inhibition of NMDA receptors, leads to a substantial increase in neuronal superoxide production through Nox2 (61). It is likely that a similar process occurs in hypoxia-ischemia, where the ischemia-induced rise in extracellular glutamate also produces overactivation of glutamate receptors.

Among the questions that remain is the relative contribution of mitochondrial vs. NADPH oxidase-derived superoxide production in brain. Recent work from our laboratory studying synaptosomes and utilizing electrochemical detection of oxygen consumption coupled with EPR spectroscopy to measure superoxide production suggests that, at least at the synapse, superoxide generation by NADPH oxidase may be as much as ~5–10-fold greater than that from mitochondria even though rates of oxygen consumption are comparable [(61); Ali and Dugan, unpublished] (Scheme 12.2). While our studies were not performed in hypoxia-ischemia directly, they were carried out after induction of Nox2 by inflammatory cytokines, and therefore could be relevant to ischemia. Our findings together suggest that induction and activation of Nox(s) could be a dominant source of ROS in brain, and specifically in neurons, although further studies to address this are needed.

12.2.3 Nitric Oxide Synthase (NOS)

In CNS parenchyma, there are three isoforms of NOS: neuronal NOS (nNOS), a constitutive isoform localized in neurons, inducible NOS (iNOS), induced in microglia/macrophages, astrocytes, and endothelial cells, and constitutive endothelial NOS (eNOS), localized in the endothelial cells. Studies using transgenic and knockout mice indicate that NO produced by nNOS and iNOS contribute to ischemic brain injury, while NO from eNOS is neuroprotective because of its vasodilatory effects. Inflammatory cytokines, IL1β and TNFα, as well as superoxide radical itself, can stimulate NO production, as well as induce expression of iNOS in astrocytes and microglia. Under normal circumstances, NOS produces exclusively nitric oxide (NO). However, when the enzyme is depleted of tetrahydrobiopterin or FAD cofactors, as occurs under hypoxic conditions, NOS can produce superoxide (62) and H_2O_2 (46). Importantly, both from NOS and $O_2^{\cdot-}$ from NADPH oxidase can be produced from activated macrophages. The combination of NO and $O_2^{\cdot-}$ produces peroxynitrite (OONO−), and microglia in the nervous system can generate significant amounts of peroxynitrite when activated by injury. Expression of iNOS after ischemia is known to peak 24–48 h after ischemia, and occurs in both infiltrating neutrophils and cerebral vascular endothelial cells, suggesting that these are later sources of RNS and oxidative stress after ischemia. Studies in cell cultures exposed to oxygen-glucose deprivation indicate that induction of iNOS, and

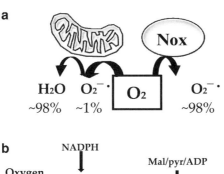

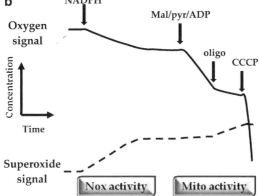

Scheme 12.2 Relative contributions of mitochondria and NADPH oxidase to synaptic superoxide production. NADPH oxidase produces substantially higher superoxide concentration than mitochondria in brain synaptosomes. (**a**) Mitochondria leak small amounts of superoxide during oxidative phosphorylation, estimated to be 1% of oxygen consumed physiologically, and up to 4% under pathological conditions. Nox stoichiometrically converts nearly 100% of oxygen consumed into superoxide radical. (**b**) Illustrative traces of oxygen consumption (solid trace) and parallel superoxide production (broken trace). Addition of NADPH activates Nox-dependent oxygen consumption and superoxide production. Mitochondrial respiration and superoxide production is initiated by the NAD$^+$-linked substrates, malate plus pyruvate. Oligomycin (ATPase inhibitor) results in reduction of the electron transport chain, increasing superoxide generation by mitochondria. Finally, the uncoupler, CCCP, dissipates the mitochondrial membrane potential and abolishes superoxide production. Despite the fact that O_2 consumption by NADPH oxidase and mitochondria is relatively similar, the rate of superoxide production by NADPH oxidase is substantially higher.

increased peroxynitrite production, as determined by formation of 3-nitrotyrosine, causes endothelial cell apoptosis, suggesting that iNOS expression directly promotes ischemic neuronal death.

12.2.4 Other Sources of ROS

In addition to the aforementioned sources of ROS, superoxide in response to anoxia-reoxygenation can be produced by several other sites. Xanthine oxidase and enzymes involved in eicosanoid and docosanoid production, cyclooxygenase(s) and lipoxygenase(s), may also be sources of ROS production in hypoxia-ischemia (63).

12.3 Effects of ROS on Blood–Brain Barrier Integrity and Cerebral Microvasculature

One key target of ROS-mediated injury is the blood–brain barrier (BBB) [see (57) for a comprehensive recent review]. Damage to the BBB occurs gradually following ischemia-reperfusion, and reflects the vulnerability of the cellular components of the BBB to direct ischemic injury, and as an indirect consequence of brain parenchymal responses. Damage to microvascular endothelial cells promotes adhesion of platelets and leukocytes, and maneuvers in turn that limit the postischemic adhesion of inflammatory cells to cerebral microvessels will blunt the subsequent inflammatory response, and protect the integrity of the BBB (64). Several mechanisms promote leukocyte adhesion to ischemia microvessels, including ICAM-1, which enhances adhesion of leukocytes to endothelial cells. ICAM-1(−/−) mice are less vulnerable to ischemic injury and have less BBB breakdown than controls. Matrix metalloproteinases (MMPs) also contribute to disruption of the BBB [reviewed in (65)]. MMP-2 and MMP-9, which degrade the perivascular extracellular matrix following ischemia, can compromise BBB function. Inhibition of MMPs pharmacologically or genetically has been shown to preserve BBB integrity, and to decrease brain edema following ischemia (66). More recently, MMPs have been suggested to have a role in promoting hemorrhagic transformation after stroke, especially in stroke patients who received tissue plasminogen activator (tPA) as thrombolytic therapy (67).

Damage to the endothelial cell or astrocytic components of the BBB not only enhances leukocyte and platelet adhesion, but it may also allow entry of blood constituents such as cytokines and inflammatory cells to the brain parenchyma. Agents that protect endothelial cells, such as antioxidants, can help preserve BBB integrity following ischemia-reperfusion (68). The response of parenchymal cells to the ischemic insult also affects BBB function. For example, provided above NMDA receptor antagonists, which reduce neuronal death and tissue infarction after focal ischemia (see later), decrease BBB disruption, possibly by blocking neuronal production of ROS triggered by NMDA receptor activation.

The accumulation of circulating leukocytes (primarily neurotrophils and monocytes) in the cerebral microvasculature and their subsequent extravasation into brain parenchyma is a well-described event following ischemia (60, 69). The mechanisms promoting the early increase in leukocyte–endothelial interactions in the cerebral circulation await further elucidation, but the elaboration of particular cytokines, NFκB-mediated increases in endothelial cell activation and adhesion molecule expression, and loss of the anti-inflammatory effects of nitric oxide may all contribute. On fact, one potential mechanism by which Nox-dependent superoxide could promote leukocyte adhesion would be through reaction with nitric oxide, thus decreasing nitric oxide levels near endothelial cells.

ROS may be involved in the inflammatory response to ischemia by multiple mechanisms: (a) ROS can activate proinflammatory signaling cascades such as NF-κB. NF-κB and other proinflammatory cytokines, in turn, will induce gene expression of inflammatory mediators including TNF-α, selectin, ICAM, and Nox(s) (59); (b) oxidative changes to the structure and function of proteins, lipids, and nucleic acids can induce a stress response, which also includes activation of inflammatory processes; (c) ROS-mediated damage to macromolecules including cell membrane lipids and other cellular components can trigger an inflammatory response. Even though there is substantial evidence indicating involvement of ROS in inflammation after ischemia, the source and the mechanism by which ROS contribute to injury remain to be fully defined.

12.4 Effects of ROS on Neuronal Circuits and Synaptic Function in Hypoxia-Ischemia

While elevated superoxide production by brain in response to injury can cause oxidative damage to macromolecules, necrotic and apoptotic cell death, and tissue infarction, it will also lead to more subtle effects on nervous system function. A delicate balance of superoxide production in brain seems to be required for normal cognitive processes, where high levels that do not cause overt cell damage may still impair LTP and memory, and inadequate levels will also cause loss of LTP and cognitive impairment (Scheme 12.2), reviewed in (70). For example, behavioral and hippocampal LTP studies show that *gp91phox*-deficient mice exhibit mild impairments in cognition that are not related to overt changes in brain morphology (71). Similarly, mice overexpressing the cytosolic SOD-1 or EC-SOD show deficits in LTP (44, 70, 72–74), but this differed by brain region and age of the mice (44).

Inactivation of synaptic proteins through oxidation is a well-described phenomenon, and is considered to be behind many of the derangements of the nervous system observed in disease states (75–77). Regulatory redox sites have been found in many proteins that are key to glutamatergic neurotransmission including, serine-racemase (77), which is responsible for the synthesis of the endogenous modulator of the glycine site in NMDA receptors; glutamine synthase (78), which is responsible for glutamate synthesis; and the glial excitatory amino acid transporter (GLT-1 or EAAT2) (79) that, together with glutamine synthase, is involved in the regulation of extraneuronal levels of glutamate. Finally, the NMDA receptor itself is tightly regulated by oxidation–reduction reactions through its redox-sensitive site (80). In particular, receptors composed of NR1:NR2A subunits are known to have a highly reversible and rapid current potentiation by sulfhydryl redox agents, including glutathione, acting on a specific redox site in NR2A (81). The oxidation status of this redox site can affect the regulation of these receptors by spermine and protons, as well as the inhibition mediated by the high-affinity Zn^{2+} site (82). Recently, SNAP25, another key synaptic protein, was shown to be the target for the presynaptic depressant actions of reactive oxygen species on neurotransmitter release (83). Oxidation of receptors and transporters does not invariably lead to inactivation. Indeed, while oxidation of GABA(A) receptors was shown to decrease its activity (84, 85), oxidation of glycine or acetylcholine receptors enhances their activity (84–86). With regard to GABA(A) receptors, those containing alpha1 subunits appear to be most sensitive to this type of redox modulation (86).

12.5 Antioxidant Strategies to Test the Contribution of ROS to CNS Alterations and Injury After HI

Interventions, either genetic or pharmacological, have been used to determine to what extent ROS contributes to injury processes involved in hypoxia-ischemia. Studies in mice overexpressing key antioxidant enzymes have consistently shown significant protection against ischemic brain injury. Transgenic mice, which overexpress SOD1, extracellular SOD (EC-SOD), or GSHPx1, show reduced infarct volume, and other indices of reduced injury. Conversely, reduction of cellular antioxidant defenses potentiates ischemic injury. These studies are extensively reviewed in (3, 4, 87).

Treatment with small-molecule antioxidants has also been shown to provide some reduction in ischemic brain injury. Ebselen, a selenocompound that mimics gluthathione peroxidase, inhibits DNA damage and lipid peroxidation after focal ischemic injury in animal studies, and has been shown to reduce functional deficits after stroke in humans. Another compound, NXY-059, a nitrone spin trap with antioxidant properties, also decreased ischemic damage.

12.6 Summary

Despite a clear role for ROS in initiating and contributing to nervous system injury and dysfunction in cerebral hypoxia-ischemia, translation of antioxidant therapeutic approaches to the clinical arena has been disappointing. However, recent advances in our understanding of the complexities of how ROS are generated in the brain, their importance to normal nervous system physiology and function, and the interplay between redox-dependent regulation of synaptic activity and inflammatory processes may allow new therapeutic targets to be identified. Studies into dysregulation of ROS and redox status during hypoxia-ischemia are also likely to provide greater insights into their actions in normal nervous system function.

References

1. Finkel, T. and Holbrook, N.J., Oxidant, oxidative stress and the biology of ageing, Nature, 2000, 408(6809): 239–247.
2. Halliwell, B., *Reactive oxygen species and the central nervous system.* J Neurochem, 1992. 59(5): 1609–23.

3. Sugawara, T., et al., *Neuronal death/survival signaling pathways in cerebral ischemia.* NeuroRx, 2004. 1(1): 17–25.
4. Sugawara, T. and P.H. Chan, *Reactive oxygen radicals and pathogenesis of neuronal death after cerebral ischemia.* Antioxid Redox Signal, 2003. 5(5): 597–607.
5. Piantadosi, C.A. and J. Zhang, *Mitochondrial generation of reactive oxygen species after brain ischemia in the rat.* Stroke, 1996. 27(2): 327–31; discussion 332.
6. Dugan, L.L., et al., *Mitochondrial production of reactive oxygen species in cortical neurons following exposure to N-methyl-D-asparate.* J Neurosci, 1995. 15: 6377–88.
7. Reynolds, I.J. and T.G. Hastings, *Glutamate induces the production of reactive oxygen species in cultured forebrain neurons following NMDA receptor activation.* J Neurosci, 1995. 15 (5, Part 1): 3318–27.
8. Nicholls, D.G. and S.L. Budd, *Mitochondria and neuronal glutamate excitotoxicity.* Biochim Biophys Acta, 1998. 1366(1–2): 97–112.
9. Sims, N.R., *Selective impairment of respiration in mitochondria isolated from brain subregions following transient forebrain ischemia in the rat.* J Neurochem, 1991. 56(6): 1836–44.
10. Schutz, H., et al., *Brain mitochondrial function after ischemia and hypoxia II. Normotensive systemic hypoxemia.* Arch Neurol, 1973. 29(6): 417–19.
11. Hillered, L., B.K. Siesjo, and K.E. Arfors, *Mitochondrial response to transient forebrain ischemia and recirculation in the rat.* J Cereb Blood Flow Metab, 1984. 4(3): 438–46.
12. Sims, N.R. and W.A. Pulsinelli, *Altered mitochondrial respiration in selectively vulnerable brain subregions following transient forebrain ischemia in the rat.* J Neurochem, 1987. 49(5): 1367–74.
13. Sims, N.R., et al., *Brain mitochondrial responses to postischemic reperfusion: a role for calcium and hydrogen peroxide?* Dev Neurosci, 2000. 22(5–6): 366–75.
14. Sims, N.R., et al., *Impairment of brain mitochondrial function by hydrogen peroxide.* Brain Res Mol Brain Res, 2000. 77(2): 176–84.
15. Hoehn-Berlage, M., et al., *Changes of relaxation times (T1, T2) and apparent diffusion coefficient after permanent middle cerebral artery occlusion in the rat: temporal evolution, regional extent, and comparison with histology.* Magn Reson Med, 1995. 34(6): 824–34.
16. Rehncrona, S., L. Mela, and B.K. Siesjo, *Recovery of brain mitochondrial function in the rat after complete and incomplete cerebral ischemia.* Stroke, 1979. 10(4): 437–46.
17. Zaidan, E. and N.R. Sims, *The calcium content of mitochondria from brain subregions following short-term forebrain ischemia and recirculation in the rat.* J Neurochem, 1994. 63(5): 1812–19.
18. Feng, Z.C., T.J. Sick, and M. Rosenthal, *Oxygen sensitivity of mitochondrial redox status and evoked potential recovery early during reperfusion in post-ischemic rat brain.* Resuscitation, 1998. 37(1): 33–41.
19. Sun, D. and D.D. Gilboe, *Ischemia-induced changes in cerebral mitochondrial free fatty acids, phospholipids, and respiration in the rat.* J Neurochem, 1994. 62(5): 1921–8.
20. Wardlaw, J.M., et al., *Studies of acute ischemic stroke with proton magnetic resonance spectroscopy: relation between time from onset, neurological deficit, metabolite abnormalities in the infarct, blood flow, and clinical outcome.* Stroke, 1998. 29(8): 1618–24.
21. De Reuck, J., et al., *Cobalt-55 positron emission tomography of ipsilateral thalamic and crossed cerebellar hypometabolism after supratentorial ischaemic stroke.* Cerebrovasc Dis, 1999. 9(1): 40–4.
22. Stevens, H., et al., *Cobalt-57 as a SPET tracer in the visualization of ischaemic brain damage in patients with middle cerebral artery stroke.* Nucl Med Commun, 1998. 19(6): 573–80.
23. Stevens, H., et al., *55Cobalt (Co) as a PET-tracer in stroke, compared with blood flow, oxygen metabolism, blood volume and gadolinium-MRI.* J Neurol Sci, 1999. 171(1): 11–18.
24. De Reuck, J., et al., *The significance of cobalt-55 positron emission tomography in ischemic stroke.* J Stroke Cerebrovasc Dis, 1999. 8(1): 17–21.
25. Cardell, M., T. Koide, and T. Wieloch, *Pyruvate dehydrogenase activity in the rat cerebral cortex following cerebral ischemia.* J Cereb Blood Flow Metab, 1989. 9(3): 350–7.

26. Zaidan, E. and N.R. Sims, *Selective reductions in the activity of the pyruvate dehydrogenase complex in mitochondria isolated from brain subregions following forebrain ischemia in rats.* J Cereb Blood Flow Metab, 1993. 13(1): 98–104.
27. White, R.J. and I.J. Reynolds, *Mitochondrial depolarization in glutamate-stimulated neurons: an early signal specific to excitotoxin exposure.* J Neurosci, 1996. 16(18): 5688–97.
28. Nicholls, D.G., *Mitochondria and calcium signaling.* Cell Calcium, 2005. 38(3–4): 311–17.
29. Thornberry, N.A. and Y. Lazebnik, *Caspases: enemies within.* Science, 1998. 281(5381): 1312–16.
30. Green, D.R. and J.C. Reed, *Mitochondria and apoptosis.* Science, 1998. 281(5381): 1309–12.
31. Merry, D.E. and S.J. Korsmeyer, *Bcl-2 gene family in the nervous system.* Annu Rev Neurosci, 1997. 20: 245–67.
32. Perez-Pinzon, M.A., et al., *Cytochrome C is released from mitochondria into the cytosol after cerebral anoxia or ischemia.* J Cereb Blood Flow Metab, 1999. 19(1): 39–43.
33. Fujimura, M., et al., *Cytosolic redistribution of cytochrome c after transient focal cerebral ischemia in rats.* J Cereb Blood Flow Metab, 1998. 18(11): 1239–47.
34. Andreyev, A.Y., B. Fahy, and G. Fiskum, *Cytochrome c release from brain mitochondria is independent of the mitochondrial permeability transition.* FEBS Lett, 1998. 439(3): 373–6.
35. Wallace, D.C., *Mitochondrial diseases in man and mouse.* Science, 1999. 283: 1482–8.
36. Tsuji, M., et al., *Reduction of cytochrome aa3 measured by near-infrared spectroscopy predicts cerebral energy loss in hypoxic piglets.* Pediatr Res, 1995. 37(3): 253–9.
37. Shiino, A., et al., *Poor recovery of mitochondrial redox state in CA1 after transient forebrain ischemia in gerbils.* Stroke, 1998. 29(11): 2421–4; discussion 2425.
38. Matsumoto, H., et al., *Does the redox state of cytochrome aa3 reflect brain energy level during hypoxia? Simultaneous measurements by near infrared spectrophotometry and 31P nuclear magnetic resonance spectroscopy.* Anesth Analg, 1996. 83(3): 513–18.
39. Skulachev, V.P., *Uncoupling: new approaches to an old problem of bioenergetics.* Biochim Biophys Acta, 1998. 1363(2): 100–24.
40. Nicholls, D.G., *Commentary on: 'Old and new data, new issues: The mitochondrial Deltapsi' by H. Tedeschi.* Biochim Biophys Acta, 2005. 1710(2–3): 63–5.
41. Serrano, F., et al., *NADPH oxidase immunoreactivity in the mouse brain.* Brain Res, 2003. 988(1–2): 193–8.
42. Infanger, D.W., R.V. Sharma, and R.L. Davisson, *NADPH oxidases of the brain: distribution, regulation, and function.* Antioxid Redox Signal, 2006. 8(9–10): 1583–96.
43. Bedard, K. and K.H. Krause, *The NOX family of ROS-generating NADPH oxidases: physiology and pathophysiology.* Physiol Rev, 2007. 87(1): 245–313.
44. Hu, D., E. Klann, and E. Thiels, *Superoxide dismutase and hippocampal function: age and isozyme matter.* Antioxid Redox Signal, 2007. 9(2): 201–10.
45. Tejada-Simon, M.V., et al., *Synaptic localization of a functional NADPH oxidase in the mouse hippocampus.* Mol Cell Neurosci, 2005. 29(1): 97–106.
46. Vallet, P., et al., *Neuronal expression of the NADPH oxidase NOX4, and its regulation in mouse experimental brain ischemia.* Neuroscience, 2005. 132(2): 233–8.
47. Finkel, T., *Signal transduction by reactive oxygen species in non-phagocytic cells.* J Leukoc Biol, 1999. 65(3): 337–40.
48. Lambeth, J.D., et al., *Novel homologs of gp91phox.* Trends Biochem Sci, 2000. 25(10): 459–61.
49. Geiszt, M. and T.L. Leto, *The Nox family of NAD(P)H oxidases: host defense and beyond.* J Biol Chem, 2004. 279(50): 51715–18.
50. Kahles, T., et al., *NADPH oxidase plays a central role in blood–brain barrier damage in experimental stroke.* Stroke, 2007. 38(11): 3000–6.
51. Martyn, K.D., et al., *Functional analysis of Nox4 reveals unique characteristics compared to other NADPH oxidases.* Cell Signal, 2006. 18(1): 69–82.
52. Walder, C.E., et al., *Ischemic stroke injury is reduced in mice lacking a functional NADPH oxidase.* Stroke, 1997. 28(11): 2252–8.

53. Green, S.P., et al., *Induction of gp91-phox, a component of the phagocyte NADPH oxidase, in microglial cells during central nervous system inflammation.* J Cereb Blood Flow Metab, 2001. 21(4): 374–84.
54. Zhan, G., et al., *NADPH oxidase mediates hypersomnolence and brain oxidative injury in a murine model of sleep apnea.* Am J Respir Crit Care Med, 2005. 172(7): 921–9.
55. Tang, L.L., et al., *Apocynin attenuates cerebral infarction after transient focal ischaemia in rats.* J Int Med Res, 2007. 35(4): 517–22.
56. Wang, Q., et al., *Apocynin protects against global cerebral ischemia-reperfusion-induced oxidative stress and injury in the gerbil hippocampus.* Brain Res, 2006. 1090(1): 182–9.
57. Yenari, M.A., et al., *Microglia potentiate damage to blood–brain barrier constituents: improvement by minocycline in vivo and in vitro.* Stroke, 2006. 37(4): 1087–93.
58. Suh, S.W., et al., *Hypoglycemic neuronal death is triggered by glucose reperfusion and activation of neuronal NADPH oxidase.* J Clin Invest, 2007. 117(4): 910–18.
59. Anrather, J., G. Racchumi, and C. Iadecola, *NF-kappaB regulates phagocytic NADPH oxidase by inducing the expression of gp91phox.* J Biol Chem, 2006. 281(9): 5657–67.
60. Wang, Q., X.N. Tang, and M.A. Yenari, *The inflammatory response in stroke.* J Neuroimmunol, 2007. 184(1–2): 53–68.
61. Behrens, M.M., et al., *Ketamine-induced loss of phenotype of fast-spiking interneurons is mediated by NADPH-oxidase.* Science, 2007. 318(5856): 1645–1647.
62. Pou, S., et al., *Generation of superoxide by purified brain nitric oxide synthase.* J Biol Chem, 1992. 267(34): 24173–6.
63. Schreiber, J., T.E. Eling, and R.P. Mason, *The oxidation of arachidonic acid by the cyclooxygenase activity of purified prostaglandin H synthase: spin trapping of a carbon-centered free radical intermediate.* Arch Biochem Biophys, 1986. 249(1): 126–36.
64. Welch, K.M.A., et al., *Primer on Cerebrovascular Diseases.* 1997, San Diego: Academic Press.
65. Jian Liu, K. and G.A. Rosenberg, *Matrix metalloproteinases and free radicals in cerebral ischemia.* Free Radic Biol Med, 2005. 39(1): 71–80.
66. Gasche, Y., et al., *Matrix metalloproteinase inhibition prevents oxidative stress-associated blood–brain barrier disruption after transient focal cerebral ischemia.* J Cereb Blood Flow Metab, 2001. 21(12): 1393–400.
67. Lo, E.H., J.P. Broderick, and M.A. Moskowitz, *tPA and proteolysis in the neurovascular unit.* Stroke, 2004. 35(2): 354–6.
68. Hall, E.D., et al., *Neuroprotective efficacy of microvascularly-localized versus brain-penetrating antioxidants.* Acta Neurochir Suppl, 1996. 66: 107–13.
69. Hartl, R., et al., *Experimental antileukocyte interventions in cerebral ischemia.* J Cereb Blood Flow Metab, 1996. 16(6): 1108–19.
70. Kishida, K.T. and E. Klann, *Sources and targets of reactive oxygen species in synaptic plasticity and memory.* Antioxid Redox Signal, 2007. 9(2): 233–44.
71. Kishida, K.T., et al., *Synaptic plasticity deficits and mild memory impairments in mouse models of chronic granulomatous disease.* Mol Cell Biol, 2006. 26(15): 5908–20.
72. Hu, D., et al., *Aging-dependent alterations in synaptic plasticity and memory in mice that overexpress extracellular superoxide dismutase.* J Neurosci, 2006. 26(15): 3933–41.
73. Thiels, E. and E. Klann, *Hippocampal memory and plasticity in superoxide dismutase mutant mice.* Physiol Behav, 2002. 77(4–5): 601–5.
74. Hu, D., et al., *Hippocampal long-term potentiation, memory, and longevity in mice that overexpress mitochondrial superoxide dismutase.* Neurobiol Learn Mem, 2007. 87(3): 372–84.
75. Satoh, T. and S.A. Lipton, *Redox regulation of neuronal survival mediated by electrophilic compounds.* Trends Neurosci, 2007. 30(1): 37–45.
76. Rowan, M.J., et al., *Synaptic memory mechanisms: Alzheimer's disease amyloid beta-peptide-induced dysfunction.* Biochem Soc Trans, 2007. 35 (Part 5): 1219–23.
77. Mustafa, A.K., et al., *Nitric oxide S-nitrosylates serine racemase, mediating feedback inhibition of D-serine formation.* Proc Natl Acad Sci USA, 2007. 104(8): 2950–5.
78. Pinteaux, E., et al., *Modulation of oxygen-radical-scavenging enzymes by oxidative stress in primary cultures of rat astroglial cells.* Dev Neurosci, 1996. 18(5–6): 397–404.

79. Volterra, A., et al., *Glutamate uptake inhibition by oxygen free radicals in rat cortical astrocytes.* J Neurosci, 1994. 14(5, Part 1): 2924–32.
80. Herin, G.A. and E. Aizenman, *Amino terminal domain regulation of NMDA receptor function.* Eur J Pharmacol, 2004. 500(1–3): 101–11.
81. Kohr, G., et al., *NMDA receptor channels: subunit-specific potentiation by reducing agents.* Neuron, 1994. 12(5): 1031–40.
82. Lipton, S.A., et al., *Cysteine regulation of protein function as exemplified by NMDA-receptor modulation.* Trends Neurosci, 2002. 25(9): 474–80.
83. Giniatullin, A.R., et al., *SNAP25 is a pre-synaptic target for the depressant action of reactive oxygen species on transmitter release.* J Neurochem, 2006. 98(6): 1789–97.
84. Pan, Z.H., et al., *Differential modulation by sulfhydryl redox agents and glutathione of GABA- and glycine-evoked currents in rat retinal ganglion cells.* J Neurosci, 1995. 15(2): 1384–91.
85. Pan, Z.H., X. Zhang, and S.A. Lipton, *Redox modulation of recombinant human GABA(A) receptors.* Neuroscience, 2000. 98(2): 333–8.
86. Leszkiewicz, D.N. and E. Aizenman, *Reversible modulation of GABA(A) receptor-mediated currents by light is dependent on the redox state of the receptor.* Eur J Neurosci, 2003. 17(10): 2077–83.
87. Sugawara, T., et al., *Overexpression of copper/zinc superoxide dismutase in transgenic rats protects vulnerable neurons against ischemic damage by blocking the mitochondrial pathway of caspase activation.* J Neurosci, 2002. 22(1): 209–17.

Chapter 13
Postnatal Hypoxia and the Developing Brain: Cellular and Molecular Mechanisms of Injury

Robert M. Douglas

Abstract Exposure to chronic constant hypoxia during the early postnatal period, as occurs during a number of disease states, can have devastating effects on the development of the mammalian CNS. Severe hypoxia can induce either apoptotic or necrotic cell death in the CNS. However, cells in the CNS are variably susceptible to hypoxia. Hypoxic exposure activates multiple signaling pathways which presumably serve to protect the organism from hypoxia-induced damage; however the final outcome depends on the interplay among the various pro-survival and pro-death signals induced during this period. This review highlights recent advances in our understanding of the cellular and molecular mechanisms that are involved in the response to hypoxic stress during the early postnatal period.

Keywords: Hypoxia; Central Nervous System; Neonate; Development; HIF-1α; ROS; NFκB

13.1 Introduction

The occurrence of hypoxic episodes during development such as in early postnatal life can have devastating effects on the growth and maturation of the central nervous system (CNS). However, hypoxia that occurs postnatally can also have adverse effects on the maturation and functional development of the CNS. Significant brain growth occurs postnatally and environmental stresses such as hypoxia can impede or significantly alter the continued development and maturation of the CNS. Diseases that impair adequate oxygen delivery and usage in the neonatal period have both short-term and long-term consequences including the development of neurocognitive morbidity that may persist into adulthood. This review addresses the pathophysiological responses and adaptations of the CNS to postnatal chronic or long-standing hypoxia.

R.M. Douglas
Department of Pediatrics, University of California, San Diego School of Medicine,
9500 Gilman Drive, 0735 La Jolla, CA 92093, USA
e-mail: rdouglas@ucsd.edu

Postnatal hypoxia can occur in several disease or environmental settings. While there are several other components involved in life at high altitude, including exposure to cold, ultraviolet radiation and decreased antioxidant capacity, life at high altitude has at its core an exposure to hypoxia. Over the last few decades or so, it has come to be appreciated that much of the pathology of high-altitude living is due to hypoxemia (1, 2). Other disease states that lead to chronic hypoxia in the infant and child include sleep-disordered breathing states such as obstructive sleep apnea (OSA), sickle cell anemia, and the apnea of prematurity (3, 4). Chronic hypoxia in neonates can also be seen in clinically relevant disorders such as severe lung disease, cyanotic congenital heart disease, coronary artery disease (CAD), myocardial infarction, and in rare cases of congenital central hypoventilation syndrome (5–7).

The extent of injury that occurs during chronic hypoxia is determined by several factors including duration, magnitude, and the pattern of hypoxic exposure (e.g., constant or intermittent). However, an extremely important parameter in the pathogenesis of hypoxia is the stage of development at which the organism is exposed to this stress. Significant advances have been made recently in the understanding of the mechanisms by which the exposure to oxygen deprivation leads to cell injury or cell death. Therefore, a focus of this review is to detail our understanding of the molecular and cellular mechanisms involved in such injury.

13.2 Normal Brain Development

The fetal brain develops in discrete steps involving cell division, migration and differentiation of neurons (8). The fetus has twice as many neurons as the adult brain, and the 2-year-old infant has twice as many synapses in the cerebral cortex as does the adult. Excess synapses are pruned until about the age of 16–20 years to achieve adult levels (9). Therefore, during brain development, cell replication and differentiation are accompanied by naturally occurring cell death (10), and programmed cell death plays a prominent role in normal CNS development (11). In the postnatal period, there is continued restructuring of the brain to achieve adult size and complexity. Normal brain development requires the formation of layered structures, fiber pathways, and morphologically defined synapses. As the brain develops, postmitotic neurons and glia, which are derived from pluripotent precursors, migrate to appropriate loci before elaborating processes to interact with other neurons. Intrinsic mechanisms such as graded gene expression and extrinsic mechanisms such as target innervation mediate the creation of connections among various cells in the CNS (12). Neurons are produced in excess but are subsequently pruned to achieve matched target and projection populations; this is probably achieved through the actions of various neurotrophins such as nerve growth factor (NGF), brain-derived neurotrophic factor (BDNF), and neurotrophins 3 and 4 [reviewed in (13)]. It is believed that neurotrophins are also involved in activity-dependent strengthening and elimination of synapses [reviewed in (14)]. Therefore, hypoxic exposure during the postnatal period can have severe adverse neurological consequences.

13.3 Altered Brain Development During Hypoxia

Postnatal hypoxia can irreversibly alter key developmental programs in the neonate and lead to permanent damage and dysfunction of the CNS due to improper cognitive and functional maturation. Postnatal hypoxia retards growth, reduces brain weight and size, and increases CNS ventricular volume (15). Neonatal mice and rats exposed to chronic hypoxia also demonstrate ventriculomegaly, changes in white matter, neuronal number, and connectivity, and long-term behavioral modifications (16, 17). Hypoxia can also lead to changes in neurite outgrowth in the cerebral cortex that are still evident after a return to normoxia. Additionally, neonatal hypoxia leads to smaller neuronal cell bodies and disorganized neurofilaments that persist into adulthood (18).

Postnatal hypoxia leads to upregulation of genes subserving presynaptic function and downregulates genes involved in synaptic maturation, postsynaptic function, and neurotransmission. At the transcriptional level, genes that affect synaptic function, maturation, and neural transmission such as snapin, synapsin-1, and reelin are altered by postnatal exposure to sublethal hypoxia. Other gene systems altered by postnatal hypoxia include those involved in glial maturation (19), vasculogenesis, and portions of the cytoskeleton. Hypoxia induces a loss of synchronization in the rate of synapse formation in different brain regions. There is also a downregulation of genes marking oligodendrocyte differentiation such as myelin-associated glycoprotein (MAG) during hypoxia, indicating a loss of mature oligodendrocytes and thus induction of demyelination during hypoxia (19). Cytoskeletal proteins involved in neurite outgrowth, synapse formation, and vesicular transport are also adversely affected. Apoptotic genes, both pro- and antiapoptotic, are variously altered as are genes within the hypoxia inducible factor (HIF) pathway [reviewed in (20)]. The result of these changes is a dysregulation or desynchronization of the maturational programs in the hypoxic developing brain (15).

13.4 Hypoxia and the Stage of Development

One must distinguish prenatal hypoxia from postnatal hypoxia. Hypoxic-ischemic injury during the prenatal period tends to lead to white matter injury and subsequent loss in gray matter during maturation [reviewed in (21)]. On the other hand, hypoxia-ischemia in the term infant, i.e., neonatal or postnatal exposure, leads preferentially to damage in the deep gray matter and parasagittal regions of the CNS [reviewed in (22)]. Periventricular white matter injury in the preterm infant appears to be a consequence of excitotoxicity (23) and oxidative stress. The preterm brain has more blood vessels, higher water content, less myelin, a developing cerebral cortex, and a prominent germinal matrix. This leads to a situation where the preterm brain is highly susceptible to periventricular (PVH) and intraventricular (IVH) hemorrhage (8, 24).

Hypoxic-ischemic damage in the early postnatal period induces a biphasic neurodegenerative response characterized by early cell death in the forebrain within the first 3h and late cell death at 6 days in the striatum and thalamus. This has been termed the *apoptotic-necrotic continuum* (25). We have also observed demyelination and increased capillary density in response to postnatal hypoxia (26). Other pathophysiological aspects of postnatal hypoxia will be discussed in the section on cellular and molecular mechanisms of injury.

13.5 Neurocognitive Effects of Postnatal Hypoxia

Hypoxia during the early postnatal period may disrupt the CNS developmental program. For example, hypobaric hypoxia such as that exists at high altitude leads to cognitive and mental dysfunction as well as memory deficits, motor impairment, and hypophagia (27–31). Impaired cognitive performance in humans has been demonstrated as a function of duration of exposure and increasing altitude. Cognitive performance was significantly impaired in 8 out of 10 performance measures, with the most sensitive to hypoxia being diminished short-term memory (working memory) and loss of critical judgement. Additionally, breathing-disordered states such as obstructive sleep apnea (OSA), which are intermittent in nature, have been postulated to induce neurocognitive morbidity in both pediatric and adult populations that suffer from these diseases (6, 32–36). Experimental hypobaric hypoxia leads to impairments in short-term or working memory (37), and lowered intellectual function seen in sickle cell disease patients is partly explained by chronic hypoxia (38). Adult rats exposed to hypobaric hypoxia for 3 days also demonstrate deficits in working memory in the Morris water maze but no change in reference memory, which requires several weeks for consolidation (37). Deficits in working memory in response to hypoxia are likely due to changes in neurotransmitter utilization and/or quantity produced (39). Hippocampal CA1 and CA3 neurons are instrumental for learning and memory in rodents, nonhuman primates, and humans (40–46). In another study, chronic hypoxia imposed on neonatal rats for 3 weeks led to decreased cell counts in the hippocampus and increased aggression and hyperactivity; however, it did not lead to significant changes in Morris water maze performance (47). The authors concluded that adaptive processes were responsible for the lack of an affect on learning and memory and that erythropoietin might play a protective role in this process (48).

13.6 Hypoxia and Altitude: Lessons from Life at High Altitude

Travel to high altitude can lead to exposure to chronic or intermittent hypoxia. Hypoxia at altitude can lead to acute mountain sickness (AMS), as well as high-altitude cerebral edema (HACE) and high-altitude pulmonary edema (HAPE). AMS, HACE, and HAPE are suggested to be due to altered blood–brain barrier permeability caused by hypoxia-induced free radical formation. Additionally, hypoxia is believed to be the trigger for signaling that leads to adaptation to high altitude (2). Adaptation to life at high altitude includes increased erythropoiesis, changes in respiratory responsiveness, remodeling of the vasculature, increased sympathetic drive, and increased exhaled nitric oxide (49).

Children from sea level demonstrate symptoms of AMS including headache, vomiting, insomnia, and cognitive morbidity when exposed to high altitude (50). There is also evidence of increased infant mortality, stunted growth, and sleep disturbance in indigenous populations living at high altitude. Indeed, oxidative stress and some of the cumulative effects of hypoxic exposure may persist for some time after a return to sea level (2). There are multiple factors at altitude that can contribute to oxidative stress generated by hypobaric hypoxia. These include energy expenditure, cold stress, ultraviolet (UV) radiation, lack of dietary antioxidants, increased catecholamine auto-oxidation, increased hypoxanthine and xanthine oxidase, as well as increased reductive stress that all lead to increased free radical production (51). Excess reactive oxygen species (ROS) production at high altitude can lead to impaired muscle function, reduced capillary perfusion, and neurological and pulmonary crises (52).

Experimental hypobaric hypoxia that simulates high-altitude exposure has also been demonstrated to lead to an increase in lactate dehydrogenase activity, free radical generation, and lipid peroxidation in animals with a maximum occurring between 3 and 7 days of exposure (37). Conversely, hypoxia led to a decrease in antioxidant levels and in glutathione reductase and superoxide dismutase activities with a concurrent decrease in reduced glutathione (GSH) levels and increased oxidized glutathione (GSSG). There were also increases in glutamate dehydrogenase activity and vesicular glutamate transporter expression, implicating glutamate toxicity in this model. Therefore, increased oxidative stress is implicated in hypoxia-induced CNS injury. Interestingly, there appeared to be acclimatization to the hypoxic stress by 14 days of exposure. It is likely that these animals were able to adapt to continued hypoxic stress using homeostatic modifiers, such as HIF-1, and were able to respond adequately to the stress to bring the system closer to baseline.

13.7 Cellular and Molecular Mechanisms of Hypoxic Cell Injury and Death

Hypoxia induces a series of cellular, molecular, and physiological responses that result in either adaptation and survival or cell death. Originally, hypoxia alone was not thought to be capable of producing severe brain injury (53). However, in vivo and in vitro studies soon demonstrated that hypoxia alone, if severe enough, can lead to cell death in the developing brain (54–59). Transient exposure to severe hypoxia in the neonatal period leads to cell death in the rat CA1 region of the hippocampus (60). In this study, it was demonstrated that neuronal death at postnatal day (P) 1 appeared to be apoptotic while it was necrotic in P7 animals. Blockade of NMDA receptors with MK801 did not prevent apoptotic cell death at P1 but rescued cells from necrosis in the P7 animals, implying that NMDA receptor activation is not involved in hypoxia-mediated cell death in the very young but becomes important at older ages (61). This is no doubt due to the maturation of the glutamate system.

13.7.1 Cell Type-Specific Responses

Depending on the developmental stage when a stress is applied, different cell types and brain regions will undergo injury at different rates, and different mechanisms of injury will be activated (62). In general, neurons are considered to be the cell type most sensitive to injury, followed by oligodendrocytes and astrocytes; the least vulnerable are endothelial cells (63). Neurons die during hypoxia because a switch to, or more reliance on, anaerobic glycolysis is insufficient to keep pace with the energetic demands of neuronal function. In adult mammals, the critical O_2 tension at which ATP levels begins to decline is ~25–40 mmHg (64). When ATP levels fall to ~35–40%, neuronal depolarization and sodium and water uptake occur causing cell death (65, 66). However, neurons in the CNS are not universally susceptible to O_2 deprivation but are differentially sensitive (67, 68). For example, brief global ischemia induces a selective loss of neurons in the hippocampal CA1 region but a loss of small and intermediate-sized neurons in the striatum with sparing of large neurons. Neurons of the CA1 region of the hippocampus tend to be more susceptible to injury as compared with CA3 and dentate gyrus neurons. Glia are also vulnerable to hypoxic damage (69), while oligodendrocytes are particularly sensitive to hypoxic insult (17).

13.7.2 Brain Region-Specific Responses

Regions of the brain are differentially sensitive to hypoxic stress. Some neurons such as those from the striatum, pons, and brainstem are relatively resistant to hypoxia (70, 71). The hippocampus, which is an important site for learning and memory, is well known to be sensitive to hypoxic/ischemic stress, with the CA1 region being particularly sensitive (63, 72–74). The hippocampus has an abundance of glutamatergic neurons and is therefore highly vulnerable to hypobaric hypoxia (75). Both the hippocampus and cerebellum have been shown to be morphologically altered by hypoxic stress during hypobaric hypoxia (28, 76). In general, it is thought that the differential sensitivity of different brain regions is due to several factors including differences in synaptic input, neurotransmitters released, neurotransmitter receptors, antioxidant capabilities, and signaling pathways (77).

13.7.3 Ion Channel and Transporter Mechanisms

Hypoxia can affect a variety of channels and transporters in the CNS. The initial response to hypoxia or ischemia is a reduction in membrane channel conductance by altering potassium, sodium, and chloride fluxes. This reduces the energy requirements of neurons during periods of O_2 lack and is referred to as *channel arrest*. Hypoxia increases the persistent Na^+ current in hippocampal cells (78), and O_2-sensitive K^+ channels are reversibly inhibited by hypoxia (79). Both Ca^{2+} and K^+ channels can be regulated directly by hypoxia (80–82).

However, prolonged or severe hypoxia can lead to a collapse of the ionic gradients across the plasma membrane and ultimately cause large increases in intracellular Na^+, Ca^{2+}, Cl^-, and H^+ and a decrease in intracellular K^+. Increased intracellular Ca^{2+} activates lipases and proteases that damage membranes, which causes the release of free fatty acids and proteins. Ultimately, this leads to activation of endonucleases, which damage DNA, glutamate efflux leading to excitotoxicity, and pertubation of the cytoskeleton (64). Furthermore, the Na^+/K^+ ATPase pump is the major ATP-consuming process in neurons; a lack of ATP during severe or prolonged hypoxia can cause a collapse of the Na^+ gradient, causing the Na^+/glutamate transporter to reverse direction and can also lead to glutamate excitotoxicty.

Other transporters are also affected by exposure to hypoxia. Glucose transporters (Glut) 3 and 4 are activated by hypoxia to increase the delivery of glucose to cells that are now relying on glycolysis for their energy production (83, 84). Hypoxia can also increase the activity of acid–base transporters such as the amiloride-sensitive Na^+/H^+ exchanger (NHE) as well as the Na^+/Ca^{2+} exchanger in nerve cells (81, 85). Under the influence of HIF, NHE-1 and -6 protein expression are upregulated during hypoxia and the activity of NHE-1 is increased during hypoxia (86). This can eventually lead to increased Na^+ influx, which can lead to necrosis and increased intracellular Ca^{2+} that can result in cell death, either apoptotic or necrotic (78). Another recent report indicates that postnatal hypoxia reduces the protein expression of two electroneutral Na^+-coupled bicarbonate (HCO_3^-) acid–base transporters: sodium-bicarbonate cotransporter 1 (NBCn1) and sodium-coupled chloride-bicarbonate exchanger (NCBE) in mouse brain, possibly as part of an energy-saving process (87). Additionally, Xue et al. (88) demonstrated a prosurvival role for NHE1 and an antisurvival role for HCO_3^--dependent transporters during prolonged hypoxia in primary neuronal cultures. These studies provide evidence for the critical role that acid–base transporters play in determining neuronal fate during prolonged hypoxia (89).

13.7.4 Hypoxia and Metabolic Arrest

Chronic hypoxia leads to a decrease in cell repair, tissue growth, and differentiation (68). Hypoxia inhibits protein synthesis in an O_2-dependent fashion to conserve energy utilization (90). For example, hypoxia- and cold-tolerant organisms use this strategy to survive periods of prolonged hypoxia/anoxia or cold exposure. Metabolic depression can be achieved by either decreasing energy-consuming processes or by increasing the efficiency of energy production (91, 92). Protein synthesis and ion-motive ATPases such as Na$^+$/K$^+$ ATPase and Ca^{2+} ATPase are the dominant consumers of cellular energy (93); therefore, reducing these activities would reduce energy turnover. Decreased membrane conductance or *channel arrest* has been observed in response to hypoxia that decreases the energy consumption of the ion-balancing ATPases (94). Inhibition of protein synthesis is considered to be part of the metabolic arrest induced by hypoxia to match energy demands with the prevailing low O$_2$ availability (95, 96). Hypoxia appears to decrease protein synthesis by inhibiting the mammalian target of rapamycin (mTOR) pathway in an HIF1-independent fashion (97). Hypoxia can also inhibit translation through the unfolded protein response and ER stress pathways (98).

Another aspect of metabolic arrest during hypoxia may be mediated in part by an overall decrease in mitochondrial respiration. In P20 mice exposed to hypobaric hypoxia for 21 days, there were decreases in mitochondrial respiration as measured by state 3 and 4 respiration and respiratory ratios (99). Additionally, NADH CoQ reductase (complex I) and cytochrome *c* oxidase (complex IV) activities were reduced in hypoxia-exposed mouse mitochondria. In this study, neonatal rats exposed to chronic hypoxia for 7 days demonstrated decreased lactate dehydrogenase and increased hexokinase activities but no change in α-ketoglutarate dehydrogenase complex (KGDHC) and citrate synthase activities (100). This is likely mediated in part by HIF-1α (101). This would aid in the transition from oxidative phosphorylation to glycolysis predicted to occur during hypoxia. However, in animals maintained in hypoxia for 30 days, there was a marked decrease in KGDHC activity, which would imply decreased activity of the TCA cycle during prolonged hypoxia. It is possible that ROS-induced damage to mitochondria or direct effects on the expression of proteins within the electron transport chain that have been shown to be modified by hydrogen peroxide (102) and decreased by chronic hypoxia (103) are responsible for the decrease in oxidative phosphorylation activity. It is possible that reduced mitochondrial activity during hypoxia is an adaptive mechanism aimed at reducing energy costs during periods of insufficient O$_2$. This is referred to as regulated metabolic depression or hypometabolism (104). Therefore, hypoxia can induce an increase in glycolytic activities but a decrease in oxidative phosphorylation in the brain and this suggests that some of the pathophysiology seen in clinically related diseases may be due in part to decreased energy metabolism.

13.7.5 Hypoxia and Apoptosis: Glutamate Excitotoxicity

In the developing brain, transient hypoxia may lead to delayed neuronal death or apoptosis if the insult is severe enough or of sufficient duration (59). In severe hypoxia, it is generally thought that excitotoxic mechanisms are involved in cell death. Both hypoxia and ischemia lead to increased release of glutamate and subsequent activation of NMDA receptors and result in excitotoxicity, which may lead to programmed cell death (105, 106). Excessive glutamate-induced neuronal depolarization leads to Ca^{2+} influx via voltage-gated Ca^{2+} channels and NMDA receptors. Collapse of the Na^+ gradient leads to reversal of the sodium-glutamate and Na^+-Ca^{2+} transporters, further increases intracellular Ca^{2+}, and activates Ca^{2+}-dependent proteases and calpains as well as nitric oxide, which eventually leads to either apoptosis or necrosis (107–109). Caspases can be activated by NMDA-induced calpain activation, and NMDA receptors are heavily implicated in cell death during severe hypoxia (110). Increased expression of the NR1 subunit of the NMDA receptor has been reported in the hippocampus of rats exposed to hypobaric hypoxia (37).

Caspase 3 plays a more important role in the neonate as compared with adult, consistent with data that caspase 3 mRNA and protein expression and activity are higher in the perinatal period (111, 112). Severe hypoxia leads to increased expression of proapoptotic proteins such as p53, p21, Bax and caspases and to decreased expression of antiapoptotic proteins such as Bcl-2 and heat-shock proteins (113). In the immature brain, severe hypoxia upregulates the expression of p53, which promotes cell cycle arrest and apoptosis via transactivation of the cell cycle inhibitor, p21 (Waf1; Cip1) and the death-inducing Bcl-2 family member Bax, respectively (59). The tumor suppressor p53 transactivates genes such as *p21*, growth arrest and DNA damage-inducible gene 45 (*GADD45*), murine double minute-2 (*mdm2*), and the proapoptotic gene *Bax* and represses the antiapoptotic gene, *Bcl-2* (114, 115). Similarly, increased expression of the executioner caspases such as caspase 1(interleukin 1β converting enzyme; ICE) and caspase 3 (Yama; CPP32) has been noted in response to severe hypoxia [reviewed in (114, 116)].

13.7.6 Hypoxia and ROS

Free radicals such as superoxide, hydroxyl, and peroxynitrite are continuously generated by cells, and the toxic capability of ROS at high levels has been well-documented (117, 118). Cellular sources of ROS production include the mitochondrial electron transport chain, NADPH oxidases, hypoxanthine/xanthine oxidase, and cytochrome P450 of the endoplasmic reticulum [reviewed in (119–122)]. Aging and its neurocognitive and pathological sequelae are correlated with the accumulation of ROS (theory of aging) (123–125). In addition, ROS are implicated in several neurodegenerative disorders such as Alzheimer's and Parkinson's diseases (126, 127). Injury normally results when the antioxidant capacity of the cell is overwhelmed. Recent evidence indicates that both hyperoxia and hypoxia can lead to ROS generation (128–132). The neonatal brain is vulnerable to oxidative damage due to its high content of unsaturated fatty acids, high rate of O_2 consumption, low amount of antioxidant molecules, and increased availability of free Fe^{2+} (133, 134). Hypoxia paradoxically stimulates or induces ROS release from mitochondria, and these species can then modulate transcriptional and posttranslational responses to low oxygen (135, 136). Hypoxia per se is able to generate increased ROS due to the lower availability of O_2 to act as an electron sink/acceptor, which allows excess electrons to form superoxide with available O_2. This ultimately leads to increased hydroxyl and peroxynitrite radicals (70, 131, 137, 138). It is believed that ROS are generated at complex I and complex III of the mitochondrial electron transport chain during hypoxia. Experimental evidence for increased ROS during hypoxia includes oxidation of fluorescent probes, decreases in the antioxidant reduced glutathione, increased free radical production as detected by electron paramagnetic resonance spectroscopy (ESR/EPR), and decreased reduced cysteine (139). Oxidant stress increases under hypoxia and can activate multiple signaling pathways. It has also been proposed that ROS may be involved in HIF1α protein stabilization and gene regulation during hypoxia but not during anoxia (140).

Experimental hypobaric hypoxia of different durations (3 and 7 days) induced oxidative stress in different regions of the CNS; these results indicate that the hippocampus is more vulnerable to hypoxia as compared with the cerebral cortex or striatum (141). The authors suggest that memory impairments during hypoxia are the result of increased free radical production due to low O_2 availability and the subsequent increased formation of superoxide, hydrogen peroxide, and hydroxyl radicals. ROS have a high propensity for the attack on membrane lipids (142), especially in the brain, which is rich in polyunsaturated fatty acids, and leads to lipid peroxidation, membrane damage, and impaired ionic conductance (137, 142, 143). Experimental hypobaric hypoxia also leads to decreased antioxidant capacity. Hypobaric hypoxia can lead to decreases in GSH levels with a concomitant increase in GSSG, perhaps due to increased utilization of GSH by the hypoxia-induced increased ROS. It also leads to decreased GSH synthesis as evidenced by decreased glutathione reductase activity.

Antioxidant defenses include enzymatic (catalases, endoperoxidases, and dismutases) and nonenzymatic (glutathione, cholesterol, ascorbic acid, and tocopherol) mechanisms [see review from (144)]. Hypoxic stress may also lead to impairment of antioxidant defense mechanisms by alterations in the activities of glutathione reductase and glutathione peroxidase (145), which can impair mitochondrial activity at complexes I and III (146). Extracellular glutamate has been noted in situations of ischemic stress (147), and glutamate inhibits cysteine uptake, which is a precursor to glutathione (148). ROS inactivate antioxidant molecules (149), and antioxidants such as reduced glutathione appear to be decreased during hypoxic stress (150). Ischemia has also been shown to promote free radical formation and there is evidence of DNA damage during hypoxia in humans (151). Additional evidence for the participation of ROS in neuronal injury is that N-acetyl-cysteine, an antioxidant, has been reported to be neuroprotective during hypoxia/ischemia (152), and other free radical quenchers and antioxidant precursors have been shown to enhance cell viability in hypoxic or ischemic stress (153).

Activation of NMDA receptors during hypoxia/ischemia has been reported to cause free radical production and calcium overload (154). Severe hypoxia also leads to calcium overload via extracellular calcium influx and release from intracellular stores, which can lead to free radical production by phospholipases and xanthine oxidase (75, 155). This may result in mitochondrial depolarization (155–157), increased mitochondrial ROS production (158), and ROS-induced damage to mitochondrial membranes that would lead to cytochrome c release during hypoxia (159). Cytochrome c release from mitochondria would lead to the initiation of the apoptotic cascade (160), with activation of proapoptotic proteins such as Bax, caspase-3, Smac/Diablo, Bid, Bad, and Apaf (161). Another radical that is produced during hypoxia/reoxygenation is nitric oxide (NO), which is derived from arginine by the enzyme nitric oxide synthase (NOS). There are three isoforms of the enzyme including constitutive neuronal NOS-1, endothelial NOS-3, and the inducible NOS-2 (162). It is postulated that nitric oxide (NO) generated by NOS during hypoxia can react with superoxide to produce peroxynitrite and precipitate membranal and cellular damage.

13.7.7 Hypoxia and the Immune System

Recently, hypoxia has been proposed to act as a neuroinflammatory agent in that it can activate microglial cells and lead to the production of inflammatory mediators (163). Hypoxia can lead to the activation of the innate immune response either through hypoxia-induced tissue damage or through signaling cascades possibly involving HIF and nuclear factor kappa B (NFκB) interactions. The ubiquitously expressed NFκB transcription factor comprises five family members (p65, p105/p50, c-Rel, RelB, and p52) and controls the expression of numerous genes involved in inflammatory and immune responses and cellular proliferation (164–166). NFκB rapidly enhances the transcription of various inflammatory cytokines, adhesion molecules, and chemokines (167, 168). Two major families that serve to activate NFκB signaling are the toll-like receptors (TLRs), which recognize lipopolysaccharide (LPS), and the tumor necrosis factor receptors (TNFRs) that are activated by their ligand, TNFα (169). It has been demonstrated that NFκB activation is preceded by increased IKK activation, IκB phosphorylation, and IκB degradation (170). NFκB activation is controlled by IκB kinases (IKK) alpha, beta, and gamma, which induce phosphorylation-induced degradation of IκBα in response to infection and inflammation (171). IKKβ has recently been shown to be activated in hypoxic cell cultures when prolyl hydroxylases that suppress its activation are inhibited (170). Hypoxia-stimulated IKK activity can lead to activation of NFκB (172).

NFκB activation has been shown to be both proapoptotic (173–175) and antiapoptotic (176, 177). There is some controversy in that various labs have reported that hypoxia-induced increased NFκB activation is due to tyrosine phosphorylation (178), reactive species generation (179), and p42/44 ERK and PI3 kinase activation (180). It has also been suggested that transient activation of NFκB may be neuroprotective while prolonged activation may lead to neuronal cell death (181). Drugs such as aspirin and indomethecin inhibit the IKKs that normally allow nuclear translocation of NFκB; use of these drugs, as well as the NFκB inhibitor SN50, prevented cell death induced by excitotoxic mechanisms (182, 183). Free radicals have also been proposed to activate NFκB (181, 184), and it was shown that the antioxidant LY231617 was neuroprotective in a global ischemia model (181). Cytokines can have direct cerebral cytotoxicity, inhibit oligodendrocyte differentiation, and induce oligodendrocyte apoptosis (24). Cytokines such as TNFα and interleukin 1β can impair glutamate transporter function and thereby lead to excitotoxicity, and local cytokine production has been demonstrated in adult stroke (185).

13.7.8 Hypoxia and Gene Transcription

Even though gene transcription is generally repressed during hypoxia, there are certain genetic programs that are upregulated, especially the universal hypoxic regulator HIF-1. Hypoxia activates multiple gene systems, mainly via HIF activity, and leads to increased expression of erythropoietin, enzymes involved in neurotransmitter synthesis, proteins involved in angiogenesis, glucose transporters, glycolytic enzymes, and other transcriptional regulators such as NFκB, activator protein 1 (AP1), and the early growth response (Egr-1)-dependent pathway. Another signaling pathway altered by hypoxic exposure is the nutrient pathway, PI3K/Akt/mTOR [reviews: (186, 187)]. The Akt protein kinase can stimulate glycolysis through increased glucose transport and mobilization of hexokinase (188). The transcription factor, hypoxia inducible factor 1 (HIF-1), is central to the response of tissues to hypoxia (189). HIFs are members of the basic helix-loop-helix, Per/ARNT/Sim (HLH-PAS) protein family and consist of three O_2-regulated alpha chains (HIF-1α, -2α, and -3α) and a constitutive beta (β) chain [HIF-1β, aryl hydrocarbon receptor nuclear translocator (ARNT)] (190, 191). HIF-1α was first identified by Wang and Semenza (189) and the HIF-1α homologue HIF-2α, originally termed endothelial PAS protein 1 (EPAS-1), which shares similar functional and regulatory features but has different roles, was later described (192).

HIF-1α is unstable under normoxic conditions due to the action of prolyl hydroxylases (PHD1, 2, and 3) that predispose to its ubiquitination by the E3 ligase complex that includes the von Hippel-Lindau tumor suppressor protein (pVHL) and degradation by the proteosome (193). Under nomoxic conditions, the pVHL binds to the oxygen-dependent degradation domain (ODD) in the carboxy terminus of HIF-1α and hydroxylates prolines 402 and 564, which allows polyubiquitination of HIF-1α. Additionally, transactivation of HIF-1 by p300 is prevented by another hydroxylase, the asparaginyl hydroxylase or factor-inihibiting HIF-1 (FIH-1), which represses activity of HIF-1 (194). These hydroxylases require molecular O_2 as a substrate and oxyglutarate as a cosubstrate as well as iron (Fe^{2+}) to hydroxylate these specific proline residues. Therefore, in the absence of O_2 or during iron depletion as with desferroxamine, HIF-1α can no longer be hydroxylated and degraded and now accumulates within cells. Hypoxia leads to a widespread accumulation of HIF-1α in virtually all tissues with subsequent tissue-specific target gene activation

(195). HIF-1α and its target genes are also upregulated in the penumbra of brain infarcts (196).

Upon dimerizing of HIF-1α with HIF-1β, HIF-1 translocates to the nucleus. HIF-1 binds a consensus sequence within the hypoxia response element (HRE) in the promoter region of O_2-responsive genes and recruits the coactivator, acetyltransferase CBP/p300 (197). This transcriptional complex regulates the expression of more than 100 prosurvival genes including erythropoietin (EPO), vascular endothelial growth factor (VEGF), and glycolytic enzymes. HIF-1 target genes play essential roles in development, angiogenesis, erythropoiesis, glucose transport, glycolysis, iron transport, and cell proliferation/survival. Neuroprotection provided by HIF-1 appears to be mediated by upregulation of the *EPO* gene that reduces the extent of apoptosis via cross talk between the Jak-2 and NFκB signaling pathways (198). Studies indicate that HIF-1 induces VEGF expression under hypoxic stress, which can lead to increased paracellular permeability (199). However, severe hypoxia can lead to HIF-1-mediated apoptotic cell death (which can be either adaptive or pathological) that requires p53 (200–202). Nitric oxide can be induced during hypoxia and may serve a neuroprotective function by stabilizing HIF-1α in addition to acting as a potent vasodilator (203).

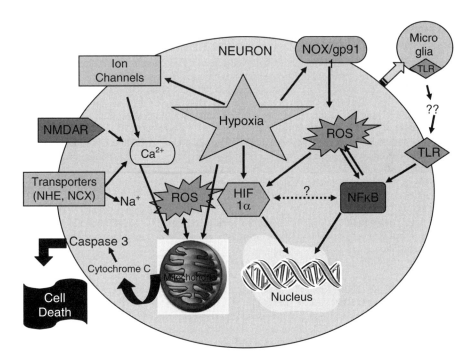

Fig. 13.1 Multiple pathways are activated or altered by exposure to chronic constant hypoxia during the early postnatal period. *HIF1α* hypoxia inducible factor 1α, *NCX* sodium-calcium exchanger, *NFκB* nuclear factor kappa B, *NHE* sodium hydrogen exchanger, *NMDAR* N-methyl-D-aspartate receptor, *NOX/gp91* NADPH oxidase, *ROS* reactive oxygen species, *TLR* toll-like receptor (*See Color Plates*)

13.8 Summary

In summary, exposure to chronic constant hypoxia during the early postnatal period, as occurs during a number of disease states, can have devastating effects on the development of the mammalian CNS. Hypoxic exposure activates multiple signaling pathways in an attempt to protect the organism from hypoxia-induced damage; however, the final outcome depends on the interplay among the various prosurvival and prodeath signals induced during this period (Fig. 13.1). The HIF transcription factor plays perhaps the major role in this phenomenon but contributions from the innate immune system may modulate this response. Hypoxia impacts ion channels, transporters, glutamate and other receptors, innate immune modulators such as NFκB, and ROS-producing and scavenging enzymes such as NADPH oxidase and glutathione, respectively. Resultantly, there are many potential points of intervention provided by this complex interplay of signaling events during postnatal hypoxia that can be utilized in developing therapeutic strategies to prevent or ameliorate cell injury and death.

References

1. Roach RC, Hackett PH. Frontiers of hypoxia research: acute mountain sickness. J Exp Biol 2001;204 (Part 18):3161–70.
2. Hornbein TF. The high-altitude brain. J Exp Biol 2001;204 (Part 18):3129–32.
3. Wiegand L, Zwillich CW. Obstructive sleep apnea. Dis Mon 1994;40(4):197–252.
4. Neubauer JA. Invited review: physiological and pathophysiological responses to intermittent hypoxia. J Appl Physiol 2001;90(4):1593–9.
5. Weese-Mayer DE, Silvestri JM, Menzies LJ, Morrow-Kenny AS, Hunt CE, Hauptman SA. Congenital central hypoventilation syndrome: diagnosis, management, and long-term outcome in thirty-two children. J Pediatr 1992;120(3):381–7.
6. Owens J, Spirito A, Nobile C, Arrigan M. Incidence of parasomnias in children with obstructive sleep apnea. Sleep 1997;20(12):1193–6.
7. Corno AF, Milano G, Samaja M, Tozzi P, von Segesser LK. Chronic hypoxia: a model for cyanotic congenital heart defects. J Thorac Cardiovasc Surg 2002;124(1):105–12.
8. Badr Zahr LK, Purdy I. Brain injury in the infant: the old, the new, and the uncertain. J Perinat Neonatal Nurs 2006;20(2):163–75; quiz 76–7.
9. Johnston MV. Brain plasticity in paediatric neurology. Eur J Paediatr Neurol 2003;7(3): 105–13.
10. Johnson EM, Jr., Deckwerth TL. Molecular mechanisms of developmental neuronal death. Annu Rev Neurosci 1993;16:31–46.
11. Blaschke AJ, Staley K, Chun J. Widespread programmed cell death in proliferative and postmitotic regions of the fetal cerebral cortex. Development 1996;122(4):1165–74.
12. Lopez-Bendito G, Molnar Z. Thalamocortical development: how are we going to get there? Nat Rev Neurosci 2003;4(4):276–89.
13. Bonhoeffer T. Neurotrophins and activity-dependent development of the neocortex. Curr Opin Neurobiol 1996;6(1):119–26.
14. Berardi N, Pizzorusso T, Maffei L. Critical periods during sensory development. Curr Opin Neurobiol 2000;10(1):138–45.
15. Curristin SM, Cao A, Stewart WB, et al. Disrupted synaptic development in the hypoxic newborn brain. Proc Natl Acad Sci USA 2002;99(24):15729–34.

16. Ment LR, Schwartz M, Makuch RW, Stewart WB. Association of chronic sublethal hypoxia with ventriculomegaly in the developing rat brain. Brain Res Dev Brain Res 1998; 111(2):197–203.
17. Weiss J, Takizawa B, McGee A, et al. Neonatal hypoxia suppresses oligodendrocyte Nogo-A and increases axonal sprouting in a rodent model for human prematurity. Exp Neurol 2004; 189(1):141–9.
18. Fagel DM, Ganat Y, Silbereis J, et al. Cortical neurogenesis enhanced by chronic perinatal hypoxia. Exp Neurol 2006;199(1):77–91.
19. Ganat Y, Soni S, Chacon M, Schwartz ML, Vaccarino FM. Chronic hypoxia up-regulates fibroblast growth factor ligands in the perinatal brain and induces fibroblast growth factor-responsive radial glial cells in the sub-ependymal zone. Neuroscience 2002;112(4):977–91.
20. Johnston MV, Nakajima W, Hagberg H. Mechanisms of hypoxic neurodegeneration in the developing brain. Neuroscientist 2002;8(3):212–20.
21. Jensen FE. Developmental factors regulating susceptibility to perinatal brain injury and seizures. Curr Opin Pediatr 2006;18(6):628–33.
22. McQuillen PS, Ferriero DM. Selective vulnerability in the developing central nervous system. Pediatr Neurol 2004;30(4):227–35.
23. Matute C, Alberdi E, Ibarretxe G, Sanchez-Gomez MV. Excitotoxicity in glial cells. Eur J Pharmacol 2002;447(2–3):239–46.
24. du Plessis AJ, Volpe JJ. Perinatal brain injury in the preterm and term newborn. Curr Opin Neurol 2002;15(2):151–7.
25. Martin LJ, Al-Abdulla NA, Brambrink AM, Kirsch JR, Sieber FE, Portera-Cailliau C. Neurodegeneration in excitotoxicity, global cerebral ischemia, and target deprivation: A perspective on the contributions of apoptosis and necrosis. Brain Res Bull 1998;46(4):281–309.
26. Kanaan A, Farahani R, Douglas RM, Lamanna JC, Haddad GG. Effect of chronic continuous or intermittent hypoxia and reoxygenation on cerebral capillary density and myelination. Am J Physiol Regul Integr Comp Physiol 2006;290(4):R1105–R1114.
27. Baumgartner RW, Eichenberger U, Bartsch P. Postural ataxia at high altitude is not related to mild to moderate acute mountain sickness. Eur J Appl Physiol 2002;86(4):322–6.
28. Shukitt-Hale B, Kadar T, Marlowe BE, et al. Morphological alterations in the hippocampus following hypobaric hypoxia. Hum Exp Toxicol 1996;15(4):312–19.
29. Singh SB, Selvamurthy W. Effect of intermittent chronic exposure to hypoxia on feeding behaviour of rats. Int J Biometeorol 1993;37(4):200–2.
30. Bahrke MS, Shukitt-Hale B. Effects of altitude on mood, behaviour and cognitive functioning. A review. Sports Med 1993;16(2):97–125.
31. Hamilton AJ, Trad LA, Cymerman A. Alterations in human upper extremity motor function during acute exposure to simulated altitude. Aviat Space Environ Med 1991;62(8):759–64.
32. Urschitz MS, Eitner S, Guenther A, et al. Habitual snoring, intermittent hypoxia, and impaired behavior in primary school children. Pediatrics 2004;114(4):1041–8.
33. Noda A, Okada T, Yasuma F, Nakashima N, Yokota M. Cardiac hypertrophy in obstructive sleep apnea syndrome. Chest 1995;107(6):1538–44.
34. Winnicki M, Shamsuzzaman A, Lanfranchi P, et al. Erythropoietin and obstructive sleep apnea. Am J Hypertens 2004;17(9):783–6.
35. Halbower AC, Degaonkar M, Barker PB, et al. Childhood obstructive sleep apnea associates with neuropsychological deficits and neuronal brain injury. PLoS Med 2006;3(8):e301.
36. Gozal D. Sleep-disordered breathing and school performance in children. Pediatrics 1998; 102(3, Part 1):616–20.
37. Barhwal K, Singh SB, Hota SK, Jayalakshmi K, Ilavazhagan G. Acetyl-L-carnitine ameliorates hypobaric hypoxic impairment and spatial memory deficits in rats. Eur J Pharmacol 2007;570(1–3):97–107.
38. Hogan AM, Pit-ten Cate IM, Vargha-Khadem F, Prengler M, Kirkham FJ. Physiological correlates of intellectual function in children with sickle cell disease: hypoxaemia, hyperaemia and brain infarction. Dev Sci 2006;9(4):379–87.
39. Freeman GB, Nielsen P, Gibson GE. Monoamine neurotransmitter metabolism and locomotor activity during chemical hypoxia. J Neurochem 1986;46(3):733–8.

40. Zola-Morgan S, Squire LR, Rempel NL, Clower RP, Amaral DG. Enduring memory impairment in monkeys after ischemic damage to the hippocampus. J Neurosci 1992;12(7):2582–96.
41. Tsien JZ, Huerta PT, Tonegawa S. The essential role of hippocampal CA1 NMDA receptor-dependent synaptic plasticity in spatial memory. Cell 1996;87(7):1327–38.
42. Tonegawa S, Tsien JZ, McHugh TJ, Huerta P, Blum KI, Wilson MA. Hippocampal CA1-region-restricted knockout of *NMDAR1* gene disrupts synaptic plasticity, place fields, and spatial learning. Cold Spring Harb Symp Quant Biol 1996;61:225–38.
43. Rempel-Clower NL, Zola SM, Squire LR, Amaral DG. Three cases of enduring memory impairment after bilateral damage limited to the hippocampal formation. J Neurosci 1996;16(16): 5233–55.
44. Dugan LL, Creedon DJ, Johnson EM, Jr., Holtzman DM. Rapid suppression of free radical formation by nerve growth factor involves the mitogen-activated protein kinase pathway. Proc Natl Acad Sci USA 1997;94(8):4086–91.
45. Alvarez P, Zola-Morgan S, Squire LR. Damage limited to the hippocampal region produces long-lasting memory impairment in monkeys. J Neurosci 1995;15(5, Part 2):3796–807.
46. McHugh TJ, Blum KI, Tsien JZ, Tonegawa S, Wilson MA. Impaired hippocampal representation of space in CA1-specific NMDAR1 knockout mice. Cell 1996;87(7):1339–49.
47. Mikati MA, Zeinieh MP, Kurdi RM, et al. Long-term effects of acute and of chronic hypoxia on behavior and on hippocampal histology in the developing brain. Brain Res Dev Brain Res 2005;157(1):98–102.
48. Maiese K, Li F, Chong ZZ. Erythropoietin in the brain: can the promise to protect be fulfilled? Trends Pharmacol Sci 2004;25(11):577–83.
49. Kirkham FJ, Datta AK. Hypoxic adaptation during development: relation to pattern of neurological presentation and cognitive disability. Dev Sci 2006;9(4):411–27.
50. Virues-Ortega J, Garrido E, Javierre C, Kloezeman KC. Human behaviour and development under high-altitude conditions. Dev Sci 2006;9(4):400–10.
51. Lieberman P, Protopapas A, Reed E, Youngs JW, Kanki BG. Cognitive defects at altitude. Nature 1994;372(6504):325.
52. Askew EW. Work at high altitude and oxidative stress: antioxidant nutrients. Toxicology 2002; 180(2):107–19.
53. Pearigen P, Gwinn R, Simon RP. The effects in vivo of hypoxia on brain injury. Brain Res 1996;725(2):184–91.
54. Tamatani M, Mitsuda N, Matsuzaki H, et al. A pathway of neuronal apoptosis induced by hypoxia/reoxygenation: roles of nuclear factor-kappaB and Bcl-2. J Neurochem 2000;75(2): 683–93.
55. Matsuoka Y, Kitamura Y, Fukunaga R, et al. In vivo hypoxia-induced neuronal damage in dentate gyrus of rat hippocampus: changes in NMDA receptors and the effect of MK-801. Neurochem Int 1997;30(6):533–42.
56. Dell'Anna ME, Calzolari S, Molinari M, Iuvone L, Calimici R. Neonatal anoxia induces transitory hyperactivity, permanent spatial memory deficits and CA1 cell density reduction in developing rats. Behav Brain Res 1991;45(2):125–34.
57. Chihab R, Bossenmeyer C, Oillet J, Daval JL. Lack of correlation between the effects of transient exposure to glutamate and those of hypoxia/reoxygenation in immature neurons in vitro. J Neurochem 1998;71(3):1177–86.
58. Bossenmeyer C, Chihab R, Muller S, Schroeder H, Daval JL. Hypoxia/reoxygenation induces apoptosis through biphasic induction of protein synthesis in cultured rat brain neurons. Brain Res 1998;787(1):107–16.
59. Banasiak KJ, Haddad GG. Hypoxia-induced apoptosis: effect of hypoxic severity and role of p53 in neuronal cell death. Brain Res 1998;797(2):295–304.
60. Grojean S, Pourie G, Vert P, Daval JL. Differential neuronal fates in the CA1 hippocampus after hypoxia in newborn and 7-day-old rats: effects of pre-treatment with MK-801. Hippocampus 2003;13(8):970–7.
61. Aitken PG, Balestrino M, Somjen GG. NMDA antagonists: lack of protective effect against hypoxic damage in CA1 region of hippocampal slices. Neurosci Lett 1988;89(2):187–92.

62. Zhu C, Wang X, Xu F, et al. The influence of age on apoptotic and other mechanisms of cell death after cerebral hypoxia-ischemia. Cell Death Differ 2005;12(2):162–76.
63. Pulsinelli WA. Selective neuronal vulnerability: morphological and molecular characteristics. Prog Brain Res 1985;63:29–37.
64. Erecinska M, Silver IA. Tissue oxygen tension and brain sensitivity to hypoxia. Respir Physiol 2001;128(3):263–76.
65. Knickerbocker DL, Lutz PL. Slow ATP loss and the defense of ion homeostasis in the anoxic frog brain. J Exp Biol 2001;204 (Part 20):3547–51.
66. Hansen AJ. Effect of anoxia on ion distribution in the brain. Physiol Rev 1985;65(1):101–48.
67. Nilsson GE. Surviving anoxia with the brain turned on. News Physiol Sci 2001;16:217–21.
68. Hochachka PW, Lutz PL. Mechanism, origin, and evolution of anoxia tolerance in animals. Comp Biochem Physiol B Biochem Mol Biol 2001;130(4):435–59.
69. Back SA, Han BH, Luo NL, et al. Selective vulnerability of late oligodendrocyte progenitors to hypoxia-ischemia. J Neurosci 2002;22(2):455–63.
70. Nieber K. Hypoxia and neuronal function under in vitro conditions. Pharmacol Ther 1999; 82(1):71–86.
71. Donnelly DF, Jiang C, Haddad GG. Comparative responses of brain stem and hippocampal neurons to O2 deprivation: in vitro intracellular studies. Am J Physiol 1992;262(5, Part 1):L549–L554.
72. Kadar T, Dachir S, Shukitt-Hale B, Levy A. Sub-regional hippocampal vulnerability in various animal models leading to cognitive dysfunction. J Neural Transm 1998;105(8–9):987–1004.
73. Gibson GE, Pulsinelli W, Blass JP, Duffy TE. Brain dysfunction in mild to moderate hypoxia. Am J Med 1981;70(6):1247–54.
74. Cervos-Navarro J, Diemer NH. Selective vulnerability in brain hypoxia. Crit Rev Neurobiol 1991;6(3):149–82.
75. Erecinska M, Silver IA. Calcium handling by hippocampal neurons under physiologic and pathologic conditions. Adv Neurol 1996;71:119–36.
76. Ramanathan L, Gozal D, Siegel JM. Antioxidant responses to chronic hypoxia in the rat cerebellum and pons. J Neurochem 2005;93(1):47–52.
77. Xu L, Sapolsky RM, Giffard RG. Differential sensitivity of murine astrocytes and neurons from different brain regions to injury. Exp Neurol 2001;169(2):416–24.
78. Hammarstrom AK, Gage PW. Inhibition of oxidative metabolism increases persistent sodium current in rat CA1 hippocampal neurons. J Physiol 1998;510 (Part 3):735–41.
79. Jiang C, Haddad GG. Oxygen deprivation inhibits a K$^+$ channel independently of cytosolic factors in rat central neurons. J Physiol 1994;481 (Part 1):15–26.
80. Peers C. Oxygen-sensitive ion channels. Trends Pharmacol Sci 1997;18(11):405–8.
81. Haddad GG, Jiang C. O2-sensing mechanisms in excitable cells: role of plasma membrane K$^+$ channels. Annu Rev Physiol 1997;59:23–42.
82. Lopez-Barneo J. Oxygen-sensing by ion channels and the regulation of cellular functions. Trends Neurosci 1996;19(10):435–40.
83. Semenza GL. Regulation of mammalian O2 homeostasis by hypoxia-inducible factor 1. Annu Rev Cell Dev Biol 1999;15:551–78.
84. Schofield CJ, Ratcliffe PJ. Oxygen sensing by HIF hydroxylases. Nat Rev Mol Cell Biol 2004;5(5):343–54.
85. Chidekel AS, Friedman JE, Haddad GG. Anoxia-induced neuronal injury: role of Na$^+$ entry and Na$^+$-dependent transport. Exp Neurol 1997;146(2):403–13.
86. Shimoda LA, Fallon M, Pisarcik S, Wang J, Semenza GL. HIF-1 regulates hypoxic induction of NHE1 expression and alkalinization of intracellular pH in pulmonary arterial myocytes. Am J Physiol Lung Cell Mol Physiol 2006;291(5):L941–L949.
87. Chen LM, Choi I, Haddad GG, Boron WF. Chronic continuous hypoxia decreases the expression of SLC4A7 (NBCn1) and SLC4A10 (NCBE) in mouse brain. Am J Physiol Regul Integr Comp Physiol 2007;293(6):R2412–R2420.
88. Xue J, Zhou D, Yao H, Haddad GG. Role of transporters and ion channels in neuronal injury under hypoxia. Am J Physiol Regul Integr Comp Physiol 2008;294(2):R451–R457.

89. Prentice HM. Key contributions of the Na$^+$/H$^+$ exchanger subunit 1 and HCO3- transporters in regulating neuronal cell fate in prolonged hypoxia. Am J Physiol Regul Integr Comp Physiol 2008;294(2):R448–R450.
90. Heacock CS, Sutherland RM. Enhanced synthesis of stress proteins caused by hypoxia and relation to altered cell growth and metabolism. Br J Cancer 1990;62(2):217–25.
91. Hochachka PW, Buck LT, Doll CJ, Land SC. Unifying theory of hypoxia tolerance: molecular/metabolic defense and rescue mechanisms for surviving oxygen lack. Proc Natl Acad Sci USA 1996;93(18):9493–8.
92. Storey KB, Storey JM. Metabolic rate depression in animals: transcriptional and translational controls. Biol Rev Camb Philos Soc 2004;79(1):207–33.
93. Rolfe DF, Brown GC. Cellular energy utilization and molecular origin of standard metabolic rate in mammals. Physiol Rev 1997;77(3):731–58.
94. Hochachka PW. Defense strategies against hypoxia and hypothermia. Science 1986;231(4735): 234–41.
95. Raley-Susman KM, Murata J. Time course of protein changes following in vitro ischemia in the rat hippocampal slice. Brain Res 1995;694(1–2):94–102.
96. Raley-Susman KM, Lipton P. In vitro ischemia and protein synthesis in the rat hippocampal slice: the role of calcium and NMDA receptor activation. Brain Res 1990;515(1–2):27–38.
97. Arsham AM, Howell JJ, Simon MC. A novel hypoxia-inducible factor-independent hypoxic response regulating mammalian target of rapamycin and its targets. J Biol Chem 2003;278(32):29655–60.
98. Koumenis C, Naczki C, Koritzinsky M, et al. Regulation of protein synthesis by hypoxia via activation of the endoplasmic reticulum kinase PERK and phosphorylation of the translation initiation factor eIF2alpha. Mol Cell Biol 2002;22(21):7405–16.
99. Chavez JC, Pichiule P, Boero J, Arregui A. Reduced mitochondrial respiration in mouse cerebral cortex during chronic hypoxia. Neurosci Lett 1995;193(3):169–72.
100. Lai JC, White BK, Buerstatte CR, Haddad GG, Novotny EJ, Jr., Behar KL. Chronic hypoxia in development selectively alters the activities of key enzymes of glucose oxidative metabolism in brain regions. Neurochem Res 2003;28(6):933–40.
101. Papandreou I, Cairns RA, Fontana L, Lim AL, Denko NC. HIF-1 mediates adaptation to hypoxia by actively downregulating mitochondrial oxygen consumption. Cell Metab 2006; 3(3):187–97.
102. Chinopoulos C, Adam-Vizi V. Depolarization of in situ mitochondria by hydrogen peroxide in nerve terminals. Ann N Y Acad Sci 1999;893:269–72.
103. Murphy BJ, Robin ED, Tapper DP, Wong RJ, Clayton DA. Hypoxic coordinate regulation of mitochondrial enzymes in mammalian cells. Science 1984;223(4637):707–9.
104. Boutilier RG. Mechanisms of cell survival in hypoxia and hypothermia. J Exp Biol 2001;204 (Part 18):3171–81.
105. Rothman SM, Olney JW. Glutamate and the pathophysiology of hypoxic-ischemic brain damage. Ann Neurol 1986;19(2):105–11.
106. Nicholls D, Attwell D. The release and uptake of excitatory amino acids. Trends Pharmacol Sci 1990;11(11):462–8.
107. Choi DW. Possible mechanisms limiting N-methyl-D-aspartate receptor overactivation and the therapeutic efficacy of N-methyl-D-aspartate antagonists. Stroke 1990;21(11 Suppl): III20–III22.
108. Lipton P. Ischemic cell death in brain neurons. Physiol Rev 1999;79(4):1431–568.
109. Lee JM, Zipfel GJ, Choi DW. The changing landscape of ischaemic brain injury mechanisms. Nature 1999;399(6738 Suppl):A7–A14.
110. Puka-Sundvall M, Hallin U, Zhu C, et al. NMDA blockade attenuates caspase-3 activation and DNA fragmentation after neonatal hypoxia-ischemia. Neuroreport 2000;11(13):2833–6.
111. Wang X, Karlsson JO, Zhu C, Bahr BA, Hagberg H, Blomgren K. Caspase-3 activation after neonatal rat cerebral hypoxia-ischemia. Biol Neonate 2001;79(3–4):172–9.
112. Hu BR, Liu CL, Ouyang Y, Blomgren K, Siesjo BK. Involvement of caspase-3 in cell death after hypoxia-ischemia declines during brain maturation. J Cereb Blood Flow Metab 2000;20(9):1294–300.

113. Bossenmeyer-Pourie C, Lievre V, Grojean S, Koziel V, Pillot T, Daval JL. Sequential expression patterns of apoptosis- and cell cycle-related proteins in neuronal response to severe or mild transient hypoxia. Neuroscience 2002;114(4):869–82.
114. Blomgren K, Leist M, Groc L. Pathological apoptosis in the developing brain. Apoptosis 2007;12(5):993–1010.
115. Miyashita T, Reed JC. Tumor suppressor p53 is a direct transcriptional activator of the human *bax* gene. Cell 1995;80(2):293–9.
116. Tamatani M, Ogawa S, Tohyama M. Roles of Bcl-2 and caspases in hypoxia-induced neuronal cell death: a possible neuroprotective mechanism of peptide growth factors. Brain Res Mol Brain Res 1998;58(1–2):27–39.
117. Forman HJ, Zhou H, Gozal E, Torres M. Modulation of the alveolar macrophage superoxide production by protein phosphorylation. Environ Health Perspect 1998;106 Suppl 5: 1185–90.
118. Cadenas E. Biochemistry of oxygen toxicity. Annu Rev Biochem 1989;58:79–110.
119. Kulkarni AC, Kuppusamy P, Parinandi N. Oxygen, the lead actor in the pathophysiologic drama: enactment of the trinity of normoxia, hypoxia, and hyperoxia in disease and therapy. Antioxid Redox Signal 2007;9(10):1717–30.
120. St-Pierre J, Brand MD, Boutilier RG. Mitochondria as ATP consumers: cellular treason in anoxia. Proc Natl Acad Sci USA 2000;97(15):8670–4.
121. Capdevila J, Chacos N, Werringloer J, Prough RA, Estabrook RW. Liver microsomal cytochrome P-450 and the oxidative metabolism of arachidonic acid. Proc Natl Acad Sci USA 1981;78(9):5362–6.
122. Babior BM. NADPH oxidase: an update. Blood 1999;93(5):1464–76.
123. Masoro EJ. Caloric restriction and aging: an update. Exp Gerontol 2000;35(3):299–305.
124. Dugan LL, Quick KL. Reactive oxygen species and aging: evolving questions. Sci Aging Knowledge Environ 2005;2005(26):pe20.
125. Ames BN, Shigenaga MK, Hagen TM. Oxidants, antioxidants, and the degenerative diseases of aging. Proc Natl Acad Sci USA 1993;90(17):7915–22.
126. Simonian NA, Coyle JT. Oxidative stress in neurodegenerative diseases. Annu Rev Pharmacol Toxicol 1996;36:83–106.
127. Beal MF. Aging, energy, and oxidative stress in neurodegenerative diseases. Ann Neurol 1995;38(3):357–66.
128. Pouyssegur J, Mechta-Grigoriou F. Redox regulation of the hypoxia-inducible factor. Biol Chem 2006;387(10–11):1337–46.
129. Pidgeon GP, Tamosiuniene R, Chen G, et al. Intravascular thrombosis after hypoxia-induced pulmonary hypertension: regulation by cyclooxygenase-2. Circulation 2004;110(17): 2701–7.
130. Guzy RD, Schumacker PT. Oxygen sensing by mitochondria at complex III: the paradox of increased reactive oxygen species during hypoxia. Exp Physiol 2006;91(5):807–19.
131. Dawson TL, Gores GJ, Nieminen AL, Herman B, Lemasters JJ. Mitochondria as a source of reactive oxygen species during reductive stress in rat hepatocytes. Am J Physiol 1993;264 (4, Part 1):C961–C967.
132. Clanton T. Yet another oxygen paradox. J Appl Physiol 2005;99(4):1245–6.
133. Siddappa AJ, Rao RB, Wobken JD, Leibold EA, Connor JR, Georgieff MK. Developmental changes in the expression of iron regulatory proteins and iron transport proteins in the perinatal rat brain. J Neurosci Res 2002;68(6):761–75.
134. McQuillen PS, Sheldon RA, Shatz CJ, Ferriero DM. Selective vulnerability of subplate neurons after early neonatal hypoxia-ischemia. J Neurosci 2003;23(8):3308–15.
135. Schumacker PT. Current paradigms in cellular oxygen sensing. Adv Exp Med Biol 2003; 543:57–71.
136. Duranteau J, Chandel NS, Kulisz A, Shao Z, Schumacker PT. Intracellular signaling by reactive oxygen species during hypoxia in cardiomyocytes. J Biol Chem 1998;273(19):11619–24.
137. Won SJ, Kim DY, Gwag BJ. Cellular and molecular pathways of ischemic neuronal death. J Biochem Mol Biol 2002;35(1):67–86.
138. Adams JH. Brain damage caused by cerebral hypoxia. Nurs Times 1975;71(17):654–6.

139. Waypa GB, Schumacker PT. O(2) sensing in hypoxic pulmonary vasoconstriction: the mitochondrial door re-opens. Respir Physiol Neurobiol 2002;132(1):81–91.
140. Schroedl C, McClintock DS, Budinger GR, Chandel NS. Hypoxic but not anoxic stabilization of HIF-1alpha requires mitochondrial reactive oxygen species. Am J Physiol Lung Cell Mol Physiol 2002;283(5):L922–L931.
141. Maiti P, Singh SB, Sharma AK, Muthuraju S, Banerjee PK, Ilavazhagan G. Hypobaric hypoxia induces oxidative stress in rat brain. Neurochem Int 2006;49(8):709–16.
142. Watson BD, Ginsberg MD, Busto R. Macroscopic indices of lipid peroxidation in cerebral ischemia/reperfusion: validity and sensitivity enhancement in terms of conjugated diene detection. Neurochem Int 1996;29(2):173–86.
143. Li RC, Row BW, Kheirandish L, et al. Nitric oxide synthase and intermittent hypoxia-induced spatial learning deficits in the rat. Neurobiol Dis 2004;17(1):44–53.
144. Mishra OP, Delivoria-Papadopoulos M. Cellular mechanisms of hypoxic injury in the developing brain. Brain Res Bull 1999;48(3):233–8.
145. Blum J, Fridovich I. Inactivation of glutathione peroxidase by superoxide radical. Arch Biochem Biophys 1985;240(2):500–8.
146. Barker JE, Bolanos JP, Land JM, Clark JB, Heales SJ. Glutathione protects astrocytes from peroxynitrite-mediated mitochondrial damage: implications for neuronal/astrocytic trafficking and neurodegeneration. Dev Neurosci 1996;18(5–6):391–6.
147. Benveniste H, Drejer J, Schousboe A, Diemer NH. Elevation of the extracellular concentrations of glutamate and aspartate in rat hippocampus during transient cerebral ischemia monitored by intracerebral microdialysis. J Neurochem 1984;43(5):1369–74.
148. Murphy TH, Miyamoto M, Sastre A, Schnaar RL, Coyle JT. Glutamate toxicity in a neuronal cell line involves inhibition of cystine transport leading to oxidative stress. Neuron 1989;2(6):1547–58.
149. Zhang Y, Marcillat O, Giulivi C, Ernster L, Davies KJ. The oxidative inactivation of mitochondrial electron transport chain components and ATPase. J Biol Chem 1990;265(27):16330–6.
150. Singh SN, Vats P, Kumria MM, et al. Effect of high altitude (7,620 m) exposure on glutathione and related metabolism in rats. Eur J Appl Physiol 2001;84(3):233–7.
151. Moller P, Loft S, Lundby C, Olsen NV. Acute hypoxia and hypoxic exercise induce DNA strand breaks and oxidative DNA damage in humans. FASEB J 2001;15(7):1181–6.
152. Jayalakshmi K, Sairam M, Singh SB, Sharma SK, Ilavazhagan G, Banerjee PK. Neuroprotective effect of *N*-acetyl cysteine on hypoxia-induced oxidative stress in primary hippocampal culture. Brain Res 2005;1046(1–2):97–104.
153. Bailey DM, Davies B. Acute mountain sickness: prophylactic benefits of antioxidant vitamin supplementation at high altitude. High Alt Med Biol 2001;2(1):21–9.
154. Sattler R, Tymianski M. Molecular mechanisms of calcium-dependent excitotoxicity. J Mol Med 2000;78(1):3–13.
155. Dugan LL, Sensi SL, Canzoniero LM, et al. Mitochondrial production of reactive oxygen species in cortical neurons following exposure to *N*-methyl-D-aspartate. J Neurosci 1995;15(10):6377–88.
156. Stout AK, Raphael HM, Kanterewicz BI, Klann E, Reynolds IJ. Glutamate-induced neuron death requires mitochondrial calcium uptake. Nat Neurosci 1998;1(5):366–73.
157. Luetjens CM, Bui NT, Sengpiel B, et al. Delayed mitochondrial dysfunction in excitotoxic neuron death: cytochrome *c* release and a secondary increase in superoxide production. J Neurosci 2000;20(15):5715–23.
158. Sengpiel B, Preis E, Krieglstein J, Prehn JH. NMDA-induced superoxide production and neurotoxicity in cultured rat hippocampal neurons: role of mitochondria. Eur J Neurosci 1998;10(5):1903–10.
159. Shimizu S, Eguchi Y, Kamiike W, et al. Involvement of ICE family proteases in apoptosis induced by reoxygenation of hypoxic hepatocytes. Am J Physiol 1996;271(6, Part 1):G949–G958.
160. Kluck RM, Bossy-Wetzel E, Green DR, Newmeyer DD. The release of cytochrome *c* from mitochondria: a primary site for Bcl-2 regulation of apoptosis. Science 1997;275(5303):1132–6.

161. Green DR, Reed JC. Mitochondria and apoptosis. Science 1998;281(5381):1309–12.
162. Facchinetti F, Dawson VL, Dawson TM. Free radicals as mediators of neuronal injury. Cell Mol Neurobiol 1998;18(6):667–82.
163. Ock J, Jeong J, Choi WS, et al. Regulation of toll-like receptor 4 expression and its signaling by hypoxia in cultured microglia. J Neurosci Res 2007;85(9):1989–95.
164. Karin M, Ben-Neriah Y. Phosphorylation meets ubiquitination: the control of NF-[kappa]B activity. Annu Rev Immunol 2000;18:621–63.
165. Ghosh S, Karin M. Missing pieces in the NF-kappaB puzzle. Cell 2002;109 Suppl: S81–S96.
166. Ben-Neriah Y. Regulatory functions of ubiquitination in the immune system. Nat Immunol 2002;3(1):20–6.
167. Siebenlist U, Franzoso G, Brown K. Structure, regulation and function of NF-kappa B. Annu Rev Cell Biol 1994;10:405–55.
168. Baeuerle PA, Baltimore D. NF-kappa B: ten years after. Cell 1996;87(1):13–20.
169. Cario E, Podolsky DK. Intestinal epithelial TOLLerance versus inTOLLerance of commensals. Mol Immunol 2005;42(8):887–93.
170. Cummins EP, Berra E, Comerford KM, et al. Prolyl hydroxylase-1 negatively regulates IkappaB kinase-beta, giving insight into hypoxia-induced NFkappaB activity. Proc Natl Acad Sci USA 2006;103(48):18154–9.
171. Hacker H, Karin M. Regulation and function of IKK and IKK-related kinases. Sci STKE 2006;2006(357):re13.
172. Rius J, Guma M, Schachtrup C, et al. NF-kappaB links innate immunity to the hypoxic response through transcriptional regulation of HIF-1alpha. Nature 2008;453:807–811.
173. Kitajima I, Nakajima T, Imamura T, et al. Induction of apoptosis in murine clonal osteoblasts expressed by human T-cell leukemia virus type I tax by NF-kappa B and TNF-alpha. J Bone Miner Res 1996;11(2):200–10.
174. Kessler JA, Ludlam WH, Freidin MM, et al. Cytokine-induced programmed death of cultured sympathetic neurons. Neuron 1993;11(6):1123–32.
175. Grilli M, Pizzi M, Memo M, Spano P. Neuroprotection by aspirin and sodium salicylate through blockade of NF-kappaB activation. Science 1996;274(5291):1383–5.
176. Beg AA, Baltimore D. An essential role for NF-kappaB in preventing TNF-alpha-induced cell death. Science 1996;274(5288):782–4.
177. Barger SW, Horster D, Furukawa K, Goodman Y, Krieglstein J, Mattson MP. Tumor necrosis factors alpha and beta protect neurons against amyloid beta-peptide toxicity: evidence for involvement of a kappa B-binding factor and attenuation of peroxide and Ca^{2+} accumulation. Proc Natl Acad Sci USA 1995;92(20):9328–32.
178. Koong AC, Chen EY, Giaccia AJ. Hypoxia causes the activation of nuclear factor kappa B through the phosphorylation of I kappa B alpha on tyrosine residues. Cancer Res 1994;54(6): 1425–30.
179. Chandel NS, McClintock DS, Feliciano CE, et al. Reactive oxygen species generated at mitochondrial complex III stabilize hypoxia-inducible factor-1alpha during hypoxia: a mechanism of O2 sensing. J Biol Chem 2000;275(33):25130–8.
180. Zampetaki A, Mitsialis SA, Pfeilschifter J, Kourembanas S. Hypoxia induces macrophage inflammatory protein-2 (*MIP-2*) gene expression in murine macrophages via NF-kappaB: the prominent role of p42/ p44 and PI3 kinase pathways. FASEB J 2004;18(10):1090–2.
181. Clemens JA, Stephenson DT, Yin T, Smalstig EB, Panetta JA, Little SP. Drug-induced neuroprotection from global ischemia is associated with prevention of persistent but not transient activation of nuclear factor-kappaB in rats. Stroke 1998;29(3):677–82.
182. Yin MJ, Yamamoto Y, Gaynor RB. The anti-inflammatory agents aspirin and salicylate inhibit the activity of I(kappa)B kinase-beta. Nature 1998;396(6706):77–80.
183. Grilli M, Goffi F, Memo M, Spano P. Interleukin-1beta and glutamate activate the NF-kappaB/Rel binding site from the regulatory region of the amyloid precursor protein gene in primary neuronal cultures. J Biol Chem 1996;271(25):15002–7.
184. Schmitz ML, Baeuerle PA. Multi-step activation of NF-kappa B/Rel transcription factors. Immunobiology 1995;193(2–4):116–27.

185. Sairanen T, Carpen O, Karjalainen-Lindsberg ML, et al. Evolution of cerebral tumor necrosis factor-alpha production during human ischemic stroke. Stroke 2001;32(8):1750–8.
186. Acker T, Acker H. Cellular oxygen sensing need in CNS function: physiological and pathological implications. J Exp Biol 2004;207 (Part 18):3171–88.
187. Brahimi-Horn MC, Chiche J, Pouyssegur J. Hypoxia signalling controls metabolic demand. Curr Opin Cell Biol 2007;19(2):223–9.
188. Plas DR, Thompson CB. Akt-dependent transformation: there is more to growth than just surviving. Oncogene 2005;24(50):7435–42.
189. Wang GL, Semenza GL. Purification and characterization of hypoxia-inducible factor 1. J Biol Chem 1995;270(3):1230–7.
190. Ratcliffe PJ. HIF-1 and HIF-2: working alone or together in hypoxia? J Clin Invest 2007; 117(4):862–5.
191. Maxwell PH. Hypoxia-inducible factor as a physiological regulator. Exp Physiol 2005; 90(6):791–7.
192. Wiesener MS, Turley H, Allen WE, et al. Induction of endothelial PAS domain protein-1 by hypoxia: characterization and comparison with hypoxia-inducible factor-1alpha. Blood 1998;92(7):2260–8.
193. Kaelin WG, Jr. The von Hippel-Lindau protein, HIF hydroxylation, and oxygen sensing. Biochem Biophys Res Commun 2005;338(1):627–38.
194. Lando D, Peet DJ, Gorman JJ, Whelan DA, Whitelaw ML, Bruick RK. FIH-1 is an asparaginyl hydroxylase enzyme that regulates the transcriptional activity of hypoxia-inducible factor. Genes Dev 2002;16(12):1466–71.
195. Stroka DM, Burkhardt T, Desbaillets I, et al. HIF-1 is expressed in normoxic tissue and displays an organ-specific regulation under systemic hypoxia. FASEB J 2001;15(13):2445–53.
196. Sharp FR, Bergeron M, Bernaudin M. Hypoxia-inducible factor in brain. Adv Exp Med Biol 2001;502:273–91.
197. Arany Z, Huang LE, Eckner R, et al. An essential role for p300/CBP in the cellular response to hypoxia. Proc Natl Acad Sci USA 1996;93(23):12969–73.
198. Digicaylioglu M, Lipton SA. Erythropoietin-mediated neuroprotection involves cross-talk between Jak2 and NF-kappaB signalling cascades. Nature 2001;412(6847):641–7.
199. Dvorak HF, Brown LF, Detmar M, Dvorak AM. Vascular permeability factor/vascular endothelial growth factor, microvascular hyperpermeability, and angiogenesis. Am J Pathol 1995;146(5):1029–39.
200. Rapino C, Bianchi G, Di Giulio C, et al. HIF-1alpha cytoplasmic accumulation is associated with cell death in old rat cerebral cortex exposed to intermittent hypoxia. Aging Cell 2005;4(4):177–85.
201. Halterman MW, Federoff HJ. HIF-1alpha and p53 promote hypoxia-induced delayed neuronal death in models of CNS ischemia. Exp Neurol 1999;159(1):65–72.
202. Carmeliet P, Dor Y, Herbert JM, et al. Role of HIF-1alpha in hypoxia-mediated apoptosis, cell proliferation and tumour angiogenesis. Nature 1998;394(6692):485–90.
203. Hagen T, Taylor CT, Lam F, Moncada S. Redistribution of intracellular oxygen in hypoxia by nitric oxide: effect on HIF1alpha. Science 2003;302(5652):1975–8.

Chapter 14
Hypoxia-Inducible Factor 1

Gregg L. Semenza

Abstract Hypoxia-inducible factor 1 (HIF-1) mediates adaptive responses to hypoxia by activating the transcription of hundreds of target genes. The expression and activity of HIF-1 are oxygen-regulated, which provides a direct mechanism for transducing changes in cellular oxygenation to changes in gene expression. HIF-1 regulates the expression of genes encoding proteins that play key roles in mediating the switch to glycolytic metabolism, the induction of angiogenesis, and the production of survival factors that block ischemia-induced apoptosis. Induction of HIF-1 activity or administration of survival factors encoded by HIF-1 target genes may be of therapeutic benefit in patients with acute stroke.

Keywords: angiogenesis; carotid body; erythropoietin; glycolysis; hypoxia; ischemia; oxygen; preconditioning; stroke

14.1 Oxygen Homeostasis and Its Impact on Evolution, Development, and Disease

All metazoan species are dependent upon a continuous supply of O_2 in order to survive. The principal requirement for O_2 is as the ultimate electron acceptor in the process of oxidative phosphorylation, which generates sufficient ATP from the catabolism of glucose and fatty acids to maintain the complex nature of vertebrate structure and function. However, the need for O_2 as metabolic substrate is opposed by the inherent risk of oxidative damage to cellular macromolecules. Because of the dual character of O_2 as an essential but potentially toxic molecule, its concentration within cells of a healthy organism is maintained within a narrow range that optimally balances supply and demand.

G.L. Semenza
Vascular Program, Institute for Cell Engineering, The Johns Hopkins University
School of Medicine, Broadway Research Building, Suite 671, 733 North Broadway,
Baltimore, MD 21205, USA
e-mail: gsemenza@jhmi.edu

The roundworm *Caenorhabditis elegans* consists of less than 1,000 cells, which receive O_2 by simple diffusion from the air. In the fruit fly *Drosophila melanogaster*, the presence of multiple cell layers necessitates the existence of a specialized system of tracheal tubes to conduct air to all cells of the organism. In vertebrates, the much larger body size requires an even more complex strategy of O_2 delivery involving a respiratory system, which provides a large pulmonary alveolar surface area for gas exchange, and a circulatory system, consisting of blood, heart, and blood vessels to deliver O_2 to all cells of the body. Although the circulatory system also functions to deliver nutrients and remove toxic metabolic wastes from the tissues, its principal function is to maintain oxygen homeostasis by precisely modulating O_2 delivery to meet the demands imposed by cells within every tissue. Thus, the requirement to maintain oxygen homeostasis represented an important driving force for the evolution of organisms with increasing biological complexity.

In mammals, the requirement for a functioning circulatory system occurs early in development when diffusion from blood vessels in the surrounding uterine tissues becomes insufficient to supply adequate O_2 to all cells of the growing embryo. In mice, failure to establish a functioning circulatory system by embryonic day 10 results in lethality. In humans, the major disease causes of mortality (cardiovascular disease, cancer, stroke, and chronic lung disease) involve changes in tissue O_2 delivery. Thus, oxygen homeostasis is an organizing principle for understanding evolution, development, and disease pathophysiology. This review will examine the molecular mechanisms underlying responses to hypoxia and ischemia from this physiological perspective.

14.2 Molecular Mechanisms of Oxygen Sensing

The master regulator of oxygen homeostasis is hypoxia-inducible factor 1 (HIF-1), which is a heterodimeric transcription factor that is composed of a constitutively expressed HIF-1β subunit and an oxygen-regulated HIF-1α subunit (1, 2). The HIF-1α subunit is continually synthesized and degraded within adequately perfused cells with normal oxygenation. Under hypoxic conditions, the degradation of HIF-1α is inhibited, leading to its accumulation, dimerization with HIF-1β, DNA binding, recruitment of coactivators, and transcriptional activation of target genes.

HIF-1α is subjected to O_2-dependent ubiquitination that is initiated by the binding of the von Hippel–Lindau tumor suppressor protein (VHL) and its recruitment of an E3 ubiquitin–ligase complex (3, 4) that contains elongin C, elongin B, cullin 2, and RBX1 (Fig. 14.1). The binding of VHL is dependent upon the hydroxylation of proline residue(s) 402 and/or 564 of HIF-1α (5–7). The HIF-1α prolyl hydroxylases that perform this reaction are dioxygenases that utilize O_2 as a substrate (8). One oxygen atom is inserted into the HIF-1α prolyl residue to form a 4-hydroxyl group and the other oxygen atom is inserted into the cosubstrate α-ketoglutarate, forming succinate and CO_2 as side products. Under hypoxic conditions, the hydroxylase activity is inhibited and the half-life of HIF-1α increases as a result of decreased hydroxylation, ubiquitination, and degradation. Although three HIF-1α

Fig. 14.1 HIF-1 activity is regulated by O_2-dependent hydroxylation. Under normoxic conditions, HIF-1α is synthesized, but is rapidly subjected to prolyl hydroxylation (*red*) by PHD2. OS9 promotes hydroxylation by binding to both PHD2 and HIF-1α. FIH-1 mediates asparaginyl hydroxylation (*blue*), which blocks the binding of the coactivator p300 (or the related protein CBP). Prolyl hydroxylation of HIF-1α is required for binding of VHL, which together with SSAT2 recruits elongin C, which in turn recruits a ubiquitin-protein ligase complex containing elongin B (B), ring box protein 1 (RBX1), cullin 2 (CUL2), and an E2 ubiquitin-conjugating enzyme. Ubiquitination of HIF-1α targets the protein for degradation by the 26*S* proteasome (*See Color Plates*)

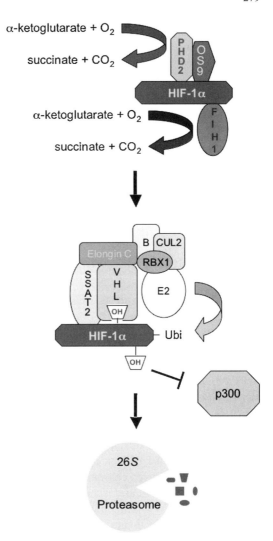

prolyl hydroxylases have been identified, the activity of PHD2 determines the basal levels of HIF-1α under aerobic conditions. Changes in O_2 concentration are very rapidly transduced to changes in HIF-1α protein levels. The speed and precision of this O_2-dependent regulation appears to reflect the involvement of multiprotein complexes in hydroxylation and ubiquitination. OS-9 binds to both HIF-1α and PHD2 and is required for efficient hydroxylation (9), whereas SSAT2 binds to HIF-1α, VHL, and elongin C and is required for efficient ubiquitination (10).

HIF-1α is also subject to O_2-dependent hydroxylation of asparagine residue 803 in the transactivation domain by factor inhibiting HIF-1 (FIH-1), which is another dioxygenase that utilizes O_2 and α-ketoglutarate (11). Hydroxylation of asparagine-803 prevents the interaction of HIF-1α with the coactivators p300 and

CBP (Fig. 14.1). Thus, both the half-life and transcriptional activity of HIF-1α are regulated by O_2-dependent hydroxylation events that provide a direct mechanism by which changes in O_2 concentration can be transduced to the nucleus as changes in the activity of HIF-1.

In addition to HIF-1α, a structurally and functionally related protein designated HIF-2α, which is the product of the *EPAS1* gene, can also heterodimerize with HIF-1β (12). HIF-1α:HIF-1β and HIF-2α:HIF-1β heterodimers appear to have overlapping but distinct target gene specificities (13, 14). Unlike HIF-1α, HIF-2α is not expressed in all cell types and when expressed can be inactive as a result of cytoplasmic sequestration (15). A third protein, designated HIF-3α, has also been identified (16). Its role has not been well defined although a splice variant, designated IPAS, has been shown to bind to HIF-1α and inhibit its activity (17, 18).

14.3   Regulation of Erythropoietin Production by HIF-1

HIF-1 was originally identified through an analysis of the cis-acting sequences and trans-acting factors required for O_2-regulated expression of the human *EPO* gene, which encodes erythropoietin (19). EPO is a glycoprotein hormone produced primarily by the kidney that is secreted into the blood stream and stimulates the proliferation and survival of red blood cell progenitors in bone marrow. The levels of EPO in the peripheral blood determine the blood O_2-carrying capacity and thus the delivery of O_2 to body tissues. In keeping with this critical role, it appears that the *EPO* gene manifests the greatest fold induction in response to hypoxia of any known HIF-1 target gene. The hypoxia-induced expression of EPO by the kidney appears to be regulated by both HIF-1α (20, 21) and HIF-2α (22).

More recently, it has been recognized that in addition to the endocrine role of EPO in regulation of red cell mass, paracrine expression of EPO may regulate survival of many tissues under ischemic conditions, due to the expression of the EPO receptor (EPOR) by endothelial cells, neurons, cardiomyocytes, and several other cell types. As described later, a large body of evidence shows that EPO is a survival factor that increases cellular tolerance to injury by blocking apoptosis.

14.4   Regulation of Angiogenesis by HIF-1

HIF-1α expression is required for proper vascularization of the mouse embryo (23, 24). HIF-1α is necessary and sufficient for the expression of multiple angiogenic growth factors including vascular endothelial growth factor (VEGF), placental growth factor (PLGF), angiopoietin 1 (ANGPT1), ANGPT2, platelet-derived growth factor B (PDGFB), stromal-derived factor 1 (SDF1), and stem cell factor (SCF; also known as kit ligand) in cultured cells and in vivo (25–27). Administration of

AdCA5, an adenovirus encoding an engineered form of HIF-1α that is constitutively active due to the presence of a deletion and point mutations that block O_2-dependent degradation of the protein, promotes the recovery of blood flow following limb ischemia by stimulating increased angiogenesis and arteriogenesis (27, 28).

Whereas transgene-directed expression of VEGF in the skin induces both hypervascularization and hyperpermeability (29), expression of a constitutively active form of HIF-1α from the same transgene construct induces hypervascularization without hyperpermeability (30). It is likely that this difference is due to the HIF-1-mediated expression of additional angiogenic factor(s) that suppress the effect of VEGF on vascular permeability. Administration of VEGF is not sufficient to induce vascularization in the superficial capillary layer of the retina, whereas administration of AdCA5 induces robust vascularization, which is associated with increased expression of mRNAs encoding VEGF, PLGF, ANGPT1, ANGPT2, and PDGFB (25). These results are consistent with the conclusion from earlier studies that physiological angiogenesis occurs through the action of multiple angiogenic growth factors (31–33). Remarkably, these factors appear to be coordinately regulated by HIF-1.

In addition to its role in stimulating angiogenic growth factor expression in hypoxic tissues, HIF-1 plays important cell-autonomous roles in endothelial cells (34–36). HIF-1 is also required for the expression of VEGFR1 in bone marrow-derived mesenchymal stem cells (MSCs) and for the chemotactic migration of MSCs toward a gradient of VEGF or PLGF (37).

14.5 HIF-1 Is Required for Carotid Body-Mediated Responses to Continuous Hypoxia

Hif1a$^{-/-}$ mouse embryos, which are homozygous for a null (knockout) allele at the *Hif1a* locus, completely lack HIF-1α expression and die between embryonic day 9.5 and 10.5 with cardiac malformations, neural tube defects, vascular regression, and massive cell death (23, 24, 38, 39). In contrast, heterozygous-null *Hif1a*$^{+/-}$ mice, which have a partial deficiency of HIF-1α, develop normally but manifest impaired responses to hypoxia or ischemia. A particularly dramatic example is the loss of O_2 sensing in the carotid body of *Hif1a*$^{+/-}$ mice (40). Although the carotid bodies are anatomically and histologically normal and depolarize normally in response to cyanide application, they show essentially no response to hypoxia. Thus, partial HIF-1α deficiency in the carotid body results in a complete loss of the ability to sense and/or respond to changes in the arterial PO_2 by stimulation of the CNS cardiorespiratory centers. The HIF-1 target genes that are critical for O_2 sensing and/or efferent responses by the carotid body have not been identified. Remarkably, in the intact animal, other chemoreceptors are less sensitive to *Hif1a* gene dosage and compensate for the loss of carotid body activity in *Hif1a*$^{+/-}$ mice (40).

14.6 HIF-1 Is Required for Carotid Body-Mediated Responses to Intermittent Hypoxia

Sleep-disordered breathing with recurrent apnea is a major cause of morbidity and mortality in the US population, affecting an estimated 18 million people. In this condition, transient repetitive episodes of apnea (cessation of breathing) result in periodic hypoxemia (decreased PO_2 in arterial blood). In severely affected patients, the frequency of apnea may exceed 60 episodes per hour and O_2 saturation of blood hemoglobin can be reduced to as low as 50%. Patients with sleep apnea have a greatly increased risk for the development of systemic hypertension and its sequelae. Sleep apnea results in both chronic intermittent hypoxia (CIH) and chronic intermittent hypercapnia. An important advance in the field of sleep apnea research was the demonstration that exposure of rats to CIH was sufficient to induce systemic hypertension (41). Studies in humans and rodents suggested that the carotid body, which is located at the bifurcation of the common carotid artery and is the primary chemoreceptor for detecting changes in arterial PO_2, mediates reflex increases in the activity of the sympathetic nervous system that result in elevated blood pressure (42). Rats in which the carotid bodies were surgically denervated or in which the sympathetic nervous system was inhibited by administration of 6-hydroxydopamine showed no increase in blood pressure in response to CIH (43, 44).

To investigate the physiological significance of CIH-induced HIF-1 transcriptional activity in intact animals, wild type and heterozygous $Hif1a^{+/-}$ littermate male mice were exposed to either CIH (15 s of hypoxia followed by 5 min of normoxia × 9 episodes/h × 8 h/day × 10 days) or to normoxia (controls). Wild-type mice exposed to CIH exhibited the following: augmented hypoxic ventilatory response; long-term facilitation (LTF) of breathing; enhanced carotid body response to graded hypoxia and sensory LTF; increased diastolic, systolic, and mean arterial blood pressures; and elevated plasma norepinephrine levels. In striking contrast, in $Hif1a^{+/-}$ mice exposed to CIH, carotid body responses to hypoxia were absent and all measured cardiorespiratory responses were either absent or markedly attenuated (45).

Immunoblot analysis of cerebral cortical tissue lysates prepared from normoxic wild type and heterozygous $Hif1a^{+/-}$ littermate mice revealed that the heterozygotes manifested a partial deficiency in the expression of HIF-1α protein as expected. Analysis of samples that were prepared from mice exposed to CIH revealed a marked increase in HIF-1α protein expression in wild-type mice (45). In contrast, $Hif1a^{+/-}$ littermate mice showed no increase in HIF-1α protein levels in response to CIH. Thus, the virtually complete absence of ventilatory and cardiovascular responses to CIH in $Hif1a^{+/-}$ mice could be attributed to the failure to induce HIF-1α protein expression in these mice.

In addition to the development of systemic hypertension, individuals with intermittent hypoxia due to sleep apnea develop a metabolic syndrome consisting of insulin resistance, hypercholesterolemia, and hypertriglyceridemia. These responses are also observed in wild-type mice exposed to CIH for 5 days, whereas in *Hifla*[+/-] mice the development of hypertriglyceridemia was impaired and was associated with impaired activation of sterol response element binding protein 1, a key activator of triglyceride synthesis (46).

14.7 Role of HIF-1 In Cerebral Preconditioning Phenomena

When adult rats are subjected to permanent middle cerebral artery occlusion, HIF-1α mRNA is induced in the penumbra or viable tissue surrounding the infarction (47). The induction of HIF-1α mRNA is temporally and spatially correlated with the expression of mRNAs encoding glucose transporter 1 and the glycolytic enzymes aldolase A, lactate dehydrogenase A, phosphofructokinase L, and pyruvate kinase M, which are all known HIF-1 target genes (23). These data suggest that induction of glycolytic metabolism may promote the survival of neurons within the penumbra. Colocalization of HIF-1α and VEGF expression has also been demonstrated in the penumbra and is associated with neovascularization (48).

When newborn rats are subjected to permanent left common carotid artery occlusion and exposed to 8% O_2, cerebral infarction occurs in the hemisphere ipsilateral to the occlusion. Rats exposed to 8% O_2 for 3h and then subjected to carotid occlusion and hypoxia 24h later are protected against cerebral infarction (49). Exposure of rats to hypoxia alone induces HIF-1α protein expression throughout the brain, whereas combined carotid occlusion and hypoxia result in decreased HIF-1α expression in the ipsilateral cortex and a striking induction within the microvasculature of the ischemic brain (50). The physiological significance of this dramatic alteration in HIF-1α expression remains to be determined.

Significant protection against brain injury can also be achieved by injecting rats with cobalt chloride or desferrioxamine, which are known inducers of HIF-1 activity (51), 24h prior to an hypoxic-ischemic insult (50). The finding that exposure of experimental animals to hypoxia or iron chelators induces HIF-1α expression in the brain and induces protection against ischemia-induced cerebral infarction has been confirmed by several subsequent studies (52–56). Given the hundreds of target genes regulated by HIF-1, the basis for its protective effects is likely to be multifactorial. HIF-1-mediated protection against ischemic injury may be attributable in part to its ability to stimulate glycolytic metabolism as a means of generating ATP under hypoxic conditions and its activation of genes encoding angiogenic growth factors to stimulate recovery of blood flow to ischemic tissue.

Studies in animal models have revealed that one of the major mechanisms by which HIF-1 promotes tissue survival in the ischemic central nervous system is through activation of the *EPO* gene (57–59). The protective effect of rhEPO was observed in gerbil brain exposed to transient bilateral carotid occlusion, even when it was administrated systemically after reperfusion (60). rhEPO can penetrate the blood–brain barrier through blood endothelial cells, and systemic administration of EPO before or up to 6h after middle cerebral artery occlusion reduces infarct volume by approximately 50–75% (61).

Binding of EPO to EPOR results in the activation of the phosphatidylinositol 3-kinase/AKT signaling pathway. The inhibition of PI3K with wortmannin blocks the effects of EPO on the recovery of cardiac function and cell apoptosis after ischemia-reperfusion (62). Systemic administration of rhEPO dramatically reduces brain infarct size and cell apoptosis through Akt activation after middle cerebral artery occlusion (63). Activated Akt phosphorylates Bcl-2 family proteins, leading to inactivation of proapoptotic Bad and activation of antiapoptotic Bcl-xL (64, 65).

A clinical trial of recombinant human EPO (rhEPO) therapy for acute stroke has also provided evidence of a protective effect (66). The advantage of rhEPO therapy is the rapid mechanism of action in contrast to induction of HIF-1 activity by treatment with iron chelators or other inhibitors of prolyl hydroxylase activity, which require several hours for induction of HIF-1 activity and the transcription and translation of protein products of HIF-1 target genes. On the other hand, HIF-1 may mediate multiple adaptive responses to ischemia that are independent of EPO. For example, HIF-1α is necessary for the acute phase of preconditioning in the isolated perfused heart (67), which is unlikely to involve EPO due to the rapidity of the protective response and the absence of EPO expression in the heart. Thus, treatment of stroke patients with a combination of one or more short acting agents, such as rhEPO and/or other survival factors, along with the administration of a low molecular weight HIF-1 inducer capable of inducing a broader adaptive response, may provide the greatest therapeutic benefit.

References

1. Wang GL, Semenza GL. Purification and characterization of hypoxia-inducible factor 1. J Biol Chem 1995;270:1230–1237.
2. Wang GL, Jiang BH, Rue EA, Semenza GL. Hypoxia-inducible factor 1 is a basic-helix-loop-helix-PAS heterodimer regulated by cellular O_2 tension. Proc Natl Acad Sci USA 1995; 92:5510–5514.
3. Salceda S, Caro J. Hypoxia-inducible factor 1α (HIF-1α) protein is rapidly degraded by the ubiquitin–proteasome system under normoxic conditions: its stabilization by hypoxia depends on redox-induced changes. J Biol Chem 1997;272:22642–22647.
4. Maxwell PH, Wiesener MS, Chang GW, Clifford SC, Vaux EC, Cockman ME, Wykoff CC, Pugh CW, Maher ER, Ratcliffe PJ. The tumour suppressor protein VHL targets hypoxia-inducible factors for oxygen-dependent proteolysis. Nature 1999;399:271–275.
5. Ivan M, Kondo K, Yang H, Kim W, Valiando J, Ohh M, Salic A, Asara JM, Lane WS, Kaelin WG Jr. HIFα targeted for VHL-mediated destruction by proline hydroxylation: implications for O_2 sensing. Science 2001;292:464–468.
6. Jaakkola P, Mole DR, Tian YM, Wilson MI, Gielbert J, Gaskell SJ, Kriegsheim Av, Hebestreit HF, Mukherji M, Schofield CJ, Maxwell PH, Pugh CW, Ratcliffe PJ. Targeting of HIF-α to the von Hippel-Lindau ubiquitylation complex by O_2-regulated prolyl hydroxylation. Science 2001;292:468–472.
7. Yu F, White SB, Zhao Q, Lee FS. HIF-1α binding to VHL is regulated by stimulus-sensitive proline hydroxylation. Proc Natl Acad Sci USA 2001;98:9630–9635.
8. Schofield CJ, Ratcliffe PJ. Signalling hypoxia by HIF hydroxylases. Biochem Biophys Res Commun 2005;338:617–626.
9. Baek JH, Mahon PC, Oh J, Kelly B, Krishnamachary B, Pearson M, Chan DA, Giaccia AJ, Semenza GL. OS-9 interacts with hypoxia-inducible factor 1α and prolyl hydroxylases to promote oxygen-dependent degradation of HIF-1α. Mol Cell 2005;17:503–512.

10. Baek JH, Liu YV, McDonald KR, Wesley JB, Hubbi ME, Byun H, Semenza GL. SSAT2 is an essential component of the ubiquitin ligase complex that regulates HIF-1α. J Biol Chem 2007;282:23572–23580.
11. Peet D, Linke S. Regulation of HIF: asparaginyl hydroxylation. Novartis Found Symp 2006;272:37–49.
12. Tian H, McKnight SL, Russell DW. Endothelial PAS domain protein 1 (EPAS1), a transcription factor selectively expressed in endothelial cells. Genes Dev 1997;11:72–82.
13. Hu CJ, Sataur A, Wang L, Chen H, Simon MC. The N-terminal transactivation domain confers target gene specificity of hypoxia-inducible factors HIF-1α and HIF-2α. Mol Biol Cell 2007;18:4528–4542.
14. Raval RR, Lau KW, Tran MG, Sowter HM, Mandriota SJ, Li JL, Pugh CW, Maxwell PH, Harris AL, Ratcliffe PJ. Contrasting properties of hypoxia-inducible factor 1 (HIF-1) and HIF-2 in von Hippel-Lindau-associated renal cell carcinoma. Mol Cell Biol 2005;25:5675–5686.
15. Park SK, Dadak AM, Haase VH, Fontana L, Giaccia AJ, Johnson RS. Hypoxia-induced gene expression occurs solely through the action of hypoxia-inducible factor 1α (HIF-1α): role of cytoplasmic trapping of HIF-2α. Mol Cell Biol 2003;23:4959–4971.
16. Gu YZ, Moran SM, Hogenesch JB, Wartman L, Bradfield CA. Molecular characterization and chromosomal localization of a third α-class hypoxia inducible factor subunit, HIF3α. Gene Exp 1998;7:205–213.
17. Makino Y, Cao R, Svensson K, Bertilsson G, Asman M, Tanaka H, Cao Y, Berkenstam A, Poellinger L. Inhibitory PAS domain protein is a negative regulator of hypoxia-inducible gene expression. Nature 2001;414:550–554.
18. Makino Y, Kanopka A, Wilson WJ, Tanaka H, Poellinger L. Inhibitory PAS domain protein (IPAS) is a hypoxia-inducible splicing variant of the hypoxia-inducible factor-3α locus. J Biol Chem 2002;277:32405–32408.
19. Semenza GL, Wang GL. A nuclear factor induced by hypoxia via de novo protein synthesis binds to the human erythropoietin gene enhancer at a site required for transcriptional activation. Mol Cell Biol 1992;12:5447–5454.
20. Yu AY, Shimoda LA, Iyer NV, Huso DL, Sun X, McWilliams R, Beaty T, Sham JS, Wiener CM, Sylvester JT, Semenza GL. Impaired physiological responses to chronic hypoxia in mice partially deficient for hypoxia-inducible factor 1α. J Clin Invest 1999;103:691–696.
21. Cai Z, Manalo DJ, Wei G, Rodriguez ER, Fox-Talbot K, Lu H, Zweier JL, Semenza GL. Hearts from rodents exposed to intermittent hypoxia or erythropoietin are protected against ischemia-reperfusion injury. Circulation 2003;108:79–85.
22. Rankin EB, Biju MP, Liu Q, Unger TL, Rha J, Johnson RS, Simon MC, Keith B, Haase VH. Hypoxia-inducible factor-2 (HIF-2) regulates hepatic erythropoietin in vivo. J Clin Invest 2007;117:1068–1077.
23. Iyer NV, Kotch LE, Agani F, Leung SW, Laughner E, Wenger RH, Gassmann M, Gearhart JD, Lawler AM, Yu AY, Semenza GL. Cellular and developmental control of O_2 homeostasis by hypoxia-inducible factor 1α. Genes Dev 1998;12:149–162.
24. Ryan HE, Lo J, Johnson RS. HIF-1α is required for solid tumor formation and embryonic vascularization. EMBO J 1998;17:3005–3015.
25. Kelly BD, Hackett SF, Hirota K, Oshima Y, Cai Z, Berg-Dixon S, Rowan A, Yan Z, Campochiaro PA, Semenza GL. Cell type-specific regulation of angiogenic growth factor gene expression and induction of angiogenesis in nonischemic tissue by a constitutively active form of hypoxia-inducible factor 1. Circ Res 2003;93:1074–1081.
26. Ceradini DJ, Kulkarni AR, Callaghan MJ, Tepper OM, Bastidas N, Kleinman ME, Capla JM, Galiano RD, Levine JP, Gurtner GC. Progenitor cell trafficking is regulated by hypoxic gradients through HIF-1 induction of SDF-1. Nat Med 2004;10:858–864.
27. Bosch-Marcé M, Okuyama H, Wesley JB, Sarkar K, Kimura H, Liu YV, Zhang H, Strazza M, Rey S, Savino L, Zhou YF, McDonald KR, Na Y, Vandiver S, Rabi A, Shaked Y, Kerbel R, Lavallee T, Semenza GL. Effects of aging and hypoxia-inducible factor-1 activity on angiogenic cell mobilization and recovery of perfusion after limb ischemia. Circ Res 2007;101:1310–1318.

28. Patel TH, Kimura H, Weiss CR, Semenza GL, Hofmann LV. Constitutively active HIF-1α improves perfusion and arterial remodeling in an endovascular model of limb ischemia. Cardiovasc Res 2005;68:144–154.
29. Thurston G, Suri C, Smith K, McClain J, Sato TN, Yancopoulos GD, McDonald DM. Leakage-resistant blood vessels in mice transgenically overexpressing angiopoietin-1. Science 1999;286:2511–2514.
30. Elson DA, Thurston G, Huang LE, Ginzinger DG, McDonald DM, Johnson RS, Arbeit JM. Induction of hypervascularity without leakage or inflammation in transgenic mice overexpressing hypoxia-inducible factor-1α. Genes Dev 2001;15:2520–2532.
31. Carmeliet P, Moons L, Luttun A, Vincenti V, Compernolle V, De Mol M, Wu Y, Bono F, Devy L, Beck H, Scholz D, Acker T, DiPalma T, Dewerchin M, Noel A, Stalmans I, Barra A, Blacher S, Vandendriessche T, Ponten A, Eriksson U, Plate KH, Foidart JM, Schaper W, Charnock-Jones DS, Hicklin DJ, Herbert JM, Collen D, Persico MG. Synergism between vascular endothelial growth factor and placental growth factor contributes to angiogenesis and plasma extravasation in pathological conditions. Nat Med 2001;7:575–583.
32. Luttun A, Tjwa M, Moons L, Wu Y, Angelillo-Scherrer A, Liao F, Nagy JA, Hooper A, Priller J, De Klerck B, Compernolle V, Daci E, Bohlen P, Dewerchin M, Herbert JM, Fava R, Matthys P, Carmeliet G, Collen D, Dvorak HF, Hicklin DJ, Carmeliet P. Revascularization of ischemic tissues by PlGF treatment, and inhibition of tumor angiogenesis, arthritis and atherosclerosis by anti-Flt1. Nat Med 2002;8:831–840.
33. Cao R, Brakenhielm E, Pawliuk R, Wariaro D, Post MJ, Wahlberg E, Leboulch P, Cao Y. Angiogenic synergism, vascular stability and improvement of hind-limb ischemia by a combination of PDGF-BB and FGF-2. Nat Med 2003;9:604–613.
34. Tang N, Wang L, Esko J, Giordano FJ, Huang Y, Gerber HP, Ferrara N, Johnson RS. Loss of HIF-1α in endothelial cells disrupts a hypoxia-driven VEGF autocrine loop necessary for tumorigenesis. Cancer Cell 2004;6:485–495.
35. Manalo DJ, Rowan A, Lavoie T, Natarajan L, Kelly BD, Ye SQ, Garcia JG, Semenza GL. Transcriptional regulation of vascular endothelial cell responses to hypoxia by HIF-1. Blood 2005;105:659–666.
36. Calvani M, Rapisarda A, Uranchimeg B, Shoemaker RH, Melillo G. Hypoxic induction of an HIF-1α-dependent bFGF autocrine loop drives angiogenesis in human endothelial cells. Blood 2006;107:2705–2712.
37. Okuyama H, Krishnamachary B, Zhou YF, Nagasawa H, Bosch-Marcé M, Semenza GL. Expression of vascular endothelial growth factor receptor 1 in bone marrow-derived mesenchymal cells is dependent on hypoxia-inducible factor 1. J Biol Chem 2006;281:15554–15563.
38. Kotch LE, Iyer NV, Laughner E, Semenza GL. Defective vascularization of HIF-1α-null embryos is not associated with VEGF deficiency but with mesenchymal cell death. Dev Biol 1999;209:254–267.
39. Compernolle V, Brusselmans K, Franco D, Moorman A, Dewerchin M, Collen D, Carmeliet P. Cardia bifida, defective heart development and abnormal neural crest migration in embryos lacking hypoxia-inducible factor-1alpha. Cardiovasc Res 2003;60:569–579.
40. Kline DD, Peng YJ, Manalo DJ, Semenza GL, Prabhakar NR. Defective carotid body function and impaired ventilatory responses to chronic hypoxia in mice partially deficient for hypoxia-inducible factor 1α. Proc Natl Acad Sci USA 2002;99:821–826.
41. Fletcher EC, Lesske J, Qian W, Miller CC III, Unger T. Repetitive, episodic hypoxia causes diurnal elevation of blood pressure in rats. Hypertension 1992;19:555–561.
42. Peng YJ, Prabhakar NR. Effect of two paradigms of chronic intermittent hypoxia on carotid body sensory activity. J Appl Physiol 2004;96:1236–1242.
43. Fletcher EC, Lesske J, Behm R, Miller CC III, Stauss H, Unger T. Carotid chemoreceptors, systemic blood pressure, and chronic episodic hypoxia mimicking sleep apnea. J Appl Physiol 1992;72:1978–1984.
44. Fletcher EC, Lesske J, Culman J, Miller CC, Unger T. Sympathetic denervation blocks blood pressure elevation in episodic hypoxia. Hypertension 1992;20:612–619.

45. Peng YJ, Yuan G, Ramakrishnan D, Sharma SD, Bosch-Marcé M, Kumar GK, Semenza GL, Prabhakar NR. Heterozygous HIF-1α deficiency impairs carotid body-mediated cardiorespiratory responses and reactive oxygen species generation in mice exposed to chronic intermittent hypoxia. J Physiol 2006;577:705–716.
46. Li J, Bosch-Marcé M, Nanayakkara A, Savransky V, Fried SK, Semenza GL, Polotsky VY. Altered metabolic responses to intermittent hypoxia in mice with partial deficiency of hypoxia-inducible factor-1α. Physiol Genomics 2006;25:450–457.
47. Bergeron M, Yu AY, Solway KE, Semenza GL, Sharp FR. Induction of hypoxia-inducible factor-1 (HIF-1) and its target genes following focal ischaemia in rat brain. Eur J Neurosci 1999;11:4159–4170.
48. Marti HJ, Bernaudin M, Bellail A, Schoch H, Euler M, Petit E, Risau W. Hypoxia-induced vascular endothelial growth factor expression precedes neovascularization after cerebral ischemia. Am J Pathol 2000;156:965–976.
49. Gidday JM, Fitzgibbons JC, Shah AR, Park TS. Neuroprotection from ischemic brain injury by hypoxic preconditioning in the neonatal rat. Neurosci Lett 1994;168:221–224.
50. Bergeron M, Gidday JM, Yu AY, Semenza GL, Ferriero DM, Sharp FR. Role of hypoxia-inducible factor-1 in hypoxia-induced ischemic tolerance in neonatal rat brain. Ann Neurol 2000;48:285–296.
51. Wang GL, Semenza GL. Desferrioxamine induces erythropoietin gene expression and hypoxia-inducible factor 1 DNA-binding activity: implications for models of hypoxia signal transduction. Blood 1993;82:3610–3615.
52. Chavez JC, LaManna JC. Activation of hypoxia-inducible factor-1 in the rat cerebral cortex after transient global ischemia: potential role of insulin-like growth factor-1. J Neurosci 2002;22:8922–8931.
53. Demougeot C, Van Hoecke M, Bertrand N, Prigent-Tessier A, Mossiat C, Beley A, Marie C. Cytoprotective efficacy and mechanisms of the liposoluble iron chelator 2,2-dipyridyl in the rat photothrombotic ischemic stroke model. J Pharmacol Exp Ther 2004;311:1080–1087.
54. Hamrick SE, McQuillen PS, Jiang X, Mu D, Madan A, Ferriero DM. A role for hypoxia-inducible factor-1α in desferoxamine neuroprotection. Neurosci Lett 2005;379:96–100.
55. Siddiq A, Ayoub IA, Chavez JC, Aminova L, Shah S, LaManna JC, Patton SM, Connor JR, Cherny RA, Volitakis I, Bush AI, Langsetmo I, Seeley T, Gunzler V, Ratan RR. Hypoxia-inducible factor prolyl 4-hydroxylase inhibition. A target for neuroprotection in the central nervous system. J Biol Chem 2005;280:41732–41743.
56. Prass K, Ruscher K, Karsch M, Isaev N, Megow D, Priller J, Scharff A, Dirnagl U, Meisel A. Desferrioxamine induces delayed tolerance against cerebral ischemia in vivo and in vitro. J Cereb Blood Flow Metab 2002;22:520–525.
57. Cerami A, Brines ML, Ghezzi P, Cerami CJ. Effects of epoetin alfa on the central nervous system. Semin Oncol 2001;28 (Suppl 8):66–70.
58. Prass K, Scharff A, Ruscher K, Löwl D, Muselmann C, Victorov I, Kapinya K, Dirnagl U, Meisel A. Hypoxia-induced stroke tolerance in the mouse is mediated by erythropoietin. Stroke 2003;34:1981–1986.
59. Liu R, Suzuki A, Guo Z, Mizuno Y, Urabe T. Intrinsic and extrinsic erythropoietin enhances neuroprotection against ischemia and reperfusion injury in vitro. J Neurochem 2006;96:1101–1110.
60. Calapai G, Marciano MC, Corica F, Allegra A, Parisi A, Frisina N, Caputi AP, Buemi M. Erythropoietin protects against brain ischemic injury by inhibition of nitric oxide formation. Eur J Pharmacol 2000;401:349–356.
61. Brines ML, Ghezzi P, Keenan S, Agnello D, de Lanerolle NC, Cerami C, Itri LM, Cerami A. Erythropoietin crosses the blood–brain barrier to protect against experimental brain injury. Proc Natl Acad Sci USA 2000;97:10526–10531.
62. Cai Z, Semenza GL. Phosphatidylinositol-3-kinase signaling is required for erythropoietin-mediated acute protection against myocardial ischemia/reperfusion injury. Circulation 2004;109:2050–2053.

63. Sirén AL, Fratelli M, Brines M, Goemans C, Casagrande S, Lewczuk P, Keenan S, Gleiter C, Pasquali C, Capobianco A, Mennini T, Heumann R, Cerami A, Ehrenreich H, Ghezzi P. Erythropoietin prevents neuronal apoptosis after cerebral ischemia and metabolic stress. Proc Natl Acad Sci USA 2001;98:4044–4049.
64. Kretz A, Happold CJ, Marticke JK, Isenmann S. Erythropoietin promotes regeneration of adult CNS neurons via Jak2/Stat3 and PI3K/AKT pathway activation. Mol Cell Neurosci 2005;29:569–579.
65. Zhande R, Karsan A. Erythropoietin promotes survival of primary human endothelial cells through PI3K-dependent, NF-kappaB-independent upregulation of Bcl-xL. Am J Physiol 2007;292:H2467–H2474.
66. Ehrenreich H, Hasselblatt M, Dembowski C, Cepek L, Lewczuk P, Stiefel M, Rustenbeck HH, Breiter N, Jacob S, Knerlich F, Bohn M, Poser W, Rüther E, Kochen M, Gefeller O, Gleiter C, Wessel TC, De Ryck M, Itri L, Prange H, Cerami A, Brines M, Sirén AL. Erythropoietin therapy for acute stroke is both safe and beneficial. Mol Med 2002;8:495–505.
67. Cai Z, Zhong H, Bosch-Marcé M, Fox-Talbot K, Wang L, Wei C, Trush MA, Semenza GL. Complete loss of ischaemic preconditioning-induced cardioprotection in mice with partialdeficiency of HIF-1α. Cardiovasc Res 2008;77:463–470.

Chapter 15
Transcriptional Response to Hypoxia in Developing Brain

Dan Zhou

Abstract Hypoxia affects gene expression in the cell. Multiple major transcription factor families are activated by hypoxia that include HIF (hypoxia-inducible factor), the factors that contribute to HIF regulation and other transcription factors and pathways that are activated by exposure to reduced oxygen. In this chapter, we summarize the current knowledge of hypoxia-responsive transcription factors and show observations regarding biological networks and pathways that contribute to long-term hypoxia adaptation through expression profiling.

Keywords: Chronic consistent hypoxia; Chronic intermittent hypoxia; Transcription factor; Expression profiling; Brain; Development

15.1 Introduction

Oxygen is essential for metazoan life. Pathological and environmental factors that affect normal blood oxygen level in the organism at early postnatal stages can affect growth and have a long-lasting effect on the structure and function of multiple organs including brain. Many childhood diseases, such as obstructive sleep apnea (OSA) (1), brochopulmonary dysplasia (BPD) (2), congenital heart disease (CHD), asthma, and sickle cell disease (SCD), can induce oxygen deprivation. Affected infants often have long-lasting pathological outcomes accompanied with cognitive and/or behavioral disturbances (3–8). The early postnatal development of central nervous system (CNS) includes the growth of axons, the formation of specific synaptic connections, and myelination, as well as a fast proliferation of glial cells. Such rapid phase of development made this organ vulnerable to hypoxic challenge. For example, children with a cyanotic congenital heart defect exhibit various neurophysiologic dysfunctions including impaired motor function, inability to sustain attention, and low academic

D. Zhou
Department of Pediatrics, University of California, San Diego, 9500 Gilman Drive,
MC0735 La Jolla, CA 92093-0735, USA
e-mail: d2zhou@ucsd.edu

achievement (9). However, the molecular mechanisms underlying the effects of long-term hypoxia on early postnatal brain development still largely remain unknown.

During the past decade, our laboratory and those of others have investigated the effects of hypoxia on brain development and function in various animal models. The aim of this chapter is to highlight potential mechanisms that mediate the effects of long-term hypoxia on the developing brain including transcriptional regulation and the adaptive or pathological outcomes. Toward that aim, we will detail results regarding the mechanisms that regulate gene expression in hypoxia and show current observations regarding biological networks and pathways that are affected by long-term O_2 deprivation through expression profiling.

15.2 Hypoxia-Responsive Transcription Factors in the Brain

Hypoxia is a strong modulator of gene expression, affecting the expression of nearly 1.0% of the genes in the genome. Many transcription factors have been found to regulate hypoxia-induced alteration of gene expression (Fig. 15.1). Among them, the hypoxia-inducible-factor (HIF)-dependent transcriptional activation is the most extensively studied response to hypoxia. The functional HIF heterodimer consists of one constitutively expressed aryl hydrocarbon receptor nuclear translocator (ARNT or HIF-1β) and one HIF-α subunit. The HIF-α subunit is subjected to biochemical and functional regulation by hypoxia. So far, three HIF-α subunits have been described: HIF-1α, HIF-2α, and HIF-3α (40–43). Levels of HIFα subunits are regulated by oxygen-dependent proteolysis that is mediated by the product of the von Hippel–Lindau (VHL) tumor suppressor gene, pVHL (44). HIF-1α levels remain very low in normoxic conditions but increase in response to hypoxia. For example, the prolines 402 and 564 of human HIF-1α are hydroxylated by HIF prolyl hydroxylase (PHD) in

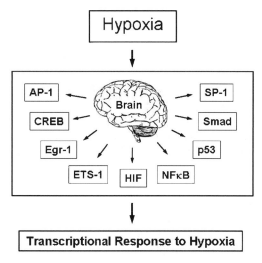

Fig. 15.1 Schematic network of hypoxia-induced transcriptional response in the brain. Multiple transcription factors are activated by hypoxia in the brain, which induces global changes in the transcriptome. This figure summarized nine transcription factors that have been found to be activated by hypoxia in the brain, which include AP-1 (10, 11), CREB (12–16), Egr-1 (17–20), ETS-1 (21), HIF (16, 22–28), NFκB (29–32), p53 (33–37), Smad (38), and SP-1 (39)

normal oxygen concentration (45, 46), and such hydroxylation promotes the interaction of HIF-1α with pVHL. pVHL is a part of a multiprotein complex that contains elongins B and C and a member of the cullin family (47–49). The interaction between pVHL and HIF-1α results in ubiquitination and proteosomal degradation of HIF-1α. A second oxygen-dependent transactivation domain of HIF is regulated in a similar manner by the asparagine hydroxylase known as factor-inhibit-HIF (FIH) (50, 51). In reduced oxygen, the hydroxylases are not active, and the oxygen-dependent proline and asparagine hydroxylation is inhibited that results in accumulation of HIF-1α. The accumulated HIF complex translocates to the nucleus, binds to consensus DNA binding elements within regulatory promoter regions (hypoxia-responsive element, HRE), and induces transcription of downstream target genes.

Recent work has demonstrated that mitochondria play an important role in regulating HIF activation. Under physiological condition, mitochondria consume approximately 90% of available oxygen to generate ATP for cellular metabolic needs (52). The remaining ~10% of oxygen is available to the cell for other processes including degradation of HIFα. In hypoxia, mitochondria act like an oxygen sink, consuming most of the available oxygen through cytochrome c oxidase, which has a high affinity to molecular oxygen (53). Thus, the viability of oxygen for oxygen-dependent HIFα hydroxylation and degradation is limited. Therefore, under hypoxic conditions, inhibition of respiration by nitric oxide (NO) or other cytochrome c oxidase inhibitors can induce HIFα degradation by redistributing available oxygen to cytosol (54, 55). Currently, accumulating evidence has shown that increase of mitochondrial reactive oxygen species (ROS) production plays a major role in HIF activation (56–58), and current efforts are directed to determine how ROS modulates PHD activity and hypoxic HIFα stabilization.

Another extensively studied hypoxia-responding transcription factor is the nuclear factor kappa-B (NFκB). As one of the central transcriptional mediators, NFκB plays a major role in innate immunity, stress responses, and cell survival and development (59–62). The family of NFκB transcription factors includes proteins that contain a highly conserved Rel homology (RH) domain. There are five members of this family: p65, cRel, and RelB, which represent the transcriptional active members, and p50 and p52, which are derived from the longer proteins p105 and p100, respectively. The NFκB family of proteins can either homodimerize or heterodimerize to form functional transcription complexes. The most common active complex is the p50–p65 dimer. In the absence of stimuli, NFκB binds to its repressor protein IκB in the cytosol. This binding of IκB protein masks the nuclear localization sequence (NLS) of NFκB and prevents the nuclear translocation of the protein. Upon stimulation, IκB is targeted for ubiquitination and degradation by specific serine phosphorylation. The NLS of NFκB is then exposed, and NFκB translocates to the nucleus where it binds to specific κB sites within the promoter regions of target genes and initiates transcription. A more dynamic model for both NFκB and IκB localization and shuttling between nucleus and cytosol has also been proposed (63). NFκB-responsive genes include those encoding inflammatory cytokines, chemokines, and cell surface adhesion molecules. A wide variety of stimuli can initiate NFκB activation including proinflammatory cytokines, bacterial products, and ultraviolet light.

The convergence point for these disparate stimuli seems to be at the level of the IκB kinases (IKKs), which are upstream of IκB phosphorylation. Hypoxia has been demonstrated to activate NFκB in a number of studies (64–68). Previous studies have identified the target genes for hypoxia-induced NFκB activation that include cyclooxygenase-2 (*COX-2*), tumor necrosis factor α (*TNFα*), interleukin-6 (*IL-6*), and macrophage inflammatory protein-2 (MIP-2). Furthermore, tyrosine phosphorylation of IκB by the Ras/Raf kinases downstream from Src has been implicated as a mechanism of NFκB activation in hypoxia (69). In addition, mitogen-activated protein kinase p42/p44 (MAPK/ERK) and PI-3 kinase have also been found to mediate hypoxia-induced transactivation of NFκB (70). Another model proposes that mitochondria-derived ROS generation in hypoxia may be responsible for NFκB activation (71). Moreover, the hypoxia-dependent transactivation of NFκB may also be coupled to redox-dependent pathways, because modulation of redox equilibrium in the cell affects the responsiveness of NFκB at the molecular level (72–74). In summary, it is clear that NFκB is a hypoxia-responsive transcription factor, and hypoxia-dependent activation can be mediated by multiple specific mechanisms.

CREB, the cyclic AMP response element binding protein, belongs to a family of leucine zipper transcription factors that are regulated by intracellular signaling mechanisms such as cAMP and Ca^{2+}. CREB regulates the expression of a diverse array of genes including the genes that are involved in inflammation, metabolism, and signal transduction. Previous studies revealed that CREB is activated through phosphorylation of Ser 133 by protein kinase A (PKA) or calmodulin kinase (CaMK) (75), and can act as an activator or a repressor of transcription for specific target genes (76). Recently, hypoxia-induced CREB-dependent gene expression has been demonstrated in vitro and in vivo, especially in ischemic diseases. For example, physiological levels of hypoxia (5% O_2, approximately 50mm Hg) rapidly induced a persistent phosphorylation of CREB on Ser133 and activated CREB-mediated transcription in PC12 cells (77). Furthermore, various studies demonstrated that cerebral ischemia triggers robust phosphorylation of CREB and CRE-mediated gene expression in neurons (78). Interestingly, hypoxia-induced CREB activation was not mediated by any of the previously known CREB kinases (e.g., PKA, CaMK, PKC, p70s6k, or MAPKAP kinase-2). Thus, the mechanism underlying hypoxia-induced CREB activation is distinct and more complex than that induced by forskolin, depolarization, or nerve growth factor. In addition, it has been shown that the strength of hypoxia-induced CREB-dependent gene expression depends on the duration and degree of hypoxic stimulus (77).

The activating protein-1 (AP-1) is another important transcription factor responding to hypoxia in the brain (79, 80). AP-1 is a pleiotropic, dimeric transcription factor that is involved in multiple neuronal functions related to neuronal development, neurodegeneration, cell death, plasticity, and repair of injury (81). AP-1 comprises members of Fos, Jun, ATF (activating transcription factors), and MAF (musculoaponeurotic fibrosarcoma) protein families that can homodimerize or heterodimerize to form the active AP-1 complex and to modulate gene expression. The complex combination of these discrete proteins provides multiple levels of gene expression control. In addition, cell type and differentiation state can dictate the phenotypic outcome, which accounts for how AP-1 can regulate apparently conflicting endpoints such as apoptosis and cell proliferation (82). Multiple factors including growth factors, proinflammatory cytokines, and UV radiation can activate AP-1. Hypoxia has also been shown to activate AP-1-dependent gene expression. Genes regulated at least partially by hypoxia-induced AP-1 activation include tyrosine hydroxylase (83), VEGF (39), and endothelial NOS (eNOS) (84). Previous works also demonstrated that AP-1 cooperates with other transcription factors such as HIF-1, GATA-2, NF-1, and NFκB to complement the activation of hypoxia-sensitive genes (83, 85–87). Thus, AP-1 may act as an important facilitator of hypoxia-induced gene expression through interaction with other transcription factors. The mechanism(s) underlying hypoxia-induced AP-1 activation is currently not fully understood. The Jun N-terminal kinase (JNK) and MAPK pathways have been proposed to mediate, at least in part, AP-1 activation under hypoxia (88, 89), but the oxygen-sensing mechanism upstream of JNK or MAPK remains unknown. Furthermore, AP-1 is a redox-sensitive transcription factor. As it has been suggested that hypoxia alters redox environment in the cell, such altered redox level may induce AP-1-dependent transcription in hypoxia (84). The other proposed mechanisms underlying hypoxia-induced AP-1 activation include nonreceptor tyrosine kinases such as Src and Ras, which propagate the hypoxic signal from G protein-coupled receptors (90).

Although a great deal of work has been done to illustrate the biological role of transcription factors in hypoxia-induced gene expression in the brain, many questions still remain. For example, further studies are required to address the diversity of oxygen-sensing mechanisms and how these sensors regulate the gene transcription machinery. It is also of particular interest to understand the mechanism(s) underlying specific hypoxia response in different cell types and regions in the brain. Furthermore, hypoxia-induced activation of transcription factors will induce alterations in global gene expression. To understand the specific role of hypoxia in brain development and function, it is important to obtain information on global changes in the transcriptome, such as which genes are activated and which are inactivated. In the following section, we will summarize results obtained from gene expression profiling of developing brain following chronic intermittent (CIH) or chronic consistent (CCH) hypoxia treatment.

15.3 Transcriptional Response to Hypoxia in Developing Brain

Hypoxia induces significant changes in gene expression in immature brain. Using long-oligo expression arrays, we have determined the changes in gene expression in the cortex and hippocampus following CIH or CCH in developing mouse brain (91). We found that, in the cortex, the expression levels of 80 transcripts following CIH (16 upregulated and 64 downregulated) and 137 transcripts following CCH (34 upregulated and 103 downregulated) were significantly altered. In the hippocampus, however, 71 transcripts following CIH (57 upregulated and 14 downregulated), and 80 transcripts following CCH (69 upregulated and 11 downregulated) were differentially expressed. In contrast to the cortex, most of the significant changes in the hippocampal region were upregulations (Fig. 15.2). Many of the genes identified in this

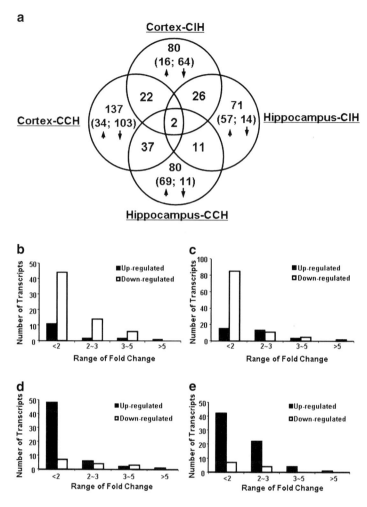

Fig. 15.2 Gene expression profiles in developing mouse brain following CCH or CIH. (**a**) Venn diagram illustrating the number of significantly altered genes and the overlaps of genes between different treatments and regions. (**b**) Gene expression profile of cortex following CIH. (**c**) Gene expression profile of cortex following CCH. (**d**) Gene expression profile of hippocampus following CIH. (**e**) Gene expression profile of hippocampus following CCH (from (91), with permission)

study had not previously been linked to hypoxia. Out of all the altered transcripts, 22 transcripts responded to both CIH and CCH in the cortex (2 upregulated and 19 downregulated) (Table 15.1). In the hippocampal region, 11 transcripts were found to be significantly altered following both CIH and CCH, and most of them were upregulated (Table 15.2). Interestingly, there were more transcripts that were significantly altered following CCH, as compared with CIH. For example, following CIH, 26 transcripts were altered in both cortical and hippocampal regions (12 upregulated and 14 downregulated) (Table 15.3), whereas, following CCH, 37 transcripts changed

Table 15.1 List of genes altered by both CIH and CCH in cortex (from (91), with permission)

		CIH		CCH	
GenBank ID	Symbol	Fold change	Diff_score	Fold change	Diff_score
NM_010380	H2-D1	5.74	45.61	4.81	41.61
NM_080445.2	B3galt6	3.17	193.92	2.85	92.55
XM_354704.1	Igh-1a	−1.57	−22.37	−1.91	−45.27
XM_130628.2	Manbal	−1.59	−67.08	−1.65	−64.47
NM_019580.3	Mir16	−1.66	−31.46	−1.75	−81.20
NM_177346.1	Gpr149	−1.75	−26.71	−1.76	−25.44
NM_152817.2	Ttc27	−1.81	−29.09	−5.40	−104.01
NM_134134.1	A630042L21Rik	−1.84	−108.16	−1.98	−116.82
NM_025542.1	2410001C21Rik	−1.85	−25.35	−2.09	−45.21
NM_181066.1	AA881470	−1.89	−98.96	−2.16	−148.51
NM_010118.1	Egr2	−1.98	−25.75	−2.03	−28.83
NM_172458.2	9030612M13Rik	−2.00	−33.41	−2.72	−108.68
XM_354710.1	Igh-6	−2.46	−32.99	−2.65	−43.37
XM_354710.1	Igh-6	−2.75	−72.18	−3.41	−137.54
XM_355058.1	LOC381142	−2.95	−106.20	−2.71	−110.71
NM_028097.2	2010300G19Rik	−2.95	−40.21	−2.13	−95.64
NM_010656.1	Sspn	−3.19	−134.32	−3.27	−137.70
NM_178592.2	Bat5	−4.40	−54.46	−1.96	−115.00

Table 15.2 List of genes altered by both CIH and CCH in hippocampus (from (91), with permission)

		CIH		CCH	
GenBank ID	Symbol	Fold change	Diff_score	Fold change	Diff_score
NM_010330.2	Emb	4.79	262.67	4.99	316.01
NM_026993.1	Ddah1	2.11	102.32	1.98	75.51
NM_026993	Ddah1	1.97	64.58	1.69	55.63
NM_011448	Sox9	1.88	24.70	1.71	43.74
NM_011708.2	Vwf	1.76	39.59	2.81	21.55
NM_007697.1	Chl1	1.73	38.74	1.69	29.63
NM_016809	Rbm3	1.69	54.07	1.78	66.14
AK036731	Trrp7	1.61	20.08	2.61	96.37
NM_025644.3	Exosc1	1.57	39.44	1.51	30.21
NM_008056	Fzd6	1.54	25.65	1.54	27.61
NM_152817.2	Ttc27	−1.69	−29.37	−2.69	−30.75
NM_010656.1	Sspn	−2.64	−50.80	−2.45	−47.77

their expression in both cortical and hippocampal regions (26 upregulated and 11 downregulated) (Table 15.4). Interestingly, two genes were found to be altered (i.e., downregulated) in *both* brain regions following *both* treatments. Both of them encode membrane-associated proteins. One encodes for sarcospan (Sspn) and the other encodes for tetratricopeptide repeat domain 27 (Ttc27). Further studies in cultured cells suggested that downregulation of Sspn induces cell death in glial cells without a profound effect on cell proliferation (91). But the specific functions and cellular expression of Sspn in the brain need to be further elucidated.

Table 15.3 List of genes altered in both cortex and hippocampus following CIH (from (91), with permission)

GenBank ID	Symbol	Cortex Fold change	Cortex Diff_score	Hippocampus Fold change	Hippocampus Diff_score
NM_010380	H2-D1	5.74	45.61	6.12	83.65
NM_080445.2	B3galt6	3.17	193.92	3.44	105.94
NM_010661.1	Krt1–12	2.80	34.97	2.06	28.19
NM_024228.1	1110015E22Rik	1.84	46.20	2.25	73.07
XM_484075	B230105J10	1.76	26.39	2.88	62.35
NM_011448	Sox9	1.71	26.43	1.88	24.70
NM_026993.1	Ddah1	1.64	38.15	2.11	102.32
NM_026993	Ddah1	1.59	65.27	1.97	64.58
NM_008112.2	Gdi3	1.55	71.79	1.72	51.79
XM_283954.2		1.54	21.43	1.90	71.65
NM_010569.2	Invs	1.51	48.16	1.53	23.23
NM_019823.2	Cyp2d22	−1.61	−49.98	−1.52	−22.78
NM_025299.1	Txnl4	−1.66	−48.44	−1.60	−22.49
NM_152817.2	Ttc27	−1.81	−29.09	−1.69	−29.37
NM_134134.1	A630042L21Rik	−1.84	−108.16	−1.66	−29.20
NM_181066.1	AA881470	−1.89	−98.96	−1.67	−30.58
NM_025390.2	Pop4	−2.26	−32.58	−2.43	−28.77
NM_025374.2	Glo1	−2.33	−46.80	−2.05	−44.72
NM_010656.1	Sspn	−3.19	−134.32	−2.64	−50.80
NM_008073.1	Gabrg2	−3.25	−183.93	−2.28	−27.41
NM_178592.2	Bat5	−4.40	−54.46	−4.31	−44.59

Based upon interactions between the significantly altered genes, the significant changes in both brain regions following CIH or CCH were further categorized into functional networks. Distinct functional networks were identified in the brain regions with different treatments (Table 15.5). In the cortical region, CIH induced significant alterations in three distinct networks, which regulate CNS function, cell death, and various neurological diseases or conditions. CCH induced significant alterations in five networks that regulate cell signaling, and CNS development and metabolism. On the other hand, in the hippocampal region, although CIH and CCH both induced an upregulation of many genes, the alterations induced by CIH were mainly in the networks that regulate inflammatory and immune responses that might lead to a variety of neurological diseases. In contrast, the alterations induced by CCH were mainly in the networks that contribute to oxygen transport and cellular rearrangement that might lead to an adaptation in the brain. These results suggest different impacts of CIH and CCH on the brain. It seems that CCH induced more of an *adaptive response* whereas CIH induced more of a *pathological response*. Indeed, CCH, but not CIH, induced a coordinated upregulation of aminolevulinic acid synthase 2 (erythroid, Alas2) and globins in both cortical and hippocampal regions, which demonstrated a synchronized upregulation of heme and globin production and hence an increase in oxygen transport capacity that can cope with oxygen deprivation in the brain (Fig. 15.3). On the other hand, glyoxalase 1 (Glo1)

Table 15.4 List of genes altered in both cortex and hippocampus following CCH (from (91), with permission)

GenBank ID	Symbol	Cortex Fold change	Cortex Diff_score	Hippocampus Fold change	Hippocampus Diff_score
NM_026218.1	Fgfr1op2	4.77	70.23	6.07	268.93
NM_139306.1	CRG-L1	3.47	117.44	3.69	171.41
NM_009063.2	Rgs5	2.97	95.19	3.63	116.20
NM_009653.1	Alas2	2.83	25.76	2.40	143.39
NM_145511.1	BC003331	2.46	36.69	3.57	89.68
NM_008218.1	Hba-a1	2.29	52.31	2.57	75.31
AK005442	Hbb-b1	2.82	32.76	2.98	60.20
AK011053	Hbb-b1	2.40	48.07	2.94	48.50
AK011069	Hbb-b1	2.34	58.08	2.90	74.30
AK011067	Hbb-b1	2.34	60.28	2.90	80.11
AK011016	Hbb-b1	2.24	60.09	2.82	43.43
AK002258	Hbb-b1	2.22	51.32	2.81	28.58
AK003096	Hbb-b1	2.21	49.51	2.71	127.11
AK010873	Hbb-b1	2.19	54.74	2.62	52.96
AK010993	Hbb-b1	2.12	49.13	2.58	60.64
NM_016956.2	Hbb-b2	1.98	20.32	2.71	54.23
NM_013655.2	Cxcl12	1.92	54.25	2.60	180.28
NM_021323.1	Usp29	1.89	49.15	2.01	74.22
NM_010738.2	Ly6a	1.87	27.77	2.18	49.32
AK036671	9830148G24Rik	1.79	23.16	2.44	23.11
NM_016809	Rbm3	1.70	20.90	1.78	66.14
NM_011708.2	Vwf	1.70	41.11	2.81	21.55
AK011092	2510042H12Rik	1.66	24.03	1.69	20.76
XM_110620.2	Slc25a18	1.61	31.14	2.29	37.96
NM_007426.2	Agpt2	1.52	26.86	1.74	42.22
NM_021388.2	Extl2	1.50	28.42	2.00	35.22
NM_198653.1	2010002H18Rik	−1.60	−22.60	−1.55	−24.98
NM_178086.2	Fa2h	−1.60	−25.02	−1.63	−28.45
NM_019580.3	Mir16	−1.75	−81.20	−1.88	−21.82
XM_355795.1	1190003M12Rik	−1.83	−40.88	−1.79	−29.84
NM_172479.1	Slc38a5	−1.89	−59.07	−1.57	−24.68
NM_009824.1	Cbfa2t3h	−1.96	−28.69	−1.72	−32.27
NM_028097.2	2010300G19Rik	−2.13	−95.64	−1.64	−26.33
NM_030707	Msr2	−3.19	−97.70	−2.65	−69.01
NM_010656.1	Sspn	−3.27	−137.70	−2.45	−47.77
NM_033080.1	D7Rp2e	−3.50	−370.77	−3.01	−78.82
NM_152817.2	Ttc27	−5.40	−104.01	−2.69	−30.75

was downregulated in both hippocampus and cortex following CIH. As the glyoxalase system plays an important role in methylglyoxal detoxification that reduces the possible occurrence of oxidative stress in cells (92–94), downregulation of this enzyme might enhance the cytotoxicity of methylglyoxal, therefore, causing CNS injury in the developing brain following CIH treatment.

Table 15.5 List of representative significant functional networks identified by Ingenuity Pathways Analysis in cortex or hippocampus following CIH or CCH treatment ($p < 0.00001$) (from (91), with permission)

Molecules in network	Focus molecules	Top functions
Cortex: CIH		
β-estradiol, CDKN2A, CTNNB1, CXCL12, CYP2D6, DCTN4, DRD5, ERBB2, FBL, GABRA1, GABRA3, GABRA4, GABRA5, GABRA6, GABRA2 (includes EG:2555), GABRG2, GABRG3, GDI2, GDPD3, GLO1, GNL2, GPER, Histone h3, HSP90AB1, IFI30, INVS, KRIT1, MAL, MAPK1, pregnenolone, RAP1A, RIOK1, TESK1, TGFB1, TNF	14	Developmental disorder, psychological disorders, neurological disease
APLP2, BRCA1, C7ORF27, CA2, CCNB1, CD8, CEBPA, CPB2, CXCL5, DBP, DHCR7, DIO2, DMBT1, EGR2, FOSL2, HABP4, HAMP, HLA-A, HMGA1, KRT12, LST1, MAG (includes EG:4099), MIR16, NFkB, PASK, PER2, Pkc(s), RGS2, ROBO1, S100A1, SAE2, SLC11A2, SLC18A3, SPP1, TNFRSF8	12	Cell signaling, molecular transport, vitamin and mineral metabolism
ATF1, ATP2A2, AXUD1, CARS, CD3EAP, CREB1, DDAH1, DHCR24, EIF4E2, EPO, GHR, GRLF1, HMGCR, HMGN2, IL1B, KRAS, MIA, MUC2, NF1, NFKBIB, PDGF BB, PKA C, POLR1B, POLR1E, prostaglandin D2, PTGDS, RGS2, RGS16, RHOA, RhoGap, Rock, SOD2, SOX9, SSPN, TFRC	9	Cellular development, cell morphology, cell death
Cortex: CCH		
Akt, CRYAB, CTSK, CXCL12, EDG2, EGR2, EGR4, G-αi, HLA-A, HLA-DQB2, HSPB1, LDLR, MAG (includes EG:4099), Mapk, MBP, Mek, MHC Class II, MIR16, NFκB, P38 MAPK, Pdgf, PDGF BB, PDGF-AA, PI3K, Pkc(s), PLC, PRKCD, Rac, RGS16, SCD, SOX10, TCR, Tgf-β, TGM2, UGT8	19	Cell-to-cell signaling and interaction, nervous system development and function, neurological disease
ACACA, BRCA1, C7ORF27, CA2, CEBPA, CRYAB, CTSK, CTSL1, ELK4, ELN, ENPP2, ENTPD2, EXTL2, FA2H, fatty acid, FCN2, IFI27, INS1, ITGA1, ITGAX, LPIN1, NMI, PDHB, PLEKHB1, Pyruvate dehydrogenase (lipoamide), retinoic acid, RGS8, SCD, SERPINE2, SLC4A1, SOX4, TGFB3, TSPAN2, UCP3, YWHAB	15	Carbohydrate metabolism, digestive system development and function, hepatic system development and function
AKT1, β-estradiol, C11ORF41, C5AR1, CA2, CD163, CD59A, CKB, CTSH, CTSL2, CYB5A, ELOVL1, ENPP2, GAL, GAL3ST1, GP5, GRIP1, HCRT, HLA-A, HLA-DQB2, IFI30, IFNG, IL6, MEF2C, MOG, MYO1D, norepinephrine, PMP22, progesterone, RYR3, SERPINB9, SHOX2, SLC6A9, SNX10, VWF	14	Neurological disease, immune response, cardiovascular system development and function
ACTA2 (includes EG:59), Caspase, CBFA2T3, CSRP1, CUGBP2, Cyclin A, DGKH, FASN, FOS, GCNT1, HAS1, HBA1, HBA2, HBB (includes EG:3043), HBB-AR, HBD, HBE1, HBG1, HBG2, HBQ1 (includes EG:3049), HBZ, Insulin, Jnk, LTBP2, MBOAT5, PDE7A, Pka, PPAP2C, RAC1, RBM3, RNH1, sorbitol, SYNJ2, TGFB1, TGFBI	12	Genetic disorder, hematological disease, molecular transport

(continued)

15 Transcriptional Response to Hypoxia in Developing Brain 299

Table 15.5 (continued)

Molecules in network	Focus molecules	Top functions
ALAS2, ANGPT2, ANLN, ARID1A, ASS1, CDKN1A, CDKN2A, CTBP2, ELL, ENPP2, EP400, EPC1, EPHA2, GLB1, HMGB2, HNRPA2B1, HTATIP, ING3, KLF7, LY6A, MAGEA2, MORF4L1, PBK, PLP1, PPFIBP1, RGS5, ROBO1, S100A4, SIM1, SOD2, SSPN, TNF, TP53, TP53INP1, VRK1	9	Cancer, cell death, respiratory disease
Hippocampus: CIH		
Aldosterone, APLP2, BAX, β-estradiol, BRCA1, C7ORF27, CCNB1, CFTR, CXCL12, CYP2A6, CYP2D6, EMB, ERBB2, GDI2, GDPD3, HLA-A, HSPB8, IFI30, LDL, MBP, MLH1, MOG, NELL2, OSBP, PDIA4, PPP3CC, RAB6A, RAP1A, RBM3, SLC4A4, TAX1BP3, TGFB1, TNFSF10, TRIM24, TYR	13	Inflammatory disease, immunological disease, neurological disease
APOA4,CD55,CDKN1A,CTNNB1,D-glucose,EIF2S3,EPHA2,EXOSC1,EXOSC2,EXOSC3,EXOSC4,EXOSC5,EXOSC6,EXOSC7,EXOSC8,EXOSC9,EXOSC10,FZD6,GLO1,GPLD1,HUS1,hydrogen peroxide,INVS,MIF4GD,MUC2,OGDH,PRKDC,pyruvaldehyde,RAPGEF5,RPA2,RPA3,SFRP1,SSPN,TCF7L2,TNF	11	Cell death, cell morphology, connective tissue development and function
ADNF, ATP, BCL2, Ca^{2+}, CALR, CANX, CD3EAP, DDAH1, F8, F9, FCN2, GABA, GABAR-A, GABRA1, GABRA5, GABRA2 (includes EG:29706), GABRB1, GABRB3, GABRG2, GP5, GRIN, GRIN2B, ITGA9 (includes EG:3680), KRAS, MASP1, MASP2, MBL2, NFKBIB, NNAT, POLR1B, POLR1E, Pp2b, SOX9, TRPM2, VWF	9	Infectious disease, developmental disorder, neurological disease
Hippocampus: CCH		
Akt, ANGPT2, Ap1, COL4A1, CXCL12, CXCR7, DAP3, DLL4, ENG, ETS, FLT1, HBA1, HBA2, HBB (includes EG:3043), HBD, HBG2, HBQ1 (includes EG:3049), HBZ, KLF3, MT2A, NCF2, NFκB, P38 MAPK, PDGF BB, PI3K, PRDX1, Rac, RYR3, SLC12A6 (includes EG:9990), SOX9, Tgf-β, TXNIP, Vegf, VEGFA, VWF	16	Genetic disorder, hematological disease, molecular transport
ACE, AGTR1, ANPEP, ATP, β-estradiol, C10ORF10, CA2, CEBPA, CSF2, CXADR, CYP3A5, DDAH1, EXTL2, F8, FLT1, GJA1, GJB1, HBD, HSP90B1, HSPA5, IL10, ITGAX, KRAS, MIR16, nitric oxide, NOSTRIN, PDIA4, PGLYRP1, RBM3, RGS2, ROCK1, TNFRSF1B, TRPM2, TXNRD1, VCAN	11	Cancer, cellular development, hematological disease
ACP5, AGTR1B, ALG1, amino acids, AXL, CBFA2T3, CDKN1A, CTNNB1, CXADR, DAP, dihydrotestosterone, EMB, ERBB2, ERG, FGF4, FOS, FZD6, GDF2, GHR, HDAC6, KCNJ16, KRT5, MAPK1, Pka, PKD1, PRKG2, prostaglandin E1, RGS5, SFRP1, SLC16A6, SNAI2, SNCG, SSPN, TCF7L2, VCAN	10	Tissue development, cancer, gene expression

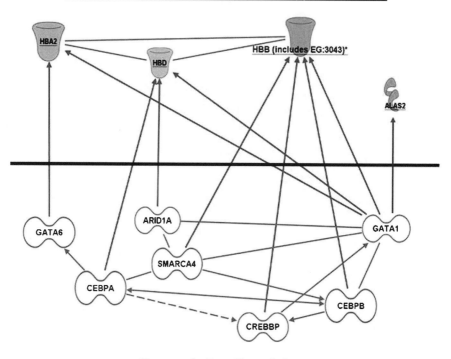

Fig. 15.3 Graphical representation of transcription factor network that regulates the expression of genes related to oxygen transport in CCH-treated mice. The relationship of transcription factors and the target genes was summarized by Network analysis using Ingenuity knowledge database (Ingenuity System, http://www.ingenuity.com). The network indicates that upregulation of oxygen delivery system requires coordination of multiple transcription factors (*See Color Plates*)

An important question is how the brain or the whole organism is trying to survive in sustained hypoxic environment. The answer to this question could be obtained, at least in part, from our studies and those of others. Note that: (a) hypoxia decreases BrdU incorporation in cells, indicating that hypoxia inhibits the ability of cells to proliferate; (b) it seems that in chronic hypoxia, especially in CCH, there is an

15 Transcriptional Response to Hypoxia in Developing Brain

increase in angiogenesis and an increase in Alas2 and globins, which leads to an increase in heme and globin production and oxygen-carrying capacity as well as the capacity of oxygen transport; and (c) it has been demonstrated that hypoxia inhibits cell differentiation and proliferation and increases the number of stem cell-like progenitor cells in CNS (95–98). Taken together, some of the overarching strategies in hypoxia would be (a) to increase oxygen-carrying capacity and transport, (b) to decrease the number of cells or cell differentiation and proliferation to decrease the overall oxygen demand, and (c) to keep cells in a hypoxia-tolerant state such as stem cell-like progenitors.

15.4 Summary

Diseases that occur during childhood, such as OSA, bronchopulmonary dysplasia, asthma, and SCD, can often have long-term pathological outcomes, accompanied by cognitive and/or behavioral disturbances. In this chapter, we summarized current studies related to hypoxia-response transcription factors and chronic hypoxia-induced transcriptomic response in developing brain. Of interest, the gene expression response to hypoxia in the developing brain seems to be region- and hypoxia paradigm-specific. For example, the majority of alterations in gene expression was essentially a downregulation in the neocortex (~80% following CIH and ~75% following CCH) but an upregulation in the hippocampus (~80% following CIH and 85% following CCH), demonstrating that hypoxia induced more suppression on gene expression in the neocortex, but more of an activation in gene expression in the hippocampus. Such dramatic differences demonstrated that distinct transcriptional regulatory mechanisms, such as different transcription factors, were involved in a tissue/cell specific manner to cope with oxygen deprivation in the brain. Furthermore, pathway analysis showed that CIH and CCH induced distinct alterations in functional networks. In general, it seems that CCH induced more of an *adaptive response* whereas CIH induced more of a *pathological response*. However, the exact mechanisms underlying transcriptional response to hypoxia and their roles in hypoxic adaptation or injury during postnatal brain development still need to be further elucidated.

Acknowledgment The author thanks Dr. Gabriel G. Haddad, Dr. Jin Xue, and Ms. Pat Spindler for helpful discussions and critical reading of the manuscript.

References

1. Bedard MA, Montplaisir J, Richer F, Rouleau I, Malo J. Obstructive sleep apnea syndrome: pathogenesis of neuropsychological deficits. J Clin Exp Neuropsychol 1991;13(6):950–64.
2. Vohr BR, Bell EF, Oh W. Infants with bronchopulmonary dysplasia. Growth pattern and neurologic and developmental outcome. Am J Dis Child 1982;136(5):443–7.

3. Bass JL, Corwin M, Gozal D, et al. The effect of chronic or intermittent hypoxia on cognition in childhood: a review of the evidence. Pediatrics 2004;114(3):805–16.
4. Gozal D. Sleep-disordered breathing and school performance in children. Pediatrics 1998;102 (3, Part 1):616–20.
5. Kirkham FJ, Datta AK. Hypoxic adaptation during development: relation to pattern of neurological presentation and cognitive disability. Dev Sci 2006;9(4):411–27.
6. O'Brien LM, Gozal D. Behavioural and neurocognitive implications of snoring and obstructive sleep apnoea in children: facts and theory. Paediatr Respir Rev 2002;3(1):3–9.
7. Rhodes SK, Shimoda KC, Waid LR, et al. Neurocognitive deficits in morbidly obese children with obstructive sleep apnea. J Pediatr 1995;127(5):741–4.
8. Wray J. Intellectual development of infants, children and adolescents with congenital heart disease. Dev Sci 2006;9(4):368–78.
9. O'Dougherty M, Wright FS, Loewenson RB, Torres F. Cerebral dysfunction after chronic hypoxia in children. Neurology 1985;35(1):42–6.
10. Chihab R, Ferry C, Koziel V, Monin P, Daval JL. Sequential activation of activator protein-1-related transcription factors and JNK protein kinases may contribute to apoptotic death induced by transient hypoxia in developing brain neurons. Brain Res Mol Brain Res 1998;63(1):105–20.
11. Gozal E, Simakajornboon N, Dausman JD, et al. Hypoxia induces selective SAPK/JNK-2-AP-1 pathway activation in the nucleus tractus solitarii of the conscious rat. J Neurochem 1999;73(2): 665–74.
12. Gao Y, Gao G, Long C, et al. Enhanced phosphorylation of cyclic AMP response element binding protein in the brain of mice following repetitive hypoxic exposure. Biochem Biophys Res Commun 2006;340(2):661–7.
13. Jin K, Mao XO, Simon RP, Greenberg DA. Cyclic AMP response element binding protein (CREB) and CREB binding protein (CBP) in global cerebral ischemia. J Mol Neurosci 2001; 16(1):49–56.
14. Mishra OP, Ashraf QM, Delivoria-Papadopoulos M. Phosphorylation of cAMP response element binding (CREB) protein during hypoxia in cerebral cortex of newborn piglets and the effect of nitric oxide synthase inhibition. Neuroscience 2002;115(3):985–91.
15. Walton M, Sirimanne E, Williams C, Gluckman P, Dragunow M. The role of the cyclic AMP-responsive element binding protein (CREB) in hypoxic-ischemic brain damage and repair. Brain Res Mol Brain Res 1996;43(1–2):21–9.
16. Zaman K, Ryu H, Hall D, et al. Protection from oxidative stress-induced apoptosis in cortical neuronal cultures by iron chelators is associated with enhanced DNA binding of hypoxia-inducible factor-1 and ATF-1/CREB and increased expression of glycolytic enzymes, p21(waf1/cip1), and erythropoietin. J Neurosci 1999;19(22):9821–30.
17. Liao H, Hyman MC, Lawrence DA, Pinsky DJ. Molecular regulation of the *PAI-1* gene by hypoxia: contributions of Egr-1, HIF-1alpha, and C/EBPalpha. FASEB J 2007;21(3):935–49.
18. Nishi H, Nishi KH, Johnson AC. Early growth response-1 gene mediates up-regulation of epidermal growth factor receptor expression during hypoxia. Cancer Res 2002;62(3):827–34.
19. Yan SF, Lu J, Zou YS, et al. Hypoxia-associated induction of early growth response-1 gene expression. J Biol Chem 1999;274(21):15030–40.
20. Zhang P, Tchou-Wong KM, Costa M. Egr-1 mediates hypoxia-inducible transcription of the *NDRG1* gene through an overlapping Egr-1/Sp1 binding site in the promoter. Cancer Res 2007;67(19):9125–33.
21. Elvert G, Kappel A, Heidenreich R, et al. Cooperative interaction of hypoxia-inducible factor-2alpha (HIF-2alpha) and Ets-1 in the transcriptional activation of vascular endothelial growth factor receptor-2 (Flk-1). J Biol Chem 2003;278(9):7520–30.
22. Bernaudin M, Nedelec AS, Divoux D, MacKenzie ET, Petit E, Schumann-Bard P. Normobaric hypoxia induces tolerance to focal permanent cerebral ischemia in association with an increased expression of hypoxia-inducible factor-1 and its target genes, erythropoietin and VEGF, in the adult mouse brain. J Cereb Blood Flow Metab 2002;22(4):393–403.
23. Chavez JC, Agani F, Pichiule P, LaManna JC. Expression of hypoxia-inducible factor-1alpha in the brain of rats during chronic hypoxia. J Appl Physiol 2000;89(5):1937–42.

24. Heidbreder M, Frohlich F, Johren O, Dendorfer A, Qadri F, Dominiak P. Hypoxia rapidly activates HIF-3alpha mRNA expression. FASEB J 2003;17(11):1541–3.
25. Jones NM, Bergeron M. Hypoxic preconditioning induces changes in HIF-1 target genes in neonatal rat brain. J Cereb Blood Flow Metab 2001;21(9):1105–14.
26. Talks KL, Turley H, Gatter KC, et al. The expression and distribution of the hypoxia-inducible factors HIF-1alpha and HIF-2alpha in normal human tissues, cancers, and tumor-associated macrophages. Am J Pathol 2000;157(2):411–21.
27. Wiener CM, Booth G, Semenza GL. In vivo expression of mRNAs encoding hypoxia-inducible factor 1. Biochem Biophys Res Commun 1996;225(2):485–8.
28. Wiesener MS, Jurgensen JS, Rosenberger C, et al. Widespread hypoxia-inducible expression of HIF-2alpha in distinct cell populations of different organs. FASEB J 2003;17(2):271–3.
29. Guo G, Bhat NR. Hypoxia/reoxygenation differentially modulates NF-kappaB activation and iNOS expression in astrocytes and microglia. Antioxid Redox Signal 2006;8(5–6):911–18.
30. Tamatani M, Che YH, Matsuzaki H, et al. Tumor necrosis factor induces Bcl-2 and Bcl-x expression through NFkappaB activation in primary hippocampal neurons. J Biol Chem 1999;274(13):8531–8.
31. Tamatani M, Mitsuda N, Matsuzaki H, et al. A pathway of neuronal apoptosis induced by hypoxia/reoxygenation: roles of nuclear factor-kappaB and Bcl-2. J Neurochem 2000;75(2):683–93.
32. Witt KA, Mark KS, Huber J, Davis TP. Hypoxia-inducible factor and nuclear factor kappa-B activation in blood–brain barrier endothelium under hypoxic/reoxygenation stress. J Neurochem 2005;92(1):203–14.
33. Culmsee C, Siewe J, Junker V, et al. Reciprocal inhibition of p53 and nuclear factor-kappaB transcriptional activities determines cell survival or death in neurons. J Neurosci 2003;23(24):8586–95.
34. Yamaguchi A, Tamatani M, Matsuzaki H, et al. Akt activation protects hippocampal neurons from apoptosis by inhibiting transcriptional activity of p53. J Biol Chem 2001;276(7):5256–64.
35. Yamaguchi A, Taniguchi M, Hori O, et al. Peg3/Pw1 is involved in p53-mediated cell death pathway in brain ischemia/hypoxia. J Biol Chem 2002;277(1):623–9.
36. Yonekura I, Takai K, Asai A, Kawahara N, Kirino T. p53 potentiates hippocampal neuronal death caused by global ischemia. J Cereb Blood Flow Metab 2006;26(10):1332–40.
37. Zhu Y, Mao XO, Sun Y, Xia Z, Greenberg DA. p38 Mitogen-activated protein kinase mediates hypoxic regulation of Mdm2 and p53 in neurons. J Biol Chem 2002;277(25):22909–14.
38. Mukerji SS, Katsman EA, Wilber C, Haner NA, Selman WR, Hall AK. Activin is a neuronal survival factor that is rapidly increased after transient cerebral ischemia and hypoxia in mice. J Cereb Blood Flow Metab 2007;27(6):1161–72.
39. Ryuto M, Ono M, Izumi H, et al. Induction of vascular endothelial growth factor by tumor necrosis factor alpha in human glioma cells. Possible roles of SP-1. J Biol Chem 1996;271(45):28220–8.
40. Ema M, Taya S, Yokotani N, Sogawa K, Matsuda Y, Fujii-Kuriyama Y. A novel bHLH-PAS factor with close sequence similarity to hypoxia-inducible factor 1alpha regulates the VEGF expression and is potentially involved in lung and vascular development. Proc Natl Acad Sci USA 1997;94(9):4273–8.
41. Flamme I, Frohlich T, von Reutern M, Kappel A, Damert A, Risau W. HRF, a putative basic helix-loop-helix-PAS-domain transcription factor is closely related to hypoxia-inducible factor-1 alpha and developmentally expressed in blood vessels. Mech Dev 1997;63(1):51–60.
42. Gu YZ, Moran SM, Hogenesch JB, Wartman L, Bradfield CA. Molecular characterization and chromosomal localization of a third alpha-class hypoxia inducible factor subunit, HIF3alpha. Gene Expr 1998;7(3):205–13.
43. Wang GL, Semenza GL. General involvement of hypoxia-inducible factor 1 in transcriptional response to hypoxia. Proc Natl Acad Sci USA 1993;90(9):4304–8.
44. Maxwell PH, Wiesener MS, Chang GW, et al. The tumour suppressor protein VHL targets hypoxia-inducible factors for oxygen-dependent proteolysis. Nature 1999;399(6733):271–5.

45. Ivan M, Kondo K, Yang H, et al. HIFalpha targeted for VHL-mediated destruction by proline hydroxylation: implications for O2 sensing. Science 2001;292(5516):464–8.
46. Jaakkola P, Mole DR, Tian YM, et al. Targeting of HIF-alpha to the von Hippel–Lindau ubiquitylation complex by O2-regulated prolyl hydroxylation. Science 2001;292(5516):468–72.
47. Duan DR, Pause A, Burgess WH, et al. Inhibition of transcription elongation by the VHL tumor suppressor protein. Science 1995;269(5229):1402–6.
48. Kibel A, Iliopoulos O, DeCaprio JA, Kaelin WG Jr. Binding of the von Hippel–Lindau tumor suppressor protein to elongin B and C. Science 1995;269(5229):1444–6.
49. Pause A, Lee S, Worrell RA, et al. The von Hippel–Lindau tumor-suppressor gene product forms a stable complex with human CUL-2, a member of the Cdc53 family of proteins. Proc Natl Acad Sci USA 1997;94(6):2156–61.
50. Hewitson KS, McNeill LA, Riordan MV, et al. Hypoxia-inducible factor (HIF) asparagine hydroxylase is identical to factor inhibiting HIF (FIH) and is related to the cupin structural family. J Biol Chem 2002;277(29):26351–5.
51. Mahon PC, Hirota K, Semenza GL. FIH-1: a novel protein that interacts with HIF-1alpha and VHL to mediate repression of HIF-1 transcriptional activity. Genes Dev 2001;15(20):2675–86.
52. Rolfe DF, Brown GC. Cellular energy utilization and molecular origin of standard metabolic rate in mammals. Physiol Rev 1997;77(3):731–58.
53. Gnaiger E, Lassnig B, Kuznetsov AV, Margreiter R. Mitochondrial respiration in the low oxygen environment of the cell. Effect of ADP on oxygen kinetics. Biochim Biophys Acta 1998;1365(1–2):249–54.
54. Acker T, Acker H. Cellular oxygen sensing need in CNS function: physiological and pathological implications. J Exp Biol 2004;207 (Part 18):3171–88.
55. Mateo J, Garcia-Lecea M, Cadenas S, Hernandez C, Moncada S. Regulation of hypoxia-inducible factor-1alpha by nitric oxide through mitochondria-dependent and -independent pathways. Biochem J 2003;376 (Part 2):537–44.
56. Bell EL, Klimova TA, Eisenbart J, et al. The Qo site of the mitochondrial complex III is required for the transduction of hypoxic signaling via reactive oxygen species production. J Cell Biol 2007;177(6):1029–36.
57. Cash TP, Pan Y, Simon MC. Reactive oxygen species and cellular oxygen sensing. Free Radic Biol Med 2007;43(9):1219–25.
58. Guzy RD, Sharma B, Bell E, Chandel NS, Schumacker PT. Loss of the SdhB, but not the SdhA, subunit of complex II triggers reactive oxygen species-dependent hypoxia-inducible factor activation and tumorigenesis. Mol Cell Biol 2008;28(2):718–31.
59. Baichwal VR, Baeuerle PA. Activate NF-kappa B or die? Curr Biol 1997;7(2):R94–R96.
60. Kaltschmidt B, Baeuerle PA, Kaltschmidt C. Potential involvement of the transcription factor NF-kappa B in neurological disorders. Mol Aspects Med 1993;14(3):171–90.
61. Lenardo MJ, Baltimore D. NF-kappa B: a pleiotropic mediator of inducible and tissue-specific gene control. Cell 1989;58(2):227–9.
62. Malek R, Borowicz KK, Jargiello M, Czuczwar SJ. Role of nuclear factor kappaB in the central nervous system. Pharmacol Rep 2007;59(1):25–33.
63. Birbach A, Gold P, Binder BR, Hofer E, de Martin R, Schmid JA. Signaling molecules of the NF-kappa B pathway shuttle constitutively between cytoplasm and nucleus. J Biol Chem 2002;277(13):10842–51.
64. Cummins EP, Comerford KM, Scholz C, Bruning U, Taylor CT. Hypoxic regulation of NF-kappaB signaling. Methods Enzymol 2007;435:479–92.
65. Greenberg H, Ye X, Wilson D, Htoo AK, Hendersen T, Liu SF. Chronic intermittent hypoxia activates nuclear factor-kappaB in cardiovascular tissues in vivo. Biochem Biophys Res Commun 2006;343(2):591–6.
66. Hedtjarn M, Mallard C, Hagberg H. Inflammatory gene profiling in the developing mouse brain after hypoxia-ischemia. J Cereb Blood Flow Metab 2004;24(12):1333–51.
67. Hu X, Nesic-Taylor O, Qiu J, et al. Activation of nuclear factor-kappaB signaling pathway by interleukin-1 after hypoxia/ischemia in neonatal rat hippocampus and cortex. J Neurochem 2005;93(1):26–37.

68. Michiels C, Minet E, Mottet D, Raes M. Regulation of gene expression by oxygen: NF-kappaB and HIF-1, two extremes. Free Radic Biol Med 2002;33(9):1231–42.
69. Koong AC, Chen EY, Mivechi NF, Denko NC, Stambrook P, Giaccia AJ. Hypoxic activation of nuclear factor-kappa B is mediated by a Ras and Raf signaling pathway and does not involve MAP kinase (ERK1 or ERK2). Cancer Res 1994;54(20):5273–9.
70. Zampetaki A, Mitsialis SA, Pfeilschifter J, Kourembanas S. Hypoxia induces macrophage inflammatory protein-2 (MIP-2) gene expression in murine macrophages via NF-kappaB: the prominent role of p42/p44 and PI3 kinase pathways. FASEB J 2004;18(10):1090–2.
71. Chandel NS, Trzyna WC, McClintock DS, Schumacker PT. Role of oxidants in NF-kappa B activation and TNF-alpha gene transcription induced by hypoxia and endotoxin. J Immunol 2000;165(2):1013–21.
72. D'Angio CT, Finkelstein JN. Oxygen regulation of gene expression: a study in opposites. Mol Genet Metab 2000;71(1–2):371–80.
73. Haddad JJ. Oxygen-sensing mechanisms and the regulation of redox-responsive transcription factors in development and pathophysiology. Respir Res 2002;3:26.
74. Haddad JJ, Olver RE, Land SC. Antioxidant/pro-oxidant equilibrium regulates HIF-1alpha and NF-kappa B redox sensitivity. Evidence for inhibition by glutathione oxidation in alveolar epithelial cells. J Biol Chem 2000;275(28):21130–9.
75. Mayr B, Montminy M. Transcriptional regulation by the phosphorylation-dependent factor CREB. Nat Rev Mol Cell Biol 2001;2(8):599–609.
76. Euskirchen G, Royce TE, Bertone P, et al. CREB binds to multiple loci on human chromosome 22. Mol Cell Biol 2004;24(9):3804–14.
77. Beitner-Johnson D, Millhorn DE. Hypoxia induces phosphorylation of the cyclic AMP response element-binding protein by a novel signaling mechanism. J Biol Chem 1998;273(31):19834–9.
78. Kitagawa K. CREB and cAMP response element-mediated gene expression in the ischemic brain. FEBS J 2007;274(13):3210–17.
79. Laderoute KR. The interaction between HIF-1 and AP-1 transcription factors in response to low oxygen. Semin Cell Dev Biol 2005;16(4–5):502–13.
80. Simakajornboon N, Kuptanon T. Maturational changes in neuromodulation of central pathways underlying hypoxic ventilatory response. Respir Physiol Neurobiol 2005;149(1–3):273–86.
81. Raivich G, Behrens A. Role of the AP-1 transcription factor c-Jun in developing, adult and injured brain. Prog Neurobiol 2006;78(6):347–63.
82. Mechta-Grigoriou F, Gerald D, Yaniv M. The mammalian Jun proteins: redundancy and specificity. Oncogene 2001;20(19):2378–89.
83. Millhorn DE, Raymond R, Conforti L, et al. Regulation of gene expression for tyrosine hydroxylase in oxygen sensitive cells by hypoxia. Kidney Int 1997;51(2):527–35.
84. Hoffmann A, Gloe T, Pohl U. Hypoxia-induced upregulation of *eNOS* gene expression is redox-sensitive: a comparison between hypoxia and inhibitors of cell metabolism. J Cell Physiol 2001;188(1):33–44.
85. Salnikow K, Kluz T, Costa M, et al. The regulation of hypoxic genes by calcium involves c-Jun/AP-1, which cooperates with hypoxia-inducible factor 1 in response to hypoxia. Mol Cell Biol 2002;22(6):1734–41.
86. Shi Q, Le X, Abbruzzese JL, et al. Cooperation between transcription factor AP-1 and NF-kappaB in the induction of interleukin-8 in human pancreatic adenocarcinoma cells by hypoxia. J Interferon Cytokine Res 1999;19(12):1363–71.
87. Yamashita K, Discher DJ, Hu J, Bishopric NH, Webster KA. Molecular regulation of the endothelin-1 gene by hypoxia. Contributions of hypoxia-inducible factor-1, activator protein-1, GATA-2, AND p300/CBP. J Biol Chem 2001;276(16):12645–53.
88. Michiels C, Minet E, Michel G, Mottet D, Piret JP, Raes M. HIF-1 and AP-1 cooperate to increase gene expression in hypoxia: role of MAP kinases. IUBMB Life 2001;52(1–2):49–53.
89. Minet E, Michel G, Mottet D, et al. *c-JUN* gene induction and AP-1 activity is regulated by a JNK-dependent pathway in hypoxic HepG2 cells. Exp Cell Res 2001;265(1):114–24.

90. Premkumar DR, Adhikary G, Overholt JL, Simonson MS, Cherniack NS, Prabhakar NR. Intracellular pathways linking hypoxia to activation of c-fos and AP-1. Adv Exp Med Biol 2000;475:101–9.
91. Zhou D, Wang J, Zapala MA, Xue J, Schork NJ, Haddad GG. Gene expression in mouse brain following chronic hypoxia: role of sarcospan in glial cell death. Physiol Genomics 2008; 32(3):370–9.
92. Amicarelli F, Colafarina S, Cattani F, et al. Scavenging system efficiency is crucial for cell resistance to ROS-mediated methylglyoxal injury. Free Radic Biol Med 2003;35(8):856–71.
93. Creighton DJ, Zheng ZB, Holewinski R, Hamilton DS, Eiseman JL. Glyoxalase I inhibitors in cancer chemotherapy. Biochem Soc Trans 2003;31 (Part 6):1378–82.
94. Rose IA, Nowick JS. Methylglyoxal synthetase, enol-pyruvaldehyde, glutathione and the glyoxalase system. J Am Chem Soc 2002;124(44):13047–52.
95. Morrison SJ, Csete M, Groves AK, Melega W, Wold B, Anderson DJ. Culture in reduced levels of oxygen promotes clonogenic sympathoadrenal differentiation by isolated neural crest stem cells. J Neurosci 2000;20(19):7370–6.
96. Studer L, Csete M, Lee SH, et al. Enhanced proliferation, survival, and dopaminergic differentiation of CNS precursors in lowered oxygen. J Neurosci 2000;20(19):7377–83.
97. Yun Z, Lin Q, Giaccia AJ. Adaptive myogenesis under hypoxia. Mol Cell Biol 2005;25(8): 3040–55.
98. Zaidi AU, Bessert DA, Ong JE, et al. New oligodendrocytes are generated after neonatal hypoxic-ischemic brain injury in rodents. Glia 2004;46(4):380–90.

Chapter 16
Acute Stroke Therapy: Highlighting the Ischemic Penumbra

Hang Yao

Abstract The ultimate goal of acute stroke therapy is to rescue the affected brain tissues that are salvageable and to avoid or minimize the potential neurological outcome. The current available thrombolytic treatment is limited by the narrow therapeutic time window due mainly to the progressively reduced rescuable tissues (ischemic penumbra) after the onset of stroke. Therefore, preserving penumbral tissue can potentially widen the therapeutic time window and improve the quality of acute stroke therapy. Hence, ischemic penumbra is the key to stroke pathology and treatment. In this chapter, the basic concept of ischemic penumbra is examined and this is followed by the discussion of the identification and evolution of penumbral tissue in animal stroke models and stroke patients. Finally, the possible mechanism of penumbral cell death and the importance of preserving penumbral tissue in the acute stroke therapy are discussed.

Keywords: Strokes, Cerebral Ischemia, Infarct, Penumbra

16.1 Introduction

Ischemic stroke produces severe reduction of blood flow to the affected brain region, which leads to neurological consequences. Current available pharmacologic intervention is related to thrombolysis afforded by a recombinant tissue-type plasminogen activator (rt-PA), the only FDA approved drug for stroke therapy (1). Although intravenous rt-PA has proven to be effective, the narrow therapeutic window (i.e., 3h) and the strict inclusion criteria make it only accessible to a small fraction of stroke patients (2, 3). This is mainly due to the significant loss of rescuable tissues and several fold increased risk of hemorrhagic transformation

H. Yao
Department of Pediatrics, University of California, San Diego, 9500 Gilman Drive, 0735 La Jolla, CA 92093-0735, USA
e-mail: hyao@ucsd.edu

to individual patient when used beyond 3 h after the onset of stroke (4). Indeed, neuroimaging analysis has demonstrated that stroke patients with better preserved penumbral tissues often show improved neurological outcomes following rt-PA treatment. Therefore, the improvement of acute stroke therapy relies on developing not only better thrombolytic agent but effective neuroprotectant to preserve the salvageable penumbral tissue.

Our knowledge about stroke pathology has been dramatically advanced due to the development of modern imaging technologies and biological research techniques. Based on the concept of ischemic penumbra, several neuroimaging modalities have been employed to study the rescuable ischemic tissue in stroke patients and animal models. As a result, the existence and the evolution of ischemic penumbra toward irreversible infarction are further confirmed in focal ischemia. Nevertheless, the ischemic penumbra does not remain viable for extended periods of time before it evolves into necrotic infarction if no appropriate intervention is given. Indeed, due to the limited cerebral blood flow (CBF) and the initiation of the ischemic cascade in the penumbra, this unique brain region clearly has a limited life span and appears to undergo irreversible damage within hours unless reperfusion is initiated. Therefore, the preservation of penumbral tissue would not only improve the quality of acute stroke therapy by thrombolysis but extend its therapeutic time window.

The residual blood flow in the penumbral region provides the possibility for the delivery of neuroprotectant to this region. Therefore, delivering neuroprotectant can be an effective way to preserve penumbral tissue before recanalization treatment is available. The practical way to extend the survival of penumbral tissue following stroke is to intervene with the major pathophysiological events that can potentially lead to cell death. In the past few decades, extensive studies have been conducted to search for drugs that can block the cascade initiated by the excessive release of glutamate in the ischemic brain region. Many compounds, such as N-methyl-D-aspartate (NMDA) receptor antagonists and calcium channel blockers, have been developed as potential therapeutic means for acute stroke, but clinical trials for these compounds have been so far disappointing. The cause of this failure has been multifactorial, but perhaps better understanding of the cellular and molecular mechanisms of cell death in the ischemic penumbra is crucial.

16.2 The Concept of Ischemic Penumbra

It has been demonstrated that following focal cerebral ischemia, extremely low CBF in certain brain region can lead to the formation of a necrotic infarct core where energy deprivation and ion pump failure have developed. Immediately around this dense ischemic center is a region that has viable brain tissues with moderately reduced CBF. Neuronal cells in this region are functionally impaired with suppressed electrical activities. It is believed that these electrically impaired cells may not be necessarily destined to die if reperfusion or other appropriate intervention is given within a certain time window. Indeed, CBF threshold for energy failure, a state in which cells can die irreversibly within a short period of time, is lower than that for the electrical failure. For example, using the abnormal elevation of extracellular K^+ concentration as a marker of functional impairment of Na^+/K^+-ATPases, investigators have found that the CBF threshold for energy failure is lower than that for electrical failure. Therefore, the ischemic region with a CBF range between the higher end of electrical failure and the lower end of energy failure is the so called *ischemic penumbra*, which is described as an ischemic region around the infarct core in which the brain tissues are electrically silent but with preserved ion homeostasis and transmembrane electrical potentials (5, 6).

Several revised versions of the definition for ischemic penumbra have been developed over time and they are largely based on the unique physiological, biochemical, or genetic changes in this hypoperfused brain region (7). One of the definitions, which is more straightforward and clinically relevant, is that the penumbra is the ischemic region evolving into ischemic infarction over time, but is rescuable if timely reperfusion or appropriate intervention is ensued (8). In addition, the penumbra is also described as a region with hypoperfusion but in which energy metabolism is preserved (9). Based on the viability threshold of CBF, a more general description for ischemic penumbra is to use the CBF threshold. For example, the ischemic region with less than 20% of normal perfusion level is the infarct core while the surrounding region with the blood supply between 20 and 40% of normal level is the penumbral tissue.

16.3 The Existence and the Evolution of Ischemic Penumbra

Since stroke is the neurological consequence of disrupted blood supply, any initial event seen in the stroke patients or animal stroke models is closely related to the perfusion levels. With respect to the physiological changes during the decrease of CBF, out of the most sensitive processes seen in the ischemic brain tissue is the suppression of protein synthesis that usually starts when CBF drops below normal (0.5–0.6 mL/g/min) and is completely inhibited at about 0.35 mL/g/min. Further decrease of CBF would stimulate the anaerobic glycolysis, which usually occurs at a flow rate between 0.35 and 0.25 mL/g/min. Pronounced acidosis and the decline of both PCr and ATP also take place at a flow rate below 0.25 mL/g/min. When the flow rate decreases to below 0.1 mL/g/min, the disruption of energy metabolic homeostasis and ionic pump failure occur (7).

Based on these data and the concept of ischemic penumbra, several methods have been developed to discriminate penumbral tissue from the infarct core. In these experiments, ATP depletion in the ischemic core was detected by the bioluminescent imaging technique. Furthermore, the whole ischemic region can be delineated by the biological events that occur within the CBF range in the penumbral region, such as the suppression of protein synthesis and the activation of anaerobic glycolisis (can be detected by examining the tissue acidosis in the whole ischemic region). Therefore, the difference between the identified whole ischemic area and the infarct core area is the ischemic penumbra (7). At molecular level, some genes were found exclusively expressed in the penumbral tissue. For example, hsp70 mRNA was found selectively expressed in the penumbra and peaked at 3 h after the vascular occlusion in a permanent middle cerebral artery occlusion (MCAO) model. The presence of this gene can be used as a marker for identifying the ischemic penumbra (10).

Now that the major target of acute stroke therapy is the ischemic penumbra, determining the existence and the volume of this salvageable tissue becomes critical for clinicians to make therapeutic decision. In fact, the identification of the penumbra may allow restricting rt-PA use to those patients with large penumbra and small infarct core even beyond the currently used 3-h time window. Noninvasive neuroimaging technologies that have been extensively studied in recent years are more clinically relevant. The development of these methods is based on the CBF threshold for different biological events, which is tightly related to the viability of brain tissues. In focal ischemia, brain tissues can be differentiated into four different classes (starting from the most peripheral to the most central), i.e., healthy brain tissues with normal CBF rate, oligemic tissues that are alive but with reduced CBF that is automatically reversible, penumbral tissues that show reduced CBF with electrical silence but salvageable if CBF is restored within a certain time period, and finally the ischemic infarct core that is composed of irreversibly damaged cells and formed shortly after the restriction of CBF. This area suffers from extensive deprivation of energy substrates due to the extremely low blood supply (11).

Using positron emission tomography (PET), magnetic resonance imaging (MRI), and computed tomography (CT), penumbral tissue can be discriminated from the irreversibly damaged infarct core. PET can provide information on CBF, cerebral blood volume (CBV), oxygen consumption ($CMRO_2$), and oxygen extraction fraction (OEF) (12). Based on the hemodynamic criteria for penumbra or infarct core, the brain region with a reduced CBF but normal $CMRO_2$, i.e., an increased OEF, is the potentially salvageable penumbral tissue. Although PET can provide relatively accurate information for the identification of ischemic penumbra, currently it is not widely used for clinical practice due to its high cost and time-consuming procedures. Compared with PET, diffusion-perfusion MRI is available at many medical facilities and currently widely used for this purpose. In stroke patients, the hypoperfused region can be detected on perfusion-weighted imaging (PWI) and irreversibly injured abnormal brain region can be detected on diffusion-weighted imaging (DWI). The mismatched region between these two areas is the approximation of the hypoperfused brain tissue without being irreversibly damaged, i.e., the ischemic penumbra (13). Although the use of DWI/PWI imaging to approximate penumbra has its limitations (since the abnormal area detected by DWI could potentially include some reversible brain tissues while the abnormal area detected by PWI could include oligemic tissues), the modified approaches can identify the penumbra with increased accuracy (14, 15). Another imaging modality tested for potentially detecting ischemic penumbra is perfusion CT, which is becoming increasingly available to patients. Perfusion CT can detect the status of CBF and CBV in the ischemic region. Since CBV reflects the autoregulation of hemodynamics in the ischemic region, the significant reduction (<0.7) of CBV is indicative of irreversible damage of the affected tissue, which distinguishes infarct core from the hypoperfused penumbral tissue (16). Therefore, a region of abnormal CBF with normal CBV would be the ischemic tissue with viable cells, i.e., the ischemic penumbra.

The basis of acute stroke therapy is to cease the growth of infarct and rescue the salvageable penumbral tissue. It is well accepted that lesion growth is clinically important and is associated with poorer neurological outcome. Ischemic infarct expansion is well known to occur in both stroke patients and experimental animal models (17). In the mouse permanent MCAO model, the infarct expands and approaches its final size by 1 day after occlusion. In this same experimental model, the final size of the infarct is precisely predicted by the initial suppression of protein synthesis, suggesting that the most reliable parameter heralding the expansion of brain infarction is the inhibition of protein synthesis. Indeed, the ischemic region with the inhibition of protein synthesis correlates better with the final size of ischemic injury than any of the other investigated variables (10). Histopathological studies from a rat MCAO model demonstrated that the ischemic area covered 10% of the hemisphere at 1 h, 22% at 2 h, 38% at 6 h, and 50% at 12 h postocclusion (18). Similar observations have been made in human stroke patients using neuroimaging approaches (19–21). About 38–156% infarct growth is estimated in stroke patients by the *mismatch* of perfusion and diffusion on MRI (22). The fact that the penumbral tissue evolves into infarction instead of dying all at once suggests a heterogeneous nature of the penumbral tissue, i.e., cells in the infarct rim, a territory residing in between the infarct core and the penumbral tissues, die earlier than those in the distal part of the penumbra. This unique area is of critical importance for understanding the mechanisms underlying the expansion of the infarct core and may be studied at cellular and molecular level with in vitro experimental models (23).

16.4 Possible Mechanisms of the Penumbral Cell Death

Although the ischemic penumbral tissue itself suffers from moderate hypoperfusion, cell fate in the penumbra may largely rely on the physical and biochemical changes in the infarct core since the two regions have an intimate relationship. The relation between the infarct core and the surrounding penumbra stems from at least three ideas: first, penumbral cells are directly exposed to the irreversibly damaged cells and challenged by the core milieu (conceptually a portion of the alive penumbral cells is next to the dead infarct cells); second, a portion of the penumbral cells and the cells in the infarct core are connected by synapses and gap junctions; third, penumbral cells are affected by electrical activities generated in the core, such as spreading depression. This intimate relationship between the penumbra and infarct core suggests that the changes occurring in the infarct core can have profound impact on the fate of the adjacent penumbral cells.

16.4.1 Cell Death in the Infarct Core

In the past decades, research has advanced our knowledge about the pathophysiology of cerebral focal ischemia. The disruption of blood supply to a certain brain area can cause irreversible damage to the affected region in a fairly short period of time. This is mainly due to the extremely reduced supply of oxygen and glucose to the affected cells. Oxygen and glucose deprivation rapidly lead to demise. The failure of the Na^+-K^+ pump results in an accumulation of intracellular Na^+ and Ca^{2+} and significant elevation of extracellular K^+ concentration. The disruption of ionic homeostasis would lead to the movement of water from interstitial space into the cell and causes cell swelling and edema in the affected brain region. Necrosis occurs in this region and the irreversible cell death leads to the formation of the dense infarct core. Timely reperfusion to this region can avoid further expansion of the lesion but cell damage in this area is generally not reversible.

16.4.2 Challenges of Penumbral Cells by Direct Exposure to the Core Milieu

As a result of cell death in the infarct core, dramatic biochemical changes in this area can have profound impact on adjacent penumbral cells. Indeed, due to the extremely low oxygen and glucose supply, the ischemic core can acquire an interstitial fluid that is characterized by large increases in K^+, H^+, and glutamate and decreases in Na^+, Ca^{2+}, and Cl^- (24, 25). Cells in the penumbra (especially those in the infarct rim) are exposed to such a milieu in addition to hypoxia and hypoglycemia. Many studies have demonstrated that the experimental conditions that simulate the biochemical changes of ischemic infarct core could induce damage in both neuronal and glial cells. Major factors that could contribute to the penumbral cell death may include acidosis, ionic shifts, and glutamate toxicity.

16.4.2.1 Acidosis

In contrast to global ischemia (complete constriction of blood supply to the brain), focal ischemia features an incomplete constriction of blood supply to the ischemic brain region. Therefore, the leakage of glucose to the hypoxic brain tissue promotes glycolysis, which generates excessive lactic acid. Lactate, in turn, titrates the intracellular HCO_3^- and increases CO_2, which diffuses into interstitial space. Similarly, extracellular HCO_3^- is titrated by raised CO_2. Therefore, excessive lactate production can exhaust the bicarbonate buffering system in the ischemic brain tissue although acid–base transporters and other intracellular buffering systems can maintain intracellular pH for a period of time. For example, in the early stage of focal ischemia (within 1 h), the lactic acid production causes an interstitial pH drop (approximately half unit) but intracellular pH remains close to normal level. Intracellular pH drops to a similar level later when the pH regulatory mechanisms and most of the buffering systems are compromised. Interstitial pH drops about 1 unit if focal ischemia occurs in hyperglycemic animals due to the higher amount of lactic acid production (26). Although interstitial HCO_3^- can keep extracellular pH at a constant level if the lactate amount is within a certain range, excessive lactate would exhaust the HCO_3^- and cause an acidotic environment in the affected brain regions (27). There is no doubt that lactic acidosis plays a role in the early stages of the ischemic core formation and most importantly, it poses a threat to the adjacent penumbral cells in the later stages.

Lactic acidosis has been proposed as a factor that may affect the outcome of ischemic cells (28). Previous studies have suggested that acidosis plays a role in neuronal and glial injury/death following hypoxia/ischemia. In vitro studies have demonstrated that lactic acid induces astrocytic injury, as does HCl (29). Such studies have shown that the effect of pH on cell fate depends on many factors, namely severity of acidosis, cell type, and experimental conditions such as the presence of low O_2 or alterations in major ionic components. As has been reported, acidosis or hypoxia alone has only a modest impact on astrocytic injury. However, the combination of acidosis and hypoxia could kill astrocytes rapidly by reducing ATP levels (30). When both hypoxia and ion shifts were introduced to mimic the ischemic condition, the acidosis created an even worse damaging effect to brain cells (23, 31). On the other hand, extracellular acidosis can damage cells by creating an intracellular acidification. Indeed, low extracellular pH can inhibit Na^+/H^+ exchanger and cause intracellular H^+ accumulation. Intracellular acidification has been found in penumbral cells during focal ischemia and may contribute to the lesion expansion following focal ischemia (32). Significant decrease in intracellular pH can be deleterious since intracellular acidosis could lead to the generation of free radicals, disruption of mitochondrial function, and activation of pH-sensitive DNA endonuclease, etc. These events are deleterious and can lead to necrosis and apoptosis in brain cells.

16.4.2.2 Ionic Disturbances

Loss of transmembrane ionic gradients accompanies brain ischemia, hypoxia, or anoxia (24). In focal ischemia, the changes in major ion species in interstitial milieu are mostly confined to brain regions with severely decreased blood supply. For example, after 0.5–3 h of MCAO in the rat, extracellular K^+ in the ischemic infarct core increased from baseline (about 3 mM) to 50–60 mM (33, 34). Also in rat MCAO model, extracellular Ca^{2+} in the ischemic core significantly decreased from baseline (1.2 mM) to nearly 0.1 mM (35). In a rat global cerebral ischemia model, a sustained elevation of extracellular K^+ (75 mM) and a decrease in extracellular Ca^{2+} (0.06 mM) were also observed (36). In this same study, after 5-min ischemia, extracellular Na⁺ and Cl⁻ in the cortex decreased significantly from a baseline of 150 to 50 mM and from 130 to 72 mM, respectively.

In contrast to the ischemic core, the penumbral tissue has sufficient energy supply to maintain its ionic homeostasis. Nevertheless, the ionic disturbances in the core pose a threat to the neighboring penumbra cells. One such threat comes from peri-infarct depolarization (PID) that is generated by the elevated K⁺ and/or glutamate in the core and propagates repeatedly among brain tissues. As aforementioned, the penumbral tissue per se is hypoperfused (with 20–40% of normal blood supply), and the residual blood supply provides only for basic needs in the penumbral cells for temporary survival. Therefore, any biological event that requires extra energy consumption would put such penumbral cells at risk. Indeed, during focal ischemia, PID spreads across the penumbral territory repetitively (37). What is important is that PID propagates the ionic disturbances among brain regions and the recovery of ionic disruption requires energy-consuming ion pumps. Therefore, PID may not be harmful to brain tissues with normal CBF and energy homeostasis. However, in the penumbra cells with compromised energy supply, PID would cause cell damage by increasing energy demands. Experiments from both mouse and rat MCAO models have shown that the number of PID is well correlated with the size of the infarction, indicating that an intervention of reducing PID could potentially preserve the penumbral tissue (38, 39).

In addition to PID, a direct exposure of penumbral cells to the ischemic core milieu may also disrupt ionic homeostasis in the adjacent penumbral tissue. Unlike PID, which propagates ionic disturbances along penumbral tissue periodically, a direct exposure of core milieu changes the extracellular environment of the penumbral cells in a relatively constant manner. Ionic disruption imposed on the penumbral cells can be an important factor that contributes to lesion expansion. Four major ions, i.e., Ca^{2+}, Na⁺, K⁺, and Cl⁻ have been extensively studied in their role of ischemic cell death.

Ca^{2+} plays a major role in ischemic cell damage. Strong activation of NMDA receptors by glutamate in the ischemic core results in massive influx of Ca^{2+} and dramatic reduction of interstitial Ca^{2+} (40). Besides the excessive and lethal Ca^{2+} accumulation in the core cells, the depletion of extracellular Ca^{2+} poses a significant impact on the adjacent penumbral cells as well. For example, low extracellular Ca^{2+} can disinhibit acid-sensing ion channels, which are activated in acidic environment during ischemia (41, 42). The fall of extracellular Ca^{2+} may also disinhibit TRPM7-mediated channels and activate calcium-sensitive nonselective channel. Therefore, the decrease of Ca^{2+} will depolarize neurons and enhance the relief of voltage-dependent blockage of NMDA receptors, leading to an influx of not only Ca^{2+} but Na⁺ in penumbral cells, which is detrimental (43). Indeed, Ca^{2+} overloading has long been known to be involved in neuronal and glial cell injury. Ca^{2+} influx leads to the activation of deleterious enzyme activities, such as persistent PKC activation that can lead to the formation of ROS. Overloading of Ca^{2+} also causes DNA fragmentation by endonuclease activity, calpain-induced cytoskeletal breakdown, and depolymerization of microtubules, which contributes to the cell injury (44).

Na⁺ ion is the major cation in the interstitial space under physiological conditions. During focal ischemia, massive Na⁺ influx takes place in the ischemic infarct core. Although still under debate, the routes for Na⁺ entry during the hypoxia/ischemia condition can include voltage-gated Na⁺ channels, Na⁺/H⁺ exchangers, Na⁺/Ca²⁺ exchangers, and Na⁺/HCO₃⁻ cotransporters, etc. Overloading with intracellular Na⁺ can be detrimental by enhancing cellular energy demand and impairing mitochondrial functions. However, as a result of the massive influx of Na⁺ in the infarct core region, Na⁺ concentration in the interstitial fluid is threefold lower than normal levels. Therefore, the neighboring penumbral cells are exposed to a very low Na⁺ core milieu. Low extracellular Na⁺ has been proven to be beneficial to neurons under simulated ischemic conditions (23). Indeed, low extracellular Na⁺ may reduce low oxygen-induced massive Na⁺ influx, which could lead to cell death in neurons (45).

K⁺ homeostasis is critical to cell fate in ischemic penumbra. Because of the impairment of Na⁺/K⁺-ATPases, the diminished K⁺ uptake mechanism in the glial cell, and the endfoot siphoning capability, interstitial K⁺ increases dramatically in the ischemic infarct core (46, 47). Therefore, neighboring penumbral cells are exposed to a milieu with a nearly 20-fold increased K⁺. Prolonged exposure to high K⁺ can be deleterious to neurons and glial cells. Indeed, high K⁺ can make intracellular pH alkaline by activating Na⁺/CHO₃⁻ cotransporter in astrocytes, which, however, bring excessive Na⁺ into the cell and lead to cell damage. High K⁺ depolarizes membrane potential that opens voltage-gated Na⁺ channels. This, together with the alkalinizing mechanisms, would promote intracellular Na⁺ accumulation, which can cause the increase of energy demand of Na⁺/K⁺-ATPases and lead to cell damage by reversal of glutamate uptake mechanism. In vivo application of high K⁺ can cause significant brain injury due to the increased glutamate secretion (48). On the other hand, high K⁺ has also been reported to be beneficial to brain injury during the stress challenge. It has been known that the efflux of intracellular K⁺ can be a trigger of caspase activation that leads to apoptosis. High extracellular K⁺ and the blocker of K⁺ channels can inhibit apoptosis by reducing K⁺ efflux (49). High K⁺ in an in vitro ischemia model that mimics the infarct rim shows a significant protective effect in hippocampal slices (23, 50).

Cl⁻ is the major anion in the interstitial space and plays an important role in both physiological and pathological conditions. The role of Cl⁻ ion homeostasis in cell death has been an emerging topic in recent years. It is known that Cl⁻ influx happens following ischemia, which causes necrotic cell death mostly in the infarct core in focal cerebral ischemia. This Cl⁻ influx, however, leads to a significant decrease of interstitial Cl⁻ concentration, which could have impact on the neighboring penumbral cells. The reduction of Cl⁻ concentration in the ischemic penumbral territory is about half of the normal level. The detrimental effect of low extracellular Cl⁻ has been observed in human lymphoid U937 and epithelial Hela cells. Death in these cells occurs due to the activation of anion channels and the enhancement of Cl⁻ efflux (51). In an in vitro ischemic slice model where ionic shifts are included, the normalization of Cl⁻ concentration alleviates the cell death while the decrease of Cl⁻ alone worsens the astrocytic cell death. These observations suggest that reduction of intracellular Cl⁻ caused by low Cl⁻ poses a major impact on the cell fate during focal ischemia (50). It is known that isotonic cell shrinkage is an important

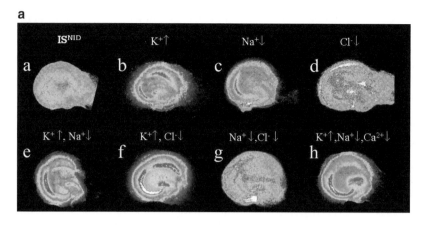

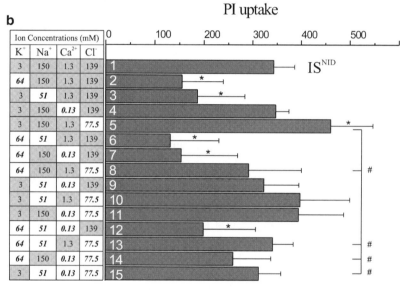

Fig. 16.1 Effects of ion shifts on cell injury in astrocyte-rich area in ischemic solution (IS)-treated hippocampal slices. (**a**) pseudocolor images of propidium iodide (PI) fluorescence acquired from eight different hippocampal slices treated with non-ion disturbed IS (IS^{NID}) (a.a), and IS^{NID} with different combinations of ion changes as shown above each slice (a.a-a.h). Pseudocolor represents the PI fluorescence intensity with purple indicating the least and white the most intense level. Suffix '↑' or '↓' of each labeling indicates that the concentration of that ion in modified IS^{NID} was adjusted to the one used in IS. (**b**) Bar graph showing PI uptake in astrocyte-rich area under different experimental conditions. Different combinations of ion changes are shown in a table to the left of the bar graph. Each row in the table presents ion concentrations of major species used in the IS^{NID} and the gray bars right next to the row show the PI uptake under each experimental condition. $^*p < 0.05$, vs. IS^{NID} group (bar 1), $^\#p < 0.05$, vs. IS^{NID} with low Cl$^-$ group (bar 5), $n = 12$ for each group, Student's t test (*See Color Plates*)

sign of apoptosis (52, 53) and the apoptotic volume decrease is accomplished by release of intracellular osmolytes, especially Cl⁻ and K⁺ (54, 55). Therefore, low extracellualar Cl⁻ may lead to apoptotic cell death by volume decrease.

Using an ischemic solution (IS) that mimics the pathological conditions in the border zone of an ischemic infarct (23), we have studied the role of major ions in the penumbra cell death during the infarct expansion in focal ischemia. Figure 16.1 summarizes these experiments done in an astrocytes-rich area (ARA) in organotypic hippocampal slice cultures. It is suggested that the increased extracellular Cl⁻ is deleterious and may contribute to the infarct growth. However, the increase of K⁺ and decrease of Na⁺ are beneficial and may protect penumbral cells from injury (23, 50).

16.4.2.3 Glutamate Toxicity

Massive release of glutamate into the interstitial space has been observed in both focal and global ischemia. In focal ischemia, the elevation of extracellular glutamate concentration can come from enhanced exocytosis or spontaneous release together with the compromised uptake mechanism. Studies have previously demonstrated that the level of glutamate is about 1–5 µM in normal brain tissue but rapidly increases to 50–90 µM in the infarct core, which can last for many hours in the prolonged focal ischemia. In contrast, glutamate concentration in the penumbral tissue only rises to 30–50 µm and falls back to baseline within 1 h. Furthermore, a delayed elevation of glutamate in the penumbra during reperfusion period has been observed, and this makes glutamate a more important factor for cell death since the damaging effect of glutamate on the penumbral cells can be exacerbated due to the presence of oxygen (56). Therefore, the impact of glutamate toxicity on the fate of penumbral cells can come from both the intrinsic elevation of glutamate in the penumbral tissue and the penetration of core fluid that contains higher amount of glutamate. Excessive extracellular accumulation of glutamate can activate NMDA receptors that are believed to be an important pathway leading to the massive influx of both Ca^{2+} and Na^+ during ischemia. In stroke animal models, several NMDA receptor antagonists have shown to reduce the infarct size, indicating that the activation of this receptor by glutamate is indeed involved in the lesion expansion during focal ischemia.

16.5 Penumbra and the Acute Stroke Therapy

The fundamental concept of acute stroke therapy is to minimize the ischemic infarct volume, which would presumably translate into improved functional outcome (22, 57–61). Since the cause of ischemic infarction is the blockade of blood supply to the affected brain region, recanalization is the basis for acute stroke treatment. However, the prerequisite for this treatment is the presence of rescuable penumbral tissue, which evolves into infarction over time. Therefore, acute stroke therapy relies on the development of not only effective means for reperfusion of the ischemic brain region but the preservation of salvageable brain tissues.

Recanalization is a straightforward approach for acute stroke therapy. Studies from animal models and stroke patients suggest that temporal profile for ischemic penumbra is heterogeneous among different species. For example, the existence of penumbra is 3–4 h in rodent, about 12 h in primate, and up to 48 h in stroke patients. This raises the possibility for clinicians to treat the stroke patients within this time window by recanalizing the blocked blood vessel and reperfusing the affected region. Approaches for recanalizition include thrombolysis by drugs and mechanical removal of clot or the combination of these two approaches. Although intravenous administration of rt-PA has been an effective way for the acute stroke therapy, the risk of hemorrhage is still a major concern, which limits the use of rt-PA in a majority of stroke patients. Therefore, more fibrin-specific thrombolytic drugs are needed to specifically target the clot and produce less impact on the blood vessels. Besides thrombolysis, one other way for recanalization is to use devices to remove the identified clot to get the ischemic tissue reperfused within the therapeutic time window. Furthermore, the combination of thrombolysis with mechanical recanalization may yield better clinical outcome.

Neuroprotection is an important approach to prolong the presence of penumbral tissue. Neuroprotection can be achieved through a number of ways, such as hypothermia, administration of neuroprotectant, etc. Hypothermia is a plausible approach to attenuate neuronal damage in the stroke patients and has been used to protect tissue during cardiovascular and neurovascular procedures for several decades. Moderate hypothermia has been shown to be neuroprotective in both clinical and animal cerebral ischemia studies (58). In focal ischemia, hypothermia can reduce the infarct size and hinder the expansion of infarcted tissue into the penumbra area (62, 63). It is known that mild hypothermia can decrease the rate of glucose, PCr, and ATP breakdown (64). Therefore, with hypothermia, the slowed or reduced breakdown of ATP may also exert its beneficial role by ameliorating the intracellular acidosis, which could contribute to penumbral cell death. Indeed, significantly reduced cell death was seen when temperature was dropped to 31°C in an in vitro model that simulates the penumbral cell death (23). Similar effects of hypothermia were also observed in other in vitro ischemia models (65, 66).

Neuroprotection by drugs that target major deleterious events is an important strategy for the preservation of penumbra tissue. In the penumbra, the diminished blood supply, the biochemically disturbed core milieu, and the periodically propagated PID can together cause irreversible cell damage in this otherwise salvageable ischemic tissue. There are a number of pathophysiological events that are potential pharmacological targets. For example, the release of excessive glutamate in both the core and the penumbral tissue can not only act on the NMDA receptors to lead to the overloading of intracellular Ca^{2+}, but can generate PID. Both of these events can be potential mechanisms underlying penumbral cell death and hence, contribute to the ischemic lesion expansion. Studies in searching for proper glutamate antagonists and Ca^{2+} channel blockers have been extensive and have demonstrated a significant protective effect in animal models. The other example is the intervention in cellular signaling pathways for apoptotic cell death. Indeed, apoptosis occurs in the ischemic penumbra (67, 68). Therefore, developing antiapoptotic strategies is another important aspect for preserving ischemic penumbral tissue. In apoptosis, a biochemical cascade activates caspases, the major executioners in the apoptotic pathway, which initiate the death program. Studies have shown that caspase inhibitors are neuroprotective in animal models. Taken together, experimental evidences have indicated that the preservation of penumbral tissue with neuroprotectant may play an important role in stroke therapy, and some compounds have shown promise in the treatment of acute stroke in clinical trials (69).

16.6 Conclusion

The concept of ischemic penumbra is the basis for understanding the stroke pathology and developing therapeutic strategy. Among the many challenges we are facing in improving the current acute stroke therapy, it is clear that developing a neuroprotectant that can effectively preserve the penumbral tissue is of critical importance.

References

1. NINDS. (1995) Tissue plasminogen activator for acute ischemic stroke. The National Institute of Neurological Disorders and Stroke rt-PA Stroke Study Group. N Engl J Med, 333, 1581–7.
2. Lyden PD. (1999) Thrombolysis for acute stroke. Prog Cardiovasc Dis, 42, 175–83.
3. Khaja AM, Grotta JC. (2007) Established treatments for acute ischaemic stroke. Lancet, 369, 319–30.
4. Brott T, Bogousslavsky J. (2000) Treatment of acute ischemic stroke. N Engl J Med, 343, 710–22.
5. Symon L, Branston NM, Strong AJ, Hope TD. (1977) The concepts of thresholds of ischaemia in relation to brain structure and function. J Clin Pathol Suppl (R Coll Pathol), 11, 149–54.
6. Astrup J, Siesjo BK, Symon L. (1981) Thresholds in cerebral ischemia – the ischemic penumbra. Stroke, 12, 723–5.
7. Hossmann KA. (2006) Pathophysiology and therapy of experimental stroke. Cell Mol Neurobiol, 26(7–8):1057–83.
8. Hakim AM. (1987) The cerebral ischemic penumbra. Can J Neurol Sci, 14, 557–9.
9. Hossmann KA. (1994) Viability thresholds and the penumbra of focal ischemia. Ann Neurol, 36, 557–65.
10. Hata R, Maeda K, Hermann D, Mies G, Hossmann KA. (2000) Dynamics of regional brain metabolism and gene expression after middle cerebral artery occlusion in mice. J Cereb Blood Flow Metab, 20, 306–15.
11. Lee DH, Kang DW, Ahn JS, Choi CG, Kim SJ, Suh DC. (2005) Imaging of the ischemic penumbra in acute stroke. Korean J Radiol, 6, 64–74.
12. Heiss WD. (2003) Best measure of ischemic penumbra: positron emission tomography. Stroke, 34, 2534–5.
13. Warach S, Dashe JF, Edelman RR. (1996) Clinical outcome in ischemic stroke predicted by early diffusion-weighted and perfusion magnetic resonance imaging: a preliminary analysis. J Cereb Blood Flow Metab, 16, 53–9.

14. Parsons MW, Yang Q, Barber PA, Darby DG, Desmond PM, Gerraty RP, Tress BM, Davis SM. (2001) Perfusion magnetic resonance imaging maps in hyperacute stroke: relative cerebral blood flow most accurately identifies tissue destined to infarct. Stroke, 32, 1581–7.
15. Kidwell CS, Alger JR, Saver JL. (2003) Beyond mismatch: evolving paradigms in imaging the ischemic penumbra with multimodal magnetic resonance imaging. Stroke, 34, 2729–35.
16. Rohl L, Ostergaard L, Simonsen CZ, Vestergaard-Poulsen P, Andersen G, Sakoh M, Le Bihan D, Gyldensted C. (2001) Viability thresholds of ischemic penumbra of hyperacute stroke defined by perfusion-weighted MRI and apparent diffusion coefficient. Stroke, 32, 1140–6.
17. Fisher M. (2006) The ischemic penumbra: a new opportunity for neuroprotection. Cerebrovasc Dis, 21 (Suppl 2), 64–70.
18. Garcia JH, Yoshida Y, Chen H, Li Y, Zhang ZG, Lian J, Chen S, Chopp M. (1993) Progression from ischemic injury to infarct following middle cerebral artery occlusion in the rat. Am J Pathol, 142, 623–35.
19. Baird AE, Benfield A, Schlaug G, Siewert B, Lovblad KO, Edelman RR, Warach S. (1997) Enlargement of human cerebral ischemic lesion volumes measured by diffusion-weighted magnetic resonance imaging. Ann Neurol, 41, 581–9.
20. Karonen JO, Vanninen RL, Liu Y, Ostergaard L, Kuikka JT, Nuutinen J, Vanninen EJ, Partanen PL, Vainio PA, Korhonen K, Perkio J, Roivainen R, Sivenius J, Aronen HJ. (1999) Combined diffusion and perfusion MRI with correlation to single-photon emission CT in acute ischemic stroke. Ischemic penumbra predicts infarct growth. Stroke, 30, 1583–90.
21. Parsons MW, Barber PA, Chalk J, Darby DG, Rose S, Desmond PM, Gerraty RP, Tress BM, Wright PM, Donnan GA, Davis SM. (2002) Diffusion- and perfusion-weighted MRI response to thrombolysis in stroke. Ann Neurol, 51, 28–37.
22. Back T, Hemmen T, Schuler OG. (2004) Lesion evolution in cerebral ischemia. J Neurol, 251, 388–97.
23. Yao H, Shu Y, Wang J, Brinkman BC, Haddad GG. (2007) Factors influencing cell fate in the infarct rim. J Neurochem, 100, 1224–33.
24. Hansen AJ. (1985) Effect of anoxia on ion distribution in the brain. Physiol Rev, 65, 101–48.
25. Siesjo BK. (1988) Mechanisms of ischemic brain damage. Crit Care Med, 16, 954–63.
26. Nedergaard M, Kraig RP, Tanabe J, Pulsinelli WA. (1991) Dynamics of interstitial and intracellular pH in evolving brain infarct. Am J Physiol, 260, R581–R588.
27. Kraig RP, Pulsinelli WA, Plum F. (1986) Carbonic acid buffer changes during complete brain ischemia. Am J Physiol, 250, R348–R357.
28. Siesjo BK, Katsura KI, Kristian T, Li PA, Siesjo P. (1996) Molecular mechanisms of acidosis-mediated damage. Acta Neurochir Suppl, 66, 8–14.
29. Nedergaard M, Goldman SA, Desai S, Pulsinelli WA. (1991) Acid-induced death in neurons and glia. J Neurosci, 11, 2489–97.
30. Swanson RA, Farrell K, Stein BA. (1997) Astrocyte energetics, function, and death under conditions of incomplete ischemia: a mechanism of glial death in the penumbra. Glia, 21, 142–53.
31. Chesler M. (2005) Failure and function of intracellular pH regulation in acute hypoxic-ischemic injury of astrocytes. Glia, 50, 398–406.
32. Anderson RE, Tan WK, Martin HS, Meyer FB. (1999) Effects of glucose and PaO_2 modulation on cortical intracellular acidosis, NADH redox state, and infarction in the ischemic penumbra. Stroke, 30, 160–70.
33. Gido G, Kristian T, Siesjo BK. (1997) Extracellular potassium in a neocortical core area after transient focal ischemia. Stroke, 28, 206–10.
34. Sick TJ, Feng ZC, Rosenthal M. (1998) Spatial stability of extracellular potassium ion and blood flow distribution in rat cerebral cortex after permanent middle cerebral artery occlusion. J Cereb Blood Flow Metab, 18, 1114–20.
35. Kristian T, Gido G, Kuroda S, Schutz A, Siesjo BK. (1998) Calcium metabolism of focal and penumbral tissues in rats subjected to transient middle cerebral artery occlusion. Exp Brain Res, 120, 503–9.
36. Hansen AJ, Zeuthen T. (1981) Extracellular ion concentrations during spreading depression and ischemia in the rat brain cortex. Acta Physiol Scand, 113, 437–45.

37. Nedergaard M. (1988) Mechanisms of brain damage in focal cerebral ischemia. Acta Neurol Scand, 77, 81–101.
38. Mies G, Iijima T, Hossmann KA. (1993) Correlation between peri-infarct DC shifts and ischaemic neuronal damage in rat. Neuroreport, 4, 709–11.
39. Shimizu-Sasamata M, Bosque-Hamilton P, Huang PL, Moskowitz MA, Lo EH. (1998) Attenuated neurotransmitter release and spreading depression-like depolarizations after focal ischemia in mutant mice with disrupted type I nitric oxide synthase gene. J Neurosci, 18, 9564–71.
40. Yao H, Haddad GG. (2004) Calcium and pH homeostasis in neurons during hypoxia and ischemia. Cell Calcium, 36, 247–55.
41. Yano S, Brown EM, Chattopadhyay N. (2004) Calcium-sensing receptor in the brain. Cell Calcium, 35, 257–64.
42. Siesjo BK, Katsura K, Kristian T. (1996) Acidosis-related damage. Adv Neurol, 71, 209–33; discussion 234–6.
43. MacDonald JF, Xiong ZG, Jackson MF. (2006) Paradox of Ca^{2+} signaling, cell death and stroke. Trends Neurosci, 29, 75–81.
44. Budd SL. (1998) Mechanisms of neuronal damage in brain hypoxia/ischemia: focus on the role of mitochondrial calcium accumulation. Pharmacol Ther, 80, 203–29.
45. Friedman JE, Haddad GG. (1994) Removal of extracellular sodium prevents anoxia-induced injury in freshly dissociated rat CA1 hippocampal neurons. Brain Res, 641, 57–64.
46. Leis JA, Bekar LK, Walz W. (2005) Potassium homeostasis in the ischemic brain. Glia, 50, 407–16.
47. Sick TJ, Xu G, Perez-Pinzon MA. (1999) Mild hypothermia improves recovery of cortical extracellular potassium ion activity and excitability after middle cerebral artery occlusion in the rat. Stroke, 30, 2416–21; discussion 2422.
48. Fujikawa DG, Kim JS, Daniels AH, Alcaraz AF, Sohn TB. (1996) In vivo elevation of extracellular potassium in the rat amygdala increases extracellular glutamate and aspartate and damages neurons. Neuroscience, 74, 695–706.
49. Yu SP, Canzoniero LM, Choi DW. (2001) Ion homeostasis and apoptosis. Curr Opin Cell Biol, 13, 405–11.
50. Yao H, Sun XL, Gu X, Wang J, Haddad GG. (2007) Cell death in an ischemic infarct rim model. J Neurochem, 103(4):1644–53.
51. Maeno E, Shimizu T, Okada Y. (2006) Normotonic cell shrinkage induces apoptosis under extracellular low Cl conditions in human lymphoid and epithelial cells. Acta Physiol, 187, 217–22.
52. Wyllie AH, Kerr JF, Currie AR. (1980) Cell death: the significance of apoptosis. Int Rev Cytol, 68, 251–306.
53. Kerr JF, Wyllie AH, Currie AR. (1972) Apoptosis: a basic biological phenomenon with wide-ranging implications in tissue kinetics. Br J Cancer, 26, 239–57.
54. Okada Y, Maeno E, Shimizu T, Manabe K, Mori S, Nabekura T. (2004) Dual roles of plasmalemmal chloride channels in induction of cell death. Pflugers Arch, 448, 287–95.
55. Bortner CD, Cidlowski JA. (1998) A necessary role for cell shrinkage in apoptosis. Biochem Pharmacol, 56, 1549–59.
56. Dubinsky JM, Kristal BS, Elizondo-Fournier M. (1995) An obligate role for oxygen in the early stages of glutamate-induced, delayed neuronal death. J Neurosci, 15, 7071–8.
57. Fisher M. (2004) The ischemic penumbra: identification, evolution and treatment concepts. Cerebrovasc Dis, 17 (Suppl 1), 1–6.
58. Ginsberg MD. (2003) Adventures in the pathophysiology of brain ischemia: penumbra, gene expression, neuroprotection: the 2002 Thomas Willis Lecture. Stroke, 34, 214–23.
59. Hakim AM. (1998) Ischemic penumbra: the therapeutic window. Neurology, 51, S44–S46.
60. Dirnagl U, Iadecola C, Moskowitz MA. (1999) Pathobiology of ischaemic stroke: an integrated view. Trends Neurosci, 22, 391–7.
61. Smith WS. (2004) Pathophysiology of focal cerebral ischemia: a therapeutic perspective. J Vasc Interv Radiol, 15, S3–S12.
62. Ridenour TR, Warner DS, Todd MM, McAllister AC. (1992) Mild hypothermia reduces infarct size resulting from temporary but not permanent focal ischemia in rats. Stroke, 23, 733–8.

63. Yanamoto H, Nagata I, Niitsu Y, Zhang Z, Xue JH, Sakai N, Kikuchi H. (2001) Prolonged mild hypothermia therapy protects the brain against permanent focal ischemia. Stroke, 32, 232–9.
64. Erecinska M, Thoresen M, Silver IA. (2003) Effects of hypothermia on energy metabolism in mammalian central nervous system. J Cereb Blood Flow Metab, 23, 513–30.
65. Frantseva MV, Carlen PL, El-Beheiry H. (1999) A submersion method to induce hypoxic damage in organotypic hippocampal cultures. J Neurosci Methods, 89, 25–31.
66. McManus T, Sadgrove M, Pringle AK, Chad JE, Sundstrom LE. (2004) Intraischaemic hypothermia reduces free radical production and protects against ischaemic insults in cultured hippocampal slices. J Neurochem, 91, 327–36.
67. Li Y, Chopp M, Jiang N, Zaloga C. (1995) In situ detection of DNA fragmentation after focal cerebral ischemia in mice. Brain Res Mol Brain Res, 28, 164–8.
68. Charriaut-Marlangue C, Margaill I, Represa A, Popovici T, Plotkine M, Ben-Ari Y. (1996) Apoptosis and necrosis after reversible focal ischemia: an in situ DNA fragmentation analysis. J Cereb Blood Flow Metab, 16, 186–94.
69. Zivin JA. (2007) Clinical trials of neuroprotective therapies. Stroke, 38, 791–3.

Chapter 17
Genes and Survival to Low O_2 Environment: Potential Insights from Drosophila

Gabriel G. Haddad

17.1 Introduction: Why Flies?

Although flies have been used for over a hundred years, since the days of T.H. Morgan, work on flies as model systems for human disease started only in the past couple of decades. At present, a considerable amount of research in Drosophila is tied to the understanding of behavioral, biochemical, or genetic processes at the molecular level with a clear bent toward the understanding of disease processes in mammals or humans. Examples in point, for instance, relate to the effort to enhance our understanding of aging, tumor formation, alcohol intoxication, neurodegeneration, and memory.

We have also been interested in a variety of questions that span from O_2 sensing to the cellular and molecular responses to hypoxia and to injury due to O_2 deprivation. We have recently discovered, for example, that *Drosophila* is very resistant to O_2 deprivation (1). This opened major avenues for us since *Drosophila* has been used so effectively for so many relevant human disease research areas.

In spite of many advances in monitoring and regulating oxygen intake and oxygenation in humans, there is still considerable morbidity and mortality arising from conditions with tissue O_2 deprivation leading to hypoxic/ischemic cell damage, especially, in brain. Part of this failure is related to the complexity of the cascade of events that ensue after celluar hypoxia. Hence, we have used *Drosophila* in our laboratory to solve some of the questions related to tolerance or susceptibility to hypoxia. In this chapter, the role and importance of genetic models, such as *Drosophila melanogaster*, are discussed and this provides with an example illustrating how to harness the power of *Drosophila* genetics. In addition, I will briefly detail approaches that have been used in general to better understand hypoxic brain injury and then detail ours in flies. This does not necessarily mean that biological processes in flies and man are the same in relation to hypoxic injury. Indeed, it is possible that the injury itself is very different. However, it is probable that the pathobiology and the genetic predisposition or susceptibility to low O_2 is not very

---

G.G. Haddad, MD
University of California and Rady Children's Hospital, San Diego, CA
e-mail: ghaddad@ucsd.edu

different and, keeping in mind that most human disease genes are present in flies, novel insights can be obtained from work on Drosophila. I will then detail some of the results that are very interesting and that have taught us an important lesson in relation to hypoxic/ischemic brain injury, a lesson that could be helpful in future designs of better therapeutic regimens.

17.2 Strategies for Understanding Hypoxic Injury in the CNS

Many approaches have been taken to study questions regarding nerve cell injury due to O_2 deprivation. Investigators, including ourselves, have resorted to in vitro techniques and others to more in vivo approaches. Some investigators have used acute settings and mostly electrophysiologic techniques to examine ionic homeostasis but others have relied on morphometric and anatomic approaches (2). Still others have focused, almost exclusively, on molecular approaches, especially in settings in which the stress is modest and cells and tissues withstood prolonged periods of hypoxia (3–5). Some of the more recent studies in our laboratory as well as in others, using molecular and genetic approaches, have provided evidence that there are genes that can protect against or predispose to cell injury and death when nerve cells are exposed to O_2 deprivation (6–10). Some investigators have studied the effect of hypoxia alone but others have combined it to glucose deprivation as well. There are laboratories that have focused on global ischemia and others have studied focal ischemia. It is of no surprise therefore that the results in the literature have often been confusing and sometimes inconsistent. This is also compounded by the fact that experimentally the PO_2 in the hypoxia system delivered is often not measured and that there can be major effects of even small PO_2 changes, especially at the lower levels of O_2. Most importantly, these studies are often biased in that the investigators focus on a pathway or a molecule that they are most familiar with or most interested in, trying to find a role for that particular pathway or molecule in hypoxia or ischemia. Clearly, this is not necessarily faulty but it illustrates the idea that the work that has been previously published, until a few years ago, was often focused on themes that might not have been the most important in understanding normative or pathobiological processes.

17.3 Our Specific Focus: Insights from Microarray Analysis

In our laboratory we have used two approaches in the Drosophila to investigate the genetic basis of the responsiveness to hypoxia. One of them is a forward genetic approach and this relies on a specific phenotype in order to discover a gene, often using positional cloning. The cloned gene, in this approach, has to be proven to be responsible for the abnormal phenotype. The hallmark of this approach is to develop a mutagenesis screen that identifies mutants with a loss-of-function or a gain-of-function mutation based on an assay (or assays), mapping the mutations, and then using molecular biologic techniques to identify and isolate the mutated gene. In flies, mutations can be induced under laboratory conditions using chemicals or radiation. More recently advances have been made to be able to test mutated flies using insertional mutagenesis (e.g., P-elements) in spite of the fact that there are *hot* or *cold* spots for these insertions. Individual flies are then checked for the presence of mutations (screen). It is, therefore, necessary to develop a test (or tests) to identify flies that are mutated or manifest an abnormal phenotype. We have developed a number of such tests or assays to identify flies harboring mutations that make them sensitive (or resistant) to low levels of oxygen.

Another approach is a reverse genetic approach. The idea behind this approach is to identify those genes that are differentially expressed during a condition or a stress. In our case, we use a variety of gases (see later) as stressors in *Drosophila* and assess whether there are mRNA species that are up- or downregulated in organisms subjected to hypoxia in comparison to controls that are unexposed. For example, recently, we have used microarrays to study all of the approximately 13,000 known *Drosophila,* demonstrating that many genes are affected during hypoxia, with some down- and others upregulated.

In the past 5–6 years, we have adopted a similar approach as the one described earlier, except that it is based on a laboratory selection experiment. In this chapter, we will detail our experiments pertaining to this strategy only. The idea was to *develop* a *Drosophila melanogaster* strain that is very resistant to very low environmental O_2 by using selection pressure, i.e., reducing O_2 levels with every 3–5 generations. At that particular time we did not know whether we would succeed with such a phenotype and whether such a phenotype would be based on a physiologic adaptation or on a genetic/epigenetic change. There were two issues that we needed to address at that juncture: (a) to provide allelic variation for this laboratory selection and (b) to decide from what O_2 level to start the selection since we were interested in making sure that all the life cycle stages would need to be able to resist a certain level of O_2 at the outset. To provide some variation, we used a number of isogenic fly lines: male and virgin female flies ($n = 20$) from each of these lines were collected and pooled in a chamber and maintained at room temperature with standard food medium. Embryos from this pooled fly population were then subjected to either long-term (over many generations) hypoxia (selection experiment) or normoxia (control experiment). To decide the O_2 concentration to start with, we tested the feasibility and tolerance of the F1 progenies of the parental cross to different O_2 concentrations (i.e., 8, 6, or 4% O_2) and quantitated the survival of the various life stages at each level of O_2. There was a dramatic decrease in the percentage of embryos reaching adult flies in 6% O_2 (<10%), and no adult flies were actually obtained in 4% O_2 (embryonic lethal). Under 8% O_2, however, the majority of the embryos (>80%) completed their development and reached the adult stage. The selection was started therefore at 8% O_2 and this concentration was gradually decreased by 1% each 3–5 generations to keep the selection pressure. The hypoxia level was first reduced to 7% and then lowered further. By the 13th generation, we obtained flies that were able to complete their development and perpetually live at 5% O_2. These hypoxia-adapted flies (AF flies) showed >50% survival rate during the first generation at 5% O_2 (the 13th generation), and this survival rate increased to more than 80% in following generations in 5% O_2. More recently, we have obtained AF flies that can even tolerate 4% of O_2 perpetually after 32 generations of selection. We hypothesized that this is, at least partially, due to newly occurring mutations or recombination of favorable alleles in the selected population. Our selection paradigm therefore allowed us to obtain flies that can complete their development and tolerate perpetually 5% and even 4% O_2, conditions that are lethal to NF (naïve flies). To test the hypothesis that AF flies are a result of selection of favorable genetic allelic variants, a subset of embryos obtained from the AF flies at the 18th generation was collected and cultured under *normoxic* condition for several consecutive generations. After eight generations cultured in normoxia, they were reintroduced to a 5% or 4% O_2 environment, and again, the majority (>80%) of the flies completed their development and could be maintained in this extreme condition perpetually. This result demonstrated that the selection, indeed, resulted in a heritable trait. Embryos, third instar larvae and adult flies were collected from each generation and stored for subsequent DNA- or RNA-based analyses.

One of the important questions that we also addressed in this paradigm is to identify genes that were altered in a microarray-based analysis and that could be at the basis of the inherited trait of hypoxia tolerance. This analysis was applied on these adapted flies, and the differentially expressed genes in the AF samples were identified by comparison of 24 arrays containing 12 replicates of the AF or the NF chambers, respectively, at the 18th generation. The reason for using F18 for the microarray analysis is that this generation displayed a phenotypic breakthrough whereby flies had been surviving at an O_2 level for more than four generations that was lethal to naïve flies. Direct comparison of the hybridizations between AF and NF samples revealed that 498 genes (~4.0% of the tested genes) had significantly altered their levels of expression, with 279 genes being upregulated and 219 genes downregulated. Several gene families were also found to be significantly altered in the AF flies ($p < 0.05$) (11). For example, antibacterial peptides (12 genes), cytochrome P450 (e.g., cytochrome P450-6 sub-family, 8 genes), and protein kinase C (2 genes) were upregulated, and proteases (26 genes), phosphatases (5 genes), and triacylglycerol lipases (6 genes) were downregulated families.

17.4 Role of Single Genes in an Inherited Complex Trait

As with any microarray analysis, the major question that emanates at the end of the analysis is whether the genes that have altered expression have something to do with the phenotype of interest. Although hypoxia tolerance is likely to be a complex trait, which involves coordinated action of many genes, individual gene expression profiling can provide clues about genes that control multigenic traits (12, 13). However, it is likely that only some of the differentially expressed individual genes are directly responsible for hypoxia tolerance in the selected AF flies. To further identify the genes that functionally contributed to the tolerance of the lethal level of hypoxia (5% O_2), we adopted a strategy that uses P-elements. These were first utilized for genes that were downregulated in the microarray experiments (single P-element insertion *D. melanogaster* lines). Out of the 82 downregulated genes tested, 26 had more than one P-element allele available. Six of these 26 genes had multiple P-element alleles that showed a remarkably greater survival as compared with *yw* or any of the parental lines. The transcripts of all alleles were, indeed, downregulated. In addition, to further confirm that survival in these severe O_2 conditions was related to the P-element insertion in these particular genes, we excised the P-element alleles of one gene, namely *sec6*, and found that the precise (molecularly) excision had less than 10% of eclosion in 5% O_2, a finding that is similar to naïve controls. Therefore, the precise excision of these P-elements reversed the hypoxia tolerance phenotype (Fig. 17.1). The *sec6* gene encodes a protein that is homologous to a mammalian sec6 protein, and it is predicted to be involved in synaptic vesicle recycling (15). This presents the first evidence that this gene is

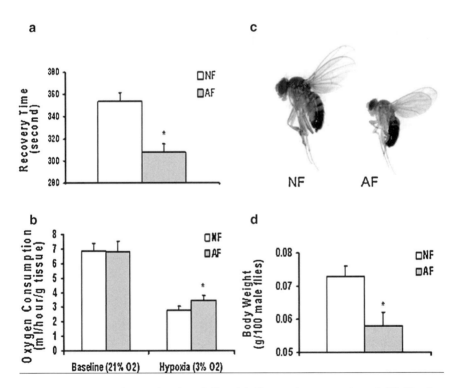

Fig. 17.1 Phenotype of hypoxia-selected flies. (**a**) Shortened recovery time of AF flies from anoxic stupor. Recovery time of each fly from the time of anoxic stupor to that of arousal following reintroduction of room air was recorded. (**b**) O_2 consumption rate was measured under normoxia (21% O_2, baseline) and hypoxia (3% O_2, hypoxia). Although both AF and NF flies decreased their O_2 consumption when switched to hypoxic condition, AF flies reduced less than NF flies. (**c, d**) AF flies had decreased body weight and size [from (14)] (*See Color Plates*)

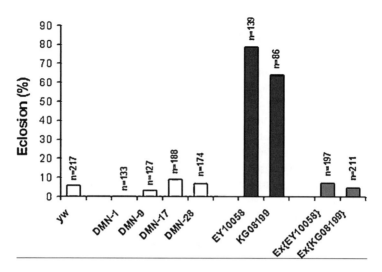

Fig. 17.2 Single genes are important in the phenotype of hypoxia survival. Survival of precise excision lines for *P*-element insertion targeting gene *sec6* was determined in hypoxic condition. Total and eclosed pupae were counted, and the ratio of eclosion for each allele was compared to that of *yw*, NF controls, and original *P*-element alleles [from (14)]

involved in conferring hypoxia tolerance in *Drosophila*, possibly through regulation of neurotransmitter release (Fig. 17.2).

Although the phenotype of hypoxia survival is, in all likelihood, complex and a number of genes may be acting in a coordinated way to affect hypoxia survival, what is surprising in our studies is that single gene alteration was able to affect the survival of flies to very low O_2 environment! Clearly this does not indicate that multiple genes are not involved, singly or in combination, and that even better survival might be achievable if a number of genes were altered at the same time. Our data indicate, however, that even single genes can be important in modulating this impressive phenotype and that a strategy like that of ours, which involved single P-element fly studies, has a chance in dissecting the importance of single genes in this phenotype of hypoxic survival.

17.5 Is Hypoxia Survival Related to Various Genetic Pathways?

The gene families obtained from the microarray experiments are interesting with respect to the phenotype of hypoxia tolerance and experimental selection since some genes would not have been necessarily predicted. The fact that antibacterial peptides were upregulated in AF is remarkable, as such genes are not generally associated with a hypoxia phenotype. On the other hand, protein phosphorylation and dephosphorylation are so fundamental to cellular processes that it is not surprising that kinases or phosphatases are observed to be up- or downregulated in our microarrays. Previous data have indeed implicated such enzymes in tissue protection during hypoxia (16–21). Furthermore, metabolic enzymes such as lipases were downregulated and it is interesting to speculate that, since this downregulation has also been found in hypoxia in mammals, the lipase has an important role to play in the accumulation of triglycerides, which, in turn, help in the hypoxia-tolerant phenotype (22–24). Although the exact molecular mechanisms underlying hypoxia tolerance are currently unknown, it seems that the experimental selection has favored the genetic inheritance of pathways that are important in cell signaling, cell cycle, and cell fate determination. For example, gene *lin19* encodes a protein component of SCF ubiquitin ligase complex that is involved in regulating cell cycle and cell proliferation (25). Interestingly, *Cul1*, the mammalian homologue of *lin19*, encodes a protein that belongs to the Skp1-Cdc53/Cul1-F-box (SCF)-like protein complex, which targets specific proteins for ubiquitination and proteolysis under the regulation of pVHL, and the pVHL-targeted proteins include transcription factor regulators (25). It is important to emphasize that the disparity in the functions of these genes likely reflects the complex physiological regulation of hypoxic responses in tissues. From our results and the function of the characterized genes obtained (i.e., *Best1, br, CG7102, dnc, lin19,* and *sec6*), we found that these genes encode proteins that play a role not only in ubiquitination (e.g., *lin19*) (24, 25) but also in transcriptional regulation (e.g., *broad*) (26, 27), in signal transduction pathways (e.g., *dunce*) (28), and in membrane transport of ions or neurotransmitters (e.g., *sec6* and *Best1*) (15, 29).

In spite of the complexity of the phenotype of hypoxia tolerance, *single* genes make a sizeable difference in this phenotype. It is completely possible, however, that such single genes, as has been shown for the HIF-1 transcription factor (30), are **master switches**. Whether one or more of the genes obtained in this study is another *switch* that can activate or inactivate a large number of genes that are relevant to the hypoxia tolerance is unknown at present.

17.6 Single Gene Vs. Multigenic Diseases

Although we have often distinguished single gene diseases from multigenic diseases with a good justification since multigenic diseases are seemingly much more complex from a genetic/phenotypic point of view as well as from the point of view of understanding its pathophysiology and cell biology, I would like to challenge the idea that this is necessarily the case. Almost every single gene disease has been more complex than what appears at the outset since often the cell biology or physiology of the disease that is related to that particular gene is not generally understood unless a whole interactive pathway is uncovered. In other words, there is so much interaction between a particular disease-inducing single gene mutation and a number of other molecules and pathways that a single disease gene becomes rather quickly a number of molecules and pathways and ceases to be a single molecule that we need to be concerned with regarding the phenotype of interest. Hence, while it is important to make a distinction between a single disease gene and a multigene disease, it is important to put in perspective that a single disease gene is in reality a multigene disease as well from the point of view of disease pathophysiology and biology!

17.7 Fly Genes and Relevance to Mammalian Hypoxic Brain Injury

On the surface, the skeptic can always argue that fly genes have little to do with mammalian genes and with mammalian stroke or hypoxic brain injury in humans. It is important to realize, however, that there are now many examples that argue to the contrary, i.e., that fly genes can give us important information that could be used as hypotheses and prove the relevance of such genes to mammalian physiology and biology. To name a few examples, fruit flies have been used to understand the basis for organ development, ageing, memory and learning, circadian rhythm, alcohol intoxication, and cancer. It should be recognized that some of the major discoveries in physiology and medicine have been made first in model systems. Examples abound. Consider, for instance, the giant squid axon and the discovery of the basis of action potential generation by Hodgkin and Huxley; *Aplysia* and the discovery of the basis for long-term potentiation and memory by Kandel and colleagues; fruit flies and the genes that control embryonic development by both Nusslein-Volhard and Wieschaus; *Caenorhabditis elegans* and the genes that control programmed cell death by Horwitz and his group; and finally the discovery of the K^+ channel structure in bacteria by Rod MacKinnon and colleagues. Hence, we believe that arguing against the potential role of fly genes in mammalian physiology can be a narrow view and that work in *Drosophila* can enhance our ability to understand biological normative processes and most probably human diseases. I illustrate this idea here with an example, which is intended to be taken as a proof of concept.

One of the questions that we asked recently was whether trehalose, which is a glucose dimer, can play an important role in protecting flies against anoxic stress. We had discovered trehalose, by accident, using nuclear magnetic resonance (NMR) experiments and found that trehalose was present in abundance in *Drosophila* heads. We first cloned the gene for trehalose-6-phosphate synthase (*tps1*), which synthesizes trehalose, and then asked whether *tps1* overexpression or mutation has an effect on the resistance of *Drosophila* to anoxia (32, 33). Upon induction of *tps1*, trehalose increased, and this was associated with increased tolerance to anoxia. A P-element inserted into an intron of the *tps1* gene resulted in an embryonic lethal fly (32). To determine whether trehalose could protect against anoxic injury in mammalian cells, we transfected the *Drosophila tps1* gene (*dtps1*) into human embryonic kidney cells (33). Glucose starvation in culture showed that HEK 293 cells transfected with pcDNA3.1 (–) *dtps1* (HEK-*dtps1*) did not metabolize intracellular trehalose and, interestingly, these cells accumulated intracellular trehalose during hypoxic exposure (Fig. 17.3). In contrast to HEK 293 cells transfected with pcDNA3.1 (–) (HEK-v), cells with trehalose were more resistant

17 Genes and Survival to Low O₂ Environment

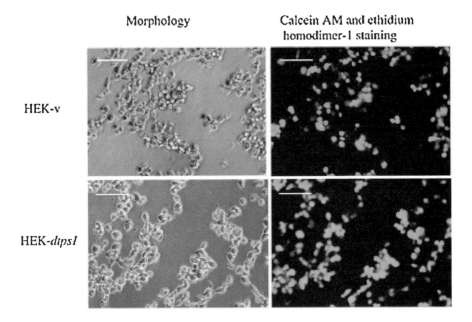

Fig. 17.3 Morphology, calcein, and PI of HEK cells in culture. Four panels, *left* (differential interference contrast (DIC) to show morphology) and *right* (calcein staining green, and propidium iodide (PI) staining red) of HEK cells in culture. *Top* two panels show cells subjected to hypoxia and nontransfected or control (transfected with only vector, v). In the bottom two panels, cells were exposed to hypoxia but also transfected with v-*dtps1*. Note the difference in cell killing between the top (not transfected with the *dtps1* gene) and bottom panels. There are many more cells stained in red in the top panel than in the bottom one (*See Color Plates*)

to low oxygen stress (1% O_2). Insoluble proteins were three times more abundant in HEK-v than in HEK-*dtps1* after 3 days of exposure to low O_2. Ubiquitinated proteins increased dramatically in HEK-v cells but not in HEK-*dtps1* cells over the same period. Our results indicated that increased trehalose in mammalian cells following transfection by the *Drosophila tps1* gene protects cells from hypoxic injury. This example illustrates that it is possible not only to learn from Drosophila but also to potentially utilize certain strategies that invertebrate organisms use to protect themselves in mammalian species, even in man, and be able to develop strategies that could be therapeutically useful.

17.8 Summary

The ability to survive in very low O_2 environments has been an enigma. How certain tissues do survive but not others, why the newborn does better than more mature differentiated cells, and what are the mechanisms underlying anoxia resistance in some anoxia-tolerant animals have been important questions for many years. The down- and upregulation of genes seen in our work and the fact that some play a crucial role in hypoxia tolerance and survival in extremely low O_2 conditions is very encouraging in our quest to understand the basic and fundamental mechanisms for hypoxia survival. In spite of the fact that we elicited the importance of these genes using an experimental selection protocol over generations, these genes may also be critical in hypoxia tolerance in physiological or pathological conditions that are characterized by low O_2 over shorter periods of time. The fact that inhibition of gene activity in this model system leads to a remarkably higher tolerance to extreme low O_2 environments implies that inhibition of the mammalian homolog of the candidate genes in mammals may also be an avenue toward ameliorating the effects of hypoxia.

References

1. Krishnan SN, Sun Y-A, Mohsenin A, Wyman RJ, and Haddad GG. Behavioral and electrophysiologic responses of *Drosophila melanogaster* to prolonged periods of anoxia. J Insect Physiol, 43(3): 203–210, 1997.
2. Haddad GG and Jiang C. O_2 deprivation in the central nervous system: on mechanisms of neuronal response, differential sensitivity and injury. Prog Neurobiol, 40: 277–318, 1993.
3. Banasiak KJ and Haddad GG. Hypoxia-induced apoptosis: effects of severity of hypoxia and role of p53 in neuronal cell death. Brain Res, 797: 295–304, 1998.
4. Banasiak BJ, Cronin T, and Haddad GG. bcl-2 Prolongs neuronal survival in graded hypoxia. Mol Brain Res, 72: 214–225, 1999.
5. Banasiak KJ, Xia Y, and Haddad GG. Mechanisms underlying hypoxia-induced neuronal apoptosis. Prog Neurobiol, 62(3): 215–249, 2000.
6. Ma E and Haddad GG. Anoxia regulates gene expression in the central nervous system of *Drosophila melanogaster*. Mol Brain Res, 46: 325–328, 1997.
7. Ma E, Xu T, and Haddad GG. Gene regulation by O_2 deprivation: an anoxia-regulated novel gene in *Drosophila melanogaster*. Mol Brain Res, 63: 217–224, 1999.
8. Ma E and Haddad GG. A drosophila Cdk5α-like molecule and its possible role in response to O_2 deprivation. Biochem Biophys Res Commun, 261: 459–463, 1999.
9. Ma E and Haddad GG. Isolation and characterization of the hypoxia-inducible factor 1β in *Drosophila melanogaster*. Mol Brain Res, 73: 11–16, 1999.
10. Ma E, Gu XQ, Wu X, Xu T, and Haddad GG. Mutation in RNA editase markedly attenuates neuronal tolerance to O_2 deprivation in *Drosophila melanogaster*. J Clin Invest, 107(6): 685–693, 2001.
11. Dahlquist KD, Salomonis N, Vranizan K, Lawlor SC, and Conklin BR. GenMAPP, a new tool for viewing and analyzing microarray data on biological pathways. Nat Genet 31: 19–20, 2002.
12. Toma DP, White KP, Hirsch J, and Greenspan RJ. Identification of genes involved in *Drosophila melanogaster* geotaxis, a complex behavioral trait. Nat Genet 31: 349–353, 2002.
13. White KP. Functional genomics and the study of development, variation and evolution. Nat Rev Genet 2: 528–537, 2001.
14. Zhou D, Xue J, Chen J, Morcillo P, Lambert JD, White KP, and Haddad GG. Experimental selection for Drosophila survival in extremely low O_2 environment. PLoS ONE 2(5): e490, 2007.
15. Lloyd TE, Verstreken P, Ostrin EJ, Phillippi A, Lichtarge O, et al. A genome-wide search for synaptic vesicle cycle proteins in *Drosophila*. Neuron 26: 45–50, 2000.

16. Greenway SC and Storey KB. Mitogen-activated protein kinases and anoxia tolerance in turtles. J Exp Zool 287: 477–484, 2000.
17. Vartiainen N, Keksa-Goldsteine V, Goldsteins G, and Koistinaho J. Aspirin provides cyclin-dependent kinase 5-dependent protection against subsequent hypoxia/reoxygenation damage in culture. J Neurochem 82: 329–335, 2002.
18. Raval AP, Dave KR, Mochly-Rosen D, Sick TJ, and Perez-Pinzon MA. Epsilon PKC is required for the induction of tolerance by ischemic and NMDA-mediated preconditioning in the organotypic hippocampal slice. J Neurosci 23: 384–391, 2003.
19. Jones NM and Bergeron M. Hypoxia-induced ischemic tolerance in neonatal rat brain involves enhanced ERK1/2 signaling. J Neurochem 89: 157–167, 2004.
20. Carini R, Grazia De Cesaris M, Splendore R, Baldanzi G, Nitti MP, et al. Role of phosphatidylinositol 3-kinase in the development of hepatocyte preconditioning. Gastroenterology 127: 914–923, 2004.
21. Neckar J, Markova I, Novak F, Novakova O, Szarszoi O, et al. Increased expression and altered subcellular distribution of PKC-delta in chronically hypoxic rat myocardium: involvement in cardioprotection. Am J Physiol Heart Circ Physiol 288: H1566–H1572, 2005.
22. Xi L, Ghosh S, Wang X, Das A, Anderson FP, et al. Hypercholesterolemia enhances tolerance to lethal systemic hypoxia in middle-aged mice: possible role of VEGF downregulation in brain. Mol Cell Biochem 291: 205–211, 2006.
23. Bruder ED, Lee PC, and Raff H. Metabolic consequences of hypoxia from birth and dexamethasone treatment in the neonatal rat: comprehensive hepatic lipid and fatty acid profiling. Endocrinology 145: 5364–5372, 2004.
24. Alberghina M, Viola M, and Giuffrida AM. Changes in enzyme activities of glycerolipid metabolism of guinea-pig cerebral hemispheres during experimental hypoxia. J Neurosci Res 7: 147–154, 1982.
25. Hatakeyama S, Kitagawa M, Nakayama K, Shirane M, Matsumoto M, et al. Ubiquitin-dependent degradation of IkappaBalpha is mediated by a ubiquitin ligase Skp1/Cul1/F-box protein FWD1. Proc Natl Acad Sci USA 96: 3859–3863, 1999.
26. Kiss I, Beaton AH, Tardiff J, Fristrom D, and Fristrom JW. Interactions and developmental effects of mutations in the Broad-Complex of *Drosophila melanogaster*. Genetics 118: 247–259, 1988.
27. Bayer CA, von Kalm L, and Fristrom JW. Relationships between protein isoforms and genetic functions demonstrate functional redundancy at the Broad-Complex during *Drosophila* metamorphosis. Dev Biol 187: 267–282, 1997.
28. Davis RL and Kiger JA. Dunce mutants of *Drosophila melanogaster*: mutants defective in the cyclic AMP phosphodiesterase enzyme system. J Cell Biol 90: 101–107, 1981.
29. Sun H, Tsunenari T, Yau KW, and Nathans J. The vitelliform macular dystrophy protein defines a new family of chloride channels. Proc Natl Acad Sci USA 99: 4008–4013, 2002.
30. Semenza GL and Wang GL. A nuclear factor induced by hypoxia via de novo protein synthesis binds to the human erythropoietin gene enhancer at a site required for transcriptional activation. Mol Cell Biol 12: 5447–5454, 1992.
31. Chen Q-F, Ma E, Behar KL, Xu T, and Haddad GG. Role of trehalose phosphate synthase in anoxia tolerance and development in *Drosophila melanogaster*. J Biol Chem 277(5): 3274–3279, 2002.
32. Chen Q-F, Behar KL, Xu T, Fan C, and Haddad GG. Expression of Drosophila trehalose phosphate synthase in HEK-293 cells increases hypoxia tolerance. J Biol Chem 49: 49113–49118, 2003.

Index

A

ABT-737, in cell death, 139
Acidosis, 27, 34, 310, 312–313, 335
Acidosis-induced neuronal injury
    ASIC1a activation in, 34–35
    role of ASIC activation in, 335
Acid-sensing ion channels, 27–28
    in acidosis-induced neuronal injury, 335
    activation of, 33–34
    developmental change of, 35
    modulation of, 31–33
    pharmacology of, 29–31
Acute intermittent hypoxia, 200
Acute mountain sickness (AMS), 258
Acute PVL, coagulative necrosis in, 341
Adaptor protein-2, 60
Adenine nucleotide transporter, 120
Adenosine, 45, 201, 215
Agrin, receptor for, 65–66
AICAR. *See* 5-aminoimidazole-4-
        carboxamide ribonucleoside
AIH. *See* Acute Acute intermittent hypoxia
Akt protein kinase, 266
Alzheimer disease (AD), Na+/K+-ATPase role
        in, 67
Alzheimer's disease, 226
Amiloride, 29, 30
5-Aminoimidazole-4-carboxamide
        ribonucleoside, 59
AMP-kinase (AMPK), 59
Angiogenesis process, 220–221
Anoikis, 230
A-317567, nonselective ASIC blocker, 30
ANT. *See* Adenine nucleotide transporter
Anthopleura elegantissima, 30
Antiapoptotic BCL-2 proteins, 122
Antioxidant defenses, 264
AP-2. *See* Adaptor protein-2
APETx2, 30

Aplysia, 330
Apocynin, Nox inhibitor, 245
ApoE. *See* Apolipoprotein E
Apolipoprotein E, 197
Apoptosis
    K+ homeostasis and, 61–62
    Na+/K+-ATPase and hybrid cell death,
        63–65
Apoptosis and potassium efflux, 103
Apoptotic genes, 257
Apoptotic-necrotic continuum, 258
Apoptotic volume decrease, 80
    AQP regulation and, 86–88
AQP. *See* Aquaporins
AQP water channel, structural model of, 337
Aquaporins, 79
    colocalization of, 88–90
    hypoxia and ischemia, 83–85
    properties of, 80–82
    regulation and AVD, 86–88
    role and significance of, 85–86
Arachidonic acid, and brain ischemia, 32
Aryl hydrocarbon receptor nuclear translocator
        (ARNT), 290
Ascorbic acid, 241
ASIC. *See* Acid-sensing ion channels
ASIC1a activation, 34–35
Astrocytes, in postischemic neuronal
        pathology, 60–61
Astrocytes-rich area (ARA), 317
ATP-dependent potassium, 107
atx1-HAH gene, 241
AVD. *See* Apoptotic volume decrease

B

BAD proteins, role of, 141–142
Bax proteins, 122, 123
    endogenous death channel activity of, 126

BBB. *See* Blood brain barrier
BCL-2 family proteins
  function as ion channels, 123–124
  in ischemic neuronal damage, 138–139
  in programmed cell death, 121–122
  regulation of action, 122–123
  role of, 118
  and VDAC, interaction of, 126
  in vivo recordings of, 124–126
BCL-xL proteins, 122, 123
  and VDAC, interactions of, 127–129
b588, flavocytochrome, 244
BH3-only proteins, 122
Bipolar disorder (BD), Na$^+$/K$^+$-ATPase role in, 68–69
Blood–brain barrier, 83, 85, 234, 245, 248
Brain acidosis, activating acid-sensing ion channels
  acidosis induces neuronal injury, 27
  ASICs, 27–28
  brain acidosis, 26–27
  tissue distribution and properties of ASICs, 28–29
Brain, AQP in, 82
Brain damage, 4, 172, 225, 229, 233, 236
Brain-derived neurotrophic factor (BDNF), 256
Brain development, 256
  during hypoxia, 257
Brain edema, model of AQP4-mediated resolution of, 85, 337
Brain injury in premature infant, pathophysiology of, 153–154
Brain stem in hypoxic response, role of, 215–216
Brain to acute injury, postnatal maturation
  increases in vulnerability, 7–8
  vulnerability to
    excitotoxicity, 5–7
    hypoxia-ischemia changes, 3–4
Brochopulmonary dysplasia (BPD), 289
3-Bromo-7-nitroindazole (3br7NI), 228
β subunits of Na$^+$/K$^+$-ATPase, structure and function of, 55

C

Caenorhabditis elegans, 278, 330
Calcium accumulation in mitochondria, energy dependence of, 130–131
Calcium channel blockers, 308
Calcium influx activation, 137
cAMP-response element binding protein, 193
Cancer, Na$^+$/K$^+$-ATPase therapeutics for, 70
Caspase 3, role in apoptosis, 263

Catalase, 173, 240, 241, 264
Catecholaminergic systems, dysregulation of, 197
CBF. *See* Cerebral blood flow
Cdk5. *See* Cyclin-dependent kinase 5
Cell death and Na$^+$/K$^+$-ATPase
  K$^+$ homeostasis and apoptosis, 61–62
  Na$^+$/K$^+$-ATPase, apoptosis and hybrid cell death, 63–65
  regulation of, 59–61
Cell death control, BCL-2 family proteins in, 122–123
Cell death pathway and intracellular release of Zn^{2+}, 104–105
Cellular Zn^{2+} influx and neuronal damage, 98
Central nervous system (CNS), 26, 29, 71, 175, 255, 289, 296, 301, 324
Cerebral blood flow (CBF), 164, 218–220, 308
Cerebral ischemia in premature infants, causes of, 166–169
Cerebral white matter vulnerability and hypoxia-ischemia, 173–174
Cerebrospinal fluid (CSF), 82, 170, 173
Channel forming integral membrane protein of 28 kDa, 80
Chinese hamster ovary, 102
CHIP28. *See* Channel forming integral membrane protein of 28 kDa
CHO. *See* Chinese hamster ovary
Chronic episodic. *See* Intermittent hypoxia
Chronic granulomatous disease (CGD), 245
Chronic hypoxia, 217–221. *See also* Hypoxia, effects of
Chronic intermittent hypoxia, 202
Chronic neurodegeneration, Zn^{2+} in, 107
CIH. *See* Chronic intermittent hypoxia
Cl-anion in homeostasis, 315
Cl-influx, 315
CNS diseases and Na$^+$/K$^+$-ATPase, 66–69
Coagulative necrosis in, acute PVL, 341
Coagulative necrosis, in PVL, 169–170
Cognition and OSA, 188–189
Computed tomography (CT), 310
Congenital heart disease (CHD), 289
Coronary artery disease (CAD), 256
COX-2. *See* Cyclooxygenase 2
CREB. *See* cAMP-response element binding protein
CSF. *See* Cerebrospinal fluid
Cul1 gene, 329
Cyclin-dependent kinase 5, 45
Cyclooxygenase 2, 195, 291
Cysteine switch, 226
Cytochrome *c,* and apoptotic cascade, 265

Cytoskeletal proteins, 257
Cytosolic calcium level, in ischemia, 136

D
2,3-Diaminonaphthalene (DAN), 229
Diffuse white matter gliosis, 153
2,2′-Dithiodipyridine, 100
DJ-1 gene, 68
DNA damage, 250
DNA damage-inducible gene 45 *(GADD45)*, 263
Dopamine, 67, 196
Dopaminergic systems, dysregulation of, 196–197
Drosophila melanogaster, 278, 323. *See also* Flies harboring mutations and studies
DTDP. *See* 2,2′-dithiodipyridine
DWI/PWI imaging, 310
DWMG. *See* Diffuse white matter gliosis

E
Ebselen, 250
Electron paramagnetic resonance spectroscopy (ESR/EPR), 264
Endogenous death channel activity, of Bax proteins, 126
Endogenous ouabain-like substances, 57
Endothelial NOS (eNOS), 246, 293
Endothelial PAS protein 1 (EPAS-1), 266
EOLS. *See* Endogenous ouabain-like substances
Episodic hypoxia in animals, neurocognitive consequences of, 189
EPO gene, 267, 283
EPSC. *See* Excitatory postsynaptic current
EPSPs. *See* Excitatory postsynaptic potential
ERK-mediated cell death pathway, 105
Erythropoietin (EPO), 267
Erythropoietin, regulation of, 280. *See also* Hypoxia-inducible factor 1 (HIF-1)
Ether-a-go-go (EAG), 103
Excitatory postsynaptic current, 10, 44
Excitatory postsynaptic potential, 44
Excitotoxicity, 4, 5, 10, 14, 173, 257, 261, 262
and ischemia, 44–47
Extracellular matrix (ECM), 226

F
Familial hemiplegic migraine type II, 69
Fenton reaction, 240
Fetal brain development, 256

FHMII. *See* Familial hemiplegic migraine type II
Flavonoids, 241
FL BCL-xL. *See* Full length BCL-xL protein
Flies harboring mutations and studies
culture under *normoxic* condition, AF flies, 326
DNA-or RNA-based analyses, 326
fly genes and hypoxic brain injury
glucose starvation in culture and hypoxic exposure, 331
tps1 overexpression, 330
ubiquitinated proteins, 331
hybridizations, between AF and NF samples, 326
selection paradigm, 326
sensitivity to low levels of oxygen, 325
single genes in inherited complex trait, 326
hypoxia-selected flies, phenotype of, 327
sec6 gene, in synaptic vesicle recycling, 327–328
uses of P-elements, in single gene role in, 327
vs. multigenic diseases, 329–330
fMRI. *See* Functional magnetic resonance imaging studies
Full length BCL-xL protein, 124
Functional magnetic resonance imaging studies, 214

G
GABA(A) receptors, 249–250
Gelatin zymography, 228, 229
GFAP. *See* Glial fibrillary acidic protein
Glial fibrillary acidic protein, 155
Global ischemia, 217
Glucose and brain ischemia, 33
Glucose and oxygen delivery, mechanism of, 214
Glutamate excitotoxicity, 26
Glutamate toxicity, 6, 12, 29, 312, 317
Glutamine synthase, 249
Glutathione peroxidase, 240, 241, 264
Glutathione reductase, 259
Glycine/acetylcholine receptors, oxidation of, 249
Glyoxalase 1 (Glo1), 297
gp91[phox.] enzyme complex, 244
γ subunits of Na^+/K^+-ATPase, structure and function of, 55–56
GTP-bound Rac, 245

H

Haber–Weiss reaction, 240
Heat-shock proteins, 263
Hemopexin domain, 226, 229
Hexokinase and VDAC, relationship of, 120–121
HIF. *See* Hypoxia inducible factor
HIF-1 activity, 347
HIF-α subunit, 290
HIF-1 transcription factor, 329
Highaltitude cerebral edema (HACE), 258
Hippocampal CA1 neurons, large channels in, 139–140
HIV-associated dementia, 226
HVR. *See* Hypoxic ventilatory response
Hybrid cell death, blocking Na$^+$/K$^+$-ATPase effects, 64–65
6-Hydroxydopamine, 282
Hypercholesterolemia, 282
Hypertriglyceridemia, 282
Hypobaric hypoxia, 258–260
Hypoglycemia, 162, 214, 312
Hypophagia, 258
Hypoxia
  blood flow change, in brain stem in, 344
  and ischemia, aquaporins in, 83–85
Hypoxia-adapted flies (AF flies), 325, 326
Hypoxia at altitude, 258
Hypoxia, effects of, 194
  on central nervous system
    brain stem, role of, 215–216
    cerebral vasodilation and blood flow, 214–215
Hypoxia, in developing brain, transcriptional response to
  BrdU incorporation and hypoxia, 300
  gene expression
    in brain, following CIH or CCH, 293–294
    following CIH or CCH, 293
  genes altered
    by both CIH and CCH in cortex, 295
    by both CIH and CCH in hippocampus, 295
    in both cortex and hippocampus following CCH, 297
    in both cortex and hippocampus following CIH, 296
  glyoxalase system, role in, 297
  increase expression of Alas2 and globins, 301
  significant functional networks identified by, 298–299
  transcription factor network, regulating expression of genes, 300
Hypoxia-inducible factor 1 (HIF-1), 60, 257, 277–280
  for carotid body-mediated responses to continuous hypoxia, 280–281
  in cerebral preconditioning phenomena, 283–284
  dependent transcriptional activation, 290
  regulation of angiogenesis by, 280–281
  regulation of erythropoietin production by, 280
  required for carotid body-mediated responses to intermittent hypoxia, 282–283
Hypoxia-ischemia in PPP1, role of, 163–165. *See also* Periventricular leukomalacia
Hypoxia response element (HRE), 267, 291
Hypoxia-responsive transcription factors, 290–293
  activating protein-1 (AP-1), 292–293
  CREB regulation by, 292
  factor-inhibit-HIF (FIH), 290–291
  HIF-α subunit, 290
  mitochondria, role in, 291
  nuclear factor kappa-B (NF?B), 291–292
  oxygen-sensing mechanisms, 293
Hypoxic and ischemic regulation, of Na$^+$/K$^+$-ATPase, 59–61
Hypoxic cell injury and death, 259
  cellular and molecular mechanisms of, 259
    brain region-specific responses, 260
    cell type-specific responses, 260
    hypoxia and apoptosis, 262
    hypoxia and gene transcription, 266
    hypoxia and immune system, 265
    hypoxia and metabolic arrest, 261
    hypoxia and ROS, 263
    ion channel and transporter mechanisms, 261
Hypoxic injury, in CNS, 324
Hypoxic-ischemic injury, during prenatal period, 257
Hypoxic ventilatory response, 198

I

IFN-γ. *See* Interferon-γ
IH. *See* Intermittent hypoxia
IH-induced cognitive deficits, pathophysiology of, 192
Immunofluorescent staining, 229

Inactivated proMMP-9 protein, effect on neuron, 230
Inducible nitric oxide synthase, 170
iNOS. *See* Inducible nitric oxide synthase
Interferon-γ, 164
Intermittent hypoxia, 187
  cognitive and behavioral effects of, 189
  effect on respiratory control, 199
  genetic and environmental risk factors in, 197
Intracellular Zn^{2+} liberation, ROS/RNS in, 101
Iodoacetamide alkylation, 231
Ionic disturbances, 313–317
Ion shifts, effect on cell death, 316
Ischemia
  excitotoxicity and, 44–47
  voltage-dependent potassium (Kv) channels and, 47–49
Ischemia/reperfusion-induced production of NO, 229
Ischemic neuronal damage
  BCL-2 family proteins in, 138–139
  events in, 135–136
  excitotoxicity, cellular events in, 136–137
  oxygen-free radicals in, 137–138
  zinc–potassium continuum in, 104
Ischemic penumbra, 308–309
  existence and evolution of, 309–311
Ischemic stroke, 307

K

K-ATP. *See* ATP-dependent potassium
α-Ketoglutarate dehydrogenase complex (KGDHC), 262
K^+ homeostasis, 315
  and apoptosis, 61–62
  pathways for, 336

L

Lactate and brain ischemia, 32–33
Lactic acidosis, 313
lin19 gene, 329
Lipid peroxidation, 100, 196, 241, 250, 259, 264
12-Lipoxygenase (12-LOX), 100
Long-term depression (LTP), 11
Long-term facilitation, 200
Long-term potentiation (LTP), 345

M

MAC. *See* Mitochondrial apoptosis-induced channel
Magnetic resonance imaging (MRI), 155, 162, 310, 311
Major intrinsic protein, 80
MALDI-TOF analysis, 231
Mammalian brain, glucose and oxygen in, 213–214
Mass spectra, 231
Matrix metalloproteinase enzymes, 225, 226
MCC channel, in mitoplast preparations, 133
MCVA. *See* Medullary cerebrovasodilator area
Medullary cerebrovasodilator area, 216
Metallothionein and Zn^{2+}, affinity for, 99–100
1-Methyl-4-phenyl-1,2,3,6-tetrahydropyridine, 107
Middle cerebral artery (MCA) occlusion, 229
Middle cerebral artery occlusion (MCAO), 310
Mild hypoxia, adaptation for, 217
MIP. *See* Major intrinsic protein
Mitochondrial apoptosis-induced channel, 126
Mitochondrial calcium accumulation, energy dependence of, 130–131
Mitochondrial calcium uniporter, 131–132
Mitochondrial inner and outer membrane ion channels, 134–135
Mitochondrial inner membrane channels, physiology of, 130
Mitochondrial ion channels. *See also* Ischemic neuronal damage; Voltage-dependent anion channel
  in cellular processes, 118
  and ischemic tolerance, 141–142
Mitochondrial megachannel, 133
Mitochondrial permeability transition pore, 132–133
Mitochondrial ROS production, 242
  complex IV, to counteract increased ROS, 244
  electron transport chain (ETC) components, function, 244
  loss of Ψ_m, 244
  metabolic imaging studies, 243
  oxidative damage, 243
  reduced status of cytochrome, 243–244
Mito KATP and KCa channels, 142
Mitoplast recording technique, 132
MMC. *See* Mitochondrial megachannel
MMP-9, activation, model of, 345
MMP-9, conversion into neurotoxin, 230
MMP gelatinolytic activity, and neuronal laminin, 233–234

MMP inhibitor GM6001, 230, 233
MMP proteolysis, inhibition
  and apoptosis, 233
  BB-94, action in, 234
  and laminin degradation, 233
  MMP-9 inhibition and hypoxia-
    ischemia, 234
  proposed reaction scheme
    of SNO-MMP, 234
  SB-3CT and SNO-MMP, role in, 234
MMP proteolysis, inhibition of, 233
MMP proteolysis-mediated neuronal
  apoptosis, 230
MMPs. See Matrix metalloproteinase enzymes
MMP-specific hydroxamate inhibitor, 230
Morris water maze performance, 258
MPTP. See 1-methyl-4-phenyl-1,2,3,
  6-tetrahydropyridine
mPTP. See Mitochondrial permeability
  transition pore
Multiple sclerosis, 226
Murine double minute-2 *(mdm2)*, 263
Myelin-associated glycoprotein (MAG), 257

N
NAA/Cr. See *N*-acetyl aspartate/creatine
Na^+/Ca^{2+} exchanger in nerve cells, 261
N-acetyl aspartate/creatine, 192
NADH and NADPH, effects on VDAC, 120
NADPH oxidase (Nox) family, 244–246
Na^+/glutamate transporter, 261
Na^+/H^+ exchanger (NHE), 261
Na^+/K^+-ATPase, 53–54
  cell death and
    K^+ homeostasis and apoptosis, 61–62
    Na^+/K^+-ATPase, apoptosis and hybrid
      cell death, 63–65
    regulation of, 59–61
  modulation of, 63, 336
  molecular structure of, 54–56, 335
  neuronal function and role of
    emerging neuronal functions of, 65–66
    Na^+/K^+-ATPase and CNS diseases,
      66–69
  physiological function of, 56–57
  physiologic regulation of, 57–59
  role as drug
    for cytoprotection, 70–71
    naturaly occurring, 69–70
    therapeutics for cancer, 70
Na^+/K^+-ATPase α subunit knockout
  phenotypes, 66
Na^+/K^+ ATPase pump, 261, 309

Na^+-K^+ pump, 312
2,3-Naphthyltriazole (NAT), 229
Nerve growth factor (NGF), 256
Neurocognitive evidences of OSA, in adults,
  188–189
Neurodegenerative disorders, 226
Neuronal apoptosis, 228
Neuronal cell death
  potassium efflux and, 103
  zinc–potassium continuum in, 104
  zinc role in, 98
Neuronal laminin, 233
Neuronal nitric oxide synthase, 202
Neuronal NOS-associated activation
  of MMP, 228
Neuroprotection, 318–319
Neurotransmitter systems alterations, in
  neurobehavioral disturbances, 196
Neurotrophins, 241
Nitric oxide (NO), 15–18, 227, 248, 265, 291
Nitric oxide synthase (NOS), 15–17, 26,
  246–247, 265
NMDA receptor, 4, 8, 12, 14, 193, 249, 260,
  264, 308, 317
NMDA receptor 2A and 2B, 45
NMDA receptor expression, developmental
  regulation of
  developmental regulation of, 9–10
  NMDA receptor desensitization, 13–14
  NR2A and NR2B subunits, contribution
    of, 10–11
  NR2A and NR2B subunits, role
    of, 11–12
  receptor subunits, 8–9
  role of synaptic and extrasynaptic, 12–13
NMDA release, 14–15
N-methyl-d-aspartate (NMDA), 3, 4, 248
N-methyld-aspartate (NMDA) receptor
  antagonists, 308
nNOS. See Neuronal nitric oxide synthase
nNOS inhibitor, 228
NO-activated MMP-9, 230
Non-MMP protease inhibitors, 233
Nonsteroid anti-inflammatory drugs, 31
NOS. See Nitric oxide synthase
Nox2-complex, proteins associated with,
  244–245
Nox2-containing NADPH oxidase, 244
Nox-dependent superoxide, 248
Nox2 expression, 245
Nox4 protein, 245
NR2A and NR2B subunits
  contribution of, 10–11
  role of, 11–12

Index

NSAIDs. *See* Nonsteroid anti-inflammatory drugs
Nuclear factor kappa B (NFκB), 265, 291
Nuclear magnetic resonance (NMR), 330
NXY-059, in ischemic damage, 250

O

Obstructive sleep apnea, 188. *See also* Intermittent hypoxia
Obstructive sleep apnea (OSA), 256
O_2 deprivation and hypoxia, 323, 324
OGD. *See* Oxygen and glucose deprivation; Oxygen glucose deprivation
OSA. *See* Obstructive sleep apnea
OSA in children, 191
Ouabain and cardiac glycosides (CGs), dual effects of, 57
Ouabain and sublethal apoptotic insults, effects of, 64
Ouabain-mediated signal transduction, 57–58
Ovarian granulosa cells, apoptosis in, 338
Oxygen and glucose deprivation, 60–61
Oxygen consumption (CMRO2), 310
Oxygen-dependent degradation domain (ODD), 266
Oxygen extraction fraction (OEF), 310
Oxygen-free radicals, in ischemic neuronal damage, 137–138
Oxygen glucose deprivation, 67, 105
Oxygen homeostasis, 277–278
Oxygen sensing, molecular mechanisms of, 278–280

P

p-Aminophenylmercuric acetate (APMA), 229
Parenchymal brain damage, 232
Parkinson's disease (PD), Na^+/K^+-ATPase role in, 67–68
Patch clamp, for mitochondrial inner membranes isolation, 132, 138
PcTX1. *See* Psalmotoxin 1
Penumbra, and acute stroke therapy, 317–319
Penumbral cell death, 311–312
  cell death in infarct core, 312
  challenges of penumbral cells by, 312
    acidosis, 312–313
    glutamate toxicity, 317
    ionic disturbances, 313–317
Perfusion-weighted imaging (PWI), 310
Peri-infarct depolarization (PID), 314

Perinatal panencephalopathy in premature infants, 153
  cause of, 154
  hypoxia-ischemia, role of, 163–165
  hypoxic-ischemic insults accumulation, role of, 177
  neuropathology of, 154
    axonal gray matter injury, 161–162
    gray matter injury in neurons, 159–161
    hemorrhage, pathologic and genetic disorders in, 162–163
    white matter injury in, 155–159
Periodic hypoxemia, 282
Periventricular leukomalacia, 153. *See also* Perinatal panencephalopathy in premature infants
  factors causing, 174–177
  hypoxia-ischemia role in, 163–165
  human clinical studies for, 166–169
  human pathologic studies for, 169–172
  pathologic studies in animal models, 172
Periventricular (PVH), 257
Peroxynitrite ($ONOO^-$), 229, 230, 246
Phospholipase A2, 26
Phospholipid hydroperoxide glutathione peroxidase (PHGPx), 241
Phrenic LTF, 201
PKA. *See* Protein kinase A
PLA2. *See* Phospholipase A2
pLTF. *See* Phrenic LTF
PO2 in hypoxia system, 324
Polyunsaturated fatty acids (PUFAs), 241
Positron emission tomography (PET), 310
Postischemic mitochondria, Zn^{2+}activated channels in, 140–141
Postnatal hypoxia, 255–257
  multiple pathways, activated/altered, 267
  neurocognitive effects of, 258
Posttranslational modifications (PTMs), 226
Potassium efflux and apoptosis, 103
Potassium role in biological processes, 97
PPPI. *See* Perinatal panencephalopathy in premature infants
PPPI, neuropathology of, 341
Prenatal hypoxia, 257
Proapoptotic BCL-2 proteins, 122
Proapoptotic proteins, expression, 263
Programmed cell death, BCL-2 family proteins in, 121–122
Prostaglandins (PGs), 31
Proteases and brain ischemia, 31–32
Protein kinase A, 49
Protein kinase regulation, of Na^+/K^+-ATPase, 58
Psalmopoeus cambridgei, 30

Psalmotoxin 1, 30
pVHL. *See* Von Hippel Lindau protein
PVL. *See* Periventricular leukomalacia

R
Rapid onset dystonia-parkinsonism, 68
Rat uterine cancer adenocarcinoma
  cell line, 87
RDP. *See* Rapid onset dystonia-parkinsonism
Reactive nitrogen species, 98, 163, 239
Reactive oxygen species, 98, 163, 239–241
  BBB integrity and cerebral
    microvasculature, effect on,
    248–249
  inflammatory response to ischemia,
    mechanisms, 248–249
  major consequence of, 239–241
Recanalization, 308, 318
Recombinant proMMP-9, 229
Recombinant tissue-type plasminogen
  activator (rt-PA), 307
Reduced glutathione (GSH), 259
Respiratory burst, 244
Respiratory control, intermittent and sustained
  hypoxia effects, 198–199
Respiratory plasticity
  AIH effects on, 200–202
  CIH effects on, 202
Reverse genetic approach, 325
RNS. *See* Reactive nitrogen species
ROS. *See* Reactive oxygen species
ROS in hypoxia-ischemia
  antioxidant strategies, to test
    contribution, 250
  neuronal circuits and synaptic function,
    effect on, 249–250
  sources of, 242
    mitochondria, 242–244
    NADPH oxidase, 244–247
    nitric oxide synthase (NOS), 246–247
    xanthine oxidase and, 247
ROS levels, critical balance of, 240
Rostral ventral lateral medulla, 216
RUCA-1. *See* Rat uterine cancer
  adenocarcinoma cell line
RUCA-1 cells, AQP 8 and 9 translocation
  in, 87, 338
RVLM. *See* Rostral ventral lateral medulla

S
SCF ubiquitin ligase complex, 329
Selenoproteins, 241

sEPSC. *See* Spontaneous miniature
  postsynaptic currents
Serine racemase, 249
SH. *See* Sustained hypoxia
Sickle cell disease (SCD), 289
Skp1-Cdc53/Cul1-F-box (SCF)-like protein
  complex, 329
Sleep apnea, 282
SNAP25 protein, 249
S-nitrosocysteine (SNOC), 229
S-nitroso-MMP-9 (SNO–MMP-9), 229
S-nitrosothiol, generation of, 229
S-nitrosylated derivative, 228
S-Nitrosylation of MMP, in vitro and in vivo
  oxidative modification, 231
    48-Da shift, mass spectrum, 232
    in-gel digestion with trypsin, 232
    MALDI-TOF analysis, 231, 232
    peptide mass fingerprinting, 231
    signature masses of human MMP-9
      fragments, 231
    S-nitrosylation of cysteine residue
      and activation of MMP-9, 232
    stroke-related processes, for MMP
      activation, 233
    via hydrolysis to form sulfenic
      acid, 232
    via NO and ROS, 231
S-Nitrosylation of recombinant MMP, 229
S-Nitroyslation of MMPs
  crystal structure model, 227
  extracellular proteolytic cascade
    and S-nitrosylation, 228
  formation of S-nitrosylated
    derivative, 227
  model of MMP-9 activation, 227
  production of RNS/ROS, 227
  reactivity of cysteine sulfur,
    enhancement, 228
  regulation of protein function
    by, 228
SNOC-treated R-proMMP-9, 229
SOD. *See* Superoxide dismutase
Sodium-coupled chloride-bicarbonate
  exchanger (NCBE), 261
Spontaneous miniature postsynaptic currents,
  65–66
Src family kinases, of Na$^+$, K$^+$-ATPase, 58–59
Src homology-3 (SH3) domains, 244
Stroke, Na$^+$/K$^+$-ATPase role in, 66–67
Subthalamic cerebrovasodilator area, 216
α subunits of Na$^+$/K$^+$-ATPase, structure and
  function of, 54–55
Superoxide dismutase, 173, 241

Index

Sustained hypoxia, 198
SVA. *See* Subthalamic cerebrovasodilator area

T
Tetrakis-(2-pyridylmethyl)ethylenediamine (TPEN), 104
Thiobarbituric acid reactive substance (TBARS), 194
Tissue plasminogen activator (tPA), 248
Transient cerebral ischemia, 43
Translocating apoptotic vesicles (TAVs), 90

V
VAH. *See* Ventilatory acclimatization to hypoxia
Vascular endothelial growth factor (VEGF), 267
VDAC. *See* Voltage-dependent anion channel
VDAC genes, 129
VDAC2, in apoptosis inhibition, 129
VDAC1 protein, 129
Ventilatory acclimatization to hypoxia, 199
Ventilatory long-term facilitation, 201
Vesicular monoamine transporter, 197
Vitamin C and E, 241
vLTF. *See* Ventilatory long-term facilitation

VMAT. *See* Vesicular monoamine transporter
Voltage-dependent anion channel
 and apoptosis, 127
 and BCL-2 family proteins, interaction of, 126, 127
 and BCL-xL interactions, 127–129
 biophysical features of, 119–120
 large channel activity, 140
 metabolism control mechanism of, 120–121
 in mitochondrial function, 118–119
Voltage-dependent potassium (Kv) channels and ischemia, 47–49
Von Hippel Lindau protein, 60
von Hippel-Lindau tumor suppressor protein (pVHL), 266

Z
Zinc in neuronal injury, 98
 cell death signaling, 100
 and chronic neurodegeneration, 106
 intracellular release of, 99–100
 mechanisms of, 99
Zinc–potassium continuum in ischemia, 104
Zinc role in biological processes, 97
Zn^{2+}–cysteine interaction, 226

Printed in the United States of America